Digital and Statistical Signal Processing

Digital and Statistical Signal Processing

Anastasia Veloni
Nikolaos I. Miridakis
Erysso Boukouvala

CRC Press
Taylor & Francis Group
Boca Raton London New York

CRC Press is an imprint of the
Taylor & Francis Group, an **informa** business

CRC Press
Taylor & Francis Group
6000 Broken Sound Parkway NW, Suite 300
Boca Raton, FL 33487-2742

First issued in paperback 2020

© 2019 by Taylor & Francis Group, LLC
CRC Press is an imprint of Taylor & Francis Group, an Informa business

No claim to original U.S. Government works

ISBN-13: 978-1-138-58006-0 (hbk)
ISBN-13: 978-0-367-73299-8 (pbk)

Library of Congress Cataloging-in-Publication Data

Names: Veloni, Anastasia, author. | Miridakis, Nikolaos, author. | Boukouvala, Erysso, author.
Title: Digital and statistical signal processing / Anastasia Veloni, Nikolaos Miridakis, and Erysso Boukouvala.
Description: Boca Raton : Taylor & Francis, a CRC title, part of the Taylor & Francis imprint, a member of the Taylor & Francis Group, the academic division of T&F Informa, plc, 2018. | Includes bibliographical references.
Identifiers: LCCN 2018027136| ISBN 9781138580060 (hardback : acid-free paper) | ISBN 9780429507526 (ebook)
Subjects: LCSH: Signal processing--Digital techniques.
Classification: LCC TK5102.9 .V45 2018 | DDC 621.382/2--dc23
LC record available at https://lccn.loc.gov/2018027136

Visit the Taylor & Francis Web site at
http://www.taylorandfrancis.com

and the CRC Press Web site at
http://www.crcpress.com

Professor Anastasia Veloni dedicates this book to her husband, Giannis.

Dr. Nikolaos I. Miridakis dedicates this book to his daughter, Ioanna,

his wife, Meni, and his parents, Ioannis and Panayota.

Dr. Erysso Boukouvala dedicates this book to George,

Dimitris, and her parents, Dimitris and Stella.

Contents

Preface...xiii
Authors...xv
List of Acronyms...xvii

Section I   Topics on Digital Signal Processing

1. Introduction..3
  1.1   Introduction...3
  1.2   Advantages of Digital Signal Processing.......................................4
  1.3   Digitization Steps of Analog Signals..5
       1.3.1   Sampling..5
       1.3.2   Quantization..7
       1.3.3   Coding..9
  1.4   Sampling and Reconstruction of Sinusoidal Signals...................10
       1.4.1   Proof of the Sampling Theorem and a Detailed Discussion....................12
  1.5   Physical Sampling..17
  1.6   Sampling and Holding...20
  1.7   Non-Accurate Reconstruction of Analog Signals.........................21
  1.8   Solved Problems...22

2. Discrete-Time Signals and Systems..45
  2.1   Discrete-Time Signals..45
  2.2   Basic Discrete-Time Signals...45
       2.2.1   Impulse Function...45
       2.2.2   Unit Step Function..47
       2.2.3   Ramp Function...47
       2.2.4   Unit Rectangular Function (Pulse Function)............48
       2.2.5   Exponential Function..48
       2.2.6   The Sinusoidal Sequence..50
  2.3   Even and Odd Discrete-Time Signals..51
  2.4   Energy and Power of a Discrete-Time Signal.................................53
  2.5   Conversion of the Independent and Dependant Variable............54
  2.6   Discrete-Time Systems..54
  2.7   Categories of Discrete-Time Systems..55
       2.7.1   Linear Discrete Systems...55
       2.7.2   Time-Invariant Discrete Systems..............................56
       2.7.3   Discrete Systems with Memory.................................57
       2.7.4   Invertible Discrete Systems.......................................57
       2.7.5   Casual Discrete Systems...57
       2.7.6   Stable Discrete Systems..58
  2.8   System Connections...58
  2.9   Convolution..59
  2.10 Deconvolution..62

2.11   Correlation — Autocorrelation...63
2.12   Difference Equations 64
2.13   Discrete-Time Systems of Finite Impulse Response.....................................65
2.14   Solved Problems...66

3. z-Transform... 113
3.1   Introduction..113
3.2   From Laplace Transform to z-Transform..113
       3.2.1   Comparison of the s- and z-Planes into the Region of Convergence..... 116
3.3   Properties of z-Transform 117
       3.3.1   Time Shift..117
       3.3.2   Linearity..118
       3.3.3   Time Reversal...118
       3.3.4   Convolution...119
       3.3.5   Differentiation in z-Plane...119
       3.3.6   Multiplication by an Exponential Sequence.....................................120
       3.3.7   Conjugation of a Complex Sequence..120
       3.3.8   Initial and Final Value Theorem...121
       3.3.9   Correlation of Two Sequences...121
3.4   Inverse z-Transform 121
       3.4.1   Method of Power Series Expansion (Division Method).......................122
       3.4.2   Method of Partial Fraction Expansion...122
       3.4.3   Method of Complex Integration...123
3.5   z-Transform in System Analysis..124
       3.5.1   Transfer Function of Discrete-Time Signal 124
       3.5.2   Causality of Discrete-Time Systems..124
       3.5.3   Stability of Discrete-Time Systems...125
       3.5.4   Transfer Function of Connected Systems...126
       3.5.5   Transfer Function of Discrete-Time Systems 127
3.6   Formula Tables...129
3.7   Solved Problems...130

4. Structures for the Realization of Discrete-Time Systems...........................179
4.1   Introduction..179
4.2   Block Diagrams 179
4.3   Realization Structures 181
       4.3.1   Implementation Structures of IIR Discrete Systems.........................183
       4.3.2   Implementation Structures of FIR Discrete Systems.........................186
4.4   Signal Flow Graphs...188
       4.4.1   Mason's Gain Formula...189
4.5   Solved Problems...190

5. Frequency Domain Analysis...211
5.1   Introduction..211
5.2   Discrete-Time Fourier Transform (DTFT)...212
5.3   Discrete Fourier Series (DFS)...214
       5.3.1   Periodic Convolution...215
       5.3.2   The Relation of the DFS Components and the DTFT over a Period.....216
5.4   Discrete Fourier Transform...216

	5.4.1	Properties of the DFT..218
		5.4.1.1 Linearity..218
		5.4.1.2 Circular Shift...218
		5.4.1.3 Circular Convolution................................219
		5.4.1.4 Multiplication of Sequences....................220
		5.4.1.5 Parseval's Theorem...............................220
5.5	Fast Fourier Transform...221
	5.5.1	FFT Equations...221
	5.5.2	Computation of the IDFT Using FFT....................228
	5.5.3	Fast Convolution..228
		5.5.3.1 Overlap and Add Method......................228
		5.5.3.2 Overlap and Save Method.....................229
5.6	Estimation of Fourier Transform through FFT.......................229
5.7	Discrete Cosine Transform 229
5.8	Wavelet Transform...231
	5.8.1	Wavelet Transform Theory..................................233
5.9	Solved Problems..236

6. Design of Digital Filters...287
6.1	Introduction 287
6.2	Types of Digital Filters..288
6.3	Digital Filter Design Specifications......................................288
6.4	Design of Digital IIR Filters..290
6.5	Indirect Methods of IIR Filter Design..................................292
	6.5.1	The Impulse Invariant Method...........................292
	6.5.2	Step Invariant Method (or z-Transform Method with Sample and Hold)...293
	6.5.3	Backward Difference Method..............................295
	6.5.4	Forward Difference Method.................................296
	6.5.5	Bilinear or Tustin Method...................................297
	6.5.6	Matched Pole-Zero Method.................................299
6.6	Direct Methods of IIR Filter Design......................................300
	6.6.1	Design of $\left\|H\left(e^{j\omega}\right)\right\|^2$ Method 300
	6.6.2	The Method of Calculating $h[n]$ 301
6.7	IIR Filter Frequency Transformations...................................301
6.8	FIR Filters...303
6.9	FIR Linear Phase Filters...304
6.10	Stability of FIR Filters...307
6.11	Design of FIR Filters..307
6.12	The Moving Average Filters 307
6.13	FIR Filter Design Using the Frequency Sampling Method....309
6.14	FIR Filter Design Using the Window Method........................311
6.15	Optimal Equiripple FIR Filter Design...................................317
6.16	Comparison of the FIR Filter Design Methods......................319
6.17	Solved Problems..320

Section II   Statistical Signal Processing

7. Statistical Models...383
   7.1   The Gaussian Distribution and Related Properties 383
         7.1.1   The Multivariate Gaussian Distribution..385
         7.1.2   The Central Limit Theorem...387
         7.1.3   The Chi-Squared RV Distribution..388
         7.1.4   Gamma Distribution..388
         7.1.5   The Non-Central Chi-Squared RV Distribution...........................389
         7.1.6   The Chi-Squared Mixed Distribution..389
         7.1.7   The Student's t-Distribution 390
         7.1.8   The Fisher-Snedecor F-Distribution..390
         7.1.9   The Cauchy Distribution 391
         7.1.10 The Beta Distribution 391
   7.2   Reproducing Distributions 392
   7.3   Fisher-Cochran Theorem 392
   7.4   Expected Value and Variance of Samples...393
   7.5   Statistical Sufficiency...395
         7.5.1   Statistical Sufficiency and Reduction Ratio.................................396
         7.5.2   Definition of Sufficient Condition...397
         7.5.3   Minimal Sufficiency...399
         7.5.4   Exponential Distributions Category 402
         7.5.5   Checking Whether a PDF Belongs to the Exponential
                 Distribution Category 404

8. Fundamental Principles of Parametric Estimation......................................405
   8.1   Estimation: Basic Components...405
   8.2   Estimation of Scalar Random Parameters 406
         8.2.1   Estimation of Mean Square Error (MSE).....................................407
         8.2.2   Estimation of Minimum Mean Absolute Error 409
         8.2.3   Estimation of Mean Uniform Error (MUE)................................. 411
         8.2.4   Examples of Bayesian Estimation..413
   8.3   Estimation of Random Vector Parameters..421
         8.3.1   Squared Vector Error...421
         8.3.2   Uniform Vector Error...422
   8.4   Estimation of Non-Random (Constant) Parameters..............................422
         8.4.1   Scalar Estimation Criteria for Non-Random Parameters.......................423
         8.4.2   The Method of Statistical Moments for Scalar Estimators....................426
         8.4.3   Scalar Estimators for Maximum Likelihood.............................429
         8.4.4   Cramer-Rao Bound (CRB) in the Estimation Variance.................433
   8.5   Estimation of Multiple Non-Random (Constant) Parameters 441
         8.5.1   Cramer-Rao (CR) Matrix Bound in the Covariance Matrix...................442
         8.5.2   Methods of Vector Estimation through Statistical Moments 446
         8.5.3   Maximum Likelihood Vector Estimation.....................................447
   8.6   Handling of Nuisance Parameters 452

9. Linear Estimation..455
   9.1   Constant MSE Minimization, Linear and Affine Estimation.............................455
         9.1.1   Optimal Constant Estimator of a Scalar RV 456

9.2 Optimal Linear Estimator of a Scalar Random Variable.......................456
9.3 Optimal Affine Estimator of a Scalar Random Variable θ...............458
    9.3.1 Superposition Property of Linear/Affine Estimators.............459
9.4 Geometric Interpretation: Orthogonality Condition and Projection Theorem......459
    9.4.1 Reconsideration of the Minimum MSE Linear Estimation....................460
    9.4.2 Minimum Affine MSE Estimation 462
    9.4.3 Optimization of the Affine Estimator for the Linear Gaussian Model....462
9.5 Optimal Affine Vector Estimator........................463
    9.5.1 Examples of Linear Estimation 464
9.6 Non-Statistical Least Squares Technique (Linear Regression)..........................467
9.7 Linear Estimation of Weighted LLS 473
9.8 Optimization of LMWLS in Gaussian Models 477

10. **Fundamentals of Signal Detection** 479
10.1 The General Detection Problem....................484
    10.1.1 Simple and Composite Hypotheses 485
    10.1.2 The Decision Function 486
10.2 Bayes Approach to the Detection Problem....................488
    10.2.1 Assign a Priori Probabilities.........................488
    10.2.2 Minimization of the Average Risk 488
    10.2.3 The Optimal Bayes Test Minimizes $\mathbb{E}[C]$...................489
    10.2.4 Minimum Probability of the Error Test......................490
    10.2.5 Evaluation of the Performance of Bayes Likelihood Ratio Test 490
    10.2.6 The Minimax Bayes Detector.....................491
    10.2.7 Typical Example 493
10.3 Multiple Hypotheses Tests....................496
    10.3.1 A Priori Probabilities.....................498
    10.3.2 Minimization of the Average Risk 498
    10.3.3 Disadvantages of Bayes Approach 501
10.4 Frequentist Approach for Detection.......................502
    10.4.1 Case of Simple Hypotheses: $\theta \in \{\theta_0, \theta_1\}$....................502
10.5 ROC Curves for Threshold Testing 506

Appendix I: Introduction to Matrix Algebra and Application to Signals and System......................517

Appendix II: Solved Problems in Statistical Signal Processing 527

Bibliography......................543

Index 545

Preface

The goal of this textbook is to support the teaching of digital and statistical signal processing in higher education. Particular attention is paid to the presentation of the fundamental theory; key topics are outlined in a comprehensible way, and all areas of the subject are discussed in a fashion that aims at simplification without sacrificing accuracy.

The book is divided into two sections.

In the first section, we aim at a deep understanding of the subject of Digital Signal Processing and provide numerous examples and solved problems often dealt with the use of MATLAB®.

The second section covers Statistical Signal Processing. The basic principles of statistical inference are discussed and their implementation in practical signal and system conditions are analyzed. The discussion is strongly supported by examples and solved problems in this section, as well.

The content of the book is developed in ten chapters and two appendices as follows:

Chapter 1 contains introductory concepts for a basic understanding on the field of Digital Signal Processing, with particular emphasis on Digitization and Reconstruction of continuous-time signals.

Chapter 2 refers to the characteristics and properties of discrete-time signals and systems.

Chapter 3 refers to z-Transform, which is a basic mathematical tool for studying discrete-time systems.

Chapter 4 analyzes the implementation of the filters FIR and IIR in various forms (Direct Form I and II, Cascade form, Parallel form).

Chapter 5 describes the Analysis of Discrete Systems in the Frequency domain. We analyze the Discrete Time Fourier Transform, Discrete Fourier Series, Discrete Fourier Transform, Fast Fourier Transform and Discrete Wavelet Transform.

Chapter 6 deals with the design of the IIR and FIR digital filters. Indirect design methods (Invariant Impulse Response, Invariant Step Response, Differential Imaging, Bilinear Transform, Pole and Zero Position Matching) and direct methods of designing IIR filters are analyzed. The design of FIR filters through sampling in the frequency domain and with the use of Windows (Rectangular, Bartlett, Hanning, Hamming, Blackman, Kaiser) is also described.

Chapter 7 describes comprehensively and develops the most important statistical models used in stochastic signal processing.

Chapter 8 deals with the concept of parametric estimation of stochastic signals when processed for both one- and multi-dimensional cases.

Chapter 9 analyzes the particular problem of linear estimation. This problem is widespread in many practical applications due to its attractively low computational complexity in processing stochastic signals.

Chapter 10 hosts the fundamental principles of stochastic signal detection. The most well-known theoretical approaches and practical techniques for detecting signals are analyzed.

Appendix I summarizes the basic principles in the Algebra of Vectors and Matrices, with a particular emphasis on their application to stochastic signals and systems.

Finally, **Appendix II** presents numerous solved problems in Statistical Signal Processing, aiming at further understanding and consolidation of the theoretical concepts of the second section of the book.

MATLAB® is a registered trademark of The MathWorks, Inc. For product information, please contact:

The MathWorks, Inc.
3 Apple Hill Drive
Natick, MA 01760-2098 USA
Tel: 508 647 7000
Fax: 508-647-7001
E-mail: info@mathworks.com
Web: www.mathworks.com

Authors

Professor Anastasia Veloni is with the University of West Attica, Department of Informatics and Computer Engineering, Greece. She has extensive teaching experience in a variety of courses in the area of automatic control and is the author/co-author of four textbooks, and her research interests lie in the areas of signal processing and automatic control.

Dr. Nikolaos I. Miridakis is with the University of West Attica, Department of Informatics and Computer Engineering, Greece, where he is currently an adjunct lecturer and a research associate. Since 2012, he has been with the Department of Informatics, University of Piraeus, where he is a senior research associate. His main research interests include wireless communications, and more specifically, interference analysis and management in wireless communications, admission control, multicarrier communications, MIMO systems, statistical signal processing, diversity reception, fading channels, and cooperative communications.

Dr. Erysso Boukouvala holds a BSc in Physics from the Department of Physics, University of Athens, an MSc in Applied Optics with distinction, and a PhD in the field of digital image restoration from the Department of Physics, University of Reading, U.K. She works at the Department of Mathematics and Physics at the Hellenic Air Force Academy as a teaching fellow. Since 2005, she has been a member of the Laboratory Teaching Staff at the Department of Informatics and Computer Engineering at the University of West Attica, Athens. She has extensive teaching and lab experience in a variety of courses such as digital signal processing, applied optics, optoelectronics, lasers, and mechanics. She has participated in research projects in the U.K. Her research focuses on the development of digital image restoration techniques.

List of Acronyms

Acronym	Technical Term
A/D	analog-to-digital
ADC	analog-to-digital converter
BIBO	bounded-input, bounded-output
BLRT	Bayes likelihood ratio test
CDF	cumulative distribution function
CLT	central limit theorem
CME	conditional mean estimator
CmE	conditional median estimator
CRB or CR	Cramer-Rao bound
CWT	continuous wavelet transform
DAC	digital-to-analog converter
DCT	discrete cosine transform
DFS	discrete Fourier series
DFT	discrete Fourier transform
DIF	decimation in frequency
DIT	decimation in time
DSP	digital signal processing
DTFT	discrete-time Fourier transform
DWT	discrete wavelets transform
FFT	fast Fourier transform
FIR	finite impulse response
FSR	full scale range
FT	Fourier transform
HPF	high-pass filter
IDCT	inverse discrete cosine transform
IDFS	inverse discrete Fourier series
IDFT	inverse discrete Fourier transform
IDWT	inverse discrete wavelets transform
IFFT	inverse fast Fourier transform
IHPF	inverse high-pass filter
IID	independent and identically distributed (RV)
IIR	infinite impulse response
ILPF	inverse low-pass filter
IUD	independent uniformly distributed random variables
IZT	inverse z-transform
LCM	least common multiple
LLS	linear least squares
LMMSE	linear minimum mean squared error
LMWLS	linear minimum weighted least squares
LPF	low-pass filter
LRT	likelihoodratio test

(Continued)

xviii

Acronym	Technical Term
LTI	linear time invariant
MAE	mean absolute error
MAP	maximum a posteriori
ML	maximum likelihood
MMSE	minimum mean squared error
MP	most powerful (test)
MP-LRT	most powerful likelihood ratio test
MSE	means quared error (estimator)
MUE	mean uniform error
OOK	On-off keying (technique)
PAM	pulse amplitude modulation
PDF	probability density function
RF	radio frequency
ROC	receiver operating characteristic (curve)
RR	reduction ratio
RV	random variable
SFG	signal flow graph
SNR	signal-to-noise ratio
STFT	short time Fourier transform
TF	transfer function
UMVU	uniform minimum variance unbiased (estimator)
WSS	wide-sense stationary
WSSE	weighted sum of squared error
ZOH	zero order hold
ZT	z-transform

Section I

Topics on Digital Signal Processing

1

Introduction

1.1 Introduction

A signal can be defined as a function that conveys information about properties, such as the state and behavior of a physical system. The signal contains this information in a form of a certain type of variation, independent of signal representation. For example, a signal can develop temporal or spatial variations. Signals can be represented as mathematical functions of one or more independent variables. Signals in which the variables of time and amplitude take discrete values are defined as digital signals. These signals can be described by a sequence of numbers.

The field of *digital signal processing* is an essential part of applied science and engineering since much of the information of our world is, today, being recorded in signals, and signals must be processed to facilitate the extraction of information. Consequently, the development of processing techniques is of great importance. These techniques are usually based on the application of transforms that convert the original signal to other forms, which are more appropriate for processing and analyzing.

The ability to proceed to digital implementation of modern signal processing systems made it possible for the field to follow the general trend of cost and size reduction and to increase the reliability of all digital systems. Digital signal processing has a vast number of applications in many different fields of modern science, such as medicine, telecommunications, geophysics, image processing, automatic control systems, etc.

In general, digital signal processing has the following objectives:

- Improvement of image quality
- Separation of undesired component in the signal
- Signal compression
- Search for periodicity
- Deconvolution of input system
- Recognition of the status of the system

Since the system conveys information about a physical quantity, then, from a mathematical point of view, it can be considered as a function. We will assume that our signal is a sequence of numbers in the memory unit of a computer of general purpose or of a specialized digital device.

Most of the signals that we encounter in theory are in analog form. This is because signals are functions of continuous variables (such as time), and their values are usually on a continuous scale. These signals can be processed by suitable analog systems such as filters,

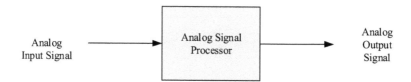

FIGURE 1.1
Analog signal processing.

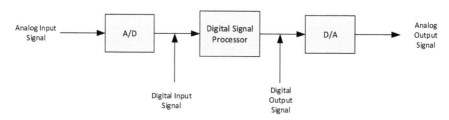

FIGURE 1.2
Digital signal processing.

frequency analyzers or frequency multipliers, in order, for example, to have their characteristics changed. In such cases, we say that the signals have been processed, as shown in Figure 1.1, where the input and the output signal are in analog form.

Digital signal processing provides an alternative method of processing an analog signal, as shown in Figure 1.2.

In order to implement digital processing, there is a need for an interface between the analog signal and the digital processor. This interface is called an *analog to digital converter* (ADC). The output of the converter is a digital signal, which, in turn, is the input to the digital signal processor.

The digital signal processor may be a computer or a microprocessor, which is programmed to perform the desired operations on the input signal. In applications where the digital output from the digital signal processor is to be given to the user in analog form, as, for example, in speech communication, another circuit has to be mediated between the digital and the analog region. A circuit that receives a digital input and converts it into analog is a *digital to analog converter* (DAC). A key feature of DACs is the accuracy in which the converter can change the output. This accuracy is a function of the number of the n bits (2^n levels of quantization) that the digital circuit of the converter can operate at.

1.2 Advantages of Digital Signal Processing

There are many reasons why digital processing of an analog signal is common practice.

- First of all, a programmable digital system allows the user to easily change the parameters of digital signal processing simply by modifying the program. Changing functions in an analog system usually requires redesigning the hardware and testing its proper operation.

- The accuracy of the desired results plays an important role in determining the form of the signal processor. Digital signal processing provides better control of the accuracy required, whereas the tolerance of an analog signal processing system makes it very difficult for the system designer to control its accuracy.
- Digital signals are easy to store on magnetic media without wear or loss of signal fidelity, so the signal can easily be transferred and processed in another laboratory space.
- The digital signal processing method allows the use of more sophisticated signal processing algorithms. It is usually very difficult to make accurate mathematical operations in an analog form. However, all these operations can be done easily by a computer.
- In some applications, digital signal processing costs less than analog. The lower cost may be due to the fact that digital equipment is cheaper, or is the result of the simplicity of adjustment of the digital application.

As a result of the aforementioned advantages, digital signal processing has been applied to a wide range of disciplines. For example, applications of the digital signal processing techniques include speech processing, signal transmission to telephone channels, image processing, seismology and geophysics, oil exploration, nuclear explosions detection, processing of space signals, etc.

However, digital signal processing has limits, too. A practical limitation is the speed of ADCs and digital signal processors. Moreover, there are signals with a wide range of frequencies, which either require very fast ADCs and digital processors or whose digital processing is beyond the capacity of any digital equipment.

1.3 Digitization Steps of Analog Signals

In *digitization*, an analog signal is converted to digital (usually binary), i.e., from a continuous function of time, the analogue signal is converted to a function of a series of discrete numerical values. The process of digitization of continuous time signals involves the following steps:

- Sampling
- Quantization
- Coding

1.3.1 Sampling

Sampling is the process in which a physical quantity that varies in time or space is depicted as a set of values at discrete locations (samples) in the domain of the physical quantity. The grid of sampling points is usually selected uniformly (*uniform sampling*). The set of samples is the discrete-time (or space) signal. In Figure 1.3, an analog signal $x(t)$ (1.3(a)) and the corresponding discrete time signal $x_s(t)$ resulting from the sampling of the analog signal (1.3(b)) with a sampling frequency, f_s, are illustrated.

We define sampling period, T_s, as the distance (in sec) between two successive samples and as sampling frequency, $f_s = 1/T_s$, the number of samples per second. The signal after sampling is not the same with the original since we only maintain the $x(kTs)$ values and

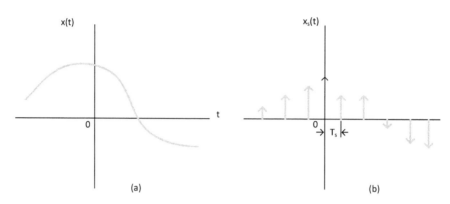

FIGURE 1.3
(a) Analog signal, (b) the signal after uniform sampling.

reject all other values. The fidelity of the discrete time signal depends on the sampling frequency, f_s. In the case of low sampling frequency, rapid changes in the original signal are not recorded, and the information contained in it is lost, resulting in degradation of the original signal. The problem is corrected by increasing the sampling frequency. Therefore, the value of the sampling frequency must meet two conflicting requirements:

- High digitization quality (high sampling frequency required)
- Digital files of a small size (low sampling frequency required)

In many practical applications of digital signal processing, the usual problem of the varying sampling rate of the signal often appears. One way to implement varying sampling rate is to convert the sampled signal to its original analog form and then sample again. Another way is to make the signal undergo new sampling in its digital form, with the advantage that no additional distortion is introduced by the successive processing of the signal by additional DACs and ADCs.

The conversion of sampling rate digitally is being implemented as follows:

1. *By reducing the sampling rate by an integer factor, e.g.,* M. Therefore, the new sampling period is $T_s' = MT_s$ and the re-sampled signal is $x_d[n] = x_a[nT_s'] = x_a[nMT_s] = x[nM]$. The procedure of reducing the sampling rate by an integer factor, M, is done by picking out each M-th sample of the sequence $x[n]$. The system that performs this procedure is called *downsampling*. Downsampling generally results in overlapping, which can be avoided if the signal, $x[n]$, is filtered prior to the process by means of a low-pass filter with a cut-off frequency, $\omega_c = \pi/M$.

2. *By increasing the sampling rate by an integer factor, e.g.,* L. In this case, it is necessary to extract the samples $x_i[n] = x_a[nT/L]$ from $x[n]$. The samples $x_i[n]$ for the values of n, which are integral multiples of L, are derived from $x[n]$ as follows: $x_i[nL] = x[n]$ The sequence, $x[n]$. The *upsampler* extends the time-scale by a factor of L by adding L-1 zeros at the intervals between the samples, $x[n]$.

3. *By converting the sampling rate by a factor that is a rational number.* A link-in-series decimator, which reduces the sampling rate by a factor, M, with an interleaver, which increases the sampling rate by a critical factor, L, leads to a system that changes the sampling rate by an rational coefficient, L/M.

Example:

Suppose that an analog signal $x_\alpha(t)$ is sampled at a frequency of 8 *KHz*. The purpose is to obtain the discrete-time signal that would have been generated if $x_\alpha(t)$ had been sampled at a frequency of 10 *KHz*. What is asked is to change the sampling rate by a factor, $\dfrac{L}{M} = \dfrac{10}{8} = \dfrac{5}{4}$. This is done by multiplying its sampling frequency by a factor of 5, filtering the frequency-amplified signal through a low-pass filter with a cut-off frequency, $\omega_c = \pi/5$, of gain 5, and, finally, dividing the frequency of the filtered signal by a factor of 4.

1.3.2 Quantization

Quantization is the conversion of continuous amplitude values into discrete amplitude values (levels). A continuous signal has a continuous amplitude domain and therefore its samples have a continuous amplitude domain, too. That is, within the finite signal domain we find an infinite number of amplitude levels. Sampling makes the analog signal discrete but if the values of the samples are to be represented in a computer, the amplitude values must also be discrete, i.e., they must be approximated by other predetermined values. The difference between the actual value of the sample and the value finally encoded is called a quantization error (Figure 1.4).

This difference introduces a distortion to the quantized signal which can be diminished if more quantization levels are used. This, however, would require the use of more bits in order to encode the sample. During the quantization process, the range of amplitude values of the continuous signal is subdivided into specific levels, and a digital code is assigned to each level. Each sample gets the digital value closest to its original value. The number of available levels depends on the number of digits of the code, n (sampling size), used to represent these values. The number of encoding levels is equal to 2^n.

Figure 1.5 illustrates the relation between the analog input and digital output for a three-input ADC or quantizer. The horizontal axis represents the analog input of a range from 0V to 10V. The vertical axis represents the encoded binary output, and, in the case of a 3-digit code, it increases from 000 to 111. In this case we have 2^n-1 analog points. In the case of a 3-digit quantizer, these analog points are 0.625, 1.875, 3.125, 4.375, 5.625, 6.875 and 8.125V. The central points of the encoded output words are 1.25, 2.50, 3.75, 5.00, 6.25, 7.50 and 8.75V.

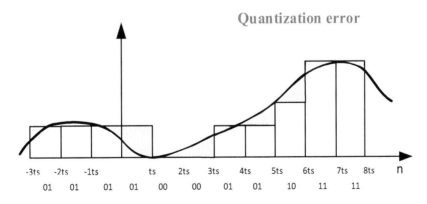

FIGURE 1.4
Quantization error.

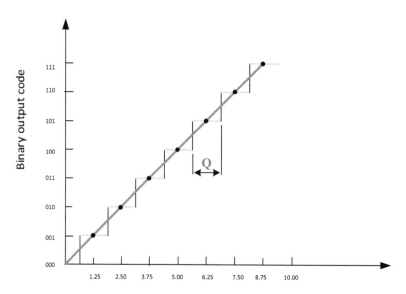

FIGURE 1.5
The input-output relation of one ADC 3 digits.

The quantization level is the step between two successive reference points and is given by the formula $Q = FSR/2^n$ where the FSR is the Full Scale Range. If the FSR is 10V, and $n = 3$, then Q equals 1.25V.

Graphically, the quantization process means that the straight line representing the relation between the input and output of a linear continuous system is replaced by a stair-step graph. The difference between the two successive discrete values is called step-size. The signals applied to a quantizer are graded at amplitude levels (steps), and all the input signals that are at the plus or minus half of a level are replaced at the output by the mid-value of this level.

Since the number of digits of a digital word is finite, the result of A/D conversion is finite. This is because the digital output must have only a finite number of levels, so an analog number should be rounded to the nearest digital level. Consequently, any A/D conversion contains a quantization error. Such errors range between 0 and $\pm\frac{1}{2}Q$. Quantization error depends on the accuracy of the quantization level and can be as small as it is desired, by increasing the number of digits, n. In practice, there is a limit to the number of the digits, n, used, and that is why there is always an error due to the quantization. The ambiguity that occurs during quantization process is called *quantization noise*. The quantization noise range is a significant measure for the quantization error because it is proportional to the average power associated with noise. The quantization error is usually described statistically, and the quantization noise is a time-invariant stochastic process, i.e., its statistical description does not change over time, and the noise probability density function is uniformly distributed over the quantization error range.

The quality of the quantized signal is measured by the signal-to-noise ratio (SNR = 6N + 1.76 (dB) - where N is the number of bits). Each bit that is added to the signal analysis increases the SNR to 6dB. The resolution, most commonly used for music and speech processing, is 16 bit and gives a SNR of 96 bit. The numerical values that will be given by the signal quantization process during digitization depend on the electrical characteristics of the converter as well as the analog signal. For optimal digital conversion, the voltage of the analog signal should be within the limits of the input voltage of the converter. This is usually accomplished by the proper adjustment of the input signal voltage through pre-amplification. Ideally, we use the entire dynamic range of the converter. If the input signal is weak due to fixed SNR, potential amplification in the digital system will also boost noise. If the signal is too strong and its electrical characteristics exceed the endurance of the converter, its extreme values will be quantized to the maximum (and minimum) value given by the specific resolution. In this way, inevitable loss of information of the original signal is introduced, which is called clipping distortion.

The quantization process, as described above, uses a uniform distance between the quantization levels. In some applications, it is preferable to use a varying distance between the quantization levels. For example, the range of voltages covered by voice signals, from the maxima of loud voice to the minima of low voice, is in the order of 1000 to 1. Using a non-uniform quantizer (with the characteristic that the step size increases as the distance from the beginning of the input-output amplitude axes increases), its last big step can include all possible fluctuations of the voice signal at high amplitude levels, which occur relatively rarely. In other words, 'weaker' steps, which require more protection, are preferred at the expense of 'stronger' ones. In this way, a percentage of uniform accuracy is achieved in most of the domain of the input signal, requiring fewer steps than those needed in the case of a uniform quantizer.

1.3.3 Coding

Coding is the representation of the quantized amplitude values of the signal samples in a bit-sequence, whereby the digital signal is generated as a series of bits. The code of each level is called a *codeword*, and the set of different codewords used for the encoding procedure is called *code*. A code is of fixed length when the set of codewords has the same size.

The main disadvantage of the digital representation of continuous media is the distortion introduced by the sampling and quantization process. On the one hand, by ignoring some analog signal values, information is lost, and on the other hand, the approximation of the actual value of the signal with one of the available levels always contains a certain amount of error. This distortion decreases as the sampling frequency and the length of the word increases. However, in this case, the volume occupied by the information also increases, and consequently larger and faster storage media, as well as faster processing units, are required.

Figure 1.6 illustrates the process of analog-to-digital signal conversion.

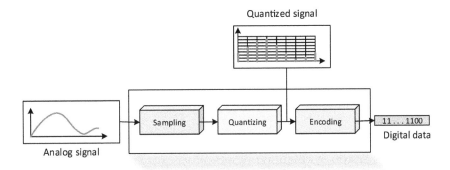

FIGURE 1.6
Analog-to-digital signal conversion.

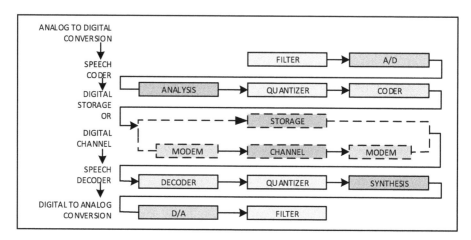

FIGURE 1.7
Digital voice coding system.

An example of the aforementioned process is voice encoding. This is a very useful process if we consider that a great amount of information is transmitted through voice and that, by compressing it, a significant reduction of the bandwidth required can be achieved (up to 4–5 times without noticeable quality degradation). In mobile telecommunication systems, bandwidth availability is an important resource as great effort is being made to serve more and more users within it. For this reason, complex voice coding algorithms are used to reduce the bandwidth of a signal without compromising quality. Most of the well-known voice coding algorithms are explicitly based on a voice production model. It is necessary to construct voice encoders that compress voice in limited bandwidth without sacrificing quality.

All the original parts of a digital voice coding system are shown in Figure 1.7. The input of the system is a continuous sound waveform, $s(t)$. This signal is filtered at low frequencies using an anti-aliasing filter and is sampled by an ADC to give the digital voice signal $s(n)$. This is the input of the encoder. Voice encoding generally consists of three parts: voice analysis, parameter quantization and parameter encoding. The input, at the stage of analysis, is the digital voice signal, while the output is the new representation of the voice signal, which will be quantized and encoded. The output at this stage may vary greatly depending on the way the voice signal is modeled. Quantization is used to reduce the information flow of the voice signal. The output of the quantizer is led to the encoder, which matches a binary code to each possible quantized representation. These binary codes are placed together for more efficient transmission or storage.

1.4 Sampling and Reconstruction of Sinusoidal Signals

Sampling is the procedure of obtaining signal values at selected instances. In the case of uniform sampling, these points are equidistant, and the distance between them is called the sampling period, T_s. At the sampling process, there is a demand of minimum amount of information loss of the continuous time signal.

Let us consider that, at the input of an ADC, there is a sinusoidal signal of the form $x(t) = A\cos(\omega t + \varphi)$. The signal at its output will be of the form $x[n] = A \cos(\omega n T_s + \varphi) = A \cos(\Omega n + \varphi)$. Consequently $\Omega = \omega T_s$ or $F = f \cdot T_s = \dfrac{f}{f_s}$.

This relation correlates the relative frequency, which expresses the regression rate of the signal $x[n]$, with the natural frequency of $x(t)$ through the sampling frequency, $f_s = 1/T_s$.

The sampling frequency f_s must be chosen to satisfy the following relation:

$$f \leq \frac{f_s}{2} \implies f_s \geq 2f \tag{1.1}$$

What Equation 1.1 describes is that when a sinusoidal signal is sampled, it must have a frequency less than half of the sampling frequency. Namely, the presence of at least two samples per cycle is necessary in order to avoid losing the information of the frequency value of the original signal.

A continuous-time signal, $x(t)$, with a maximum frequency, f_{max}, can be successfully reconstructed from its samples, provided that:

$$f_{max} \leq \frac{f_s}{2} \tag{1.2 (a)}$$

or

$$f_s \geq 2 f_{max} \tag{1.2 (b)}$$

Equations 1.2 (a) and 1.2 (b) are the mathematical expressions of the *sampling theorem* (Shannon, 1948). Frequency $f_s/2$ is called *cut-off frequency* or *Nyquist frequency*. If we choose to work with sampling frequencies f_s, that do not satisfy Shannon's criterion, then the original spectrum will not be completely recovered, and effects, such as aliasing or folding, will corrupt it significantly. Examples, proof and a detailed discussion of the sampling theorem are presented next.

Example 1:

If we digitize a signal of a 30 KHz frequency using a sampling frequency $f_s = 50$KHz, then we will get the folding effect, and the output signal will have a $f = 50-30 = 20$ KHz frequency. A sampling frequency greater than the Nyquist frequency is a waste of storage space since it creates additional samples without them being necessary for the successful reconstruction of the signal.

Example 2:

In case we want to transmit a speech signal, the frequency range 0-4000 Hz is sufficient. Therefore, we use a low-pass filter that allows the frequencies 0-4000 Hz to pass while cutting the remaining frequencies. The maximum frequency of the signal will be $f_{max} = 4000$Hz. According to the sampling theorem, we will gain all the information contained in the spectrum of a conversation session if we transmit at least 8000 samples/sec of this spectrum. Therefore, the sampling frequency should be 8000Hz.

1.4.1 Proof of the Sampling Theorem and a Detailed Discussion

Consider the model of ideal sampling with the use of impulses, as shown in Figure 1.8, where the signal $x(t)$ is to be sampled and $s(t)$ is the sampling impulse train of infinite duration.

The signal $x(t)$ is band limited in $[-B,B]$, so the range of the spectrum is $2B$. The signal $s(t)$ that has the following spectrum $S(f) = F\{s(t)\} = F\left\{\sum_{n=-\infty}^{\infty} \delta(t - nT_s)\right\}$, is periodic with a period T_s, and in order to compute its Fourier transform, we have to evaluate the coefficients, C_m, of its Fourier series, namely

$$s(t) = \sum_{n=-\infty}^{\infty} \delta(t - nT_s) = \sum_{m=-\infty}^{\infty} C_m e^{jm\omega_0 t} \text{ with } \omega_0 = \omega_s = \frac{2\pi}{T_s}$$

and

$$C_m = \frac{1}{T_s} \int_{(T_s)} s(t)e^{-jm\omega_0 t}\, dt = \frac{1}{T_s} \int_{-\frac{T_s}{2}}^{\frac{T_s}{2}} \delta(t)e^{-jm\omega_0 t}\, dt = \frac{1}{T_s}$$

Thus,

$$C_m = \frac{1}{T_s} \forall\, m \in Z$$

and

$$s(t) = \sum_{m=-\infty}^{\infty} C_m e^{jm\omega_0 t} = \frac{1}{T_w} \sum_{m=-\infty}^{\infty} e^{jm\omega_0 t}$$

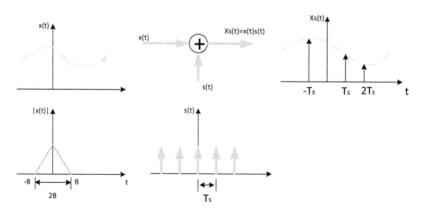

FIGURE 1.8
Ideal sampling using impulses.

The Fourier transform of *s(t)* would be:

$$S(f) = F\left\{s(t)\right\} = F\left\{\frac{1}{T_s} \sum_{m=-\infty}^{\infty} e^{jm\omega_0 t}\right\} = \frac{1}{T_s} \sum_{m=-\infty}^{\infty} F\left\{e^{jm\omega_0 t}\right\} \Rightarrow$$

$$S(f) = \frac{1}{T_s} \sum_{m=-\infty}^{\infty} \delta(f - m \underset{\underset{f_0=f_s}{\downarrow}}{f_0}) = \frac{1}{T} \sum_{m=-\infty}^{\infty} \delta\left(f - m\frac{1}{T_s}\right)$$

The result is illustrated in Figure 1.9.

For the signal $x_s(t)$, resulting after the sampling process, it is true that $x_s(t) = x(t)s(t)$, so it can be assumed that it is derived from the multiplication of the original signal, *x(t)*, and the signal *s(t)* (sampling function), which controls the state of the switch, as illustrated in Figure 1.10.

The spectrum of $x_s(t)$ can be estimated as follows:

$$X_s(f) = F\left\{x_s(t)\right\} = F\left\{x(t) \cdot s(t)\right\} = F\left\{x(t)\right\} * F\left\{s(t)\right\} =$$

$$X(f) * S(f) = X(f) * \left[\frac{1}{T_s} \sum_{m=-\infty}^{\infty} \delta\left(f - m\frac{1}{T_s}\right)\right] \Rightarrow$$

$$X_s(f) = \frac{1}{T_s} \sum_{m=-\infty}^{\infty} X\left(f - m\frac{1}{T_s}\right)$$

The spectrum $|X_s(f)|$ of $x_s(t)$ is illustrated in Figure 1.11.

As shown in Figure 1.11, it is obvious that, in order to avoid overlap between the successive copies of the original spectrum, $X(f)$, around the center frequencies, $mf_x = m\frac{1}{T_x}$, the minimum distance between the central frequencies (i.e., f_s) must be at least equal to the spectral range $2B$. That is, we must apply the rule $f_s \geq 2B$. In this case, the spectrum of the original signal is contained completely in the spectrum of the sampled signal and can be easily isolated using a low-pass filter without any distortion.

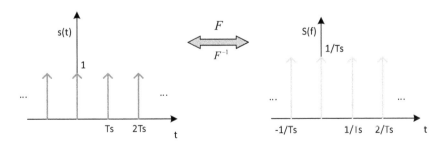

FIGURE 1.9
Signals *s(t)* and *S(f)*.

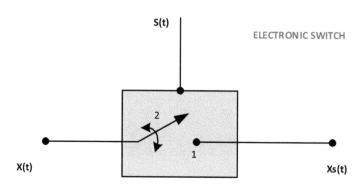

FIGURE 1.10
Sampling system.

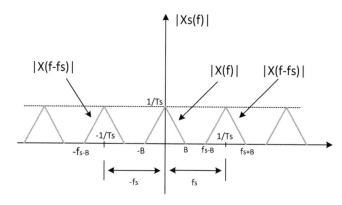

FIGURE 1.11
Spectrum $\left| X_s(f) \right|$ of $x_s(t)$.

Under the circumstances that the above condition is not satisfied, there will be an overlap between successive copies of the spectrum, $X(f)$. The effect of overlap has the following result: the original spectrum, $X(f)$, cannot be completely recovered from the spectrum, $X_s(f)$, and consequently, the original signal, $x(t)$, from the signal, $x_s(t)$. In this case, the spectra of the various components at frequencies $0, \pm f_s, \pm 2f_s, \dots\dots$ overlap, resulting in the distortion of the spectrum of the original signal. So, after the spectrum of the original signal is truncated by a filter, frequencies from the neighboring spectral component are still present within its range, too, resulting in the aforementioned spectrum distortion. In Figure 1.12 we notice that, due to the choice of sampling frequency, $f_s < 2B$, the upper spectral region of $X(f)$ is distorted.

The inverse process of sampling, which aims to the recovery of the original signal, is called reconstruction of the signal. From the spectra of the original and the sampled signal, it becomes clear that the first can only be completely recovered when all frequencies, which are greater than f_{max} in the sampled signal, are cut off.

Generally, the truncation of the spectrum $X(f)$, and consequently of the original signal, $x(t)$, is implemented by using a low-pass filter around the central spectrum, which cuts

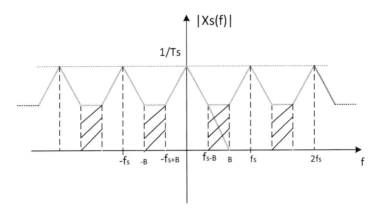

FIGURE 1.12
Spectrum $X_s(f)$ with overlaps.

off the 'remaining' spectra and preserves only the central part, with a cut-off frequency $f_c = \dfrac{f_s}{2} \geq B$ (1.3), as illustrated in Figures 1.13 (a), (b) and (c).

Thus, in order for the original signal to be recovered, it is necessary that there is no overlap between the copies of the original signal spectrum in the spectrum of the sample. This only happens when the criterion $f_s \geq 2f_{max}$ of the sampling theorem is satisfied.

The ideal low-pass filter, with a cutoff frequency $f_c = B$, has a frequency response and an impulse response given by the Equations 1.4 and 1.5, respectively:

$$H(f) = \begin{cases} T_s \, , & |f| \leq \dfrac{f_s}{2} \\ 0 \, , & |f| > \dfrac{f_s}{2} \end{cases} \tag{1.4}$$

$$h(t) = F^{-1}\{H(f)\} = \sin c\left(\pi \dfrac{t}{T_s}\right) \tag{1.5}$$

Consequently, the output of the low-pass filter is determined by using the convolution integral as follows:

$$x(t) = x_s(t) * h(t) = \left[\underbrace{\sum_{n=-\infty}^{\infty} x(nT_s)\delta(t - nT_s)}_{x_s(t)}\right] * \sin c\left(\pi \dfrac{t}{T_s}\right) \Rightarrow$$

$$x(t) = \sum_{n=-\infty}^{\infty} x(nT_s)\sin c\left[\dfrac{\pi}{T_s}(t - nT_s)\right] \tag{1.6}$$

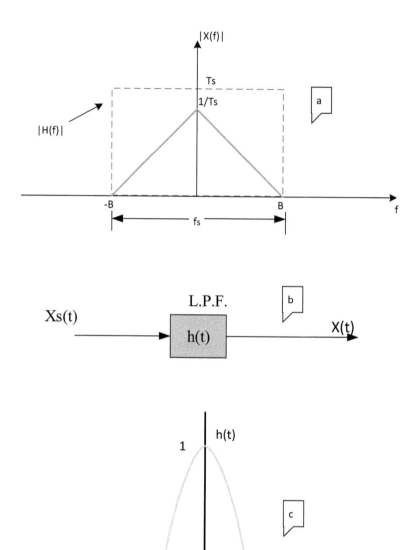

FIGURE 1.13
(a) Spectrum, (b) low-pass filter, (c) impulse response of the ideal low-pass filter.

In the case that we choose $T_s = \dfrac{1}{f_s} = \dfrac{1}{2B}$, we have:

$$x(t) = \sum_{n=-\infty}^{\infty} x\left(\frac{n}{2B}\right) \sin c\left[2\pi B\left(t - n\frac{1}{2B}\right)\right] \quad , \quad f_s = 2B \tag{1.7}$$

Equations 1.6 and 1.7 are the interpolation relations through which we can achieve the recovery of the original continuous signal, $x(t)$, from its samples, $x(nTs)$ Equation 1.7 states that ideally an analog signal (at infinite points in time) can be reconstructed when its samples $x(nT_s)$ are linearly combined with the function sinc.

1.5 Physical Sampling

The discussion in previous paragraph is only theoretical, since in practice there are several problems, namely:

1. There is no such signal (containing low harmonics) for which $X(f) = 0$, when $|f| > f_{max}$, is valid. Of course, the amplitude of the spectrum for most of the signals is greatly reduced beyond a certain frequency. That is, there is a frequency, f_{max}, up to which the power or energy of the signal sums up to at least a 95% of the total. To prevent the signal from being distorted by overlap between the remaining part of the spectrum and the following one, this remaining part can be cut off using anti-aliasing filters prior to sampling. For example, in ISDN telephony all voice frequencies above 3.4 KHz are cut off using an anti-aliasing filter with a cut-off frequency $f_c = 3.4\ KHz$, so that the voice signal is sampled at 8000 samples/second.

2. There is practically no ideal low-pass filter having the aforementioned properties, since real low-pass filters do not completely clip the frequencies, which are greater than the cut-off frequency, but simply reduce their amplitude significantly. Thus, when recovering the original signal, the frequencies of the sampled spectrum that did not exist in the spectrum of the original signal are not completely cut off. The problem is solved by selecting $f_s > 2\ f_{max}$, so that as the difference between the two frequencies increases, the space between successive copies of the original signal spectrum increases, too, resulting in a better reconstruction of the original signal.

3. The sampling function, which is a series of delta functions spaced uniformly in time with a period, T_s, i.e., $\sum\limits_{K=-\infty}^{+\infty}\delta(t-kT_s)$, cannot be implemented in practice but can be approached by rectangular periodic pulses of very short duration. If the period and the duration of the pulses are T_s and τ, respectively, then this pulse will track the original signal for a period equal to its duration, τ.

The original signal, the pulse train and the sampled signal are illustrated in Figure 1.14. The latter consists of a number of pulses whose amplitude is the amplitude of the original signal within the specified time interval. The method of this kind of sampling is called *physical sampling* and has the main disadvantage that the amplitude of the sample does not remain constant throughout the pulse.

Because the output signal does not get its maximum value instantaneously, the opening time is divided into two parts. The first part corresponds to the time the output takes to obtain the maximum amplitude value and is called *acquisition time*, τ_{aq}. The second part corresponds to the time duration that the pulse tracks the original input signal and is called *tracking time*, τ_{tr}. Figure 1.15 illustrates the form of such a pulse.

The sampling function in the case of physical sampling is described by:

$$s(t)=\begin{cases}1, & nT_s-(\tau/2)\le t\le nT_s+(\tau/2)\\1, & nT_s-(\tau/2)\le t\le nT_s+(\tau/2)\end{cases}\quad n=\pm0,\pm1,\pm2,\ldots \tag{1.8}$$

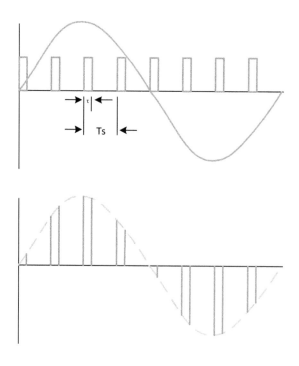

FIGURE 1.14
Physical sampling.

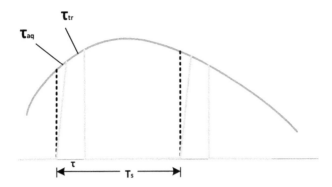

FIGURE 1.15
Pulse format of sampled signal.

Equation 1.8 can be written in Fourier series as follows:

$$s(t) = \sum_{n=-\infty}^{+\infty} C_n \exp(j2\pi n f_s t) \tag{1.9}$$

where

$$C_n = (f_s \tau) \sin c(n f_s \tau) = d \sin c(n f_s \tau) \tag{1.10}$$

Time $d = \tau/T_s$ is called a *duty cycle*, and from Equation 1.10 it follows that $C_0 = d > C_{\pm1} > C_{\pm2} > \ldots$, meaning that the amplitude of each harmonic component of the spectrum of the sampled signal $s(t)$ decreases with the increase of the order of harmonic, and consequently, the same happens to the spectrum of the sampled signal, $x_s(t)$. This makes the recovery of the original signal in physical sampling easier than the ideal case of sampling, since the signal is of low harmonic content, given the absence of ideal filters. This is true because, in physical sampling, the low-pass filter must eliminate the amplitudes of the harmonic components, which are already reduced compared to ideal sampling. Thus, these amplitudes have values that are significantly lower in the case of physical sampling. Note that, at the end of the recovery process, the amplitude of the output signal will have been reduced compared to the original since it is multiplied with the factor $d < 1$. Figure 1.16 shows the spectra of the signals $x(t)$, $s(t)$ and $x_s(t)$.

The signal-to-noise ratio (SNR) is used to act as a standard for the evaluation of the signal recovered at the receiver (i.e., whether or not it contains undesirable frequencies which do not exist in the original signal). In other words, this is the ratio of the average power of the signal to the average power of noise (i.e., the undesirable components) measured in dB and given by the relation (Eq. 1.11) with $x(t)$ and $n(t)$ being the signal and the noise, respectively.

$$S/N(db) = 10 \log_{10} \frac{\overline{x^2(t)}}{\overline{n^2(t)}} \tag{1.11}$$

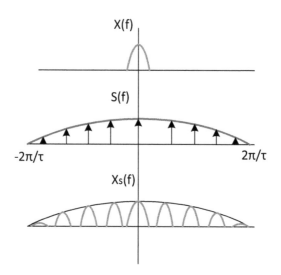

FIGURE 1.16
Signal spectra in physical sampling.

1.6 Sampling and Holding

In cases where it is necessary to process the signal further in discrete or digital form, it is necessary that the amplitude of the samples remains constant. In these cases, *the sample and holding* method is used, in which, after the end of the sample tracking-time, the amplitude of each sample is kept constant until the beginning of the next. The shape of the pulses is illustrated in Figure 1.15, where time, $T_s - \tau$, is called holding time. The amplitude of the sample is kept constant by using a memory element at the output of the sampling system that acts as a holding element.

Figure 1.17 illustrates a sampling and holding device.

The capacitance, C, of the capacitor, connected to the output, is determined by two factors: First, the product of capacitance with the resistance of the switch in position 1 must be small enough for the capacitor to reach the value of the input signal, $x(t)$, in a relatively short time. In this way, the shortest possible holding time is achieved. Secondly, the product of the capacitance with the input resistance of the device connected to the output of the sampling and holding device must be very large. Thus, the discharge time of the capacitor

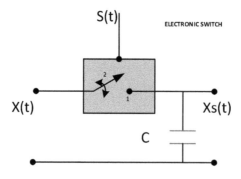

FIGURE 1.17
Sampling and holding device.

would be large enough so that the amplitude of the signal, $x_s(t)$, when the switch is in position 2, remains practically constant, i.e., holding is achieved. The pulse duration of the signal, $s(t)$, is selected rather small. The sampling procedure, as described in the previous paragraphs, is defined as Pulse Amplitude Modulation (PAM), and the sampled signal is called PAM signal.

1.7 Non-Accurate Reconstruction of Analog Signals

Through Equation 1.7, we can accurately recover the original signal from the samples if the analog signal is of finite bandwidth and the sampling frequency satisfies the Nyquist criterion. In real-time processing applications, the analog signal is continuously sampled and the samples are driven in a digital system that processes them and generates output samples at the same rate as the samples arriving to its input. The digital output is converted into analog while maintaining the same time scale as the analog signal.

In practice, most applications require real-time processing and reconstruction. The function sin c, that is defined as $\sin c(\varphi) = \dfrac{\sin \varphi\pi}{\varphi\pi}$, ranges from $-\infty$ to $+\infty$, so each sample contributes at all the reconstruction instances. Therefore, in order to generate the analog signal we have to wait until all samples are available. Consequently, the reconstruction process proposed by the sampling theorem is inappropriate, and the method described by Equation 1.12 is used, leading to an inaccurate reconstruction.

$$x_a(t) = \sum_{n=-\infty}^{+\infty} x_n\varphi(\frac{t}{T_s} - n) \tag{1.12}$$

The function $\varphi(\tau)$ is chosen to be a function of finite duration so that at each instant, t, a finite number of successive samples are involved in the reconstruction, $x_a(t)$. This property provides a real-time reconstruction. Some typical functions used in practice are the stair-step, the triangular, and the finite sinc. Because of its simplicity, the stair-step reconstruction is the most widely used.

$$\text{Step-like}: \varphi(\tau) = \begin{cases} 1, & -1/2 \leq \tau \leq 1/2 \\ 0, & elsewhere \end{cases} \tag{1.13}$$

$$\text{Triangular}: \varphi(\tau) = \begin{cases} 1-|\tau|, & -1 \leq \tau \leq 1 \\ 0, & elsewhere \end{cases} \tag{1.14}$$

$$\text{Finite Sin } c: \varphi(\tau) = \begin{cases} \dfrac{\sin \pi\tau}{\pi\tau}, & -k \leq \tau \leq k \\ 0, & elsewhere \end{cases} \tag{1.15}$$

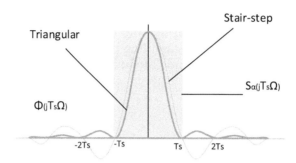

FIGURE 1.18
Step and triangular reconstruction analysis.

In Figure 1.18, the stair-step and the triangular reconstruction are illustrated in order to compare the quality of these two methods.

The functions $\varphi(\tau)$ and $\Phi(j\Omega)$ for the step method are of the form:

$$\varphi(\tau) = \begin{cases} 1, & -1/2 \le \tau \le 1/2 \\ 0, & elsewhere \end{cases} \quad , \quad \Phi(j\Omega) = \frac{\sin\dfrac{\Omega}{2}}{\dfrac{\Omega}{2}}$$

while, for the triangular method, they are:

$$\varphi(\tau) = \begin{cases} 1-|\tau|, & -1 \le \tau \le 1 \\ 0, & elsewhere \end{cases} \quad , \quad \Phi(j\Omega) = \left(\frac{\sin\dfrac{\Omega}{2}}{\dfrac{\Omega}{2}}\right)^2$$

The window function $S_\alpha(jT_s\Omega)$ is the Fourier transform of the function sinc. The triangular reconstruction creates weaker parasitic frequencies as it has smaller ripples than the step, but its central lobe is significantly different from the window function, so the distortion at the actual frequencies of the signal is greater.

1.8 SOLVED PROBLEMS

1.8.1   A digital signal, derived from analog signal sampling, is a linear combination of three sinusoidal signals at frequencies: 0.15, 0.2 and 0.4 KHz.

a. If the sampling frequency is 8 KHz, find the analog frequencies that correspond to the above three digital ones.

b. If the digital signal is reconstructed using a reconstruction period $T_s =$ 0.1 *msec*, which analog frequencies will the three digital ones correspond to in this case?

SOLUTION: A digital frequency φ corresponds to an analog one through the relation:

$$f = \varphi \cdot f_s.$$

a. If the sampling frequency is $f_s = 8$ KHz, the digital frequencies $\varphi_1, \varphi_2, \varphi_3$ correspond to the analog frequencies $f_1 = 1.2$ KHz, $f_2 = 1.6$ KHz and $f_3 = 3.2$ KHz.

b. If the digital signal is reconstructed with a reconstruction period $T_s = 0.1$ *msec* corresponding to a sampling frequency of $f_s = 10$ KHz, then the digital frequency φ_1 corresponds to $f_1 = \varphi_1 \cdot 10000 = 1500$ Hz, the digital frequency φ_2 corresponds to $f_2 = \varphi_2 \cdot 10000 = 2000$ Hz, and the digital frequency φ_3 corresponds to $f_3 = \varphi_3 \cdot 10000 = 4000$ Hz.

1.8.2 Determine the minimum sampling frequency for the following band pass signals:

a. $x_\alpha(t)$ is real with non-zero $X_\alpha(f)$ only for *9 KHz<f <12 KHz*

b. $x_\alpha(t)$ is real with non-zero $X_\alpha(f)$ only for *18 KHz<f <22 KHz*

c. $x_\alpha(t)$ is complex with non-zero $X_\alpha(f)$ only for *9 KHz<f <12 KHz*

SOLUTION:

a. For this signal, the bandwidth is $B = f_2 - f_1 = 3KHz$, and $f_2 = 4B$ is an integer multiple of B, thus the minimum sampling frequency is $f_s = 2B = 6Kz$.

b. For this signal, the bandwidth is $B = f_2 - f_1 = 4KHz$ and $f_2 = 22KHz$ is not an integer multiple of B. If we set $B' = f_2/5 = 4.4$, then f_2 is an integer multiple of the width of B', thus we can sample $x_\alpha(t)$ using the sampling frequency $f_s = 2B' = 8.8KHz$.

c. For a complex bandwidth signal with a non-zero spectrum for $f_1 < f < f_2$, the minimum sampling frequency is $f_s = f_2 = f_1 = 5KHz$.

1.8.3 Explain why the data transmission rate of telephony systems for digital voice transmission is 64 Kbit/s.

SOLUTION: The telephone channel has a bandwidth of 4 kHz. In order to convert the analog voice signal to digital, we should sample at a rate of at least 8000 samples/sec according to the sampling theorem. Digital telephony systems use 8-bit for each level of the analog signal ($2^8 = 256$ quantization levels). Therefore, the information rate of the digital signal resulting from the analog speech signal will be:

$$8000 \text{ sample/sec} \times 8 \text{ bits/sample} = 64 \text{ Kbit/sec}$$

1.8.4 An analog signal of the form

$$x(t) = 3 \cos(400\pi t) + 5 \sin(1200\pi t) + 6 \cos(4400\pi t)$$

is sampled at a frequency $f_s = 4kHz$, creating the sequence $x(n)$ Find its mathematical expression, $x(n)$.

SOLUTION: Obviously $t = nT_s = n\dfrac{1}{4000}$ sec. Thus, the signal $x[n]$ will be:

$$x[n] = 3\ cos(400\pi nT_s) + 5\ sin(1200\pi nT_s) + 6\ cos(4400\pi nT_s) \Rightarrow$$

$$x[n] = 3\ cos\left(\dfrac{\pi n}{10}\right) + 5\ sin\left(\dfrac{12\pi n}{40}\right) + 6\ cos\left(\dfrac{44\pi n}{40}\right)$$

1.8.5 An analog signal of the form

$$x_a(t) = \cos(10\pi t) + \cos(20\pi t) + \cos(40\pi t)$$

is sampled to produce a sequence, $x(n)$. Calculate the sampled sequence for the following sampling frequencies: (a) 45 Hz, (b) 40 Hz, and (c) 25 Hz.

SOLUTION: $x_a(t) = \cos(10\pi t) + \cos(20\pi t) + \cos(40\pi t)$

  a. $F_s = 45Hz$ Thus, $x[n] = \cos\left(\dfrac{2\pi n}{9}\right) + \cos\left(\dfrac{4\pi n}{9}\right) + \cos\left(\dfrac{8\pi n}{9}\right)$

  b. $F_s = 40Hz$ Thus, $x[n] = \cos\left(\dfrac{\pi n}{4}\right) + \cos\left(\dfrac{\pi n}{2}\right) + \cos(\pi n)$

  c. $F_s = 25Hz$ Thus, $x[n] = \cos\left(\dfrac{2\pi n}{5}\right) + \cos\left(\dfrac{4\pi n}{5}\right) + \cos\left(\dfrac{8\pi n}{5}\right)$

1.8.6 For a continuous-time signal, we write:

$$x(t) = 1 + \cos(2\pi f_1 t) + \cos(2\pi f_2 t),\ \text{with}\ f_2 > f_1$$

The signal is samped at a sampling frequency, f_s, which is 10 times the minimum sampling frequency. Write the expression of the sampling signal in the time domain.

SOLUTION: The maximum frequency of the signal $x(t) = 1 + \cos(2\pi f_1 t) + \cos(2\pi f_2 t)$ is $f_{max} = f_2$. Therefore, the minimum sampling frequency is $f_{s,min} = 2f_{max} = 2f_2$

The signal is sampled at frequency $f_s = 10f_{s,min} = 20f_2$ with a sampling period $T_s = \dfrac{1}{f_s} = \dfrac{1}{20f_2}$. The expression of the sampling signal in the time domain is:

$$x_s[n] = 1 + \cos(2\pi f_1 nT_s) + \cos(2\pi f_2 nT_s)$$

1.8.7 A continuous-time signal is given by:

$$x_a(t) = 3.085\cos 1000\pi t - 0.085\sin 700\pi t - 3\sin 300\pi t$$

where t is given in seconds. Find the maximum sampling period that can be used to produce the perfect reconstruction of the signal $x_a(t)$ from its samples.

SOLUTION: Obviously the maximum frequency present in the given signal is $\Omega_{max} = 1000\pi$ rad/sec. According to the Nyquist criterion, the maximum sampling period would be $T_{s,max} = \pi/\Omega_{max} = 1$ msec. In fact, because Ω_{max} is known (as it is the frequency of a sinusoid term in the signal), T_s should be strictly less than 1 msec.

1.8.8 The signal $x(t) = cos(2\pi800t)$ is sampled at 600 *samples*/sec creating the sequence $x[n]$. Find the mathematical expression describing $x(n)$.

SOLUTION: The signal after sampling is given by

$$x[n] = cos(2\pi\ 800nT_s) = cos(2\pi\ (200\ +600)nT_s) =$$
$$= cos(2\pi \cdot 200nT_s +\ 2\pi \cdot 600nT_s) = cos(2\pi \cdot 200nT_s +\ 2\pi \cdot 600 \cdot n\ /\ 600) =$$
$$= cos(2\pi \cdot 200nT_s +\ 2\pi n) = cos(2\pi \cdot 200nT_s)$$

We notice that the signal, at 800 *Hz*, has a copy at 200 *Hz* because it is sampled at 600 *Hz* < 2 800 = 1600 *Hz*.

1.8.9 The following analog signals, $x_1(t) = cos(2\pi10t)$ and $x_1(t) = cos(2\pi50t)$, that are sampled at a sampling frequency of 50 *Hz* create the sequences $x_1[n]$ and $x_2[n]$. Find the mathematical expression of $x_1[n]$ and $x_2[n]$ and comment on the conclusions.

SOLUTION: Sampling at $f_s = 50Hz$:

$$x_1[n] = cos(2\pi10/40n) = cos(n\pi/2)$$

$$x_2[n] = cos(2\pi50/40n) = cos(5n\pi/2)$$

$$\text{But } 5\pi/2 = \pi/2, \text{ thus } x_1[n] = x_2[n].$$

Consequently, $x_1[n]$ and $x_2[n]$ are identical. Therefore, frequency 50 Hz is folded back to frequency 10 Hz when sampled at 40 Hz. All frequencies (10 + 40k) Hz, k = 1, 2, ... are folded back to 10 Hz, resulting in an infinite number of sinusoidal continuous-time functions being represented, after sampling, by the same discrete-time signal.

1.8.10 For the signal $x(t)$ with spectrum

$$X(f) = cos(2\pi ft_o)\left[u\left(f + \frac{1}{4t_o}\right) - u\left(f - \frac{1}{4t_o}\right)\right], (t_o = 1/16)$$

calculate the minimum sampling frequency as well as the form of the signal $x(t)$ at the time-domain.

SOLUTION: The spectrum of the signal for $t_o = 1/16$ is written:

$$X(f) = \cos\left(\frac{\pi f}{8}\right)\left[u(f+4) - u(f-4)\right] \Rightarrow$$

$$X(f) = \cos\left(\frac{\pi f}{8}\right)rect\left(\frac{f}{8}\right) \tag{1}$$

The second term corresponds to a square pulse with a range of 8Hz (i.e., from −4Hz to 4Hz) symmetrically located at the origin of the axes, therefore the spectrum of X(f) is illustrated graphically in the following figure:

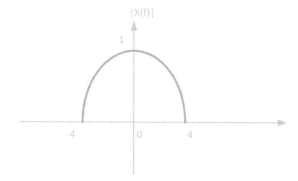

From the figure, it is obvious that $f_{max} = 4$ Hz, therefore, the minimum sampling frequency will be $F_s = 8Hz$.

It is known that:

$$\cos(2\pi f t_o)rect\left(\frac{f}{8}\right) \leftrightarrow \frac{1}{2}\left[\delta(t-t_o) + \delta(t+t_o)\right] \cdot 8\sin c(8t) =$$

$$= 4\left[\sin c\left(8(t-t_o)\right) + \sin c\left(8(t+t_o)\right)\right] \tag{2}$$

Thus, the signal x(t) at time domain for $t_o = 1/16$ is given by:

$$x(t) = 4\left[\sin c(8t - 0.5) + \sin c(8t + 0.5)\right] \tag{3}$$

1.8.11   The signal $y(t) = \sqrt{2}\cos(4\pi f_1 t) + 4\cos(\pi f_2 t)$ is sampled at a sampling frequency that is 20 times the minimum sampling frequency. Write the expressions of the sampled signal, $y_s[n]$, in the time domain and in the frequency domain, $Y_s(f)$.

SOLUTION: The maximum frequency of the signal $y(t) = \sqrt{2}\cos(4\pi f_1 t) + 4\cos(\pi f_2 t)$ is $f_{max} = 2f_1$, so the minimum sampling frequency is $f_s = 2f_{max} = 4f_1$.

For a sampling frequency of 20 times the minimum sampling frequency, the sampling period is such that the expression of the sampling signal in the time domain is:

$$y[n] = y(t)\Big|_{t=nT_\delta} = \sqrt{2}\cos(2\pi 2 f_1 nT_\delta) + 4\cos\left(2\pi \frac{f_2}{2} nT_\delta\right) \Rightarrow$$

$$y[n] = y(t)\Big|_{t=nT_\delta} = \sqrt{2}\cos\left(\frac{\pi n}{20}\right) + 4\cos\left(\frac{\pi n f_2}{80 f_1}\right) \tag{1}$$

The Fourier transform of the sampled signal is:

$$Y(f) = \frac{1}{2}\left[\sqrt{2}\left(\delta(f - 2f_1) + \delta(f + 2f_1)\right) + 4\left(\delta\left(f - \frac{f_2}{2}\right) + \delta\left(f + \frac{f_2}{2}\right)\right)\right] \tag{2}$$

Therefore, the sampled signal in the frequency domain will be:

$$X(f) = f_\delta \sum_{m=-\infty}^{\infty}\left[X(f - mf_\delta)\right] =$$

$$X_\delta(f) = 80 f_1 \sum_{m=-\infty}^{\infty} \frac{1}{2}\left[\begin{array}{l}\sqrt{2}\left(\delta(f - mf_\delta - 2f_1) + \delta(f - mf_\delta + 2f_1)\right) + \\ +4\left(\delta\left(f - mf_\delta - \frac{f_2}{2}\right) + \delta\left(f - mf_\delta + \frac{f_2}{2}\right)\right)\end{array}\right] \tag{3}$$

Replacing the sampling frequency we obtain:

$$X_\delta(f) = 80 f_1 \sum_{m=-\infty}^{\infty} \frac{1}{2}\left[\begin{array}{l}\sqrt{2}\left(\delta\left(f - m80 f_1 - 2f_1\right) + \delta(f - m80 f_1 + 2f_1)\right) + \\ 4\left(\delta\left(f - m80 f_1 - \frac{f_2}{2}\right) + \delta\left(f - m80 f_1 + \frac{f_2}{2}\right)\right)\end{array}\right] \tag{4}$$

1.8.12 Suppose that we have the signal $x(t) = a\mathrm{sinc}(b^2 t)$, $a > 0$. At which value of b does the signal have a sampling period of 100 msec?

SOLUTION: The spectrum of the signal $X(f)$ is computed next.

Using the Fourier transform property of time scaling we have:

$$\text{sinc}(t) \overset{F}{\longleftrightarrow} \text{rect}(f) \Rightarrow$$

$$\text{sinc}(B^2 t) \overset{F}{\longleftrightarrow} \frac{1}{B^2} \text{rect}\left(\frac{f}{B^2}\right) \Rightarrow A\,\text{sinc}(B^2 t) \overset{F}{\longleftrightarrow} \frac{A}{B^2} \text{rect}\left(\frac{f}{B^2}\right) \tag{1}$$

The maximum frequency that the signal $X(f)$ contains, equals $B^2/2$, so the sampling frequency is $f_s = 2B^2/2 = B^2$, and the sampling period is

$$T_s = \frac{1}{f_s} = \frac{1}{B^2} = 100ms \Rightarrow B = \pm\sqrt{10}.$$

The spectrum of the signal $X(f)$ is illustrated in the following figure.

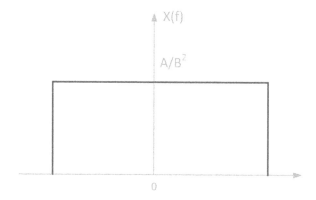

1.8.13   Consider the case of the signal $x(t) = 2\cos^2(40\pi t) + 5\sin(20\pi t)$. Starting from time t = 0, write the sequence of bits resulting from the sampling and coding of the first 5 samples (0, Ts, 2Ts, ..., 5Ts) of the signal. In order to compute the quantization levels, consider that the maximum value of the signal is 7 V and the minimum is –7 V.

SOLUTION: The expression of the discrete signal $x[nT_s]$ is $x[nT_s] = 2\cos^2(40\pi n T_s) + 5\sin(20\pi n T_s)$ (1). The interval between samples is $T_s = 1/F_s = 1/400$ s. We calculate the first five samples of the signal:

$$t = 0T_s\,,\, x[0] = 2$$

$$t = 1T_s\,,\, x[T_s] = 2\cos^2\left(\frac{40\pi}{400}\right) + 5\sin\left(\frac{20\pi}{400}\right) = 2.59$$

$$t = 2T_s\,,\, x[2T_s] = 2\cos^2\left(\frac{80\pi}{400}\right) + 5\sin\left(\frac{40\pi}{400}\right) = 2.85$$

$$t = 3T_s, x[3T_s] = 2\cos^2\left(\frac{120\pi}{400}\right) + 5\sin\left(\frac{60\pi}{400}\right) = 2.96$$

$$t = 4T_s, x[2T_s] = 2\cos^2\left(\frac{160\pi}{400}\right) + 5\sin\left(\frac{80\pi}{400}\right) = 3.13$$

In the case of 4 quantization zones, the range of each is going to be $[+7-(-7)]/4 = 3.5$ V. Therefore, from -7 to -3.5 V, we will have a codeword 00; from -3.5 to 0 V we will have a codeword 01; from 0 to 3.5 V will have a codeword 10; and finally, from 3.5 to 7 V would have a code word 11. Thus, the sequence 2, 2.59, 2.85, 2.96, 3.13 will be coded as: 10 101010 10.

1.8.14 Explain how a voltage signal is coded between -20 V and 20 V if 3 bits are used.

SOLUTION: If 3 bits are used, 8 quantization levels are required.

The quantization step is $q = (V_{max} - V_{min})/2^n = (20 - (-20))/8 = 5$. We divide the range into 8 levels of a height of 5 each, and at the center of each level we assign a value from 0 to $2^n - 1$.

The analytical calculations of the aforementioned procedure are given in Table 1.1.

1.8.15 A digital signal produced by the sampling of an analog signal, is a linear combination of three sinusoidal signals at frequencies of 0.15, 0.2 and 0.4.

  a. If the sampling frequency is 8 KHz, find which analog frequencies the above three digital frequencies correspond to.

  b. If the digital signal is reconstructed with a reconstruction period $T_s = 0.1msec$, what analog frequencies would the three digital frequencies correspond to in this case?

TABLE 1.1

Analytical Calculations

Levels	Mean Value of Level	3 Bits Encoding
-20 to 15	-17.5	000
-15 to 10	-12.5	001
-10 to 5	-7.5	011
-5 to 0	-2.5	010
0 to $+5$	2.5	100
$+5$ to $+10$	7.5	101
$+10$ to $+15$	12.5	110
15 to $+20$	17.5	111

SOLUTION:

a. The frequencies of the digital signal are $f_1 = 0.15$, $f_2 = 0.2$ and $f_3 = 0.4$. It is known, that a digital frequency, f, is assigned to an analog frequency, f_a, through the relation $f_a = f \cdot f_s$. Therefore, for the sampling frequency $f_s = 8KHz$, frequencies f_1, f_2, f_3, are assigned to $f_1 = 1.2KHz$, $f_2 = 1.6KHz$, $f_3 = 3.2KHz$.

b. If the digital signal is reconstructed using a sampling period $T_s = 0.1$ ms corresponding to a sampling frequency $f_s = 10$ KHz, then frequency f_1 is assigned to $f_{a,1} = f_1 \cdot 10000 = 1500Hz$, frequency f_2 is assigned to $f_{a,2} = f_2 \cdot 10000 = 2000Hz$ and frequency f_3 is assigned to $f_{a,3} = f_3 \cdot 10000 = 4000Hz$.

Hence, if f_1, f_2, f_3 correspond to the rate of the analog signal on which the sampling was implemented, then the answers are the three first frequencies, whereas the next three values of frequencies correspond to the reconstructed signal.

1.8.16   Find the Nyquist frequency for the signals:

$$sinc^2(100t), 0.15sinc^2(100t), sinc(100t) + 3sinc^2(60t), sinc(50t)sinc(100t).$$

SOLUTION: The signal $sinc^2(100t)$ is equal to the signal $\dfrac{1}{100} tri\left(\dfrac{\omega}{200\pi}\right)$ so the maximum frequency of the signal is 200π rad/sec. The Nyquist frequency will be $\omega_s = 400\pi$ rad/sec.

The signal 0.15 $sinc^2(100t)$ differs only in amplitude from the signal $0.15sinc^2(100t)$ so the spectrum is not affected, and the Nyquist frequency will be $\omega_s = 400\pi$ rad/sec.

The signal $sinc(100t) + sinc^2(60t)$ is equivalent to the signal $\dfrac{1}{100} rect\left(\dfrac{\omega}{200\pi}\right) + 3\dfrac{1}{60} tri\left(\dfrac{\omega}{120\pi}\right)$ so the maximum frequency of the signal is 120π rad/sec. The Nyquist frequency will be $\omega_s = 240\pi$ rad/sec.

The signal $sinc(50t)sinc(100t)$ is equivalent to the signal $\dfrac{1}{50} rect\left(\dfrac{\omega}{100\pi}\right) \cdot \dfrac{1}{100} rect\left(\dfrac{\omega}{200\pi}\right)$, so the maximum frequency of the signal is 100π rad/sec. The Nyquist frequency will be $\omega_s = 200\pi$ rad/sec.

1.8.17   Calculate the sampling frequency for the analog signal $x(t) = e^{-t}, t \geq 0$.

SOLUTION: The Fourier transform of the signal $x(t)$ is $X(j\Omega) = \dfrac{1}{1 + j\Omega}$ with amplitude $|X(j\Omega)| = \dfrac{1}{\sqrt{1 + \Omega^2}}$. For the relation $X(j\Omega) = 0$ to apply, it should be $f_{max} \to \infty$. Therefore, the sampling period must be zero, which is not possible. Let us choose as Ω_{max} the angular frequency that contains 95% of the signal energy. Then:

$$\int_0^{\Omega_{max}} |X(j\Omega)|\,d\Omega = A\int_0^{\infty} |X(j\Omega)|\,d\Omega \Rightarrow$$

$$\int_0^{\Omega_{max}} |X(j\Omega)|\,d\Omega = 0.95\int_0^{\infty} \frac{1}{\sqrt{1+\Omega^2}}\,d\Omega \tag{1}$$

Thus,

$$\tan^{-1}(\Omega)\,|_0^{\Omega max} = 0.95\tan^{-1}(\Omega)\,|_0^{\infty} = 0.95\left(\tan^{-1}(\infty) - \tan^{-1}(0)\right)0.95\frac{\pi}{2} \tag{2}$$

We solve the relation (2):

$$\tan^{-1}(\Omega)\,|_0^{\Omega max} = 0.95\frac{\pi}{2} \Rightarrow \tan^{-1}(\Omega_{max}) = 0.95\frac{\pi}{2} \Rightarrow \Omega_{max} = \tan\left(0.95\frac{\pi}{2}\right) \Rightarrow$$

$$\Omega_{max} = 12.7\ rad/sec$$

Therefore, the sampling frequency is:

$$f_s \geq 2f_{max} = 2\left(\frac{\Omega_{max}}{2\pi}\right) = \frac{12.7}{\pi} = 4.04\ samples/sec.$$

1.8.18 Consider the continuous-time signal: $x(t) = 3\cos(2000\pi t) + 5\sin(6000\pi t) + 10\cos(12000\pi t)$.

What is the discrete-time signal that is produced if we use $f_s = 5000$ samples/sec, and what is the continuous-time signal resulting from the reconstruction from its samples?

SOLUTION: The maximum frequency of the signal is $f_{max} = 6000\ Hz$ Therefore, according to the Nyquist criterion, the minimum sampling frequency that guarantees the correct reconstruction of the continuous-time signal from its samples is $f_{s,min} > 2f_{max} = 12000$ samples/sec.

The signal is sampled with frequency $f_s = 10f_{s,min} = 20f_2$, thus with a sampling period $T_s = \frac{1}{f_s} = \frac{1}{20f_2}$. The sampled signal in the time domain is: $x_s[n] = 1 + \cos(2\pi f_1 nT_s) + \cos(2\pi f_2 nT_s)$.

Sampling with sampling period $T_s = 1/f_s = 1/5000$ sec yields:

$$x[n] = x_a[nT_s] = 3\cos 2\pi \left(\frac{1}{5}\right) n + 5\sin 2\pi \left(\frac{3}{5}\right) n + 10\cos 2\pi \left(\frac{6}{5}\right) n$$

$$= 3\cos 2\pi \left(\frac{1}{5}\right) n + 5\sin 2\pi \left(1-\frac{2}{5}\right) n + 10\cos 2\pi \left(1+\frac{1}{5}\right) n$$

$$= 3\cos 2\pi \left(\frac{1}{5}\right) n + 5\sin 2\pi \left(-\frac{2}{5}\right) n + 10\cos 2\pi \left(\frac{1}{5}\right) n$$

$$= 13\cos 2\pi \left(\frac{1}{5}\right) n - 5\sin 2\pi \left(\frac{2}{5}\right) n$$

That is, the frequencies contained in the interval $(-f_s/2, f_s/2)$, not exceeding the Nyquist frequency $f_s/2$, are $f_s/5, 2 f_s/5$.

Consequently the continuous-time signal resulting from the reconstruction is: $x_a(t) = 13\cos 2000\pi t - 5\sin 4000\pi t$. We can see here the results of the aliasing resulting from the low sampling frequency.

1.8.19 A continuous-time signal $x_a(t)$ consists of a linear combination of sinusoidal signals at the frequencies 300 Hz, 400 Hz, 1.3 KHz, 3.6 KHz and 4.3 KHz. The signal is sampled with a frequency of 2 kHz, and the sampled sequence passes through an ideal low-pass filter with a cut-off frequency of 900 Hz, generating a continuous-time signal $y_a(t)$. What frequencies are present in the reconstructed signal?

SOLUTION: Since the signal $x_a(t)$ is sampled at a rate of 2 KHz, there will be multiple copies of the spectrum at the frequencies $F_i \pm 2000k$ where F_i is the frequency of the ith-sinusoid component of $x_a(t)$. Thus,

$$F_1 = 300 \ Hz,$$

$$F_2 = 400 \ Hz,$$

$$F_3 = 1300 \ Hz,$$

$$F_4 = 3600 \ Hz,$$

$$F_5 = 4300 \ Hz,$$

$$F_{1m} = 300, \ 1700, \ 2300, \ \dots \ , \ Hz$$

$$F_{2m} = 400, \ 1600, \ 2400, \ \dots \ , \ Hz$$

$$F_{3m} = 700, 1300, 3300, \ldots, Hz$$

$$F_{4m} = 400, 1600, 3600, \ldots, Hz$$

$$F_{5m} = 300, 2300, 4300, \ldots, Hz.$$

Therefore, after filtering with a low-pass filter with a cut-off frequency of 900 Hz, the frequencies present in $y_a(t)$ are 300 Hz, 400 Hz and 700 Hz.

1.8.20 The following analog signals are given:

$$x_1(t) = \begin{cases} \cos\left(\dfrac{\pi}{4m\sec}t\right), t \in [0,8]m\sec. \\ 0, t \notin [0,8]m\sec. \end{cases} \quad \text{and } x_2(t) = \begin{cases} -1, t \in [0,8]m\sec. \\ 0, t \notin [0,8]m\sec. \end{cases}$$

a. Sample the signals so that we get nine samples in the [0,8] msec interval. (Take the first sample at 0 sec, and the ninth at 8 msec.) What is the sampling frequency, f_s, and what are the exact values of the respective samples?

b. Assuming 3-bit and 4-bit rounding quantization, respectively, for level coding by arithmetic with two's complement, what are the respective values of the samples of both signals after the quantization?

SOLUTION:

a. The analog signal $x_1(t)$ is written: $x_1(t) = \cos\left(\dfrac{\pi}{4m\sec}t\right) = \cos\left(\dfrac{2\pi}{8m\sec}t\right)$ for t $\in$[0,8msec]. So the period of the signal is $T = 8m$ sec.
In order to have 9 samples in the interval [0,8msec], sampling frequency should be $T_s = \dfrac{1}{8}T = \dfrac{1}{8}\cdot 8m\sec = 1m\sec$. Thus, $f_s = \dfrac{1}{T_s} = \dfrac{1}{1m\sec} = 1kHz.$
The exact values of the samples of both signals are:

$$x_1(0) = \cos\left(\frac{\pi}{4}\cdot 0\right) = 1, x_1(1) = \cos\left(\frac{\pi}{4}\cdot 1\right) = \frac{\sqrt{2}}{2}, x_1(2) = \cos\left(\frac{\pi}{4}\cdot 2\right) = 0$$

$$x_1(3) = \cos\left(\frac{\pi}{4}\cdot 3\right) = -\frac{\sqrt{2}}{2}, x_1(4) = \cos\left(\frac{\pi}{4}\cdot 4\right) = -1, x_1(5) = \cos\left(\frac{\pi}{4}\cdot 5\right) = -\frac{\sqrt{2}}{2}$$

$$x_1(6) = \cos\left(\frac{\pi}{4}\cdot 6\right) = 0, x_1(7) = \cos\left(\frac{\pi}{4}\cdot 7\right) = \frac{\sqrt{2}}{2} \text{ και } x_1(8) = \cos\left(\frac{\pi}{4}\cdot 8\right) = 1.$$

For the second signal, we have:

$$x_2(0) = x_2(1) = x_2(2) = x_2(3) = x_2(4) = x_2(5) = x_2(6) = x_2(7) = x_2(8) = -1.$$

b. For quantization of 3 bits level $\Rightarrow 2^3 = 8$ levels

So, the two signals after the output of the 3-bit quantizer will have the following values:

$$\hat{x}_1(0) = 0.75,\ \hat{x}_1(1) = 0.75,\ \hat{x}_1(2) = 0,\ \hat{x}_1(3) = -0.75,\ \hat{x}_1(4) = -1,\ \hat{x}_1(5) = -0.75,\ \hat{x}_1(6) = 0,$$
$$\hat{x}_1(7) = 0.75\ \text{και}\ \hat{x}_1(8) = 0.75$$

whereas

$$\hat{x}_2(0) = \hat{x}_2(1) = \hat{x}_2(2) = \hat{x}_2(3) = \hat{x}_2(4) = \hat{x}_2(5) = \hat{x}_2(6) = \hat{x}_2(7) = \hat{x}_2(8) = -1.$$

For quantization of 4 bits, there are $\Rightarrow 2^4 = 16$ levels, and the quantization step is
$$q = \frac{2|x_{\max}|}{2^4} = \frac{2 \cdot 1}{2^4} = 2^{-3} = 0.125.$$

Therefore, the two signals at the output of the 4-bit quantizer will have the following values: $\hat{x}_1'(0) = 0.875,\ \hat{x}_1'(1) = 0.75,\ \hat{x}_1'(2) = 0,\ \hat{x}_1'(3) = -0.75,\ \hat{x}_1'(4) = -1,$ $\hat{x}_1'(5) = -0.75,\ \hat{x}_1'(6) = 0,\ \hat{x}_1'(7) = 0.75$ and $\hat{x}_1'(8) = 0.875,$
whereas

$$\hat{x}_2'(0) = \hat{x}_2'(1) = \hat{x}_2'(2) = \hat{x}_2'(3) = \hat{x}_2'(4) = \hat{x}_2'(5) = \hat{x}_2'(6) = \hat{x}_2'(7) = \hat{x}_2'(8) = -1.$$

1.8.21 Consider the signal processing system illustrated in the figure below. The sampling periods of the A/D and D/A converters are $T = 5ms$ and $T' = 1\ ms$, respectively. Calculate the output $y_a(t)$ of the system if the input is $x_a(t) = 3\cos100\pi t + 2\sin250\pi t$. The post-filter cuts off the frequencies above $F_s/2$.

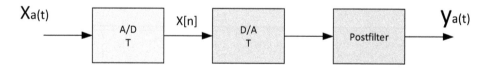

SOLUTION: The sampling frequencies of the ADCs and DACs are $F_{s1} = \dfrac{1000}{5} = 200$ samples/sec, $F_{s2} = 1000$ samples/sec.

The signal $x(n)$ is:

$$x[n] = 3\cos\frac{100\pi}{200}n + 2\sin\frac{250\pi}{200}n = 3\cos\left(\frac{\pi n}{2}\right) + 2\sin\left(\frac{5\pi n}{4}\right) \Rightarrow$$

$$x[n] = 3\cos\left(\frac{\pi n}{2}\right) + 2\sin\left(2\pi - \frac{3\pi n}{4}\right) = 3\cos\left(\frac{\pi n}{2}\right) - 2\sin\left(\frac{3\pi n}{4}\right)$$

Thus, the output $y_a'(t)$ of the system is:

$$y_a'(t) = 3\cos\left(\frac{\pi \cdot 1000}{2}\right)t - 2\sin\left(\frac{3\pi \cdot 1000}{4}\right)t + \cdots =$$
$$= 3\cos 500\pi t - 2\sin 750\pi t + \ldots\ldots$$

The post-filter filters the frequencies over 500 Hz. So, the output of the system is:

$$y_a(t) = 3\cos 500\pi t - 2\sin 750\pi t.$$

1.8.22 The continuous-time signal $x(t) = \cos(1000\pi t) + \cos(2400\pi t)$ is sampled with a frequency of 2 kHz. The resulting discrete-time signal, $x[n]$, passes through a digital signal processor with a unit amplitude response for all frequencies. An ideal analog low-pass filter with a cut-off frequency of 1 KHz is used to reconstruct the analog signal $y(t)$ from the output $y[n]$ of the digital signal processor.

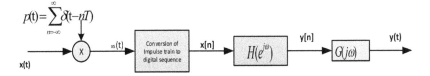

Compute the spectrums of $x(t)$ and of $x[n]$.

SOLUTION: The signal $x(t) = \cos(1000\pi t) + \cos(2400\pi t)$ is written as

$$x(t) = \frac{1}{2}(e^{j1000\pi t} + e^{-j1000\pi t}) + \frac{1}{2}(e^{j2400\pi t} + e^{-j2400\pi t}) \tag{1}$$

By using the formula $X(j\Omega) = 2\pi \sum\limits_{k=-\infty}^{\infty} a_k \delta(\Omega - \Omega_0 k)$ for a periodic signal, we derive relation (2) that describes the spectrum of the signal $x(t)$.

$$X(j\Omega) = 2\pi \left(\frac{1}{2} \delta(\Omega + 1000\pi) + \delta(\Omega - 1000\pi) \right) +$$

$$+ 2\pi \left(\frac{1}{2} \delta(\Omega + 2400\pi) + \delta(\Omega - 2400\pi) \right) \Rightarrow$$

$$X(j\Omega) = \pi \big(\delta(\Omega + 1000\pi) + \delta(\Omega - 1000\pi) \big) +$$

$$+ \pi \big(\delta(\Omega + 2400\pi) + \delta(\Omega - 2400\pi) \big) \tag{2}$$

The signal $x[n]$ resulting from the sampling is

$$x[nT] = \frac{1}{2} (e^{j1000\pi nT} + e^{-j1000\pi nT}) + \frac{1}{2} (e^{j2400\pi nT} + e^{-j2400\pi nT})$$

$$x[n/2000] = \frac{1}{2} (e^{j0.5\pi n} + e^{-j0.5\pi n}) + \frac{1}{2} (e^{j1.2\pi n} + e^{-j1.2\pi n})$$

$$or\ x[n] = \frac{1}{2} (e^{j0.5\pi n} + e^{-j0.5\pi n}) + \frac{1}{2} (e^{j1.2\pi n} + e^{-j1.2\pi n}) \tag{3}$$

So the spectrum $x[n]$ will be

$$X(e^{j\omega}) = \pi \big(\delta(\omega + 0.5\pi) + \delta(\omega - 0.5\pi) \big) +$$

$$+ \pi \big(\delta(\omega + 1.2\pi) + \delta(\omega - 1.2\pi) \big) \tag{4}$$

1.8.23   Consider the continuous-time signal $x(t) = \cos(2\pi f t)$ of frequency $f = 400 KHz$. Sketch the sampled signal (amplitude spectrum and time-waveform) for the following sampling frequencies $f_s = f$, $1.2f$, $1.5f$, $2f$, $10f$ (using MATLAB).

SOLUTION: For sampling frequency $f_s = f = 400 KHz$ the code in MATLAB is:

```
f  = 400e3;  % f_s = f
fs = 1.*f;
Nt = 100;
Ts = 1./fs;
t  = 0:Ts:(Nt-1).*Ts;
x  = cos(2.*pi.*f.*t);
z  = fft(x);
subplot(211),plot((0:1:(Nt-1)).*fs./Nt,abs(z));
subplot(212),plot(t,x)
```

We notice that the frequency of the reconstructed signal is $f = 0KHz$. We repeat the sampling procedure for all the given frequencies, from which we reach to the following conclusions:

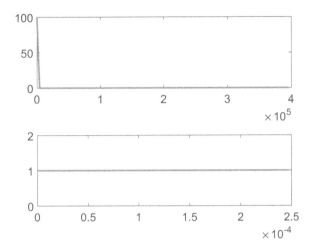

For sampling frequency $f_s = 1.2f = 480KHz$, the reconstructed signal frequency is $f = 81.6KHz$ (aliasing).

For sampling frequency $f_s = 1.5f = 600KHz$, the reconstructed signal frequency is $f = 198KHz$ (aliasing).

For sampling frequency $f_s = 2f = 800KHz$ the reconstructed signal frequency is $f = 400KHz$.

For sampling frequency $f_s = 10f = 4000KHz$ the reconstructed signal frequency is $f = 400KHz$.

1.8.24 Consider the continuous time signal $x(t) = cos(7t) + cos(23t)$. Using three different sampling periods (ts1 = 0.05 sec, ts2 = 0.1 sec and ts3 = 0.2 sec), study the effect of the choice of the sampling period on the effectiveness of the reconstruction procedure of the signal from its samples.

SOLUTION: The MATLAB code, with the corresponding comments on each case, is:

```
ts1 = 0.05; ts2 = 0.1; ts3 = 0.2;
ws1 = 2*pi/ts1; ws2 = 2*pi/ts2; ws3 = 2*pi/ts3;
w1 = 7; w2 = 23;
t = [0:0.005:2];
x = cos(w1*t)+cos(w2*t);
subplot(2,2,1)
plot(t,x),grid,xlabel('Time(s)'),ylabel('Amplitude'),
  title('ContinuousTimeSignal); x(t) = cos(7t)+cos(23t)')
% Sampling of the analog signal with sampling period ts = 0.05 s and
  ws1 = 5.5*w2
t1 = [0:ts1:2];
xs1 = cos(w1*t1)+cos(w2*t1);
subplot(2,2,2)
```

```
stem(t1,xs1);grid;hold on, plot(t,x,'r:'),hold off,
    xlabel('Time(s)'),
ylabel('Amplitude'), title('Sampled version of x(t) with ts =
    0.005s')
% Sampling of the analog signal with sampling period s = 0.1 s και
    ws2 = 2.7*w2
t2 = [0:ts2:2];
xs2 = cos(w1*t2)+cos(w2*t2);
subplot(2,2,3)
stem(t2,xs2);grid;hold on, plot(t,x,'r:'),hold off,
    xlabel('Time(s)'),
ylabel('Amplitude'), title('Sampled version of x(t) with ts = 0.1s')
% Sampling of the analog signal with sampling period ts = 0.2 s and
    ws3 = 1.37*w2<2*w2
t3 = [0:ts3:2];
xs3 = cos(w1*t3)+cos(w2*t3);
subplot(2,2,4)
stem(t3,xs3);grid;hold on, plot(t,x,'r:'),hold off,
    xlabel('Time(s)'),
ylabel('Amplitude'), title('Sampled version of x(t) with ts=0.2 s')
```

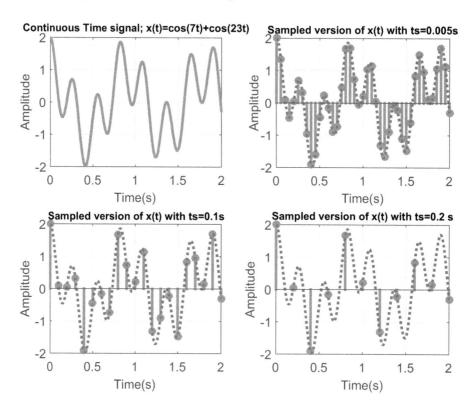

1.8.25 Let an analog signal $x_a(t) = \cos(10\pi t)$, $0 \leq t < 1$, $f = 5Hz$. The signal is sampled with the frequency $f_s = 10Hz(T_s = 0.1s)$. Sketch the signal after sampling and the reconstructed signal resulting from the relation $y(t) = \displaystyle\sum_{n=-\infty}^{\infty} x[n] \frac{\sin\left[\pi(t-nT)/T\right]}{\pi(t-nT)/T}$.

SOLUTION: For $f_s = 10Hz(T_s = 0.1s)$, the sampled signal will be $x[n] = x_a(nT) = \cos(10\pi nT) = \cos(\pi n)$.

The MATLAB code for illustrating, in figures, the signals asked is:

```
T = 0.1;
n = 0:10;
x = cos(10*pi*n*T);
stem(n,x);
dt = 0.001;
t = ones(11,1)* [0:dt:1];
n = n'*ones(1,1/dt+1);
y = x*sinc((t-n*T)/T);
hold on;
plot(t/T,y,'r')
```

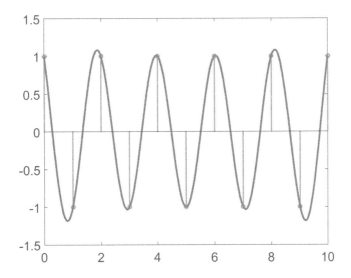

NOTE: The code in MATLAB for designing the sinc function is

```
x = linspace(-5,5);
y = sinc(x);
plot(x,y);
```

1.8.26 Let the continuous-time signals:

$$x_1(t) = \cos(6\pi t), \ x_2(t) = \cos(14\pi t), \ x_3(t) = \cos(26\pi t) \text{ and } y(t) = x_1(t) + x_2(t) + x_3(t).$$

    a. For sampling frequency of 1KHz, sketch the samples of the given signals.

    b. For sampling frequencies of 10Hz and 100Hz, sketch the samples of the given signals. Also, sketch the signal resulting from the reconstruction (based on the sampling theorem).

SOLUTION:

a. The maximum frequency of the signal $y(t)$ is 13Hz. We will sketch the samples of the given signals for sampling frequency 1000Hz, which is clearly greater than 2·13Hz. Because we chose a sampling frequency that exceeds the minimum limit set by Nyquist, we are able not only to reconstruct the original signal, but also to produce much better graphs.

    The MATLAB code follows:

```
fs = 1000; % set sampling frequency 1000Hz
ts = 1/fs; % compute sampling period
n = 0:ts:1; %set time interval
x1 = cos(6*pi*n);
x2 = cos(14*pi*n);
x3 = cos(26*pi*n);
y = x1+x2+x3;
subplot(411),plot(n,x1);
subplot(412),plot(n,x2);
subplot(413),plot(n,x3);
subplot(414),plot(n,y);title('Final signal y(t) - fs=1000Hz')
```

b. If we choose as a sampling frequency 10Hz < 2·13Hz, using the appropriate code in MATLAB, we will get a result that is going to differ from the previous one due to the aliasing effect.

```
fs1 = 10; % Set sampling frequency 10Hz
ts1 = 1/fs1; % compute sampling period
n1 = 0:ts1:1; %set time interval
x11 = cos(6*pi*n1);
x12 = cos(14*pi*n1);
x13 = cos(26*pi*n1);
y1 = x11+x12+x13;
subplot(411),plot(n1,x11);
subplot(412),plot(n1,x12);
subplot(413),plot(n1,x13);
subplot(414),plot(n1,y1);title('Final signal y1(t) - fs1=10Hz')
```

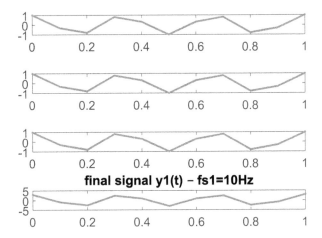

final signal y1(t) – fs1=10Hz

With the following commands we will sketch the original signal *y(t)* and the one produced from its reconstruction after setting 10*Hz* as the sampling frequency. The original signal will be reconstructed based on the sampling theorem.

Having in mind the relation $y_a(t) = \displaystyle\sum_{n=-\infty}^{\infty} y_a(nT_s)\sin c\left[f_s(t-nT_s)\right]$, where $y_a(t)$ is the analog reconstructed signal, and $y_a[nT_s]$ is the sampled signal, we get:

```
for i = 1:length(y)
yrec(i)  = sum(y1.*sinc(fs1*(n(i)-n1)));
end
subplot(211);plot(n,y);
subplot(212);plot(n,yrec);
```

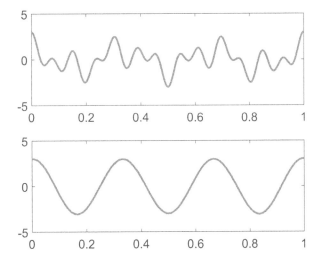

We notice that the reconstructed signal has no resemblance to the original due to overlapping. The signal resulting from the reconstruction is a cosine of 3*Hz*, with an amplitude of 3.

The spectrum of the sampled signal is formed by shifting the original spectrum by integer multiples of the sampling frequency and by adding up the results, that is, $Y_s(f) = \dfrac{1}{T_s} \displaystyle\sum_{k=-\infty}^{\infty} Y(f + kf_s)$. The function $Ys(f)$ is periodic with a period, f_s. Consider

an interval $[-f_s/2, f_s/2]$ of $Y_s(f)$, namely $[-5,5]$. We will find out what spectral components are present in this interval, and because of the periodicity we will have absolutely defined $Y_s(f)$.

Because of the component $x_1(t) = cos(6\pi t) = cos(2 \cdot pi \cdot 3 \cdot t)$ in the original signal, there will be a frequency component with a frequency of 3Hz in the sampled sig-

nal. This component corresponds to $Y_s(f) = \dfrac{1}{T_s}\displaystyle\sum_{k=-\infty}^{\infty} Y(f + kf_s)$ for $k = 0$ (zero shift), and it is in $[-5,5]$ interval.

The component $x_2(t) = cos(14\pi t) = cos(2 \cdot pi \cdot 7 \cdot t)$ is out, but its shift $cos(2 \cdot pi \cdot (10-7) \cdot t) = cos(2 \cdot pi \cdot 3 \cdot t)$ is in the interval $[-5,5]$ and is of frequency 3Hz. The component $x_3(t) = cos(26\pi t) = cos(2 \cdot pi \cdot 13 \cdot t)$ is out, but its shift $cos(2 \cdot pi \cdot (13 - 7 \cdot t) = cos(2 \cdot pi \cdot 3 \cdot t)$ is in the interval $[-5,5]$ and is of frequency 3Hz.

So the wrong sampling with $f_s = 10Hz < 2 \cdot 13Hz$ turned the frequency signals of 7Hz and of 13Hz to signals of 3Hz. That is the reason that the signal at the output is a cosine of amplitude 3 (as a result of the superposition of these three signals).

Let us choose as sampling frequency $100Hz > 2 \cdot 13Hz$.

```
fs1 = 100; % set sampling frequency 10Hz
ts1 = 1/fs1; % compute sampling period
n1 = 0:ts1:1; %set time interval
x11 = cos(6*pi*n1);
x12 = cos(14*pi*n1);
x13 = cos(26*pi*n1);
y1 = x11+x12+x13;
subplot(411),plot(n1,x11);
subplot(412),plot(n1,x12);
subplot(413),plot(n1,x13);
subplot(414),plot(n1,y1);title('Final signal y1(t) - fs1=100Hz')
The reconstructed signal will be the same as the original:
for i = 1:length(y)
yrec(i) = sum(y1.*sinc(fs1*(n(i)-n1)));
end
subplot(211);plot(n,y);
subplot(212);plot(n,yrec);
```

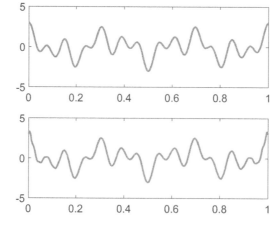

1.8.27 Sketch the quantized sinusoidal wave signal using a 3-bit quantizer (using MATLAB).

SOLUTION: The following MATLAB code is used for the 3-bit quantization of a sinusoidal waveform (8 levels of quantization).

```
clc;
close all;
Am = 4;
bit = 3;
f = 1;
fs = 30;
t = 0:1/fs:1*pi;
y = Am*sin(2*pi*f*t);
Nsamples = length(y);
quantised_out = zeros(1,Nsamples);
del = 2*Am/(2^bit);
Llow = -Am+del/2;
Lhigh = Am-del/2;
for i = Llow:del:Lhigh
for j = 1:Nsamples
if(((i-del/2)<y(j))&&(y(j)<(i+del/2)))
quantised_out(j) = i;
end
end
end
stem(t,quantised_out);
hold on;
plot(t,y,'color','r');
```

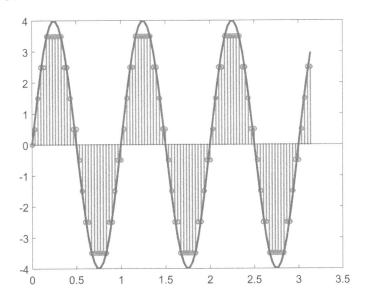

2

Discrete-Time Signals and Systems

2.1 Discrete-Time Signals

Discrete-time signals are the signals defined only for a discrete set of values of the independent variable of time, e.g., for all integers (Figure 2.1). In this case, the independent variable receives only discrete values, which are uniformly distributed. Therefore, discrete-time signals are described as sets of samples whose amplitudes can receive constant values. When each sample of a discrete-time signal is quantized (i.e., the amplitude must receive only a finite set of discrete values) and then coded, the final signal is referred to as a digital signal. The output of a digital computer is an example of a digital signal.

A signal may be discrete by nature. One example is the closing prices of the stock market index. However, let us consider another category of discrete signals: those resulting after the sampling of continuous-time signals. Sampling is being implemented at fixed intervals of time and is mathematically expressed by the relation $x(t) = [nT_s]$. The result is a sequence of numbers, which is a function of the variable n.

2.2 Basic Discrete-Time Signals

2.2.1 Impulse Function

The impulse function is defined as:

$$\delta[n] = \begin{cases} 1 & for \quad n = 0 \\ 0 & for \quad n \neq 0 \end{cases} \tag{2.1}$$

and is illustrated in Figure 2.2.

The shifted impulse function is, then, given by

$$\delta[n-k] = \begin{cases} 1 & for \quad n = k \\ 0 & for \quad n \neq k \end{cases} \tag{2.2}$$

The properties of the impulse function are given in the following, Table 2.1.

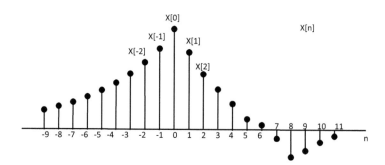

FIGURE 2.1
Discrete-time signal.

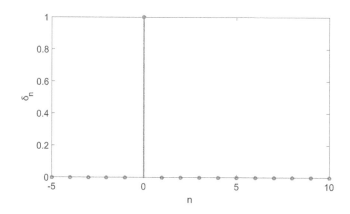

FIGURE 2.2
Impulse function.

TABLE 2.1

Properties of Impulse Function

$\displaystyle\sum_{m=-\infty}^{+\infty} x[n]\delta[n] = x[0]$	$\displaystyle\sum_{m=-\infty}^{+\infty} x[n]\delta[n-N] = x[N]$
$x[n]\cdot\delta[n] = \begin{cases} x[0] & \text{when } n = 0 \\ 0 & \text{when } n \neq 0 \end{cases}$	$x[n]\cdot\delta[n-N] = \begin{cases} x[N] & \text{when } n = N \\ 0 & \text{when } n \neq N \end{cases}$

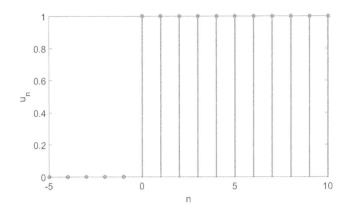

FIGURE 2.3
Unit sequence function.

2.2.2 Unit Step Function

The unit step function is illustrated in Figure 2.3 and is defined as:

$$u[n] = \begin{cases} 1, & n \geq 0 \\ 0, & n < 0 \end{cases} \tag{2.3}$$

The shifted unit step function is given by

$$u[n-k] = \begin{cases} 1, & n \geq k \\ 0, & n < k \end{cases} \tag{2.4}$$

Between the functions $\delta[n]$ and $u[n-k]$, the following relations are true:

$$\delta[n] = u[n] - u[n-1] \text{ and } u[n] = \sum_{m=-\infty}^{n} \delta[m].$$

2.2.3 Ramp Function

The ramp function (of unit slope) is illustrated in Figure 2.4, and it is defined as:

$$ramp[n] = \begin{cases} n & for & n \geq 0 \\ 0 & for & n < 0 \end{cases} \tag{2.5}$$

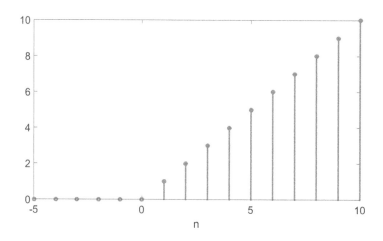

FIGURE 2.4
Ramp function.

2.2.4 Unit Rectangular Function (Pulse Function)

The unit rectangular function is illustrated in Figure 2.5 and it is defined as:

$$\text{rect}[n] = \begin{cases} 1, k_1 \le n \le k_2 \\ 0, n < k_1 \ \& \ n \vec{>} k_2 \end{cases} \tag{2.6}$$

2.2.5 Exponential Function

The unit exponential function is defined as:

$$g[n] = \begin{cases} Ca^n, \ n \ge 0 \\ 0 \quad , \ n < 0 \end{cases} \tag{2.7}$$

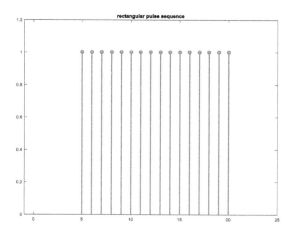

FIGURE 2.5
Unit rectangular function.

The real exponential function is illustrated in Figure 2.6, where $C \in \mathbb{R}$ and $a \in \mathbb{R}$, for $0 < \alpha < 1$ and for $0 > \alpha 1$, respectively.

In complex exponential signals, the quantities A and Ω are generally complex numbers. A complex number can be written as the product of its amplitude and its phase. Thus, $C = |C|e^{j\theta}$ and $a = |a|e^{j\Omega}$. By replacing the relation of the definition of complex exponential signals, we obtain:

$$x[n] = |C|e^{j\theta}|a|^n e^{j\Omega n} = |C| \cdot |a|^n e^{j(\Omega n + \theta)}, \text{ or}$$

$$x[n] = |C| \cdot |a|^n \cos(\Omega n + \theta) + j|C| \cdot |a|^n \sin(\Omega n + \theta) \qquad (2.8)$$

When $|a| = 1$, the real and imaginary parts of the signal are sinusoidal signals of discrete time. When $|a| < 1$, the sinusoidal signals are multiplied by an exponential number, which decreases. Finally, for $|a| > 1$, the real and imaginary parts of $x[n]$ are sinusoidal sequences whose amplitude increases exponentially with n. Figure 2.7 illustrates the complex exponential sequence:

$$x = \exp((-0.1 + j0.3)n), \quad for \ -10 \le n \le 10.$$

Period of complex exponential signals of discrete time: Let N be the period of the sequence $x[n]$. Then, according to the definition: $x[n] = x[n + N] = Ae^{j\Omega n}$ or $e^{j\Omega n} = e^{j\Omega n} \cdot e^{j\Omega n}$, which is obviously true when $e^{j\Omega n} = 1$. This relation is true when $\Omega N = k2\pi$. So,

$$\Omega = 2\pi \frac{k}{N} \qquad (2.9)$$

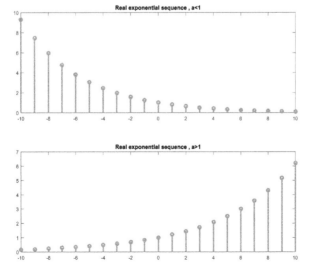

FIGURE 2.6
The real exponential function.

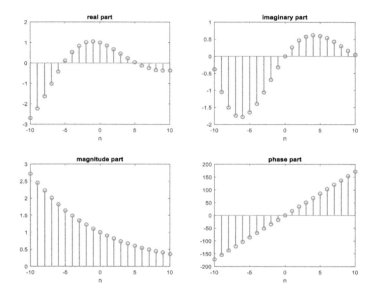

FIGURE 2.7
Complex exponential sequence.

Therefore, in order for the complex exponential sequence to be periodic, there must be integers k and λ, such that the circular frequency, Ω, can be written as $\Omega = 2\pi \dfrac{\kappa}{\lambda}$. When this is true, the period of the exponential sequence is given by:

$$\Omega N = k2\pi \Rightarrow N = k\frac{2\pi}{\Omega} \tag{2.10}$$

The fundamental period occurs for the smallest value of k, for which N becomes an integer.

2.2.6 The Sinusoidal Sequence

The sinusoidal sequence is defined as:

$$f[n] = \begin{cases} A \sin \omega_0 n, & n \geq 0 \\ 0, & n < 0 \end{cases} \tag{2.11}$$

and is illustrated in Figure 2.8.

Period of discrete-time sinusoidal signals: Let N be the period of the signal $x[n] = A \sin(\Omega n + \varphi)$; then, according to the definition of periodicity:

$$x[n] = x[n+N] \text{ or } A \sin(\Omega n + \varphi) = A \sin[\Omega(n+N) + \varphi] = A \sin[\Omega n + \Omega N + \varphi] \Rightarrow$$
$$\sin(\Omega n + \varphi) = \sin[(\Omega n + \varphi) + \Omega N].$$

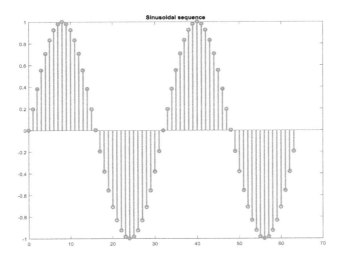

FIGURE 2.8
Discrete-time sinusoidal sequence.

Based on the periodicity of trigonometric functions, this equality is true only when $\Omega N = 2k\pi$. Thus: $\Omega = 2\pi \dfrac{k}{N}$. Therefore, in order for a sinusoidal sequence to be periodic, there must be integers μ and λ, such that the circular frequency, Ω, can be written as $\Omega = 2\pi \dfrac{\mu}{\lambda}$. When this is true, the fundamental period of the sinusoidal sequence is given by Equation 2.9, i.e., $N = k \dfrac{2\pi}{\Omega}$.

2.3 Even and Odd Discrete-Time Signals

The presence of *even and odd symmetry* in a signal is of particular interest because it reduces the amount of time required to process it with a computer. It is noteworthy that most signals are neither even nor odd. Each signal, though, can be written as the sum of an even and an odd signal.

A discrete-time signal is even when:

$$x[-n] = x[n] \tag{2.12}$$

for all values of n.

The graphs of the even signals are easily recognized by their symmetry with respect to the vertical axis (Figure 2.9).

A discrete-time signal is odd when:

$$-x[-n] = x[n] \tag{2.13}$$

for all values of n.

Odd signals are symmetric with respect to the origin of the axes (Figure 2.10). For a discrete-time signal, an even signal $x_e[n]$ and an odd signal $x_o[n]$ can be calculated,

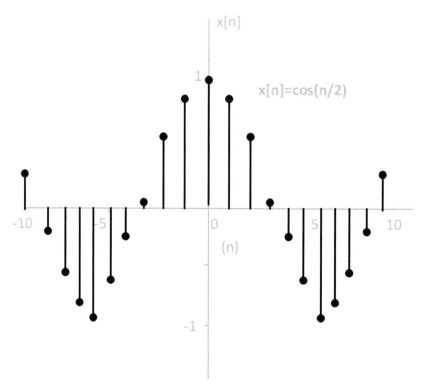

FIGURE 2.9
An even discrete-time signal.

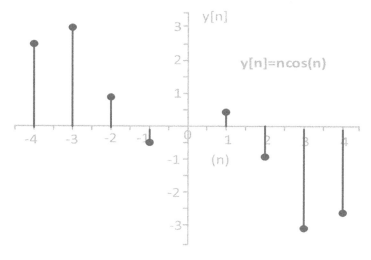

FIGURE 2.10
An odd discrete-time signal.

and then, if added together, can result in the original signal, $x[n]$. Signs $x_e[n]$ and $x_o[n]$ are often referred to as the even and odd components of $x[n]$. It is:

$$x_e[n] = \frac{x[n] + x[-n]}{2} \qquad (2.14a)$$

$$x_o[n] = \frac{x[n] - x[-n]}{2} \qquad (2.14b)$$

2.4 Energy and Power of a Discrete-Time Signal

The *energy of a discrete-time signal* is given by:

$$E = \sum_{n=-\infty}^{+\infty} |x[n]|^2 = E = \lim_{N \to \infty} \sum_{n=-N}^{+N} |x[n]|^2 \qquad (2.15)$$

The *power of a discrete-time signal* is given by:

$$P = \lim_{N \to \infty} \frac{1}{2N+1} \sum_{n=-N}^{+N} |x[n]|^2 \qquad (2.16)$$

Discrete-time signals are generally classified into two categories:

1. Finite energy signals. These are the signals for which energy is not infinite, that is $0 < E < \infty$. In the case of discrete-time signals, the average power is given by $P = \lim_{N \to \infty} \frac{E}{2N+1}$; and, since the numerator is finite, the average power will be 0.
2. Finite power signals. These are the signals for which power is not infinite, that is $0 < P < \infty$. From $P = \lim_{N \to \infty} \frac{E}{2N+1}$, it is obvious that if power gets a non-zero value, then energy must be non-finite, that is, $E = \infty$.

However, there are also signals that do not belong to any of the aforementioned categories. Such an example is the signal $x[n] = n$. Periodic signals form an important category of signals that have infinite energy and finite power.

2.5 Conversion of the Independent and Dependant Variable

There are three types of conversion of the independent variable of a signal:

1. *Time shift*: If $m > 0$, the signal $x[n - m]$ is delayed by m on n axis. That is, it shifts right on the n-axis. The signal $x[n + m]$ proceeds $x[n]$, i.e., it is shifted to the left on the n-axis.
2. *Time reversal (time reversal mirroring)*: The signal $x[-n]$ is symmetrical $x[-n]$, with respect to the vertical axis.
3. *Time scaling*: If $a > 1$, the signal $x[an]$ is compressed in time, that is, it lasts a times less than $x[n]$. If $a < 1$, the signal $x[an]$ stretches over time.

Conversion of the dependent variable causes variations of the signal along the vertical axis. For example, the signal $x[n] + 3$ results by shifting the signal $x[n]$ 3 units upwards, the signal $3x[n]$ results by multiplying the ordinate of the signal by 3 and the signal $-x[n]$ is symmetrical to $x[n]$ with respect to the horizontal axis.

2.6 Discrete-Time Systems

A system can be considered to be a physical object, which processes one or more input signals applied to it, resulting in the generation of new signals. A *discrete-time system* processes an input sequence, $x[n]$, to generate an output sequence $y[n]$. When $y[n] = x[n]$, $\forall n$, then the system is called identical. The signal $x[n]$ is the input of the system whereas the resulting signal $y[n]$ is the output of the system. The transform that the system induces to the input signal can be described either with a function or not. From a mathematical point of view, a discrete-time system can be considered as a transform, T, that transforms the signal $x[n]$ applied to the input, to a signal $y[n] = T\{x[n]\}$ at the output (Figure 2.11).

The system of Figure 2.11 is known as a single-input, single-output system or SISO. Most of the systems in the physical world have more inputs and/or outputs. A system with multiple inputs and a single output, e.g., an adder, is known as the MISO system (multiple inputs, single output). Systems with multiple inputs and multiple outputs are known as MIMO systems (Multiple Inputs, Multiple Outputs).

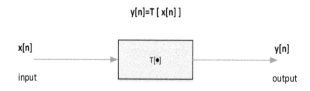

FIGURE 2.11
Discrete-time signal.

Examples of discrete time systems are: the ideal discrete time delay system $y[n] = x[n-l]$, the discrete-time moving average system, $y[n] = \dfrac{1}{M}\sum_{m=0}^{M-1} x[n+m]$, the discrete-time system, defined by the mathematical relation $y[n] = \sum_{k=-\infty}^{n} x[k]$, and it is called the accumulator, because in one sense, it sums all values from $-\infty$ up to n, etc.

Various algorithms of numerical analysis are discrete-time systems. Also, processors that work on discrete-time data can be considered discrete-time systems. A special category of discrete-time systems is that of the digital systems. These are systems whose inputs and outputs are digital signals. A large part of the modern hardware corresponds to digital systems.

In modern technology, in order to study a system, four techniques are followed:

1. System analysis
2. System modeling
3. System identification
4. Model validation

System analysis includes the study of system inputs and outputs over time and frequency. Through this study, an attempt is made to understand the complexity of the system and classify it into a category. System modeling aims to "limit the description" of the system to the basic mathematical relations that govern its behavior. These mathematical relations are expressed in terms of parameters that need to be defined to respond to the operation of the particular system at some point in time. These parameters may be constant or time-varying. System identification is intended to determine the parameters of the model so that its behavior is "as close as possible" to that of the system. Finally, the quality of the model must be checked. Different criteria are used for this purpose.

In practice, only special categories of systems are of interest. Each category is usually defined based on the number of constraints acting on transform, T. A very interesting category of systems offered for mathematical analysis is the linear time invariant systems (LTI systems).

2.7 Categories of Discrete-Time Systems

2.7.1 Linear Discrete Systems

When a signal $x_1[n]$ is applied to the input of a discrete-time system, T, and it gives rise to the output $y_1[n]$, while the input, $x_2[n]$, gives rise to the output, $y_2[n]$, then the system, T, is called *linear* if the input, $x[n] = ax_1[n] + bx_2[n]$, gives rise to the output, $y[n] = ay_1[n] + by_2[n]$, where a, b are constant, i.e., if the superposition principle is true. The signal, $x[n]$, is called linear combination of the signals $x_1[n]$ and $x_2[n]$.

Linear systems are those for which the properties of homogeneity and additivity are true, i.e., the behavior of a linear system does not change when the input is multiplied by a constant number. Two signals applied to the input of a linear system do not interact with each other (Figure 2.12).

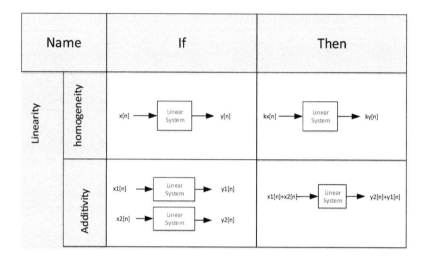

FIGURE 2.12
Homogeneity and additivity in a discrete time system.

In a linear system:

- The output may depend directly on time
- The output may not depend directly on time
- The output may result from the time shift of the input
- The output may result from the compression/decompression of the input

Examples of linear systems are those defined by mathematical relations such as: $y[n] = 2x[n]$, $y[n] = 2n \cdot x[n]$, $y[n] = 2n \cdot x[n-1]$, $y[n] = 3n \cdot x[2n]$, $y[n] = 2n \cdot x[2n-1]$, $y[n] = 2n^2x[n-2]$, etc.
Examples of non-linear systems are those defined by mathematical relations, such as: $y[n] = 2x^3[n]$, $y[n] = 2n^2x^3[n-2]$, etc.

2.7.2 Time-Invariant Discrete Systems

A system is defined as **time-invariant**, when its response to an input signal does not change with time. In time-invariant systems if the input signal is delayed by time, t_0, then the output will also be delayed by the same time, t_0. When a signal, $x[n]$, is applied to the input of the discrete-time system, T, and causes the output signal, $y[n]$, then the system, T, will be called time-invariant if the input $x[n-k]$ gives rise to the output signal, $y[n-k]$, for each k. Time-invariant is a system whose response depends on the form of the stimulation and not from the time it is applied.

A time-invariant system can be identified by the form of the input-output equation. Hence:

- Systems where the output does not depend directly on time are time-invariant.
- Systems that shift the input with respect to time are time-invariant. Systems where the input has been compressed or decompressed in time are time-varying.
- When the relation between input-output is described by a differential equation with constant coefficients — that is, the coefficients do not depend directly or indirectly on time, then the system is time-invariant.

- When the relation between input-output is described by a difference equation with constant coefficients — that is, the coefficients do not depend directly or indirectly on time, then the system is time-invariant.

Examples of time-invariant systems are those defined by the mathematical relations: $y[n] = 2x^3[n - 2]$, $y[n] = 2x^2[n - 4]$, $y[n] - y[n - 2] = 2x[n]$, $y[n + 1] + y[n - 2] = x[n + 1] - 2x[n]$, etc.
Examples of time-varying systems are those defined by the mathematical relations, such as: $y[n] = 2n \cdot x[n]$, $y[n] = x[2n]$, $y[n] = x[2n]$, $y[n] - ny[n - 2] = 2x[n]$, etc.

2.7.3 Discrete Systems with Memory

A system is called a *static* or *memory-less system* if, for each moment, its output depends only on the current input value. In all other cases, the system is called dynamic or memory-based system. In dynamic systems, the output may depend on the current input value but also on past or future input values.

LTI systems are generally memory-based systems. Memory is a fundamental property of the convolution sum. The only case, that a discrete-time LTI system is memory-less, is when it satisfies the relation $h(n) = 0$, for $n \neq 0$. Hence, $h(n) = h(0)\delta(n)$; therefore, this is a system that corresponds to a multiplication with a constant coefficient and which carries out amplification or damping of the signal. The input-output relation is: $y(n) = h(0)x(n)$.

For example, the system $y[n] = (x^2[n])^2$ is memory-less because, at a random moment, n, the output depends only on the input value, $x[n]$, at that moment. Indeed, the input-output equation does not contain any past or future input values. Instead, the system $y[n] = \dfrac{x[n+1] + x[n] + x[n-1]}{3}$ is a memory-based system because the output depends not only on the previous input, $x[n - 1]$, but also on the future one, $x[n + 1]$.

2.7.4 Invertible Discrete Systems

A system, T, is called *invertible* when, there is another system, T', which, when connected in series with the system, T, provides as its output the input signal of system, T. This procedure is implemented in applications where the effect of a system on a signal must be removed. Concluding, we could say that, if we know the system T' and can connect it in series with the original system, we end up with an identical system.

For example, the system described by the relation $y[n] = \displaystyle\sum_{k=-\infty}^{n} x[k]$ is invertible and has as an inverse system, the system with an input-output relation, $y[n] = x[n] - x[n - 1]$. Also, the system $y[n] = 6x[n]$ is reversible and has as an inverse system, $y[n] = \dfrac{1}{6}x[n]$. On the other hand, systems $y[n] = x^2[n]$ and $y[n] = 0$ are not invertible.

2.7.5 Casual Discrete Systems

A system is called *casual* when the output at any point in time n_0 depends only on the present or the previous values of the input; that is, it depends on $x[n]$ for $n \leq n_0$. In casual systems, the output does not depend on future input values. Generally, in the case of casual discrete-time systems, the output at n will depend on the input values at $n, n - 1, n - 2$, etc.,

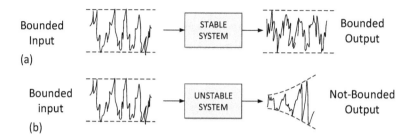

FIGURE 2.13
(a) Stable system and (b) unstable system.

and not on the input values at $n + 1$, $n + 2$, etc. So, the system $y[n] = \sum_{k=0}^{n} x[n-k]$ is casual, while the system $y[n] = \sum_{k=-n}^{n} x[n-k]$ is not casual.

All memory-less systems are casual. The reverse is not true. Indeed, for a memory-less system, the output at any time depends only on the present value of the input (it does not depend on future or past input values).

For example, the system described by the difference equation, $y[n] + y[n-1] = x[n] - 4x[n-1]$, is a system with memory because the output depends on past input values due to the presence of the term $4x[n-1]$. In addition, the system is casual because the output does not depend on future input values.

2.7.6 Stable Discrete Systems

A discrete-time system is *BIBO stable* (Bounded Input-Bounded Output stable) when, for each bounded input, the output also remains bounded for all instances. Namely, if $|x[n]| \leq k$, then $|y[n]| \leq m$, for all n. The requirement of bounded input and output signals is fulfilled when the system is stable (Figure 2.13).

For example, the system described by the relation $y[n] = nx[n]$ is unstable because, although the input, $x[n] = 2u[n]$ applied in the system, is bounded, the response of the system will be the signal $y[n] = 2nu[n] = 2r[n]$, which is not bounded.

The property of BIBO stability is expressed by the following relation:

$$S = \sum_{k=-\infty}^{\infty} |h(k)| < \infty \tag{2.17}$$

2.8 System Connections

Complex system analysis is greatly facilitated if the system is considered to be the result of the connection of simpler systems. The fundamental types of system connections are in series, parallel, mixed connection and connection with feedback (see Figures 2.14 through 2.16).

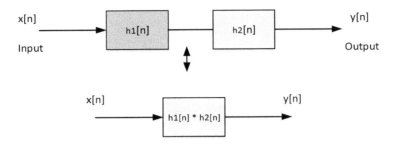

FIGURE 2.14
In series connection of discrete-time systems.

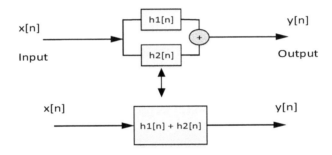

FIGURE 2.15
Parallel connection of discrete-time systems.

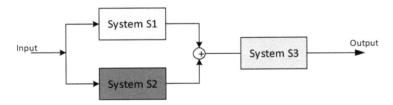

FIGURE 2.16
Mixed connection of discrete-time systems.

2.9 Convolution

Let an impulse sequence, $\delta[n]$, be applied to the input of a discrete-time system. The resulting output is called impulse response and is usually denoted by $h[n]$. If the impulse response of an LTI system is known, then the problem of deriving the response for any input is a purely a mathematical problem solved by the convolution theorem.

Consider an LTI system with impulse response, $h[n]$. If the input, $x[n]$, is applied to it, the resulting output is given by:

$$y[n] = x[n] * h[n] = \sum_{k=-\infty}^{k=+\infty} x[k]h[n-k] \tag{2.18}$$

TABLE 2.2

Properties of Convolution

Commutativity	$x[n] * h[n] = h[n] * x[n]$
Associativity	$\{x[n] * h_1[n]\} * h_2[n] = x[n] * \{h_1[n] * h_2[n]\}]$
Distributivity	$x[n] * \{h_1[n] + h_2[n]\} = x[n] * h_2[n]$
Identity	$\delta[n] * x[n] = x[n]$
	$\delta[n - k] * x[n] = x[n - k]$

Methods of convolution estimation, allow the computation of the output of an LTI system when the input and impulse response of the system are known. In Table 2.2 below, the properties of convolution are presented.

Regarding the physical concept of the commutativity property, after rewriting Equation 2.18 as $y[n] = \sum_{k=-\infty}^{k=+\infty} h[k]x[n-k]$, it is suggested that the output signal at a certain instant occurs as a linear combination of all samples of the input signal with appropriate weighting factors. The distributivity property suggests that the output of a series of connected systems is independent of the sequence to which they are connected, while the distributivity property expresses the possibility to connect systems in parallel.

There are three techniques for calculating the convolution of two finite-length signals:

1. Analytical, using Equation 2.18
2. With the use of a matrix (Figure 2.17), for $x[n] = \{x[0], x[1]\}$ and $h[n] = \{h[0], h[1], h[2], h[3]\}$.

   The order of terms (i.e., the value of n) of the signal $y[n]$ is estimated by adding the corresponding indices, n, of the signals $h[n]$ and $x[n]$. Each term of the sequence $y[n]$ is computed by adding the diagonal terms.
3. **Chunking method**: The successive steps for the calculation of $y[n]$ are:
   - The sequences $x[k]$ and $h[k]$ are designed
   - The reflection of $h[k]$ is designed, i.e., $h[-k]$

h[n]		h[0]	h[1]	h[2]	h[3]
x[n]		2	1	1	1
x[0]	1	1x2	1x1	1x1	1x1
x[1]	2	2x2	2x1	2x1	2x1

FIGURE 2.17
Computing convolution with the use of a matrix.

- $h[-k]$ is shifted to the left so that the last on the right non-zero term of $h[-k]$ coincides with the first on the left non-zero term of $x[k]$
- $h[n-k]$ is shifted to the right by one step. Then the terms of $h[n-k]$ and $x[k]$ are multiplied and the partial products are summed

4. A fourth technique, far more complex than the three aforementioned, is the graphical one. The graphical technique is the only one that can be used in the case that one of the two signals is of infinite length. From the definition of convolution, the steps for its graphical approximation are:
   - Flipping over one of the signals with respect to y-axis
   - Shifting it by n
   - Multiplying this flipped, shifted signal with the other one
   - Calculating the area under this product
   - Assigning this value to $x[n]* h[n]$ at n

Now, $x[n]$ and $h[n]$ are of N and M length, correspondingly, and furthermore let
$\begin{cases} x[n] = 0 \\ h[n] = 0 \end{cases}$ when $n < 0$. Then the convolution $y[n] = x[n] * h[n]$ is of $N + M - 1$ length and the terms of $y[n]$ are produced by the following relations:

$$y[0] = x[0]h[0]$$

$$y[1] = x[1]h[0] + x[0]h[1]$$

$$y[2] = x[2]h[0] + x[1]h[1] + x[0]h[2]$$

$$y[3] = x[3]h[0] + x[2]h[1] + x[1]h[2] + x[0]h[3]$$

$$y[4] = x[4]h[0] + x[3]h[1] + x[2]h[2] + x[1]h[3] + x[0]h[4]$$

These equations can also be written in a matrix form:

$$\begin{bmatrix} y[0] \\ y[1] \\ y[2] \\ y[3] \\ y[4] \end{bmatrix} = \begin{bmatrix} x[0] & 0 & 0 & 0 & 0 \\ x[1] & x[0] & 0 & 0 & 0 \\ x[2] & x[1] & x[0] & 0 & 0 \\ x[3] & x[2] & x[1] & x[0] & 0 \\ x[4] & x[3] & x[2] & x[1] & x[0] \end{bmatrix} \begin{bmatrix} h[0] \\ h[1] \\ h[2] \\ h[3] \\ h[4] \end{bmatrix} \qquad (2.19)$$

Lower triangular matrix

The convolution of two sequences can be written in the form of a product between a Toeplitz vector matrix and a vector.

$$
\begin{bmatrix} y[0] \\ y[1] \\ y[2] \\ y[3] \\ y[4] \\ y[5] \\ y[6] \end{bmatrix} = \begin{pmatrix} h[0] & & \\ h[1] & h[0] & \\ h[2] & h[1] & h[0] \\ h[3] & h[2] & h[1] \\ h[4] & h[3] & h[2] \\ & h[4] & h[3] \\ & & h[4] \end{pmatrix} \begin{bmatrix} x[0] \\ x[1] \\ x[2] \end{bmatrix}
\tag{2.20}
$$

$$
\text{or} \quad \begin{bmatrix} y[0] \\ y[1] \\ y[2] \\ y[3] \\ y[4] \\ y[5] \\ y[6] \end{bmatrix} = \begin{pmatrix} x[0] & & & \\ x[1] & x[0] & & \\ x[2] & x[1] & x[0] & \\ & x[2] & x[1] & x[0] \\ & & x[2] & x[1] & x[0] \\ & & & x[2] & x[1] \\ & & & & x[2] \end{pmatrix} \begin{bmatrix} h[0] \\ h[1] \\ h[2] \\ h[3] \\ h[4] \end{bmatrix}
\tag{2.21}
$$

The characteristic of the matrices formed by the terms of the impulse response or the input signal is that the elements covering the same diagonal are equal to each other and are, therefore, Toeplitz matrices.

The representation of the convolution as a product of a matrix and a vector — aside from being an additional method for its computation — is a useful tool in the search for fast algorithms serving the same purpose.

2.10 Deconvolution

Deconvolution of the system is the calculation of the impulse response of the system when its response to an input signal, other than the impulse function, is known. It is:

$$
h[0] = \frac{y[0]}{x[0]} \quad and \quad h[n] = \frac{y[n] - \displaystyle\sum_{k=0}^{n-1} h[k]\, x[n-k]}{x[0]}
\tag{2.22}
$$

where: $n = 1, 2, \ldots, (N-1)$, $x[0] \neq 0$.

2.11 Correlation — Autocorrelation

The Correlation or *Cross-Correlation* of the signals $x[n]$ and $y[n]$ is defined by the inner product:

$$r_{xy}(\ell) = \sum_{n=-\infty}^{\infty} x[n]y[n-\ell] , \qquad -\infty \le \ell \le \infty \tag{2.23}$$

The independent variable, ℓ, is called *lag*. The correlation measures the time shift between two signals. By comparing Relations 2.18 and 2.23, one can easily observe that convolution implies the reflection and shift of one of the two signals, while correlation denotes only a shift. It is true that: $x[n] * y[-n]\big|_{n=\ell} = \sum_{m=-\infty}^{\infty} x[m]y[m-\ell] = r_{xy}(\ell)$, namely, the correlation of two signals $x[n]$ and $y[n]$ is equal to the convolution of the signals $x[n]$ and $y[-n]$.

The correlation coefficient, $\rho_{xy}(l)$, is the value of correlation normalized to $r_{xx}(0)$ and $r_{yy}(0)$, which are the maximum values of $r_{xx}(l)$ and $r_{yy}(l)$, and is defined by

$$\rho_{xy}(\ell) = \frac{r_{xy}(\ell)}{\sqrt{E_x}\sqrt{E_y}} , \qquad -\infty \le \ell \le \infty \tag{2.24}$$

where $E_x = \sum_{n=-\infty}^{\infty} x^2[n]$ is the energy of $x[n]$, and $E_y = \sum_{n=-\infty}^{\infty} y^2[n]$ is the energy of $y[n]$.

Applications of correlation are encountered in various signal processing topics (for example, the search for periodicity to signals with noise and the search for delay in two identical signals), in digital communications, etc. Shifting a signal, $x[n]$, to itself changes the degree of correlation. Knowing this change can be very useful to the calculation of the spectrum of the signal.

The autocorrelation sequence is defined by:

$$r_{xx}(\ell) = \sum_{n=-\infty}^{\infty} x[n]x[n-\ell] , \qquad -\infty \le \ell \le \infty \tag{2.25}$$

The autocorrelation coefficient, $\rho_{xx}(\ell)$, is defined by:

$$\rho_{xx}(\ell) = \frac{r_{xx}(\ell)}{E_x} , \qquad -\infty \le \ell \le \infty \tag{2.26}$$

But, $E_x = r_{xx}(0)$, so Relation 2.26 becomes:

$$\rho_{xx}(\ell) = \frac{r_{xx}(\ell)}{r_{xx}(0)} , \qquad -\infty \le \ell \le \infty \tag{2.27}$$

Cross-Correlation Properties

$$r_{xy}(l) = r_{yx}(-l) \tag{2.28}$$

$$\left|\rho_{xy}(\ell)\right| \leq 1 \tag{2.29}$$

$$\left|r_{xy}(\ell)\right| \leq \sqrt{r_{xx}(0)r_{yy}(0)} \leq \frac{r_{xx}(0) + r_{yy}(0)}{2} \tag{2.30}$$

Autocorrelation Properties

$$r_{xx}(l) = r_{xx}(-l) \tag{2.31}$$

$$\left|r_{xx}(\ell)\right| \leq r_{xx}(0) \tag{2.32}$$

Let two signals, x(n) and y(n), of finite duration defined at $0\ n \leq N - 1$ domain and $0\ n \leq N - 1$ domain with $N < M$, respectively.

For $N = 3$ and $M = 5$, correlation can be written in a matrix form as:

$$\begin{bmatrix} r_{xy}(-4) \\ r_{xy}(-3) \\ r_{xy}(-2) \\ r_{xy}(-1) \\ r_{xy}(0) \\ r_{xy}(1) \\ r_{xy}(2) \end{bmatrix} = \underbrace{\begin{bmatrix} y[4] & 0 & 0 \\ y[3] & y[4] & 0 \\ y[2] & y[3] & y[4] \\ y[1] & y[2] & y[3] \\ y[0] & y[1] & y[2] \\ 0 & y[0] & y[1] \\ 0 & 0 & y[0] \end{bmatrix}}_{\text{Toeplitz matrix}} \begin{bmatrix} x[0] \\ x[1] \\ x[2] \end{bmatrix} \tag{2.33}$$

2.12 Difference Equations

Difference equations model mathematically discrete systems, in the way that differential equations model analog systems. The general form of an N-order difference equation is:

$$\sum_{k=0}^{N} b_k y[n-k] = \sum_{m=0}^{M} a_m x[n-m] \tag{2.34}$$

with initial conditions $y[-1]$, $y[-2]$, …, $y[-N]$ and $x[-1]$, $x[-2]$, …, $x[-M]$.

The general solution of Equation 2.34 can be calculated both in n domain (by methods similar to the solution of a differential equation in time domain), and in z domain by using the z-transform. In order to solve a difference equation describing an LTI system, N initial conditions are required for $y[n]$ and $x[n]$. For a casual system, the initial conditions required to calculate the first value of $y[n]$ are considered to be equal to zero, i.e., $y[-1]$, $y[-2]$, ... and $x[-1]$, $x[-2]$, ... are zero. If the system is non-causal, the solution of the difference equation includes one more term derived from the initial conditions.

The solution of the difference equation can be calculated either by application of convolution, after calculation of $h[n]$, or directly by the difference equation, writing Equation 2.34 in this form:

$$y[n] = -\sum_{k=0}^{N} \frac{b_k}{b_0} y[n-k] + \sum_{m=0}^{M} \frac{\alpha_m}{b_0} x[n-m] \tag{2.35}$$

The general difference equation in the form of Equation 2.35 is called a recursive equation because a value of the output $y[n]$ depends on previous values, $y[n-1]$, $y[n-2]$, ..., of the output, as well.

In the case that the coefficients of Equation 2.34 are zero for $k = 1, 2, \ldots, N$, the difference equation becomes:

$$y[n] = \sum_{m=0}^{M} \frac{\alpha_m}{b_0} x[n-m] \tag{2.36}$$

Equation 2.36 is non-recursive since the value of the output, $y[n]$, is described only by the input values and not by any other previous values of the output.

2.13 Discrete-Time Systems of Finite Impulse Response

If the impulse response, $h[n]$, of a discrete-time system is of finite duration then the system is called *Finite Impulse Response* (FIR).

For non-recursive systems, we have:

$$h[n] = \begin{cases} \dfrac{\alpha_n}{b_0} & , \quad n = 0, 1, \ldots, M \\ 0 & , \quad elsewhere \end{cases} \tag{2.37}$$

Thus, non-recursive discrete time systems are FIR systems.

Consider a casual system described by the equation:

$$y[n] = \sum_{m=0}^{-1} \frac{\alpha_m}{b_0} x[n-m] = \frac{\alpha_0}{b_0} x[n] + \frac{\alpha_{-1}}{b_0} x[n+1].$$

The system cannot be physically realized because, in order to calculate the term $y[n]$, the future term $x[n + 1]$ is needed, or the impulse response would be $h[n] = \begin{cases} \dfrac{\alpha_n}{b_0} & n = -1,\ 0 \\ 0 & \text{elsewhere} \end{cases}$.

That is, $h[n]$ is different from zero for $n = -1$, and the system is non-causal. It can, however, be implemented using a delay system or on a computer where causality is not forbidden. All FIR systems are stable since the relation (2.17) is valid. If the impulse response, $h[n]$, of a discrete system is of infinite duration, then the system is called IIR (*Infinite Impulse Response*). Systems which are described using recursive equations are IIR systems.

2.14 Solved Problems

2.14.1 Consider the signal $x[n]$. Sketch the signals $x[n - 3]$ and $x[n + 2]$.

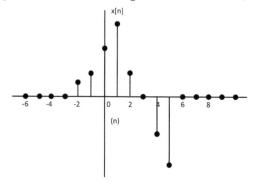

SOLUTION: The signal $x[n - 3]$ is derived by shifting the original signal by 3 units to the right. The signal $x[n - 3]$ is delayed by 3 units. The signal $x[n + 2]$ results from the time shift of the original signal by 2 units to the left. The signal $x[n + 2]$ is ahead of $x[n]$ by 2 units.

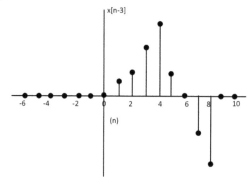

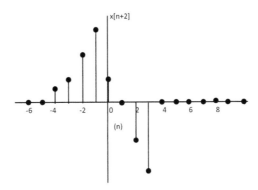

2.14.2 Consider the signal $x[n]$. Sketch the signals $x[3n]$ and $x\left[\dfrac{n}{2}\right]$.

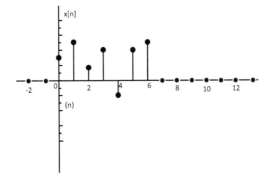

SOLUTION: The signal $x[3n]$ is obtained by dividing the abscissae of the original signal by 3. After division, only those elements corresponding to interger time intervals remain. From the original signal, only the time-terms that are multiples of 3 remain. While the original signal extends from $n = 0$ to $n = 6$, signal $x[3n]$ extends from $n = 0$ to $n = 2$.

The signal $x[n/2]$ is obtained by multiplying the abscissae of the original signal by 2. The initial signal extends from $n = 0$ to $n = 6$, while $x[n/2]$ extends from $n = 0$ to $n = 12$.

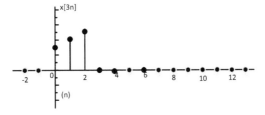

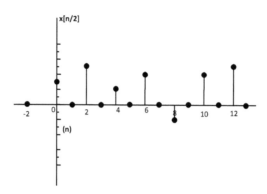

2.14.3 Let two periodic signals be $x_1[n]$ and $x_2[n]$ with fundamental periods N_1 and N_2. What is the condition for the signal $x[n] = x_1[n] + x_2[n]$ to be periodic? What is its fundamental period?

SOLUTION: Since the signals $x_1[n]$ and $x_2[n]$ are periodic, we have:

$x_1[n] = x_1[n + N_1] = x_1[n + mN_1]$ where m: positive integer and

$x_2[n] = x_2[n + N_2] = x_2[n + kN_2]$ where k: positive integer.

Substituting the above equations in $x[n] = x_1[n] + x_2[n]$, we obtain:

$$x[n] = x_1[n + mN_1] + x_2[n + kN_2] \tag{1}$$

If $x[n]$ is periodic with a fundamental period N, then

$$x[n] = x[n + N] = x_1[n + N] + x_2[n + N]$$

Relations (1) and (2) yield:

$$x_1[n + mN_1] + x_2[n + kN_2] = x_1[n + N] + x_2[n + N] \tag{3}$$

In order for (3) to be true, it must be:

$mN_1 = kN_2 = N$ or $\dfrac{N_1}{N_2} = \dfrac{k}{m}$, or the quotient of the two periods should be a rational number.

The fundamental period N of the signal $x[n]$ will be the LCM (Least Common Multiple) of the fundamental periods N_1 and N_2 given by the relation $mN_1 = kN_2 = N$, provided that k and m are prime numbers.

2.14.4 For the following signals: (a) $x[n] = 4\cos(\pi n)$, (b) $x[n] = \cos^2\left(\dfrac{\pi}{6}n\right)$, and (c) $x[n] = e^{j\frac{\pi}{4}n}$, examine which are periodic and find their fundamental period.

SOLUTION:

a. $x[n] = 4\cos(\pi n)$

The angular frequency is $\Omega = \pi$. In order for the signal to be periodic, there must be intergers κ and λ such that $\Omega = 2\pi \dfrac{\kappa}{\lambda}$. In this case, $\kappa = 1$ and $\lambda = 2$, so the signal is periodic with the fundemental period $N = 2$.

b. $x[n] = \cos^2\left(\dfrac{\pi}{6}n\right)$

Based on the trigonometric identity $\cos^2 \varphi = \dfrac{1 + \cos 2\varphi}{2}$, the signal $x[n]$ is written: $x[n] = \dfrac{1}{2} + \dfrac{1}{2}\cos\left(\dfrac{\pi}{3}n\right)$, i.e., it is the sum of the signals: $x_1[n] = \dfrac{1}{2}$ and $x_2[n] = \dfrac{1}{2}\cos\left(\dfrac{\pi}{3}n\right)$.

The signal $x_1[n]$ is written in the form $x_1[n] = \dfrac{1}{2} = \dfrac{1}{2} \cdot 1^n$, so it is periodic with period $N_1 = 1$. The signal $x_2[n]$ is also periodic because the quotient $\dfrac{2\pi}{\Omega_2} = \dfrac{2\pi}{\pi/3} = 6$ is a rational number. That is $N = 6$.

c. $x[n] = e^{j\frac{\pi}{4}n}$

The signal $x[n]$ is in the form $e^{j\Omega n}$ with $\Omega = \dfrac{\pi}{4}$. The signal is periodic because it is $\dfrac{2\pi}{\Omega} = 8$, i.e., a rational number. The fundamental period is $N = 8$.

2.14.5 Calculate the energy and power of the following discrete-time signals: (a) $x[n] = u[n]$, (b) $y[n] = (-0.2)^n u[n]$, and (c) $z[n] = Ae^{j\Omega n}$.

SOLUTION:

a. $x[n] = u[n]$

The energy of the signal $x[n] = u[n]$ is given by:

$$E = \sum_{n=-\infty}^{+\infty} |x[n]|^2 = \sum_{n=-\infty}^{+\infty} |u[n]|^2 = \sum_{n=0}^{+\infty} 1^2 = \infty$$

The power is given by:

$$P = \lim_{N \to \infty} \frac{1}{2N+1} \sum_{n=-N}^{+N} |[x[n]|^2 = \lim_{N \to \infty} \frac{1}{2N+1} \sum_{n=-N}^{+N} |[u[n]|^2 \Rightarrow$$

$$P = \lim_{N \to \infty} \frac{1}{2N+1} \sum_{n=0}^{+N} 1^2 = \lim_{N \to \infty} \left(\frac{N+1}{2N+1}\right) = \frac{1}{2}$$

b. $y[n] = (-0.2)^n u[n]$

The energy of the signal $y[n] = (-0.2)^n u[n]$ is given by:

$$E = \sum_{n=-\infty}^{+\infty} |x[n]|^2 = \sum_{n=-\infty}^{+\infty} |(-0.2)^n u[n]|^2 = \sum_{n=0}^{+\infty} |(-0.2)^n|^2 \Rightarrow$$

$$E = \sum_{n=0}^{+\infty} \left(\frac{1}{5}\right)^{2n} = \sum_{n=0}^{+\infty} \left(\frac{1}{25}\right)^n = \frac{1}{1-(1/25)} = \frac{25}{24}$$

Since the energy is finite the average power will be 0.

c. $z[n] = Ae^{j\Omega n}$

The energy of the signal $z[n] = Ae^{j\Omega n}$ is given by:

$$E = \sum_{n=-\infty}^{+\infty} |x[n]|^2 = \sum_{n=-\infty}^{+\infty} |Ae^{j\Omega n}|^2 = A^2 \sum_{n=-\infty}^{+\infty} |e^{j\Omega n}|^2 \rightarrow \infty$$

The power of the signal is given by:

$$P = \lim_{N \to \infty} \frac{1}{2N+1} \sum_{n=-N}^{+N} |x[n]|^2 = \lim_{N \to \infty} \frac{1}{2N+1} \sum_{n=-N}^{+N} |Ae^{j\Omega n}|^2 \Rightarrow$$

$$P = \lim_{N \to \infty} \frac{1}{2N+1} A^2 \sum_{n=-N}^{+N} 1 = \lim_{N \to \infty} \left(\frac{2N+1}{2N+1} A^2\right) = A^2$$

Thus, $P = \lim_{N \to \infty} \frac{1}{2N+1} A^2 \sum_{n=-N}^{+N} 1 = \lim_{N \to \infty} \left(\frac{2N+1}{2N+1} A^2\right) = A^2$

2.14.6 Calculate the even and odd component of the sequence $x[n]$.

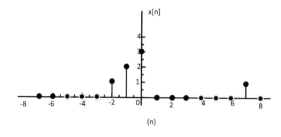

SOLUTION: Each discrete sequence can be written as a sum of an even and an odd one given respectively by the relations:

$$x_A[n] = \frac{1}{2}\{x[n] + x[-n]\} \quad (1) \quad \text{and} \quad x_O[n] = \frac{1}{2}\{x[n] - x[-n]\} \quad (2).$$

The sequence $x[n]$ can be written as a sum of sequences δ.

$x[n] = \delta[n+2] + 2\delta[n+1] + 3\delta[n] + \delta[n-7]$, so $x[x-n]$ is written:

$x[-n] = \delta[-n+2] + 2\delta[-n+1] + 3\delta[-n] + \delta[-n-7]$.

Calculation of the even part: Sustituting in (1), we obtain:

$$x_A[n] = \frac{1}{2}\{\delta[n+2] + 2\delta[n+1] + 3\delta[n] + \delta[n-7] +$$
$$+ \delta[-n+2] + 2\delta[-n+1] + 3\delta[-n] + \delta[-n-7]\} \Rightarrow$$

$$x_A[n] = \frac{1}{2}\{\delta[-n-7] + \delta[n+2] + 2\delta[n+1] + 6\delta[n]$$
$$+ \delta[-n+2] + 2\delta[-n+1] + \delta[n-7]\}$$

Thus: $x_A[n] = \left\{\dots, 0, \frac{1}{2}, 0,0,0,0, \frac{1}{2}, 1,3,1, \frac{1}{2}, 0,0,0,0, \frac{1}{2}, 0, \dots\right\}$

Calculation of the odd part: Sustituting in (2), we obtain:

$$x_O[n] = \frac{1}{2}\{\delta[n+2] + 2\delta[n+1] + 3\delta[n] + \delta[n-7] -$$
$$- \delta[-n+2] - 2\delta[-n+1] - 3\delta[-n] - \delta[-n-7]\} \Rightarrow$$

$$x_O[n] = \frac{1}{2}\{-\delta[-n-7] + \delta[n+2] + 2\delta[n+1] -$$
$$- \delta[-n+2] - 2\delta[-n+1] + \delta[n-7]\}$$

Thus: $x_O[n] = \left\{\dots, 0, -\frac{1}{2}, 0,0,0,0, \frac{1}{2}, 1,0,-1, -\frac{1}{2}, 0,0,0,0, \frac{1}{2}, 0, \dots\right\}$

2.14.7 Study the casuality, memory, temporal invariance and linearity of the systems described by the equations (a) $y[n] = 4x[n]$, (b) $y[n] = 4x[n] + 3$ and (c) $y[n] = x[4n+3]$.

SOLUTION:

a. $y[n] = 4x[n]$

At any point in time, the output depends only on the current value of the input, so the system has no memory, and all systems that have no memory are causal.

Time invariance: Assume that the signal $x_1[n]$ is implemented at the input of the system, which gives the signal $y_1 = 4x_1[n]$ (1) at the output.

As the signal $x_1[n]$ is shiftet by n_0, the output becomes: $x_2[n] = x_1[n - n_0]$ (2), and from the input-output equation, the corresponding output would be: $y_2[n] = 4x_2[n]$ (3). In order for the system to be time-invariant, it must be: $y_2[n] = y_1[n - n_0]$.

By substituting (2) in (3) we get: $y_2[n] = 4x_1[n - n_0]$. And from (1), we get: $y_1[n - n_0] = 4x_1[n - n_0]$.

Consequently, $y_2[n] = y_1[n - n_0]$. Thus, the system is time-invariant.

Linearity: When the signal $x_1[n]$ is applied to the input of the system the output is $y_1[n] = 4x_1[n]$. Respectively, when the signal $x_2[n]$ is applied to the input of the system, the output is $y_2[n] = 4x_2[n]$. The system is linear when, at an input signal $x_0[n] = ax_1[n] + bx_2[n]$, the output is $y_0 = ay_1[n] + by_2[n]$. Based on input-output relation, when an input $x_0[n]$ is applied to the system the output is:

$$y_0[n] = 4x_0[n] = 4\{ax_1[n] + bx_2[n]\} =$$

$$4ax_1[n] + 4bx_2[n] = a\{4x_1[n]\} + b\{4x_2[n]\} = 4ax_1[n] + 4bx_2[n]$$
$$= a\{4x_1[n]\} + b\{4x_2[n]\}$$

Thus, the system is linear.

b. $y[n] = 4x[n] + 3$

At any point in time, the output depends only on the current value of the input, so the system has no memory and is therefore casual.

Time-invariance: For an input $x_1[n]$, the output is $y_1[n] = 4x_1[n] + 3$ (4). If $x_1[n]$ is shifted by n_0, the new output is $x_2[n] = x_1[n - n_0]$ (5), and, respectively, the output is $y_2[n] = 4x_2[n] + 3$ (6). In order for the system to be time-invariant, it must be: $y_2[n] = y_1[n - n_0]$.

By substituting (5) in (6) we get: $y_2[n] = 4x_1[n - n_0] + 3$.

Equation (4) yields: $y_1[n - n_0] = 4x_1[n - n_0] + 3$

Hence, $y_2[n] = y_1[n - n_0]$.

So, the system is time-invariant.

Linearity: Let $x_1[n] \to y_1[n]$ and $x_2[n] \to y_2[n]$. For the system to be linear, the linear combination of the two input signals should give at the output the corresponding linear combination of the two output signals, i.e.: $ax_1[n] + bx_2[n] \to ay_1[n] + by_2[n]$.

Based on the input-output relation, when the input signal $ax_1[n] + bx_2[n]$ is applied, the output is:

$$y[n] = 4x[n] + 3 = 4\{ax_1[n] + bx_2[n]\} + 3 = a[4x_1[n] + 3] + b[4x_2[n]]$$
$$\neq ay_1[n] + by_2[n]$$

Thus, the system is not linear.

c. $y\{n\} = x[4n + 3]$

The system is not a casual. This is obvious because $y[4] = x[8]$, i.e., the output depends on future input values. The value of the output $y[n]$ depends on other values of the input but not on the current value $x[n]$, so the system has memory.

Time-invariance: When an input signal, $x_1[n]$, causes an output signal, $y_1[n]$, then, for this case, it is: $y_1[n] = x_1[2n]$ (1). Assume that the input signal is shifted by n_0 over time. This is equivalent to applying a signal $x_2 = x_1[n - n_0]$ (2) to the input, which will cause the signal $y_2[n] = x_2[2n]$ (3) to output. In order for the system to be time-invariant, it must be: $y_2[n] = y_1[n - n_0]$. Substituting (2) in Equation (3), we get: $y_2 = x_1[2n - n_0]$. Substituting in (1), we get $x_1[2(n - n_0)]$, therefore, $y_2[n] \neq y_1[n - n_0]$, thus, the system is a time-varying.

Linearity: The output results from the time compression of the input signal by a factor of 2, therefore the system is linear.

The input $x_1[n]$ gives the output signal:

$$y_1[n] = \begin{cases} (-1)^n x_1[n] & \text{when} & x_1[n] < 0 \\ 2x_1[n] & \text{when} & x_1[n] \geq 0 \end{cases} \qquad (4)$$

and the input $x_2[n] = x_1[n - n_0]$ gives the output signal:

$$y_2[n] = \begin{cases} (-1)^n x_2[n] & \text{when} & x_2[n] < 0 \\ 2x_2[n] & \text{when} & x_2[n] \geq 0 \end{cases} \qquad (5)$$

The system would be linear if the input $y_2[n] = y_1[n - n_0]$ gives the output

signal: $y_2[n] = \begin{cases} (-1)^n x_1[n - n_0] & \text{when} & x_1[n - n_0] < 0 \\ 2x_1[n - n_0] & \text{when} & x_1[n - n_0] \geq 0 \end{cases}$, where $y_0 = x_0[2n]$.

From (4) and (5) it is: $ay_1[n] + by_2[n] = ax_1[2n] + bx_2[2n]$ or $y_0[n] = x_0[2n]$. Thus, the system is linear.

2.14.8 Examine the stability of the systems described by the following equations:

$$y[n] = \begin{cases} (-1)^n x[n] & \text{when} & x[n] < 0 \\ 2x[n] & \text{when} & x[n] \geq 0 \end{cases} \quad \text{and } y[n] = (x[n])^2.$$

SOLUTION:

a. It should be considered if $|x[n]| \leq k$ for every n.

For the case that $|y[n]| = |(-1)^n x[n]| = |x[n]| \leq k$. Therefore $|y[n]| \leq k$.

For the case that $x[n] \geq 0$: $|y[n]| = |2x[n]| = 2|x[n]| = 2k|$. Therefore $|y[n]| \leq 2k$.

The two inequalities indicate that $|y[n]| \leq 2k$ for all n, thus, the system is stable.

b. Based on $|x(t)| \leq k$ and raising to the square: $|x(t)|^2 \leq k^2$ so $|y(t)| \leq k^2$, thus, the system is stable.

2.14.9 Let $h[n]$ the impulse respons of a LTI system. Find the response, $y[n]$, when $x[n]$ is the input signal.

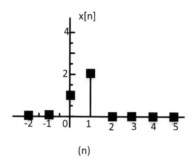

(n)

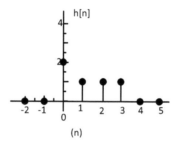

(n)

SOLUTION: We shall compute the convolution with three different ways:

    a. Solution using the analytical technique:

       From the figure, we have:

$$x[n] = \delta[n] + 2\delta[n-1] \tag{1}$$

$$h[n] = 2\delta[n] + \delta[n-1] + \delta[n-2] + \delta[n-3] \tag{2}$$

Thus,

$$x[n] * h[n] = \{\delta[n] + 2\delta[n-1]\} * h[n]$$

$$= \delta[n] * h[n] + 2\delta[n-1] * h[n]$$
$$= h[n] + 2h[n-1] \tag{3}$$

But

$$h[n] = 2\delta[n] + \delta[n-1] + \delta[n-2] + \delta[n-3] \tag{4}$$

$$2h[n-1] = 4\delta[n-1] + 2\delta[n-2] + 2\delta[n-3] + 2\delta[n-4] \tag{5}$$

So the convolution is computed as follows:

$$(3) \Rightarrow x[n] * h[n] = \overset{(4)}{h[n] + 2h[n-1]} \underset{(5)}{=}$$

$$= 2\delta[n] + 5\delta[n-1] + 3\delta[n-2] + 3\delta[n-3] + 2\delta[n-4]$$

(6)

b. Solution with the use of matrix:

We create the matrix below. Each term of the sequence $y[n]$ is calculated by adding the diagonal terms.

h[n] x[n]	h[0] 2	h[1] 1	h[2] 1	h[3] 1
x[0] 1	1x2=2	1x1=1	1x1=1	1x1=1
x[1] 2	2x2=4	2x1=2	2x1=2	2x1=2

y[0]=2   y[1]=5   y[3]=3   y[4]=2
         y[2]=3

So the signal $y[n]$ that is produced from the convolution $x[n]* h[n]$ is:

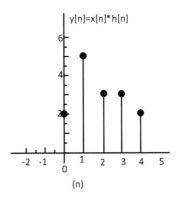

c. Solution with the use of the chunking method:

The successive steps for computing $y[n]$ are:

- The sequences $x[k]$ and $h[k]$ are designed
- The reflection of $h[k]$ is designed, i.e., $h[-k]$
- $h[-k]$ is shifted to the left so that the last on the right non-zero term of $h[-k]$ coincides with the first on the left non-zero term of $x[k]$.
- $h[n-k]$ is shifted to the right by one step. Then the terms $h[n-k]$ and $x[k]$ are multiplied, and the partial products are summed.

For $n = 0$:

$y[0] = x[k] \cdot h[-k] =$
$= 0.1 + 0.1 + 0.1 + 1.2 = 2$

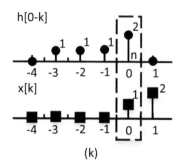

For $n = 1$:

$y[1] = x[k] \cdot h[-k] =$
$= 0.1 + 0.1 + 1.1 + 2.2 = 5$

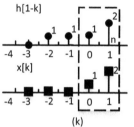

For $n = 2$:

$y[2] = x[k] \cdot h[2 - k] =$
$= 0.1 + 1.1 + 2.1 + 0.2 = 3$

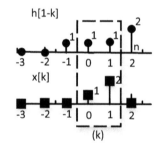

For $n = 3$:

$y[3] = x[k] \cdot h[3 - k]$
$= 1.1 + 2.1 + 0.1 + 0.2 = 3$

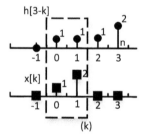

For $n = 4$:

$y[4] = x[k] \cdot h[4 - k]$
$= 1.0 + 2.1 + 0.1 + 0.1 + 0.2 = 2$

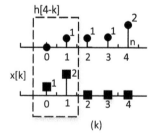

2.14.10 Let a linear, time-invariant system with the following impulse response be:

$$h[n] = \begin{cases} 1, & 0 \le n \le 4 \\ 0, & elsewhere \end{cases}$$

Compute the output $y[n]$ of the system when the input signal is:

$$x[n] = \begin{cases} \alpha^n, & -3 \le n \le 5 \\ 0, & elsewhere \end{cases}$$

where α is a non-zero coefficient. It is given that $\sum_{n=0}^{N-1} \lambda^n = \dfrac{\lambda^N - 1}{\lambda - 1}$ for $\lambda \ne 1$.

SOLUTION: It is $y[n] = 0$, $n < -3$, $n > 9$.

For $-3 \le n \le 1$:

$$y[n] = \sum_{k=-3}^{n} x[k] = \sum_{k=-3}^{n} \alpha^k = \alpha^{-3} \sum_{k=0}^{n+3} \alpha^k = \alpha^{-3} \frac{\alpha^{n+4} - 1}{\alpha - 1}.$$

For $2 \le n \le 5$:

$$y[n] = \sum_{k=n-4}^{n} x[k] = \alpha^{n-4} \sum_{k=0}^{4} \alpha^k = \alpha^{n-4} \frac{\alpha^5 - 1}{\alpha - 1}.$$

Finally, for $6 \le n \le 9$:

$$y[n] = \sum_{k=n-4}^{5} x[k] = \alpha^{n-4} \sum_{k=0}^{9-n} \alpha^k = \alpha^{n-4} \frac{\alpha^{10-n} - 1}{\alpha - 1}.$$

2.14.11 Compute the convolution, $y[n] = x[n] * v[n]$ when

$$x[n] = u[n] - u[n-4], \quad v[n] = 0.5^n u[n]$$

SOLUTION: The convolution of the two sequences is computed as follows:

$$y[n] = x[n] * v[n] = \sum_{k=-\infty}^{+\infty} x[k]v[n-k] = \sum_{k=-\infty}^{+\infty} (u[k] - u[k-4])0.5^{n-k} u[n-k] \qquad (1)$$

In order to graphically calculate the convolution, $v[-k]$ is sketched and then shifted by n. The last right non-zero element of $v[n-k]$ has n as its abscissa.

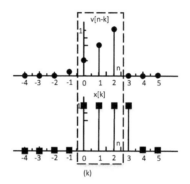

Consider the following time-intervals:

When $n < 0$, there is no overlap, so $y[n] = 0$.

When $0 \leq n \leq 3$: For $k < 0$, it is $x[k] = 0$, so the lower limit of the summation of the convolution will be 0. The dashed line shows the overlap of the two sequences that are from $k = 0$ to $k = n$. The upper limit of the summation will be n.

Thus, it will be: $x[n] * v[n-k] = \displaystyle\sum_{k=0}^{n} 0.5^{n-k} = 0.5^{n} \sum_{k=0}^{n} 0.5^{-k}$ (2)

But, $\displaystyle\sum_{k=0}^{n} 0.5^{-k} = \sum_{k=0}^{n} (0.5^{-1})^{k} = \sum_{k=0}^{n} 2^{k}$ is the sum of $(n + 1)$ first terms of geometrical progression, with the first term being the 1 and ratio 2, so

$$y[n] = x[n] * v[n-k] = \sum_{k=0}^{n} 0.5^{n-k} = 0.5^{n} \sum_{k=0}^{n} 2^{k} = 2 - 0.5^{n} \quad (3)$$

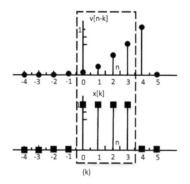

When $n \geq 4$: The dashed line shows the overlap of the two sequences, which is from $k = 0$ to $k = 3$. These are the limits of summation.

$$x[n] * v[n-k] = \sum_{k=0}^{3} 0.5^{n-k} = 0.5^n \sum_{k=0}^{3} 0.5^{-k} = 0.5^n \frac{1-2^4}{1-2} = -0.5 + 0.5^{n-4} \qquad (4)$$

$$\text{Thus, } y[n] = \begin{cases} 0 & \text{for} & n < 0 \\ 2 - 0.5^n & \text{for} & n = 0,1,2,3 \\ 0.5^{n-4} - 0.5 & \text{for} & n \geq 4 \end{cases} \qquad (5)$$

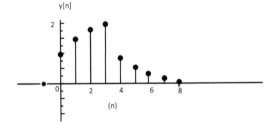

2.14.12 Find the casual system that gives the response $y[n] = \{1, 2, 3, 4, 5, 4, 2\}$ when the input signal is $x[n] = \{1, 1, 1, 2, 2\}$

SOLUTION: From Relation 2.22, we have:

$$h[0] = \frac{y[0]}{x[0]} = \frac{1}{1} = 1$$

$$h[1] = \frac{(y[1] - x[1]h[0])}{x[0]} = \frac{2 - 1 \cdot 1}{1} = 1$$

$$h[2] = \frac{(y[2] - h[0]x[2] - h[1]x[1])}{x[0]}$$

$$h[3] = \frac{4 - 1 \cdot 2 - 1 \cdot 1 - 1 \cdot 1}{1} = 0$$

$$h[4] = \frac{5 - 1 \cdot 2 - 1 \cdot 2 - 1 \cdot 1 - 0 \cdot 1}{1} = 0$$

Thus, $h[n] = \left\{ \underset{\uparrow}{1}, 1, 1, 0, 0 \right\}$.

2.14.13 Consider the system that is characterized by the impulse response $h[n] = (1/2)^n$, $n \geq 0$. Find the input signal $x[n]$ that creates the output $y[n] = \{0, 1, 4, 2, 0, 0\}$.

SOLUTION: It is: $y[0] = x[0]h[0]$, $y[1] = x[1]h[0] + x[0]h[1]$, etc.

Thus,

$$x[0] = \frac{y[0]}{h[0]} = \frac{0}{1} = 0$$

$$x[1] = \frac{(y[1] - x[0]\, h[1])}{h[0]} = \frac{1 - 0 \cdot \dfrac{1}{2}}{1} = 1$$

$$x(2) = \frac{(y[2] - (x[0]\, h[2] + x[1]\, h[1]))}{h[0]} = \frac{4 - (0 \cdot (0.5)^2 + 1 \cdot 0.5)}{1} = 3.5$$

.

So: $x[n] = \left\{\underset{\uparrow}{0},\ 1,\ 3.5, \ldots\right\}.$

2.14.14 Compute the first six samples of the impulse responses of the following difference equations using the iterative method.

  i. $y[n] = x[n] + 2x[n-1] + 3x[n-2] - 4x[n-3] + 5x[n-4]$
 ii. $y[n] = 4x[n] - 0.5y[n-1]$

SOLUTION:

  i.   For the non-recursive difference equation

$$y[n] = x[n] + 2\ x[n-1] + 3\ x[n-2] - 4\ x[n-3] + 5\ x[n-4]$$

we create the following Table 2.3 of values
Then,

$$y[n] = \{..\ 0,\ ..,\ 0,1,2,3,\ -4,\ 5,\ 0,..,0,...\}$$

TABLE 2.3

Values for the Non-Recursive Difference Equation

N	x[n]	x[n−1]	x[n−2]	x[n−3]	x[n−4]	y[n]
0	1	0	0	0	0	1
1	0	1	0	0	0	2
2	0	0	1	0	0	3
3	0	0	0	1	0	−4
4	0	0	0	0	1	5
5	0	0	0	0	0	0
:	:	:	:	:	:	:

TABLE 2.4

Values for the Recursive Difference Equation

n	x[n]	y[n–1]	y[n]
0	1	0	4
1	0	4	–2
2	0	–2	1
3	0	1	–0.5
4	0	0.5	0.25
5	0	0.25	–0.125
:	:	:	:

ii. For the recursive difference equation

$$y[n] = 4\,x[n] - 0.5\,y[n-1]$$

we create the following Table 2.4 of values

It is:

$$y[n] = \{..,0,..,\ 0,\ 4,\ -2,\ 1,\ -0.5,\ 0.25,\ -0.125,...\}$$

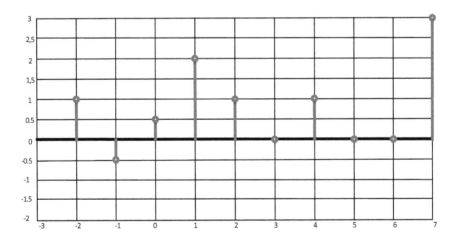

2.14.15 For the system of the following figure, it is given that $h_2[n] = u[n] - u[n-2]$ and also that its total discrete impulse response is considered known. Calculate: (a) $h_1[n]$ and (b) the response of the system for the input $x[n] = \delta[n] - \delta[n-1]$.

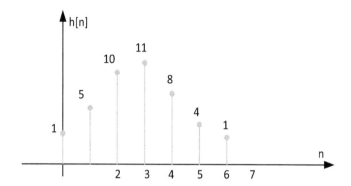

SOLUTION: Two different approaches for the solution of both parts of the problem are given next.

a. It is $h[n] = h_1[n] * h_2[n] * h_2[n]$ (1)

$$h_2[n] * h_2[n] = h_3[n] = \sum_{k=0}^{\infty} h_2[k] h_2[n-k] = h_2[0] h_2[n] + h_2[1] h_2[n-1] +$$ (2)

$$+ h_2[2] h_2[n-2] + \ldots$$

and $h_2[n] = u[n] - u[n-2]$. Thus,

$$(2) \Rightarrow h_3[0] = h_2[0] h_2[0] = 1$$
$$h_3(1) = h_2[0] h_2[1] + h_2[1] h_2[0] = 2$$
$$h_3(2) = h_2[0] h_2[2] + h_2[1] h_2[1] + h_2[2] h_2[0] = 1 \quad \Bigg\} \Rightarrow h_3[n] = \{ \underset{\uparrow}{1}, 2, 1, 0 \}$$
$$h_3(2) = h_2[0] h_2[3] + h_2[1] h_2[2] + h_2[2] h_2[1] + h_2[3] h_2[0] = 0$$

$$(1) \Rightarrow h[n] = h[n] * h[n] = h[n] * ([n] + 2\delta[n-1] + \delta[n-2]) \Rightarrow h[n] = h_1[n] + 2\, h_1[n-1] + h_1[n-2]$$ (3)

$$(3) \Rightarrow h[0] = h_1[0] + 2\, h_1[-1] + h_1[-2] \Rightarrow h[0] = 1$$

$$h[1] = h_1[1] + 2\, h[0] + h_1[-1] \Rightarrow h[1] = h[1] - 2h[0] = 3$$

$$h[2] = 10 = h_1[2] + 2\, h_1[1] + h_1[0] \Rightarrow h_1[2] = 3$$

$$h[3] = 11 = h_1[3] + 2\, h_1[2] + h_1[1] \Rightarrow h_1[3] = 2$$

$$h[4] = 8 = h_1[4] + 2\, h_1[3] + h_1[2] \Rightarrow h_1[4] = 1$$

.

.

$$h_1[n] = 0 \ \forall n \geq 5$$

Second proposed solution:

It is $h[n] = h_1[n]* h_2[n]* h_2[n]$ (1)

and $h_2[n] = u[n] - u[n-2] = \left\{\underset{\uparrow}{1},1\right\}$

$$h_2[n]*h_2[n] = \sum_{k=0}^{\infty} h_2[k]h_2[n-k] = h_2'[n]$$ (2)

$(2) \Rightarrow h_2'[0] = \sum_{k=0}^{\infty} h_2[k]h_2[-k] = h_2[0]h_2[0] = 1\cdot1 = 1$

$h_2'[1] = \sum_{k=0}^{\infty} h_2[k]h_2[1-k] = h_2[0]h_2[1] + h_2[1]h_2[0] = 1\cdot1+1\cdot1 = 2$

$h_2'[2] = \sum_{k=0}^{\infty} h_2[k]h_2[2-k] = h_2[0]h_2[2] + h_2[1]h_2[1] + h_2[2]h_2[0] = 1 \Rightarrow$

$$h_2'[n] = \left\{\underset{\uparrow}{1},2,1\right\}$$ (3)

$(1) \Rightarrow h[n] = h_1[n]* h_2'[n].$

$h_1[n]$ will be computed using deconvolution.

$$h[n] = \sum_{k=0}^{\infty} h_1[k]h_2'[n-k]$$ (4)

$(4) \Rightarrow h[0] = \sum_{k=0}^{\infty} h_1[k]h_2'[-k] = h_1[0]h_2'[0] \Rightarrow h_1[0] = \dfrac{h[0]}{h_2'[0]} = \dfrac{1}{1} = 1$

$h[1] = h_1[0]h_2'[1] + h_1[1]h_2'[0] \Rightarrow h_1[1] = \dfrac{h[1] - h_1[0]h_2'[1]}{h_2'[0]} = \dfrac{5-1\cdot2}{1} = 3$

$h[2] = \sum_{k=0}^{\infty} h_1[k]h_2'[2-k] = h_1[0]h_2'[2] + h_1[1]h_2'[1] + h_1[2]h_2'[0] \Rightarrow$

$10 = 1\cdot1 + 3\cdot2 + h_1[2]\cdot1 \Rightarrow h_1[2] = 10-7 = 3$

$$h[3] = h_1[0]h_2'[3] + h_1[1]h_2'[2] + h_1[2]h_2'[1] + h_1[3]h_2'(0) \Rightarrow$$
$$11 = 1 \cdot 0 + 3 \cdot 1 + 3 \cdot 2 + h_1[3] \cdot 1 \Rightarrow h_1[3] = 11 - 9 = 2$$

$$h[4] = h_1[0]h_2'[4] + h_1[1]h_2'[3] + h_1[2]h_2'[2] + h_1[3]h_2'[1] +$$
$$h_1[4]h_2'[0] \Rightarrow 8 = 3 \cdot 1 + 2 \cdot 2 + h_1[4] \cdot 1 \Rightarrow h_1[4] = 8 - 7 = 1$$

$$h[5] = h_1[0]h_2'[5] + h_1[1]h_2'[4] + h_1[2]h_2'[3] + h_1[3]h_2'[2] +$$
$$h_1[4]h_2'[1] + h_1[5]h_2'[0] \Rightarrow 4 = 2 \cdot 1 + 1 \cdot 2 + h_1[5] \cdot 1 \Rightarrow h_1[5] = 4 - 4 = 0$$

Thus, $h_1[n] = \left\{ \underset{\uparrow}{1}, 3, 3, 2, 1 \right\}$

b. First proposed solution

$$y[n] = x[n] * h[n] = (\delta[n] - \delta[n-1]) * h[n] = h[n] - h[n-1] \tag{5}$$

$$(5) \Rightarrow \quad \left. \begin{aligned} y[0] &= h[0] = 1 \\ y[1] &= h[1] - h[0] = 4 \\ y[2] &= h[2] - h[1] = 5 \\ y[3] &= h[3] - h[2] = 1 \\ y[4] &= h[4] - h[3] = -3 \\ etc \end{aligned} \right\} \Rightarrow$$

It is true that,

The system response will be:

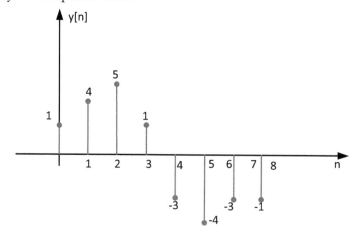

Second proposed solution

It is true that $y[n] = x[n]* h[n]$

$x[n] = \delta[n] - \delta[n-1] = \left\{ \underset{\uparrow}{1}, -1 \right\}$ and $h[n] = \left\{ \underset{\uparrow}{1}, 5, 10, 11, 8, 4, 1 \right\}$.

Hence,

$$y[0] = x[0]h[0] = 1 \times 1 = 1$$
$$y[1] = x[0]h[1] + x[1]h[0] = 5 - 1 = 4$$
$$y[2] = x[0]h[2] + x[1]h[1] + x[2]h[0] = 1 \times 10 - 1 \times 5 = 5$$
$$y[3] = x[0]h[3] + h[2] = 11 - 10 = 1$$
$$y[4] = x[0]h[4] + x[1]h[3] = 8 - 11 = -3$$
$$y[5] = x[0]h[5] + x[1]h[4] = 4 - 8 = -4$$
$$y[6] = x[0]h[6] + x[1]h[5] = 1 - 4 = -3$$
$$y[7] = x[0]h[7] + x[1]h[6] = 0 - 1 = -1$$

So, the response of the system will be:

$$y[n] = \left\{ \underset{\uparrow}{1}, 4, 5, 1, -3, -4, -3, -1 \right\}$$

2.14.16 Solve the following difference equation.

$$y[n] - ay[n-1] = z[n], \ y[-1] = b, \ z[n] = c\delta[n]$$

SOLUTION: An iterative solution of the difference equation will be given.

For $n \geq 0$, it holds:

$$y[0] = c\delta[0] + ay[-1] = c + ab$$
$$y(1) = c\delta[1] + ay[0] = a(c + ab)$$
$$y[2] = c\delta[2] + ay[1] = a^2(c + ab)$$
$$\cdot$$
$$\cdot$$
$$y[n] = a^n(c + ab) = a^{n+1}b + a^n c$$

For $n < 0$, it is true that:

$$y[n-1] = \frac{1}{a}(y[n] - x[n]) \quad , \quad y[-1] = b$$

$$y[-2] = \frac{1}{a}(y[-1] - x[-1]) = \frac{b}{a}$$

$$y[-3] = \frac{1}{a}(y[-2] - x[-2]) = \frac{b}{a^2}$$

.
.
.

$$y[-n] = \frac{b}{a^{n-1}} = b \cdot a^{1-n} \Rightarrow y[n] = b \cdot a^{n+1}$$

Thus,

$$y[n] = \begin{cases} b \cdot a^{n+1} + c \cdot a^n, & n \geq 0 \\ b \cdot a^{n+1}, & n < 0 \end{cases}$$

2.14.17 Consider the following two signals of finite length:

$$x_1[n] = -\left(\frac{n}{4}\right) \cdot (u[n] - u[n-4])$$

$$x_2[n] = \left(1 - \frac{n}{5}\right) \cdot (u[n] - u[n-5])$$

Compute the convolution and correlation of the two signals using a matrix-vector product.

SOLUTION: $x_1[n] = -\left(\frac{n}{4}\right) \cdot (u[n] - u[n-4]) \Rightarrow x_1[n] = \left\{ \ldots, 0, \underset{\uparrow}{-\frac{1}{4}}, -\frac{1}{2}, -\frac{3}{4}, 0, \ldots \right\}$ and

$$x_2[n] = \left(1 - \frac{n}{5}\right) \cdot (u[n] - u[n-5]) \Rightarrow x_2[n] = \left\{ \ldots 0, \underset{\uparrow}{1}, \frac{4}{5}, \frac{3}{5}, \frac{2}{5}, \frac{1}{5}, 0, \ldots \right\}$$

Computation of convolution:

It is $y[n] = x_1[n] * x_2[n] = \sum_{\forall k} x_1[k] x_2[n-k] = \sum_{k_1}^{k_2} x_2[k] x_1[n-k]$

where $\begin{cases} k_1 = \max(0, n-3) \\ k_2 = \min(4, n-1) \\ k_1 \leq k_2 \end{cases}$ and $1 \leq n \leq 7$

$$
\begin{bmatrix} y[1] \\ y[2] \\ y[3] \\ y[4] \\ y[5] \\ y[6] \\ y[7] \end{bmatrix}
=
\begin{bmatrix}
x_1[1] & 0 & 0 & 0 & 0 \\
x_1[2] & x_1[1] & 0 & 0 & 0 \\
x_1[3] & x_1[3] & x_1[2] & 0 & 0 \\
0 & x_1[3] & x_1[2] & x_1[1] & 0 \\
0 & 0 & x_1[3] & x_1[2] & x_1[1] \\
0 & 0 & 0 & x_1[3] & x_1[2] \\
0 & 0 & 0 & 0 & x_1[3]
\end{bmatrix}
\cdot
\begin{bmatrix} x_2[0] \\ x_2[1] \\ x_2[2] \\ x_2[3] \\ x_2[4] \end{bmatrix}
$$

$$y[n]=x_2[n]*x_1[n]$$

And substituting the elements yields:

$$
\begin{bmatrix} y[1] \\ y[2] \\ y[3] \\ y[4] \\ y[5] \\ y[6] \\ y[7] \end{bmatrix}
= -
\begin{bmatrix}
\tfrac{1}{4} & 0 & 0 & 0 & 0 \\
\tfrac{2}{4} & \tfrac{1}{4} & 0 & 0 & 0 \\
\tfrac{3}{4} & \tfrac{2}{4} & \tfrac{1}{4} & 0 & 0 \\
0 & \tfrac{3}{4} & \tfrac{2}{4} & \tfrac{1}{4} & 0 \\
0 & 0 & \tfrac{3}{4} & \tfrac{2}{4} & \tfrac{1}{4} \\
0 & 0 & 0 & \tfrac{3}{4} & \tfrac{2}{4} \\
0 & 0 & 0 & 0 & \tfrac{3}{4}
\end{bmatrix}
\cdot
\begin{bmatrix} 1 \\ \tfrac{4}{5} \\ \tfrac{3}{5} \\ \tfrac{2}{5} \\ \tfrac{1}{5} \end{bmatrix}
= -
\begin{bmatrix} \tfrac{1}{4} \\ \tfrac{7}{10} \\ \tfrac{13}{10} \\ 1 \\ \tfrac{7}{10} \\ \tfrac{2}{5} \\ \tfrac{3}{20} \end{bmatrix}
$$

$$y[n]=x_2[n]*x_1[n]$$

Computation of the correlation:

It is $r_{x_2 x_1}(\ell) = \displaystyle\sum_{n=N_1}^{N_2} x_2[n]x_1[n-\ell]$

where $\begin{cases} N_1 = \max(0, \ell+1) \\ N_2 = \min(4, \ell+3) \\ N_1 \le N_2 \end{cases}, \; -3 \le \ell \le 3$

$$
\begin{bmatrix} r_{21}(-3) \\ r_{21}(-2) \\ r_{21}(-1) \\ r_{21}(0) \\ r_{21}(1) \\ r_{21}(2) \\ r_{21}(3) \end{bmatrix}
=
\begin{bmatrix}
x_1[3] & 0 & 0 & 0 & 0 \\
x_1[2] & x_1[3] & 0 & 0 & 0 \\
x_1[1] & x_1[2] & x_1[3] & 0 & 0 \\
0 & x_1[1] & x_1[2] & x_1[3] & 0 \\
0 & 0 & x_1[1] & x_1[2] & x_1[3] \\
0 & 0 & 0 & x_1[1] & x_1[2] \\
0 & 0 & 0 & 0 & x_1[1]
\end{bmatrix}
\cdot
\begin{bmatrix} x_2[0] \\ x_2[1] \\ x_2[2] \\ x_2[3] \\ x_2[4] \end{bmatrix}
$$

Substituting the values yields,

$$r_{21}(\ell) = r_{x_2 x_1}(\ell) = -\left\{0,\ \frac{3}{4},\ \frac{11}{10},\ \frac{11}{10},\ \frac{8}{10},\ \frac{1}{2},\ \frac{1}{5},\ \frac{1}{20},\ 0,\ldots\right\}$$

2.14.18 Compute the autocorrelation of the signal: $x[n] = \cos\left(\dfrac{2\pi n}{N}\right)$.

SOLUTION: It is $r_{xx}(l) = \dfrac{1}{N}\displaystyle\sum_{n=N_L}^{n=N_H} x[n]\cdot x[n-l]$

Here, we have: $N_L = 0,\ N_H = N-1$

$$r_{xx}(l) = \frac{1}{N}\sum_{n=0}^{N-1}\sin\left(\frac{2\pi n}{N}\right)\cdot\sin\left(\frac{2\pi(n-l)}{N}\right) =$$

$$= \frac{1}{N}\sum_{n=0}^{N-1}\frac{1}{2}\left[\cos\left(\frac{2\pi n - 2\pi(n-l)}{N}\right) - \cos\left(\frac{2\pi n + 2\pi(n-l)}{N}\right)\right] =$$

$$= \frac{1}{2N}\sum_{n=0}^{N-1}\left[\cos\left(\frac{2\pi l}{N}\right) - \cos\left(\frac{4\pi n}{N} - \frac{2\pi l}{N}\right)\right] =$$

$$\frac{1}{2N}\sum_{n=0}^{N-1}\cos\frac{2\pi l}{N} - \underbrace{\frac{1}{2N}\sum_{n=0}^{N-1}\cos\left(\frac{4\pi n}{N} - \frac{2\pi l}{N}\right)}_{0} \Rightarrow$$

Consequently, the autocorrelation of the signal $x[n] = \cos\left(\dfrac{2\pi n}{N}\right)$ is:

$$r_{xx}(l) = \frac{1}{2N}N\cos\frac{2\pi l}{N} = \frac{1}{2}\cdot\cos\frac{2\pi l}{N}$$

2.14.19 Consider the systems with the following input-output relations:

System 1 (S1): $y[n] = \begin{cases} x\left[\dfrac{n}{2}\right], & n : even\ (2l) \\[2mm] 0, & n : odd\ (2l+1) \end{cases}$

System 2 (S2): $y[n] = x[n] + \tfrac{1}{2}x[n-1] + \tfrac{1}{4}x[n-2]$

System 3 (S3): $y[n] = x[2n]$

S1, S2, S3 are connected in the way illustrated in the figure below:

Find the input-output relation of this system, i.e., the difference equation.

SOLUTION: The outputs of the three systems are given respectively by the relations:

S3: $y[n] = x[2n]$

S2: $z[n] = w[n] + \dfrac{1}{2} w[n-1] + \dfrac{1}{4} w[n-2]$

S1: $w[n] = \begin{cases} x\left[\dfrac{n}{2}\right], & n = 2k \\ \\ 0, & n = 2k+1 \end{cases}$

From the above relations, we compute the output:

$$y[n] = z[2n] = w[2n] + \frac{1}{2} w[2n-1] + \frac{1}{4} w\left[2(n-1)\right] =$$

$$= x[n] + 0 + \frac{1}{4} x[n-1] = x[n] + \frac{1}{4} x[n-1]$$

Therefore, the difference equation is:

$$y[n] = x[n] + \frac{1}{4} x[n-1]$$

2.14.20 Calculate the impulse response of the discrete-time system illustrated in the figure below when: $h_1[n] = \delta[n] + \dfrac{1}{2}\delta[n-1]$, $h_2[n] = \dfrac{1}{2}\delta[n] - \dfrac{1}{4}\delta[n-1]$, $h_3[n] = 2\delta[n]$, $h_4[n] = -2\left(\dfrac{1}{2}\right)^n u[n]$.

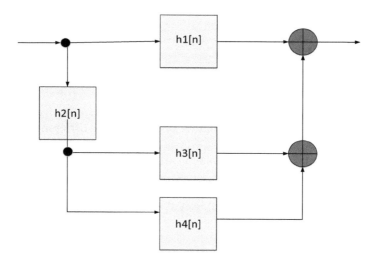

SOLUTION: Based on the figure, the systems with impulse responses $h_3[n]$ and $h_4[n]$ are connected in parallel and in series with $h_2[n]$. The total response of the system is:

$$h[n] = h_1[n] + h_2[n] * \left[h_3[n] + h_4[n] \right] = h_1[n] + h_2[n] * h_3[n] + h_2[n] * h_4[n]$$

But,

$$h_2[n] * h_3[n] = \left\{ \frac{1}{2}\delta[n] - \frac{1}{4}\delta[n-1] \right\} * 2\delta[n] = \frac{1}{2}\delta[n] * 2\delta[n] - \frac{1}{4}\delta[n-1] * 2\delta[n] = \delta[n] - \frac{1}{2}\delta[n-1]$$

$$h_2[n] * h_4[n] = \left\{ \frac{1}{2}\delta[n] - \frac{1}{4}\delta[n-1] \right\} * \left\{ -2\left(\frac{1}{2}\right)^n u[n] \right\}$$

$$= \frac{1}{2}\delta[n] * \left\{ -2\left(\frac{1}{2}\right)^n u[n] \right\} - \frac{1}{4}\delta[n-1] * \left\{ -2\left(\frac{1}{2}\right)^n u[n] \right\} = -\left(\frac{1}{2}\right)^n u[n] + \frac{1}{2}\left(\frac{1}{2}\right)^{n-1} u[n-1]$$

$$= \left(\frac{1}{2}\right)^n \left\{ u[n] - u[n-1] \right\} = -\left(\frac{1}{2}\right)^n \delta[n] = -\delta[n]$$

Thus, $h[n] = \left\{ \delta[n] + \frac{1}{2}\delta[n-1] \right\} + \left\{ \delta[n] - \frac{1}{2}\delta[n-1] \right\} + \{-\delta[n]\} = \delta[n]$

2.14.21 a. Using MATLAB, sketch the following sequences: *delta[n], u[n], a^n, r^n e^{j\omega n}, r[n],* and *x[n] = 2delta[n + 2] delta[n] + e^n(u[n + 1] − u[n − 2]).*

   b. Sketch the odd and even part of the step sequence from −20 to 20.

SOLUTION: a.1. Code for designing *delta*[*n*] = $\delta[n]$

The easiest way is the use of **gauspuls** *function*:

```
n=-3:3
d=gauspuls(n);
stem(n,d)
```

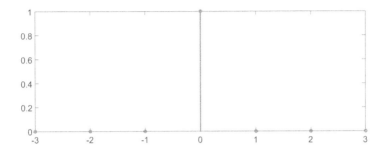

Another way to work is to set separate time intervals and place them in apposition. Then we define the values of the function $\delta[n]$ in each timeframe and plot it.

```
n1=-3:-1;
n2=0;
n3=1:3;
n=[n1 n2 n3]
d1=zeros(size(n1));
d2=1;
d3=zeros(size(n3));
d=[d1 d2 d3]
stem(n,d)
```

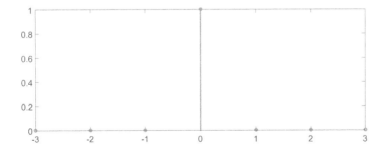

Alternatively, we can write the following code in which the sequence $\delta[n]$ is plotted over the time interval $-5 \leq n \leq 10$. The command x = (k == 0) gives result 1 when k is 0 and gives result 0 when k is other than 0.

```
k1=-5;
k2=10;
k=k1:k2;
x=(k==0);
stem(k,x)
```

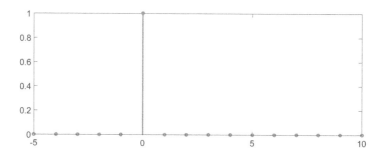

We can also create a *function* (**impseq.m**), which implements the unit impulse function δ[n]

```
function [x,n]= impseq(n0,n1,n2)
if ((n0<n1)|(n0>n2)|(n1>n2))\
error('arguments must satisfy n1 <= n0 <= n2')
end

n = [n1:n2];
x = [(n-n0) == 0];
```

After saving the function to an M-File, we run it from the command window and design for example the *δ(n-1)*.

```
[x,n]= impseq(1,-3,4);
stem(n,x)
```

2. Code for designing $u[n]$ for $-3 \leq n \leq 5$.

```
n2=0:5;
n=[n1 n2];
u1=zeros(size(n1));
u2=ones(size(n2));
u=[u1 u2];
stem(n,u)
```

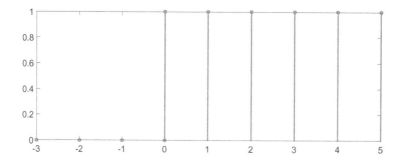

We can also create another *function* (**stepseq.m**), which implements the unit step sequence *u*[*n*].

```
function [x,n] = stepseq(n0,n1,n2)
if ((n0<n1) | (n0>n2) | (n1>n2))
error('arguments must satisfy n1 <= n0 <= n2')
end

n = [n1:n2];
x = [(n-n0) >= 0];
```

After saving the *function* **stepseq.m** to an M-File, we run it from the command window and design u(*n* − 1) for −3 ≤ 5.

```
[x,n]=stepseq(1,-3,5);
stem(n,x);
```

3. Code for designing a^n για $a = 0.8 < 1$

```
a1=0.8;
x1=a1.^n;
stem(n,x1);
```

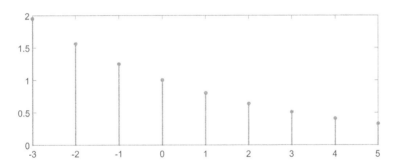

Code for designing a^n for $a = 1.2 > 1$

```
a2=1.2;
x2=a2.^n;
stem(n,x2);
```

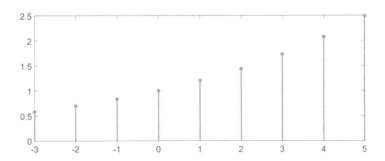

4. Code for design $r^n e^{j\omega n}$.

```
n=-10:10;
r=0.9;
w=1;
x=(r.^n).*exp(j*w*n);
stem(n,real(x)); % plot of the real part of r^n e^{jωn}
stem(n,imag(x)) % plot of the imaginary part of r^n e^{jωn}
stem(n,abs(x)) % plot of the magnitude of r^n e^{jωn}
stem(n,angle(x)) % plot of the phase of r^n e^{jωn}
```

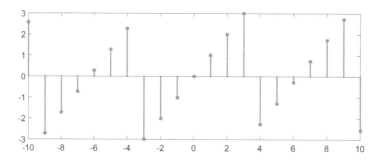

5. Code for designing $r[n]$

```
n=-3:6;
r=0.5*n.*(sign(n)+1);
stem(n,r)
grid
```

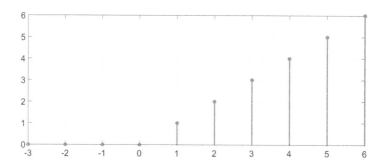

Alternatively, we can create the *function* **uramp.m** and apply it accordingly

```
function [sig,time]= uramp(n)
time=n;
sig=zeros(size(n));
sig1=find(n>=0);
sig(sig1)=n(sig1)-n(sig1(1));
stem(time,sig);
grid;
```

6. Code for designing: $x[n] = 2delta[n + 2] - delta[n] + e^n(u[n + 1]) - u[n - 2])$

```
n = -3:3;   % define time interval
% Define x[n]
x = 2.*((n >= -2) - (n >= -1)) ...          % 2delta[n+2]
    - 1.*((n >= 0) - (n >= 1)) ...            % -delta[n]
    + exp(n).*((n >= -1) - (n >= 2));    % e^n(u[n+1]-u[n-2])
stem(n,x);      % Graph x[n] with respect to n
xlabel('n');      % label on the x-axis
ylabel('x[n]'); % label on the y-axis
title('x[n] = 2delta[n+2] - delta[n]+ e^n(u[n+1]-u[n-2])');
```

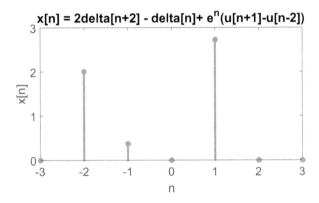

b. We shall design the sequences $x_e[n] = \frac{1}{2}(x[n] + x[-n])$ and $x_o[n] = \frac{1}{2}(x[n] - x[-n])$.

```
% partition odd - even
n=-20:20;
m=length(n);
k=find(n==0);
u=zeros(1,m);
u(k:m)=1;
ur=u(m:-1:1);
% even part xe(n)=1/2[x(n)+x(-n)]
ue=1/2*(u+ur);
% odd part xo(n)=1/2[x(n)-x(-n)]
uo=1/2*(u-ur);
% graph of signal x, x1 ,x2
subplot(3,1,1);
stem(n,u);
title('original');
axis([-20 20 -1 1.5])
subplot(3,1,2);
stem(n,ue);
title('even part');
axis([-20 20 -1 1.5])
subplot(3,1,3);
stem(n,uo);
title('odd part');
axis([-20 20 -1 1.5])
```

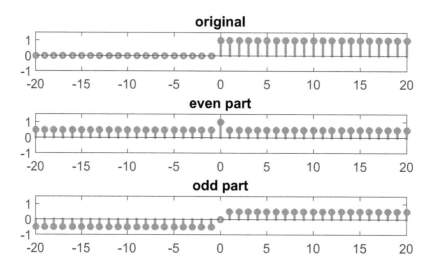

2.14.22 Let the sequence $x[n] = 0.9^n$ for $-10 \leq n \leq 10$ and $x[n] = 0$, elsewhere. Design using MATLAB the sequences $x[n-10]$, $x[n+10]$, $x[2n]$, $x[n/2]$ and $x[-n]$.

SOLUTION: The code in MATLAB for designing the signal $x[n] = 0.9^n$

```
n=-20:20;
m=length(n);
k=find(n==0);
a=0.9;
x=zeros(1,m);
x(k-10:k+10) = a.^n(k-10:k+10);
stem(n,x);
```

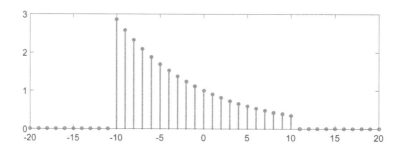

Definition and design of the signal $x[n-10]$:

```
x1=zeros(1,m);
x1(k:m)=x(k-10: k+10);
stem(n,x1);
```

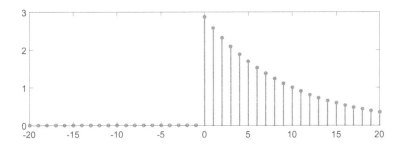

Definition and design of the signal $x[n + 10]$:

```
x2=zeros(1,m);
x2(k-20:k)=x(k-10:k+10);
stem(n,x2);
```

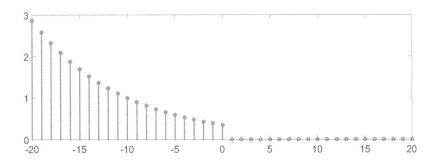

Definition and design of the signal $x[2n]$:

```
x3 =zeros(1,m);
x3(k-5:k+5)=x(k-10:2:k+10);
stem(n,x3)
```

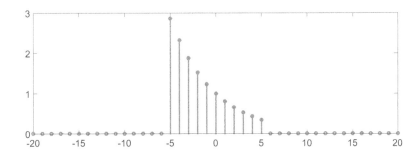

Definition and design of the signal $x[n/2]$:

```
x4=zeros(1,m);
x4(1:2:m)=x(k-10: k+10);
stem(n,x4);
```

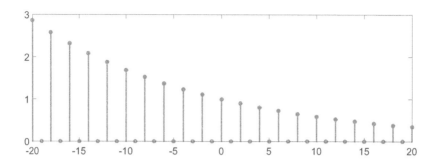

Definition and design of the signal $x[-n]$:

```
x5(1:m)=x(m:-1:1 );
stem(n,x5)
```

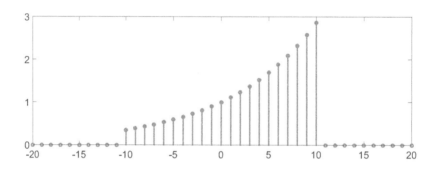

2.14.23 a. Write a *function* that implements a shift to a discrete-time signal. Implement and design the signal $y[n] = x[n-2]$ for $x[n] = [1, 2, 3, 4], 0 \leq n \leq 3$.

b. Write *functions* that implement frequency division and frequency multiplication, respectively, to a discrete-time signal.

SOLUTION:

a. We create the *function* **sigshift.m** that implements a shift to a discrete signal

```
function [y,n] = sigshift(x,m,n0)
% y(n)  = x(n-n0)
% x is the signal,  m is the time , n0 the shit
n = m+n0; y = x;
```

In order to implement and design the signal $y[n] = x[n-2]$ for $x[n] = [1, 2, 3, 4]$, $0 \leq n \leq 3$, we write the code:

```
x=[ 1 2 3 4];
m=[ 0 1 2 3];
n0=2;
```

```
[y,n]=sigshift(x,m,n0)
stem(n,y)
```

b. We create the *function* **sigfreqdiv.m** that implements the frequency division to a discrete-time signal.

```
function y=sigfreqdiv(a,x)
nl=length(x);
m=floor(nl/a);
for i=1:m
    y(i)=x(i*a);
end;
```

We create the *function* **sigfreqmul.m** that implements the frequency multiplication to a discrete-time signal.

```
function [y]=sigfreqmul(a,x)
nl=length(x);
m=nl*a;
for i=1:m
    y(i)=0;
    if mod(i, a)==0
        y(i)=x(i/a);
    end;
end;
```

2.14.24 Write a code in MATLAB to compute the linear convolution of the functions:

$$x[m] = h[n] = \begin{bmatrix} 1 & 2 & 3 & 4 & 5 & 6 \end{bmatrix}.$$

SOLUTION: The code is:

```
clc;
clear all;
close;
disp('enter the length of the first sequence m=');
m=input('');
disp('enter the length of first sequence x[m]=');
for i=1:m
    x(i)=input('');
end
disp('enter the length of the second sequence n=');
n=input('');
disp('enter the length of second sequence h[n]=');
for j=1:n
    h(j)=input('');
end
```

```
y=conv(x,h);
figure;
subplot(3,1,1);
stem(x);
ylabel ('amplitude---->');
xlabel('n---->');
title('x(n) Vs n');
subplot(3,1,2);
stem(h);
ylabel('amplitude---->');
xlabel('n---->');
title('h(n) Vs n');
subplot(3,1,3);
stem(y);
ylabel('amplitude---->');
xlabel('n---->');
title('y(n) Vs n');
disp(The linear convolution of x[m] and h[n] is y');
```

So, the linear convolution of x[m] and h[n] is:
$$y = 1\ 4\ 10\ 20\ 35\ 56\ 70\ 76\ 73\ 60\ 36$$

2.14.25 Design the signal:

$$x[n] = 2delta[n+2] - delta\big[n\big] + e\,^\wedge\,n\big(u[n+1] - u[n-2]\big)$$

SOLUTION: We write the code:

```
% signal x[n] = 2delta[n+2] - delta[n]
%                              + e^n(u[n+1]-u[n-2]).
n = -3:3; % define discrete-time variable
% define x[n]
x = 2.*((n >= -2) - (n >= -1)) ...      % 2delta[n+2]
    - 1.*((n >= 0) - (n >= 1)) ...      % -delta[n]
    + exp(n).*((n >= -1) - (n >= 2)); % e^n(u[n+1]-u[n-2])
stem(n,x); % plot x[n] vs n
xlabel('n'); % label plot
ylabel('x[n]');
title('A Combination Signal');
```

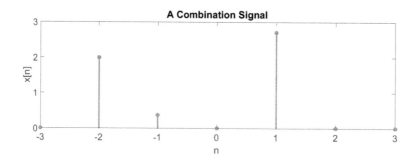

2.14.26 A basic discrete-time signal that appears quite often is: $x[n] = A \cos(\omega_0 n + \varphi)$. Write a *function* that produces a sinusoidal signal of finite-length and receives five input values: three for the signal parameters and two for determining the first and the last signal index of n. The *function* will return as the output, a column vector containing n values and a column vector containing the corresponding signal values. Then, using this function, design the signal $2\sin(\pi n/11)$ for $-20 \le n \le 20$.

SOLUTION: The code for the *function* is:

```
function [x,n]=sinusoid(A,w,phi,n1,n2)
% SINUSOID   Creates a discrete time signal
% The command   [x,n]=sinusoid(A,w,phi,n1,n2)
% creates the signal A*cos(w*n+phi) at the time interval
% n1<=n<=n2.
n=[n1:n2]';
x=A*cos(w*n+phi);
```

Now in order to create the signal $2\sin(\pi n/11)$ for $20 \le n \le 20$, we run the function as follows:

```
>>[x,n]=sinusoid(2,pi/11,pi/2,-20,20);
% Plot of the signal
>> stem(n,x)
>> grid
>> xlabel('n')
>> ylabel('sin(pi n/11)')
```

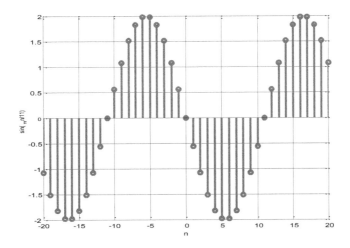

2.14.27 Consider the input sequence $x[n] = [1, 2, 3, 4, 5,]$, $0 \le n \le 4$. Plot the output of the system described by the impulse response $h[n] = [1, 2, 1]$, $-1 \le n \le 1$.

SOLUTION:

First suggested solution:

We define the two signals at the same time interval. So, we have to compute the convolution of the input signal $x[n] = [1, 2, 3, 4, 5]$, $0 \leq n \leq 4$ and the impulse response $h[n] = [1, 2, 1, 0, 0, 0]$, $-1 \leq n \leq 4$.

We plot the two signals:

```
n=-1:4;
x=[0,1,2,3,4,5];
h=[1,2,1,0,0,0];
subplot(121);
stem(n,x);
axis([-1.1 4.1 -.1 5.1]);
legend('x[n]')
subplot(122);
stem(n,h);
axis([-1.1 4.1 -.1 5.1]);
legend('h[n]')
```

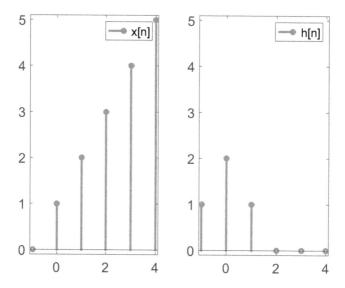

We will compute the convolution of the two discrete time signals using the **conv** function. The result of the function y = conv (x, h) is a vector, y, the length of which is M + N − 1, where M is x-length, and N is h-length. So, it holds *length(y) = length(x) + length(h) − 1*. We choose to design the output at a time interval, which is twice as that of the input and the impulse response.

```
y=conv(x,h);
stem(-2:8,y)
axis([-2.5 8.5 -.5 16.5]);
legend('y[n]')
```

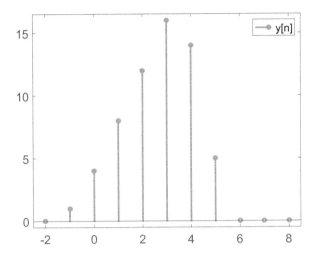

Second suggested solution:

The second method is equivalent to the first one and is based on the 2nd convolution principle: If the non-zero values of $x[n]$ are contained within the time interval $[M_x, N_x]$, and the non-zero values of $h[n]$ are contained within the time interval $[M_h, N_h]$, then the non-zero values of $y[n]$ will be contained within the time interval: $[M_x + M_h, N_x + N_h]$. The convolution of only the non-zero parts of the two signals is more convenient to be calculated. The output is then transferred to the appropriate time interval and is plotted.

The following code computes and designs the convolution of two discrete-time signals.

```
n=-10:10;
m=length(n);
k=find(n==0);
x=zeros(1,m);
h=zeros(1,m);
y=zeros(1,m);
x(k:k+4)=[1,2,3,4,5];
h(k-1:k+1)=[1 2 1];
yt=conv(x(k:k+4),h(k-1:k+1));
y(k-1:k+5)=yt;
stem(n,y);
axis([-2.5 8.5 -.5 16.5]);
legend('y[n]')
```

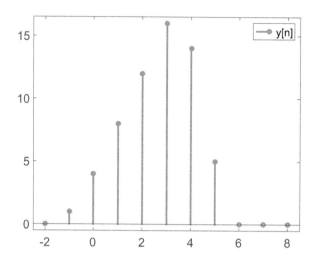

The output signals are equivalent.

2.14.28 Compute the convolution of the signals $h[k] = 0.5^k$, $0 \leq k \leq 10$ and $x[k] = u[k] - u[k-4]$.

SOLUTION: After defining both of the signals in the same time-domain, we can compute their convolution using the **conv** *function*.

```
k=0:10;
u=ones(size(k));
k1=0:3;
u4_1=zeros(size(k1));
k2=4:10;
u4_2=ones(size(k2));
u4=[u4_1 u4_2];
x=u-u4;
h=0.5.^k;
y=conv(x,h);
stem(0:20,y);
xlim([-1 21]);
legend('y[n]')
```

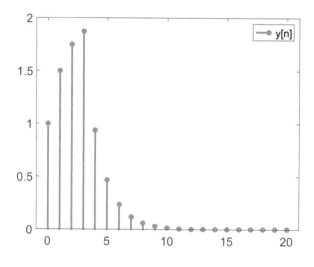

2.14.29 Consider a system described by the following difference equation: $y[n] = -y[n-1] - 0.5y[n-2] + 0.2x[n] + 0.1x[n-1] + 0.1x[n-2]$. Compute and design for $0 \leq n \leq 50$:

a. the impulse response of the system

b. the step response of the system

c. the response of the system when the input signal is: $x[n] = n \cdot 0.9^n, 0 \leq n \leq 5$.

SOLUTION:

a. The impulse response of the system will be calculated in MATLAB with the command filter.

```
a=[1 1 0.5];
b=[.2 .1 .1];
d=[1 zeros(1,50)];
h=filter(b,a,d);
stem(0:50,h);
```

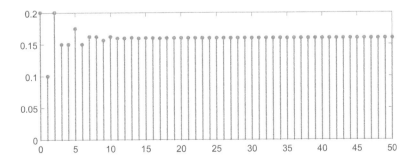

b. The step response of the system is computed and designed using the following code:

```
a=[1 1 0.5];
b=[.2 .1 .1];
u=ones(1,51);
usr=filter(b,a,u);
stem(0:50,usr);
```

c. The response of the system when the input signal is $x[n] = n \cdot 0.9^n$, $0 \le n \le 5$, is computed and plotted as follows:

```
a=[1 1 0.5];
b=[.2 .1 .1];
nx1=0:5;
x1=nx1.*(0.9.^nx1)
nx2=6:50;
x2=zeros(size(nx2))
x=[x1 x2];
y=filter(b,a,x);
stem(0:50,y);
legend('y[n]');
```

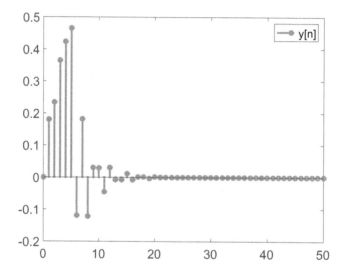

2.14.30 Based on the following figure, and given that the impulse responses of the systems S_1, S_2 and S_3 are $h_1[n] = [2, 3, 4]$, $h_2[n] = [-1, 3, 1]$ and $h_3[n] = [1, 1, -1]$ for $0 \le n \le 2$, respectively, find:

a. The total impulse response, $h[n]$, of the system
b. The output of the system when the input signal is $x[n] = u[n] - u[n - 2]$.

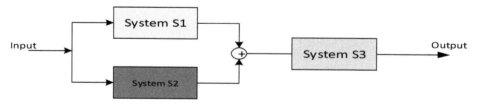

SOLUTION:

a. For S_1 and S_2 (connected in parallel), using the distributive property, it is: $h_{12} = h_1 + h_2 = [2, 3, 4] + [-1, 3, 1] = [1, 6, 5]$.

For the connection in series and using the associative property, we have: $h = h_{12} * h_3$.

Thus, using the following code, we can compute the total impulse response, $h(n)$.

```
h12=[1 6 5];
h3=[1 1 -1];
h=conv(h12,h3)
```

The total impulse response $h(n)$: h = 1   7   10   -1   -5

b. The output signal results from the convolution of the input and the total impulse response.

```
x=[1 1 0 0 0];
y=conv(x,h);
stem(0:8,y);
legend('y[n]');
xlim([-1 9]);
```

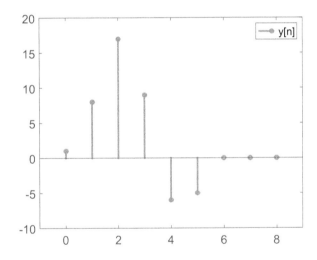

2.14.31 For the system with the difference equation, $y[n] -1.1y[n-1] + 0.5y[n-2] + 0.3y[n-4] = 0.5x[n] -0.2x[n-1]$, find:

a. The impulse response, $h[n]$, of the system for the time interval $0 \leq n \leq 10$.

b. The output of the system when the input signal is x = [5, 1, 1, ,1, 0, 0, 1, ,1, 1, 0, 0] using the MATLAB *functions* **conv** and **filter**.

SOLUTION:

a. Using the following code, the impulse response of the system is computed and designed:

```
n=0:10;
a=[1 -1.1 0.5 0 0.3];
b=[0.5 -0.2];
x=(n==0);
```

```
h=filter(b,a,x);
stem(n,h)
xlim([-.5 10.5]);
legend('h(n)')
```

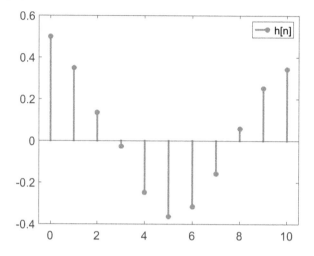

b. Since the impulse response, $h[n]$, is now known, it can be used to compute the output of the filter. We shall use the MATLAB *function* **conv**:

```
x= [5 1 1 1 0 0 1 1 1 0 0];
y1=conv(x,h);
stem(0:20,y1);
xlim([-.5 20.5]);
legend('y[n]');
```

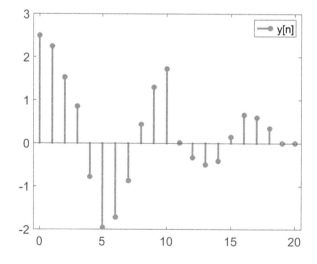

In order to get 21 output terms from the **filter** *function* (since we have 21 output terms from the **conv** *function*), we have to redefine the input signal $x[n]$ for

$0 \leq n \leq 20$, i.e., to fill it in with zeros. This is done with the command x(21) = 0. This command adds zeros up to the 21st position of the x vector. Adding zeros to x gives the **filter** command the extra information it needs in order to continue its calculations further.

```
a=[1 -1.1 0.5 0 0.3];
b=[0.5 -0.2];
x=[5 1 1 1 0 0 1 1 1 0 0];
x(21)=0;
y2=filter(b,a,x);
stem(0:20,y2);
xlim([-.5 20.5]);
legend('y[n]');
```

We notice that there is a match between the two results but only with respect to the first 11 terms. The result produced from the use of the **filter** (y2) is perfectly correct but y1 is partly incorrect. Our system has an impulse response of infinite length. However, we approximated it (we could not do anything different!) by using only 11 terms when calculating the impulse response of the system.

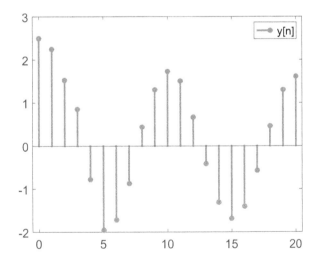

Using **conv**, 11 output terms will be computed with "absolute accuracy," but the rest will be incorrect. In other words, the **conv** *function* 'considered' the system to be of finite impulse response with length 11. It is concluded that, if we try to compute the output based on the impulse response, in the case of the systems with infinite impulse response (IIR systems), we will surely get incorrect results because of the approximation we have to do when determining the number of its terms.

In the case of finite impulse response systems (FIR systems), described by difference equations, the use of the **conv** and **filter** *functions* is equivalent when computing the output.

2.14.32 Let the input sequence:

$$x[n] = \begin{cases} 1, & 10 \le n \le 20 \\ 0, & elsewhere \end{cases}.$$

Design the output of the system when its impulse response is:

$$h[n] = \begin{cases} n, & -5 \le n \le 5 \\ 0, & elsewhere \end{cases}.$$

SOLUTION:

First suggested solution: The two signals must by defined at the same time domain, thus, $x[n] = \begin{cases} 0, & -5 \le n \le 9 \\ 1, & 10 \le n \le 20 \end{cases}$ and $h[n] = \begin{cases} n, & -5 \le n \le 5 \\ 0, & 6 \le n \le 20 \end{cases}$

The code in MATLAB is:

```
n1=-5:9;
x1=zeros(size(n1));
n2=10:20;
x2=ones(size(n2));
x=[x1 x2];
n1=-5:5;
h1=n1;
n2=6:20;
h2=zeros(size(n2));
h=[h1 h2];
y=conv(x,h);
stem(-10:40,y)
ylim([-16 16])
grid
```

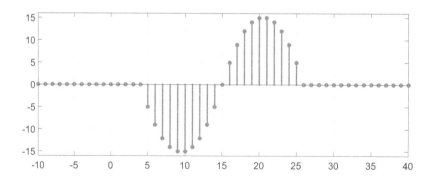

Since the input signal, $x[n]$, and impulse response, $h[n]$, are defined over the whole time interval $-5 \le n \le 20$, we must fill them in with zeros (where needed).

Second suggested solution:

If the non-zero values of $x[n]$ are contained within the time interval $[M_x, N_x]$, and the non-zero values of $h[n]$ are contained within the time interval $[M_h, N_h]$, then the non-zero values of $y[n]$ will be contained within the time interval $[M_x + M_h, N_x + N_h]$.

Therefore, the non-zero values of the output $y[n] = x[n] * h[n]$ will be limited to the interval [5,25].

The code in MATLAB is:

```
nx=10:20;
x=ones(size(nx));
nh=-5:5;
h=nh;
y=conv(x,h);
stem(5:25,y)
axis([-10 40 -16 16]);
grid;
legend('y[n]');
```

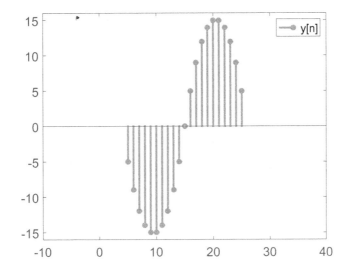

We notice that the two output figures are identical, so, the two suggested solutions are equivalent.

3

z-Transform

3.1 Introduction

Analog systems are designed and analyzed with the use of *Laplace transforms*. On the other hand, discrete-time systems are analyzed by using a similar technique, called the *z-Transform*.

The basic lines of reasoning are the same for both cases: After determining the impulse response of the system, the response of any other input signal can be extracted by simple mathematical operations. The behavior and the stability of the system can be predicted from the zeros and poles of the transfer function.

As Laplace transform converts the differential equations into algebraic terms, with respect to s, the z-transform converts the difference equations into algebraic terms, with respect to z. Both transforms are mapping a complex quantity to the complex plane. It is noteworthy that the z-plane (i.e., the domain of z-transform) is structured in a *Polar* form, while the s-plane (i.e., the domain of Laplace transform) is structured in a *Cartesian* form.

3.2 From Laplace Transform to z-Transform

z-Transform facilitates significantly the study and design of nonlinear time-varying discrete-time systems, because it transforms the difference equation that describes the system in an algebraic equation.

In Figure 3.1, the procedure followed by using the z-transform is illustrated, where there are three steps to solve the difference equation (D.E.). The direct solution of the given D.E. through higher mathematics is much more laborious.

To show that z- and Laplace transforms are two parallel techniques, the Laplace transform will be used, and making use of it, the mathematical expression of z-transform will be developed.

Laplace transform is defined as $X(s) = \displaystyle\int_{t=-\infty}^{+\infty} x(t) \cdot e^{-st}\, dt$, where s is a complex number.

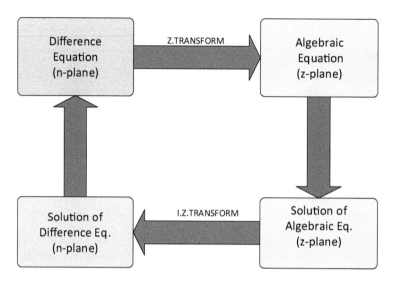

FIGURE 3.1
Solution of D.E. via z-transform.

By substituting $s = \sigma + j\omega$ one can reach to an alternative form of $X(s)$ function, which is $X(\sigma,\omega) = \displaystyle\int_{t=-\infty}^{+\infty} x(t) \cdot e^{-\sigma t} \cdot e^{-j\omega t}\, dt$. Substituting in the latter expression $e^{-j\omega t} = \cos(\omega t) - j\sin(\omega t)$ yields:

$$X(\sigma,\omega) = \int_{t=-\infty}^{+\infty} x(t) \cdot e^{-\sigma t} \cdot \big[\cos(\omega t) - j\sin(\omega t)\big]\, dt \Rightarrow$$

$$X(\sigma,\omega) = \int_{t=-\infty}^{+\infty} x(t) \cdot e^{-\sigma t} \cos(\omega t)\, dt - j \int_{t=-\infty}^{+\infty} x(t) \cdot e^{-\sigma t} \sin(\omega t)\, dt \tag{3.1}$$

According to Equation 3.1, $x(t)$ is analyzed in sine and cosine waves, whose amplitude varies exponentially in the time domain (because of the existence of the term $e^{-\sigma t}$). Each point of the complex s-plane is determined by its real and imaginary part, i.e., the parameters σ and ω. At each point of the s-plane, the complex quantity $X(\sigma,\omega)$ can be calculated.

The real part of $X(\sigma,\omega)$ arises by multiplying the signal $x(t)$ with a cosine waveform of frequency ω whose amplitude decreases exponentially with a rate, σ, and then is integrated for all instances. Thus:

$$\Re\big\{X(\sigma,\omega)\big\} \sim \int_{t=-\infty}^{+\infty} x(t) \cdot e^{-\sigma t} \cdot \cos(\omega t)\, dt \tag{3.2}$$

Likewise, the imaginary part is obtained by multiplying the $x(t)$ signal with a sine waveform of frequency ω, whose amplitude decreases exponentially with a rate, σ. Hence,

$$\Im m\{X(\sigma,\omega)\} \sim \int_{t=-\infty}^{+\infty} x(t) \cdot e^{-\sigma t} \cdot \sin(\omega t)\, dt \tag{3.3}$$

Based on the above representation of the Laplace transform, one can produce z-transform, i.e., the corresponding transform relation for discrete signals in *three steps*:

Step 1: The first step is the most obvious: By changing the signal from continuous to discrete, i.e., $x(t) \to x[n]$, and, of course, by replacing the integral with a sum.

Therefore: $\int_{t=-\infty}^{+\infty} \to \sum_{n=-\infty}^{+\infty}$ so,

$$X(\sigma,\omega) = \sum_{n=-\infty}^{+\infty} x[n] \cdot e^{-\sigma n} \cdot e^{-j\omega n} \tag{3.4}$$

Despite the fact that $x[n]$ signal is a discrete one, $X(\sigma,\omega)$ is continuous since σ and ω variables can take continuous values.

In the case of the Laplace transform, one could move up to any point (σ,ω) of the complex plane (not quite any point; the integral will not converge for points not belonging in the region of convergence) and define the real and imaginary part of $X(\sigma,\omega)$ by integrating over time, as previously explained.

If the case of the z-transform, one may, again, move up to any point of the complex plane, but instead of integration, one may work with summation.

Step 2: In the second step, polar coordinates are introduced to represent the exponential $e^{-\sigma n}$.

The exponential signal $y[n] = e^{-\sigma n}$ can be written as: $y[n] = r^{-n}$, where, apparently, the substitution $e^{\sigma} = r$ has been implemented, which yields $\sigma = \ln r$.

It is noteworthy that when:

- $y[n] = e^{-\sigma n}$, $y[n]$ increases with time when $\sigma < 0$.
- $y[n] = r^{-n}$, $y[n]$ increases with time when $r < 1$.
- $y[n] = e^{-\sigma n}$, $y[n]$ decreases with time when $\sigma > 0$.
- $y[n] = r^{-n}$, $y[n]$ decreases with time when $r > 1$.
- $y[n] = e^{-\sigma n}$, $y[n]$ remains unchanged when $\sigma = 0$.
- $y[n] = r^{-n}$, $y[n]$ remains unchanged when $r = 1$.

$$\text{Hence: } X(r,\omega) = \sum_{-\infty}^{\infty} x[n] r^{-n} \cdot e^{-j\omega n} \tag{3.5}$$

Step 3: The substitution $z = r \cdot e^{j\omega}$ is implemented, therefore the standard form of z-transform arises.

$$X(z) = \sum_{-\infty}^{\infty} x[n] \cdot z^{-n} \qquad (3.6)$$

Equation 3.6 forms the definition of bilateral z-transform.

Because there are both positive and negative z-powers, convergence is, in some cases, possible inside the region of a ring of the complex plane. A serious difficulty with this transform is that, in order to be able to return to time domain, the convergence region must also be known. For casual systems, the unilateral z- transform is defined by Equation 3.7.

It is a power series of negative z-powers and convergence is possible if $|z|$ is large enough.

$$X(z) = \sum_{n=0}^{n=\infty} x[n] \cdot z^{-n} \qquad (3.7)$$

For the convergence region it is true that:

1. A sequence of finite-length has a z-transform for which the convergence region is the entire z-plane, except perhaps from the points $z = 0$ and $z = \infty$. $z = \infty$ is included when $x[n] = 0$ for $n < 0$ is true and point $z = 0$ is included when $x[n] = 0$ for $n > 0$ is true.

2. A right-sided sequence has a z-transform for which the region of convergence (ROC) is the exterior plane of a ROC circle: $|z| a$.

3. A left-sided sequence has a z-transform for which the region of convergence (ROC) is the interior plane of a ROC circle: $|z| < b$.

z-Transform (Eq. 3.6) is a valuable tool for analyzing discrete Linear Time-Invariant (LTI) systems. It can be used for:

- Efficient calculation of the response of an LTI system (convolution in the n-plane $y(n) = x(n)*h(n)$ is computed as a product in the z-plane: $Y(z) = X(z)H(z)$, so $y(n) = IZT(Y(z))$.
- Stability analysis of an LTI system (through calculation of the ROC).
- Description of LTI system with regards to its behavior in the frequency domain (low-pass filter, band-pass filter, etc.).

3.2.1 Comparison of the s- and z-Planes into the Region of Convergence

The main differences between the s- and z-planes are presented in Figure 3.2.

The points of s-plane are described by two parameters: σ parameter, which corresponds to the real axis and determines the rate of reduction of the exponential, and ω parameter, which corresponds to the imaginary axis, which determines the oscillation frequency. Both parameters have a rectangular arrangement on the s-plane.

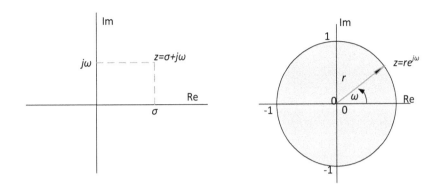

FIGURE 3.2
s-plane is Cartesian whereas *z*-plane is polar.

The resulting geometry arises from the fact that for each *s* number, the position is determined by the relation: $s = \sigma + j\omega$.

The *z*-plane is *polar*. Each complex number *z* is determined by its distance, *r*, from the origin, which corresponds to the reduction rate of the exponential (recall the expression $X(r, \omega)$), and by ω parameter, which corresponds to the angle between *r* and the positive horizontal semi-axis. The *z*-plane is structured in polar coordinates, and this is interpreted by the equations: $z = r \cdot e^{j\omega}$ or $z = re^{j\omega} = r(\cos \omega + j \sin \omega)$.

As a consequence of the above differences, the vertical lines in the *s*-plane become circles in the *z*-plane. This is due to the substitution $\sigma = \ln r$, which has previously been implemented. For instance, the imaginary axis of the *s*-plane, i.e., the line formed by setting $\sigma = 0$, will correspond to a circle with a radius $r = 1$ into the *z*-plane. Indeed, the relation $\sigma = 0$ corresponds to $\sigma = \ln r$ in the *z*-plane, which yields $r = 1$.

Lines parallel to the imaginary axis located in the left half-plane ($\sigma < 0$) correspond to concentric circles, which are located within the unit circle in the *z*-plane and, likewise, for lines located in the right *s*-half-plane.

3.3 Properties of *z*-Transform

The most important properties of the *z*-transform, commonly used for problem solving, are presented in the following sections.

3.3.1 Time Shift

If $X(z)$ denotes the *z*-transform of the casual sequence $x[n]$, then the corresponding transform of $x[n - N]$ is given by $z^{-N}X(z)$. Time shift adds or subtracts the axes' origin or infinity from the region of convergence of $X(z)$. This is one of the most important properties of the *z*-transform since it allows difference equations with zero initial conditions to be expressed as functions of *z* and to calculate the output of a system regardless of the type of input.

If...	Then...
$x[n] \leftrightarrow X(z)$	$x[n - N] \leftrightarrow z^{-N}X(z)$

In the case of non-zero initial conditions:

If...	Then...
$x[n] \leftrightarrow X(z)$	$x[n-N] \leftrightarrow z^{-N}X(z) + \sum_{i=1}^{m} x(-i)z^{-m+i}$

Proof: Let $y[n] = x[n-N]$, then $Y(z) = \sum_{n=-\infty}^{\infty} X[n-N]z^{-n}$

Assume $k = n - N$ then $n = k + N$, and substituting in the above equation, we have:

$$Y(z) = \sum_{k=-\infty}^{\infty} x[k]z^{-k-N} = z^{-N}X[z]$$

3.3.2 Linearity

Let $x[n]$ be a function, which is produced from the linear combination of two functions, $x_1[n]$ and $x_2[n]$, with regions of convergence P_1 and P_2, respectively. The region of convergence of $x[n]$ includes the intersection of $P_1 \cap P_2$.

If...	Then...
$x[n] \leftrightarrow X(z)$	$ax_1[n] + bx_2[n] \overset{ZT}{\leftrightarrow} aX_1(z) + bX_2(z)$

3.3.3 Time Reversal

If the z-transform of the $x[n]$ function is $X(z)$, with a region of convergence, P, then the transform of $x[-n]$ is $X\left(\dfrac{1}{z}\right)$ with a region of convergence $\dfrac{1}{P}$.

If...	Then...
$x[n] \leftrightarrow X(z)$	$x[-n] \overset{ZT}{\leftrightarrow} X\left(\dfrac{1}{z}\right)$

Proof: If we have $x(n) \Leftrightarrow X(z)$, then $x^*[-n] \overset{z}{\longleftrightarrow} X^*\left(1/z^*\right)$
Let $y[n] = x^*[-n]$, then

$$Y(z) = \sum_{n=-\infty}^{\infty} x^*[-n]z^{-n} = \left(\sum_{n=-\infty}^{\infty} x[-n][z^*]^{-n}\right)^* = \left(\sum_{k=-\infty}^{\infty} x[k]\left(1/z^*\right)^{-k}\right)^* = X^*\left(1/z^*\right)$$

If the ROC of $F(z)$ is $r_R < |z| < r_L$, then the ROC of $Y(z)$ is

$$r_R < \left|1/z^*\right| < r_L \quad i.e., \quad \frac{1}{r_R} > |z| > \frac{1}{r_L}$$

Moreover, it is easy to show that:

$$x[-n] \xleftrightarrow{z} X\left(1/z\right) \text{ and ROC is } \frac{1}{r_R} > |z| > \frac{1}{r_L}$$

3.3.4 Convolution

Let two functions $x_1[n]$ and $x_2[n]$ with corresponding z-transforms and regions of convergence $x_1[n] \leftrightarrow X_1(z)$ when $z \in P_1$ and $x_2[n] \leftrightarrow X_2(z)$ when $z \in P_2$. The transform of the convolution of the signals $x_1[n]$ and $x_2[n]$ is $x_1[n] * x_2[n] \leftrightarrow X_1(z) X_2(z) = X(z)$, where the region of convergence of $X(z)$ is equal or contains the intersection of the regions of convergence of $X_1(z)$ and $X_2(z)$.

If...	Then...
$x[n] \leftrightarrow X(z)$	$x_1[n] * x_2[n] \leftrightarrow X_1(z) X_2(z)$

Proof: The convolution of $x_1[n]$ and $x_2[n]$ is defined as $x[n] = x_1[n] * x_2[n] = \sum_{k=-\infty}^{\infty} x_1[k]x_2[n-k]$

The z-transform of x[n] is $X(z) = \sum_{n=-\infty}^{\infty} x[n]z^{-n} = \sum_{n=-\infty}^{\infty} \left[\sum_{n=-\infty}^{\infty} x_1[k]x_2[n-k] \right] z^{-n}$

By interchanging the order of the summation and applying the time-shift property, we obtain $X(z) = \sum_{k=-\infty}^{\infty} x_1(k) \left[\sum_{n=-\infty}^{\infty} x_2[n-k]z^{-n} \right] = X_2(z) \sum_{k=-\infty}^{\infty} x_1[k]z^{-k} = X_2(z)X_1(z)$

3.3.5 Differentiation in *z*-Plane

Let $X(z)$ be the transform of $x[n]$ function with a region of convergence, Π. Then, it is $nx[n] \leftrightarrow -z\dfrac{dX(z)}{dz}$ with the same region of convergence.

If...	Then...
$x[n] \leftrightarrow X(z)$	$nx[n] \leftrightarrow -z\dfrac{dX(z)}{dz}$

Proof:

$$x(z) = \sum_{n=-\infty}^{\infty} x[n]z^{-n}$$

$$-z\frac{dX(z)}{dz} = -z\sum_{n=-\infty}^{\infty} -nx[n]z^{-n-1} = \sum_{n=-\infty}^{\infty} -nx[n]z^{-n}$$

$$-z\frac{dX(z)}{dz} \xleftrightarrow{z} nx[n]$$

3.3.6 Multiplication by an Exponential Sequence

Let $X(z)$ be the transform of $x[n]$ function with a region of convergence, P. Then, $z_0^n x[n] \leftrightarrow X\left(\dfrac{z}{z_0}\right)$. The consequence is pole and zero locations are scaled by z_0. If the ROC of FX(z) is rR < |z| < rL, then the ROC of Y(z) is rR < |z/z0| < rL, i.e., |z0|rR < |z| < |z0| rL.

If...	Then...
$x[n] \leftrightarrow X(z)$	$z_0^n x[n] \leftrightarrow X\left(\dfrac{z}{z_0}\right)$

Proof: $Y(z) = \sum_{n=-\infty}^{\infty} z_0^n x[n]z^{-n} = \sum_{n=-\infty}^{\infty} x[n]\left(\dfrac{z}{z_0}\right)^{-n} = X\left(\dfrac{z}{z_0}\right)$

3.3.7 Conjugation of a Complex Sequence

If we have $x(n) \Leftrightarrow X(z)$, then $x^*[n] \xleftrightarrow{z} X^*(z^*)$, and ROC = R_f

If...	Then...
$x[n] \leftrightarrow X(z)$	$x^*[n] \xleftrightarrow{z} X^*(z^*)$

Proof: Let $y[n] = x^*[n]$, then

$$Y(z) = \sum_{n=-\infty}^{\infty} x^*[n]z^{-n} = \left(\sum_{n=-\infty}^{\infty} x[n][z^*]^{-n}\right)^* = X^*(z^*)$$

3.3.8 Initial and Final Value Theorem

If $x[n] = 0$ when $n < 0$, then $x[0] = \lim_{z \to \infty} X(z)$ (Initial Value Theorem)

For causal stable systems, it is true that:

$$x[\infty] = \lim_{z \to \infty} x(n) = \lim_{z \to 1}(z - 1)X(z) \text{ (Final Value Theorem)}$$

If...	Then...
$x[n] \leftrightarrow X(z)$	$x[0] = \lim_{z \to \infty} X(z)$
$x[n] \leftrightarrow X(z)$	$x[\infty] = \lim_{n \to \infty} x[n] = \lim_{z \to 1}(z - 1)X(z)$

3.3.9 Correlation of Two Sequences

If $x_1[n] \leftrightarrow X_1(z)$ and $x_2[n] \leftrightarrow X_2(z)$

$$r_{x_1 x_2}(m) = \sum_{n=-\infty}^{\infty} x_1(n)x_2(n-m) \xleftrightarrow{\;z\;} X_1(z)X_2(z^{-1})$$

The following is the correlation of two sequences.

$$rx_1 x_2(m) = \sum_{n=-\infty}^{\infty} x_1(n)x_2\big[(n-m)\big]$$

Or, rearranging the terms, we get $rx_1 x_2(m) = \sum_{n=-\infty}^{\infty} x_1(n)x_2\big[-(m-n)\big].$

Thus, the right-hand side of the second equation represents the convolution $x_1(n)$ and $x_2(-m)$ and can be written as

$$rx_1 x_2(m) = x_1(m) \otimes x_2(-m)$$

$$Z[rx_1 x_2(m)] = Z[x_1(m) \otimes x_2(-m)]$$
$$Z[rx_1 x_2(m)] = Z[x_1(m)] \otimes Z[x_2(-m)]$$
$$Z[x_1(m)] = X_1(z) \text{ and } z[x_2(m)] = X_2(z^{-1})$$
$$Z[rx_1 x_2(m)] = X_1(z)X_2(z^{-1})$$

3.4 Inverse z-Transform

Applying the z-transform results in the transfer from the discrete-time domain to the z-domain. The inverse procedure is carried out using the inverse z-transform.

The inverse z-transform is defined by

$$x[n] = Z^{-1}\left[X(z)\right] = \frac{1}{2\pi j}\oint_c X(z)z^{n-1}\,dz \tag{3.8}$$

where c is a circle with its center at the origin of the z-plane, such that all poles of $X(z)z^{n-1}$ are included in it. This integral can be derived by Cauchy's residue theorem.

Since the calculation of the integral is quite difficult, the use of matrices is suggested. In this way, the time functions of fundamental complex functions are derived. The majority of these matrices, however, can cover only a few cases. Therefore, other methods are used to calculate the inverse z-transform.

There are three methods for calculating the inverse transform of a function, $X(z)$:

  a. *Power series expansion*
  b. *Expansion partial fraction*
  c. *Complex integration* (through the residue theorem)

3.4.1 Method of Power Series Expansion (Division Method)

Using this method, specific samples of the inverse z-transform are calculated, but a corresponding analytical expression is not provided. By dividing its numerator by the denominator, function $X(z)$ forms a series with respect to z.

3.4.2 Method of Partial Fraction Expansion

Partial fraction expansion is a particularly useful method for the analysis and design of systems because the effect of any characteristic root or eigenvalue becomes obvious.

It is helpful to use $\dfrac{X(z)}{z}$ and not $X(z)$ while applying the method of partial fraction expansion. We distinguish *three cases* of partial fraction expansion for $\dfrac{X(z)}{z}$, according to the form of its poles.

- Case of distinct real poles:

  In this case, $\dfrac{X(z)}{z}$ is expanded in a fractional sum series, as follows:

$$\frac{X(z)}{z} = \frac{B(z)}{(z-p_1)...(z-p_n)} = \frac{c_1}{(z-p_1)} + ... + \frac{c_n}{(z-p_n)} \tag{3.9}$$

  The c_i coefficients are computed using the *Heaviside formula* for distinct poles, hence:

$$c_i = \lim_{z \to p_i}\left\{(z-p_i)\frac{B(z)}{(z-p_1)(z-p_2)...(z-p_n)}\right\} \tag{3.10}$$

- Case of non-distinct real poles (multiple real poles):

  In this case, $\dfrac{X(z)}{z}$ is expanded in a series of fractions, as follows:

  $$\frac{X(z)}{z} = \frac{B(z)}{(z - p_1)^n \dots (z - p_n)} = \frac{c_{11}}{(z - p_1)} + \frac{c_{12}}{(z - p_1)^2}$$

  $$+ \dots \dots \frac{c_{1n}}{(z - p_1)^n} + \frac{c_2}{(z - p_2)} \dots + \frac{c_n}{(z - p_n)} \tag{3.11}$$

  The c_{ij} coefficients are computed using the Heaviside formula for multiple poles, hence:

  $$c_{ij} = \frac{1}{(n - j)!} \lim_{z \to p_i} \left\{ \frac{d^{(n-j)}}{dz^{(n-j)}} (z - p_i)^n \frac{X(z)}{z} \right\} \tag{3.12}$$

  The remaining coefficients are computed using Equation 3.10.
- Case of complex roots:

  In this case, the coefficient of the numerator for one of the complex roots is computed using Equation 3.10 or 3.12. Therefore, the coefficient that is the numerator of the term, which has the conjugate root of the former at the denominator would be its corresponding conjugate.

3.4.3 Method of Complex Integration

This method is quite general and is being used when one or more partial fractions of the expanded $F(z)$ are not included into the lookup tables of z-transform pairs.

This method relies on the use of the definition formula of the *inverse z-transform*.

In order to use Equation 3.7, the application of the *residue theorem* is required, which is given by:

$$\oint F(z) z^{n-1} \, dz = 2\pi j \sum residues \left[F(z) z^{n-1} \right] \tag{3.13}$$

In the latter expression, also known as *Cauchy's formula*, $\sum$ represents the sum of residues of all the poles of $F(z)$, which are included in the c curve.

Combining the above expressions, we obtain:

$$f[n] = \sum residues \left[F(z) z^{n-1} \right] \tag{3.14}$$

- If there is a simple first-order pole of $F(z) \cdot z^{n-1}$ (i.e., $z = \alpha$), then its residual is given by:

  $$F(z) z^{n-1} (z - \alpha) \Big|_{z = \alpha} \tag{3.15}$$

- If there is an m-order pole of $F(z) \cdot z^{n-1}$, then its residual is given by:

$$\frac{1}{(m-1)!} \frac{d^{(m-1)}}{dz^{(m-1)}} F(z)z^{n-1}(z-\alpha)^m \bigg|_{z=\alpha} \tag{3.16}$$

3.5 z-Transform in System Analysis

z-Transform gives a significant amount of possibilities in system analysis for:

- Effective computation of the response of a discrete LTI system (the convolution in the discrete-time domain $y[n] = x[n]*h[n]$ is calculated as a product in the z-transform domain).
- Analysis of the stability of an LTI system (through the calculation of the convergence region).
- Characterization of an LTI in relation to its behavior in the frequency domain.

3.5.1 Transfer Function of Discrete-Time Signal

The transfer function is defined as the quotient of the z-transform of the output of an LTI system to the z-transformation of its input, when the initial conditions are zero, and corresponds to a relation which describes the dynamics of the particular system.

$$H(z) = \frac{Y(z)}{X(z)}\bigg|_{I.C.} = 0 \tag{3.17}$$

The transfer function is the z-transform of the impulse response. Equation 3.17 shows that the response of the discrete system can be calculated by the relation:

$$y[n] = IZT\big(X(z)H(z)\big) \tag{3.18}$$

Let the input of an LTI system be a signal described by the impulse function. The z-transform of the impulse response $h[n]$ is the transfer function of the system. The transfer function does not depend on the waveform of the input but only on the system characteristics.

In the case that the variable z is in the unit circle, i.e., when $z = e^{j\Omega}$, then the transfer function is the same as the system response frequency, provided that the unit circle belongs to the region of convergence.

Moreover, if, at the input of a discrete-time LTI system, the complex exponential sequence $x[n] = z^n$ is applied, then the output will be $H(z) \cdot z^n$. That is, the sequences of z^n form, are eigenfunctions of the system with eigenvalue $H(z)$.

3.5.2 Causality of Discrete-Time Systems

We can examine whether an LTI system is casual by studying its impulse response. According to the definition, a discrete-time LTI system is casual when the impulse response is zero, i.e., $h[n] = 0$ for each $n < 0$.

According to the condition of causality, the impulse response of a causal system is zero for each negative value of the variable n; therefore, the impulse response is a clockwise sequence. Then, the region of convergence of the transfer function will be in the form of an area outside of a circle. Based on the definition a causal sequence, $h[n]$, corresponds to a function of the form $H(z) = \sum_{n=0}^{+\infty} h[n]z^{-n}$, i.e., $H(z)$ does not include positive exponents of z, and consequently, the range of convergence includes infinity.

A discrete-time LTI system is casual if and only if the region of convergence of the transfer function is part of the complex plane exterior to the circle passing through the most distant (from the origin) pole and includes infinity.

In the case of discrete-time signals, the shift by N in the time domain corresponds to the multiplication times z^{-N} in the z-domain. That is, if $x[n] \leftrightarrow X(z)$, then $x[n-N] \leftrightarrow z^{-N}X(z)$. If function $X(z)$ is rational, then the z-transform of the displaced sequence will also be a rational function of z. From the theorem of the initial value: $h[0] = \lim_{2 \to \infty} H(z)$. If H(z) is not rational with respect to z, the latter relation is true only if the rank of the numerator is smaller than or equal to the rank of the denominator.

Thus, an LTI system with a rational transfer function is casual when:

- The region of convergence is outside of the circle defined by the origin and the most distant from the origin pole.
- The rank of the numerator is smaller than or equal to the rank of the denominator.

3.5.3 Stability of Discrete-Time Systems

An LTI discrete-time system is stable when the impulse response $h[n]$ converges, i.e., when

$$\left| \sum_{n=-\infty}^{+\infty} h[n] \right| < \infty \tag{3.19}$$

A conclusion of great practical value is derived from the comparison of the condition $\left| \sum_{n=-\infty}^{+\infty} h[n] \right| < \infty$, with the transform of the sequence $h[n]$ From the definition of the z-transform, we have: $H(z) = \sum_{n=-\infty}^{+\infty} h[n]z^{-n} \Rightarrow H(z)_{z=1} = \sum_{n=-\infty}^{+\infty} h[n]$. In other words, the requirement $\left| \sum_{n=-\infty}^{+\infty} h[n] \right| < \infty$ of stability is equivalent to the requirement for $H(z)$ to be finite when $|z| = 1$.

Or, in other words, the region of convergence should include the unit circle.

Therefore, an LTI discrete-time system with an impulse response $h[n]$ is stable if and only if the region of convergence includes the unit circle.

A causal LTI system with a transfer function $H(z)$ is stable when the poles are included in the unit circle, i.e., their magnitude is less than one. The closer to the center of the circle the poles are, the greater the relative stability of the system is going to be.

The stability of the output of an LTI system depends both on the stability of the input signal and on the stability of the system. Under certain conditions, an unstable system may produce a stable output signal.

A system is practically feasible when it is both stable and casual. Such a system will have a transfer function with a region of convergence of the form $|z| > a$, where $0 \le a \le 1$. Consequently, all of the poles of $H(z)$ must be included in the unit circle.

3.5.4 Transfer Function of Connected Systems

The transfer function of a discrete system can be considered to be produced by the connection of other simpler systems and vice versa.

1. *Systems connected in series* (illustrated in Figure 3.3)

   In this case the total transfer function of the equivalent system is equal to:

$$H(z) = H_1(z) \cdot H_2(z) \tag{3.20}$$

2. *Systems connected in parallel* (illustrated in Figure 3.4)

   In this case, the total transfer function of the equivalent systems is equal to:

$$H(z) = H_1(z) + H_2(z) \tag{3.21}$$

3. *Feedback connection of systems* (illustrated in Figure 3.5)

   In this case, the total transfer function of the equivalent system is:

$$H(z) = \frac{H_1(z)}{1 + H_1(z) \cdot H_2(z)} \tag{3.22}$$

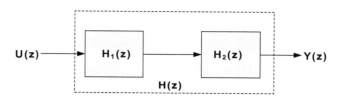

FIGURE 3.3
Systems connected in series.

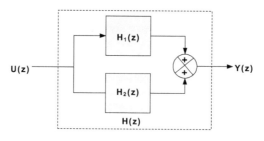

FIGURE 3.4
Systems connected in parallel.

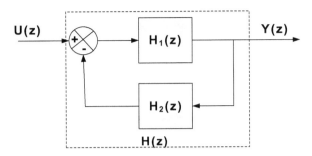

FIGURE 3.5
Feedback connection of systems.

3.5.5 Transfer Function of Discrete-Time Systems

In the digital signal analysis, a linear system (filter) is described by a difference equation, the general form of which is: $\displaystyle\sum_{k=0}^{N} a_k y[n-k] = \sum_{k=0}^{M} b_k x[n-k]$. Applying z-transform to both parts of the difference equation produces the system transfer function.

$$Y(z)\sum_{k=0}^{N} a_k\, z^{-k} = X(z)\sum_{k=0}^{M} b_k z^{-k} \tag{3.23}$$

Thus,

$$H(z) = \frac{Y(z)}{X(z)} = \frac{\displaystyle\sum_{k=0}^{M} b_k z^{-k}}{\displaystyle\sum_{k=0}^{N} a_k z^{-k}} \tag{3.24}$$

From the transfer function $H(z)$, the following can be derived:

1. The impulse response, $h[n]$, by applying the inverse z-transform.
2. The frequency response, $H(e^{j\omega})$, by substituting $z = e^{j\omega}$, where ω is the digital frequency in the interval $(0, \pi)$.

$$H(e^{j\omega}) = \frac{Y(e^{j\omega})}{X(e^{j\omega})} = \frac{\displaystyle\sum_{k=0}^{M} b_k e^{-j\omega k}}{\displaystyle\sum_{k=0}^{N} a_k e^{-j\omega k}} \tag{3.25}$$

In a discrete system, the transfer function is defined by:

$$H(e^{j\omega}) = \sum_{n=-\infty}^{\infty} h[n] e^{-j\omega n} \tag{3.26}$$

$H(e^{j\omega})$ is periodic with a period 2π and determines the effect of the system on the frequency content of the input signal.

If an input signal $x[n] = e^{j\omega n}$ is applied to a discrete system with an impulse response, $h[n]$, then the output will be:

$$y[n] = x[n] * h[n] = h[n] * x[n] = \sum_{k=-\infty}^{\infty} h[k]x[n-k] = \sum_{k=-\infty}^{\infty} h[k]e^{j\omega(n-k)}$$

$$= e^{j\omega n} \sum_{k=-\infty}^{\infty} h[k]e^{-j\omega k} \Rightarrow$$

$$y(n) = e^{j\omega n}H(e^{j\omega}) = e^{j\Omega n}\left|H(e^{j\omega})\right|\angle H(e^{j\omega}) \tag{3.27}$$

That is, the output, $y[n]$, will be equal to the input, $x[n] = e^{j\omega n}$, multiplied by a weight function, $H(e^{j\omega})$.

As already mentioned, the frequency response of an LTI discrete-time system results from the transfer function $H(z)$ by substituting $z = e^{j\omega}$. This substitution is allowed when the following are true:

- The sum $\sum_{n=-\infty}^{+\infty} |x[n]|$ must be finite, i.e., $\sum_{n=-\infty}^{+\infty} |x[n]| < \infty$
- The region of convergence must include the unit circle.

The study of the frequency response involves the calculation and plotting of its amplitude and phase. Quite often the group delay is studied, which is defined as:

$$\tau(\omega) = -\frac{d\varphi(\omega)}{d(\omega)} \tag{3.28}$$

An LTI discrete system is called of linear phase when the frequency response is of the following form:

$$H(e^{j\omega}) = \left|H(e^{j\omega})\right|e^{-j\alpha\omega} \tag{3.29}$$

where α is a real number. Linear phase systems have a constant group delay $\tau(\omega) = \alpha$. Consequently, the group delay (which indicates the delay that the input signal suffers from the system) is the same for all ω values.

3.6 Formula Tables

The *z*-transforms of the fundamental functions and the properties of *z*-transform are given in the following Tables 3.1 and 3.2, respectively.

TABLE 3.1

z-Transform of Fundamental Functions

	$x[n]$	$X(z)$	ROC
1	$\delta[n]$	1	All z
2	$u[n]$	$\dfrac{z}{z-1}$	$\lvert z \rvert > 1$
3	$-u[-n-1]$	$\dfrac{z}{z-1}$	$\lvert z \rvert < 1$
4	$\delta[n-m]$	z^{-m}	All z-plane except from 0 ($m > 0$) or ∞ ($m < 0$)
5	$a^n u[n]$	$\dfrac{z}{z-a}$	$\lvert z \rvert > \lvert a \rvert$
6	$-a^n u[-n-1]$	$\dfrac{z}{z-a}$	$\lvert z \rvert < \lvert a \rvert$
7	$na^n u[n]$	$\dfrac{az}{(z-a)^2}$	$\lvert z \rvert > \lvert a \rvert$
8	$-na^n u[-n-1]$	$\dfrac{az}{(z-a)^2}$	$\lvert z \rvert < \lvert a \rvert$
9	$(n+1)a^n u[n]$	$\left[\dfrac{z}{z-a}\right]^2$	$\lvert z \rvert > \lvert a \rvert$
10	$(\cos \Omega n)u[n]$	$\dfrac{z^2(\cos \Omega)z}{z^2 - (2\cos \Omega)z + 1}$	$\lvert z \rvert > 1$
11	$(\sin \Omega n)u[n]$	$\dfrac{(\sin \Omega)z}{z^2 - (2\cos \Omega)z + 1}$	$\lvert z \rvert > 1$
12	$(r^n\cos \Omega n)u[n]$	$\dfrac{z^2 - (r\cos \Omega)z}{z^2 - (2r\cos \Omega)z + r^2}$	$\lvert z \rvert > r$
13	$(r^n\sin \Omega n)u[n]$	$\dfrac{(r\sin \Omega)z}{z^2 - (2r\cos \Omega)z + r^2}$	$\lvert z \rvert > r$

TABLE 3.2

Properties of z-Transform

	Property	Function	Transform	ROC
		$x[n]$	$X(z)$	R
		$x_1[n]$	$X_1(z)$	R_1
		$x_2[n]$	$X_2(z)$	R_2
1	Linearity	$ax_1[n] + bx_2[n]$	$aX_1(z) + bX_2(z)$	$R' \supset R_1 \cap R_2$
2	Time shift	$x[n - N]$	$Z^{-N}X(z)$	$R' \supset R$
3	Multiplication with z_0^n	$z_0^n x[n]$	$X\left(\dfrac{z}{z_0}\right)$	$R' = \|z_0\|R$
4	Multiplication with $e^{j\Omega n}$	$e^{j\Omega n} x[n]$	$X(e^{-j\Omega}Z)$	$R' = R$
5	Time reverse	$x[-n]$	$X\left(\dfrac{1}{z}\right)$	$R' = \dfrac{1}{R}$
6	Multiplication with n	$nx[n]$	$-z\dfrac{dX(z)}{dz}$	$R' = R$
7	Sum	$\displaystyle\sum_{k=-\infty}^{n} x[n]$	$\dfrac{1}{1-z^{-1}}X(z)$	$R' \supset R \cap \{\|z\| > 1\}$
8	Convolution	$x_1[n]^* x_2[n]$	$X_1(z) \cdot X_2(z)$	$R' \supset R_1 \cap R_2$
9	Initial value	If $x[n] = 0$ when $n < 0$, then $x[0] = \lim_{z\to\infty} X(z)$		
10	Final value	For stable systems $\lim_{n\to\infty} x[n] = \lim_{z\to 1}(z-1)X(z)$		

3.7 Solved Problems

3.7.1 Compute the z-transform of unit-step function: $u[n] = \begin{cases} 1 & \text{when} & n \geq 0 \\ 0 & \text{when} & n < 0 \end{cases}$.

SOLUTION: Based on the definition of the z-transform, the following is calculated:

$$X(z) = \sum_{-\infty}^{+\infty} u[n].z^{-n} = \sum_{n=0}^{+\infty} 1.z^{-n} = \sum_{n=0}^{+\infty} (z^{-1})^n = \frac{1}{1-z^{-1}} = \frac{z}{z-1} \tag{1}$$

The unit-step function affects the limits of summation.

$\displaystyle\sum_{n=0}^{+\infty} (z^{-1})^n$ is a geometric series (with first term 1 and ratio z^{-1}). Consequently, it must be true that $|z| < 1$ or $|z| > 1$.

The condition $|z| > 1$ defines the *region of convergence* of the transform, i.e., the set of values of the complex z-plane, for which the sum of the z-transform converges.

3.7.2 Compute the *z*-transform of the function: $x[n] = a^n \cdot u[n]$.

SOLUTION: Based on the definition of the *z*-transform, the following is calculated:

$$X(z) = \sum_{n=-\infty}^{\infty} a^n u[n] \cdot z^{-n} = \sum_{n=0}^{\infty} (a \cdot z^{-1})^n$$

The unit-step function restricts the summation limits from $n = 0$ to infinity.

The term $(a \cdot z^{-1})n$ is the general expression of a geometric progression with: $a_1 = 1$

and $N(z) = \dfrac{A}{z - \dfrac{1}{2}} + \dfrac{B}{z - \dfrac{1}{3}}$.

In order for $\displaystyle\sum_{n=0}^{\infty} (az^{-1})^n$ to converge, the geometric progression should be decreasing, i.e., $|a \cdot z^{-1}| < 1$.

This constraint yields:

$$\sum_{n=-\infty}^{\infty} (a \cdot z^{-1})^{-n} = \frac{1}{1 - az^{-1}} = \frac{1}{1 - \dfrac{a}{z}} = \frac{z}{z - a} \tag{1}$$

Hence: $x[n] = a^n \cdot u[n] \leftrightarrow X(z) = \dfrac{z}{z - a}$ \hfill (2)

The above transform is true only when: $|a| \cdot |z^{-1}| < 1$ or $|a| < |z|$ or $|z| > |a|$

The condition $|z| > |a|$ defines the region of convergence of the transform, i.e., the set of values of the complex *z*-plane for which the *z*-transform converges.

3.7.3 Compute the *z*-transforms of the functions: $x_1[n] = \delta[n-2]$ and $x_2[n] = \delta[n+2]$.

SOLUTION: Functions $x_1[n]$ and $x_2[n]$ are two time-shifted impulse functions.

Function $x_1[n] = \delta[n-2]$ is equal to zero except when $n = 2$.

The *z*-transform is $X_1(z) = \displaystyle\sum_{n=-\infty}^{n=+\infty} x_1[n]z^{-n} = z^{-2}$ and has a double pole at $z = 0$.

The region of convergence is the entire complex plane except the point $(0,0)$, yet it includes infinity.

Correspondingly, for the function $x_2[n]$, it will be $X_2(z) = \displaystyle\sum_{n=-\infty}^{n=+\infty} x_2[n]z^{-n} = z^2$ with

a pole at infinity. The region of convergence is the entire complex plane including $(0,0)$, yet not including infinity.

3.7.4 Compute the *z*-transform and the region of convergence for each of the following functions: $x[n] = \left(\dfrac{5}{6}\right)^n u[n]$ and $y[n] = \left(\dfrac{5}{6}\right)^{n+5} u[n+5]$.

SOLUTION: The function $x[n]$ takes non-zero values only when $n \geq 0$, hence, it is a causal function.

The function $y[n]$ is derived by shifting $x[n]$ by 5 time units to the left $y[n] = x[n+5]$; therefore, there are some negative values of n for which $y[n] \neq 0$. Thus, $y[n]$ is a non-causal function.

The transform (5), given in the Transform Lookup Table 3.1, yields $X(z) = \dfrac{z}{z - \dfrac{5}{6}}$ with a region of convergence $|z| < \dfrac{5}{6}$.

The transform of the sequence $y[n]$ can easily be computed using the time-shift property (Property 2 – Table 3.2). $Y(z)$ is then given by:

$$Y(z) = z^{-5}\, \frac{z}{z - \dfrac{5}{6}} = \frac{z^{-4}}{z - \dfrac{5}{6}}$$

The poles and regions of convergence of $X(z)$ and $Y(z)$ functions are presented in the following Table 3.3:

3.7.5 Compute the z-transform and the region of convergence for each of the following functions: $x[n] = -\left(\dfrac{5}{4}\right)^{n} u[-n-1]$ and $y[n] = -\left(\dfrac{5}{4}\right)^{n-3} u[-n+2]$.

SOLUTION: The function $y[n]$ arises from the time shift of $x[n]$ by 3 time units to the right, $y[n] = x[n-3]$, hence, as shown in the figures of Table 3.4, there are some positive values of the variable n for which $y[n] \neq 0$. The function $y[n]$ is non-causal.

From the transform of (5), given in the Transform Lookup Table 3.1, it is true that $X(z) = \dfrac{z}{z - \dfrac{5}{4}}$ with a region of convergence $|z| < \dfrac{5}{4}$.

The transform of $y[n]$ can be easily computed by using the time-shift property (Property 2 – Table 3.2). The function is given as

$$Y(z) = \frac{z^{-2}}{z - \dfrac{5}{4}} = \frac{1}{z^{2}\left(z - \dfrac{5}{4}\right)} \tag{1}$$

The poles and regions of convergence of $X(z)$ and $Y(z)$ functions are presented in the following Table 3.4.

The computation of the z-transform of the function $y[n] = -\left(\dfrac{5}{4}\right)^{n-3} u[-n+2]$ without using the time-shift property is given next.

Based on the definition of the z-transform, we get:

$$Y(z) = \sum_{n=-\infty}^{n=+\infty} y[n]z^{-n} = = -\sum_{n=-\infty}^{n=+\infty} \left(\frac{5}{4}\right)^{n-3} u[-n+2]z^{-n} = -\sum_{n=-\infty}^{n=2} \left(\frac{5}{4}\right)^{n-3} z^{-n}$$

TABLE 3.3

The Poles and Regions of Convergence for $X(z)$ and $Y(z)$

$$x[n] = \left(\frac{5}{6}\right)^n u[n]$$

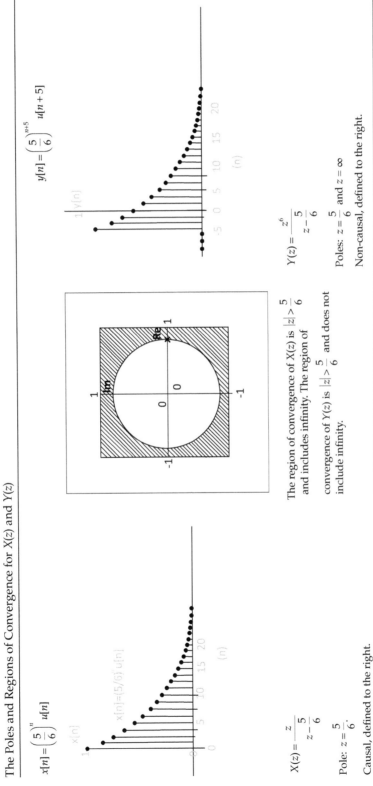

$$y[n] = \left(\frac{5}{6}\right)^{n+5} u[n+5]$$

$$X(z) = \frac{z}{z - \frac{5}{6}}$$

Pole: $z = \frac{5}{6}$.

Causal, defined to the right.

The region of convergence of $X(z)$ is $|z| > \frac{5}{6}$ and includes infinity. The region of convergence of $Y(z)$ is $|z| > \frac{5}{6}$ and does not include infinity.

$$Y(z) = \frac{z^6}{z - \frac{5}{6}}$$

Poles: $z = \frac{5}{6}$ and $z = \infty$

Non-causal, defined to the right.

TABLE 3.4

The Poles and Regions of Convergence for X(z) and Y(z)

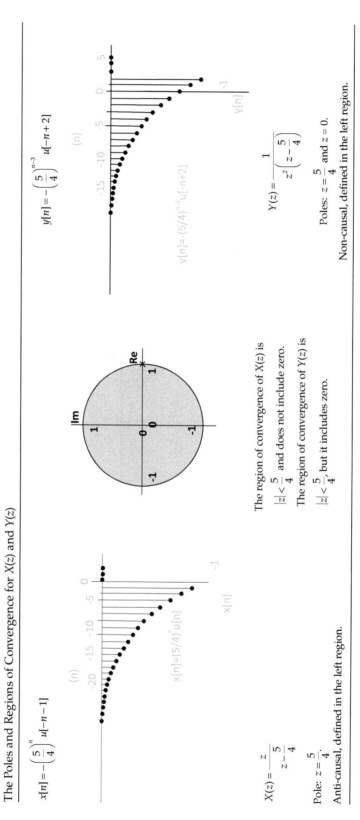

$$x[n] = -\left(\frac{5}{4}\right)^n u[-n-1]$$

$$X(z) = \frac{z}{z - \frac{5}{4}}$$

Pole: $z = \frac{5}{4}$.

Anti-causal, defined in the left region.

The region of convergence of X(z) is

$|z| < \frac{5}{4}$ and does not include zero.

The region of convergence of Y(z) is

$|z| < \frac{5}{4}$, but it includes zero.

$$y[n] = -\left(\frac{5}{4}\right)^{n-3} u[-n+2]$$

$$Y(z) = \frac{1}{z^2 \left(z - \frac{5}{4} \right)}$$

Poles: $z = \frac{5}{4}$ and $z = 0$.

Non-causal, defined in the left region.

$$= -\left(\frac{5}{4}\right)^{-3} \sum_{n=-\infty}^{n=2}\left(\frac{5}{4}\right)^n z^{-n} = -\left(\frac{5}{4}\right)^{-3} \sum_{n=-\infty}^{n=2}\left(\frac{5}{4}z^{-1}\right)^n$$

$$\text{or } Y(z) = \sum_{n=-\infty}^{n=+\infty} y[n]z^{-n} = -\left(\frac{5}{4}\right)^{-3}\sum_{n=-\infty}^{n=2}\left(\frac{5}{4}z^{-1}\right)^n \tag{2}$$

The sum $\displaystyle\sum_{n=-\infty}^{n=2}\left(\frac{5}{4}z^{-1}\right)^n$ is written as $\displaystyle\sum_{n=-\infty}^{n=2}\left(\frac{5}{4}z^{-1}\right)^n$

$$= \dots + \left(\frac{5}{4}z^{-1}\right)^{-k} + \dots\left(\frac{5}{4}z^{-1}\right)^{-2} + \left(\frac{5}{4}z^{-1}\right)^{-1} + \left(\frac{5}{4}z^{-1}\right)^{0} + \left(\frac{5}{4}z^{-1}\right)^{1} + \left(\frac{5}{4}z^{-1}\right)^{2} \tag{3}$$

A more sophisticated way to build in the relation is to set $\left(\dfrac{5}{4}z^{-1}\right)^{-1} = W$.

Then, $\displaystyle\sum_{n=-\infty}^{n=2}\left(\frac{5}{4}z^{-1}\right)^n = \dots + W^k + \dots W^2 + W^1 + W^0 + W^{-1} + W^{-2}$ or

$$\sum_{n=-2}^{n=\infty} W^n = \dots + W^k + \dots W^2 + W^1 + W^0 + W^{-1} + W^{-2} \text{ or}$$

$$\sum_{n=-2}^{n=\infty} W^n = W^{-2} + W^{-1} + W^0 + W^1 + W^2 + \dots + W^k + \dots$$

which is a series with the first term $W^{-2} = \left(\dfrac{5}{4}z^{-1}\right)^2$ and ratio $W = \dfrac{4}{5}z$.

And for decreasing geometric progression: $|W|<1$ or $\left|\dfrac{4}{5}z\right| < 1$ or $|z| < \dfrac{5}{4}$.

Substituting the first term a_1 for $W^{-2} = \left(\dfrac{5}{4}z^{-1}\right)^2$ and then $W = \dfrac{4}{5}z$ into $\dfrac{a_1}{1-W}$ yields

$$\frac{a_1}{1-W} = \frac{\left(\dfrac{5}{4}z^{-1}\right)^2}{1-\dfrac{4}{5}z} = -\frac{\left(\dfrac{5}{4}z^{-1}\right)^2}{\dfrac{4}{5}z-1} = -\frac{5}{4}\frac{\left(\dfrac{5}{4}z^{-1}\right)^2}{z-\dfrac{5}{4}} = -\left(\frac{5}{4}\right)^3\frac{z^{-2}}{z-\dfrac{5}{4}} \tag{4}$$

Then, from relation (2), $Y(z)$ is computed:

$$Y(z) = \sum_{n=-\infty}^{n=+\infty} y[n]z^{-n} = -\left(\frac{5}{4}\right)^{-3} \sum_{n=-\infty}^{n=2} \left(\frac{5}{4}z^{-1}\right)^n = \frac{z^{-2}}{z - \frac{5}{4}} = \frac{1}{z^2\left(z - \frac{5}{4}\right)} \tag{5}$$

3.7.6 Compute the z-transform and the region of convergence for the function:

$$x[n] = \left(\frac{1}{3}\right)^n u[n] + 2^n u[-n-1].$$

SOLUTION: The given function can be considered the sum of $x_1[n]$ and $x_2[n]$, where $x_1[n] = \left(\frac{1}{3}\right)^n u[n]$ and $x_2[n] = 2^n u[-n-1]$.

The transform of $x_1[n]$ directly results from the transform (5) of Table 3.1, which is $x_1[n] = \left(\frac{1}{3}\right)^n u[n] \leftrightarrow \dfrac{z}{z - \frac{1}{3}} = X_1(z)$ with a region of convergence $|z| > \frac{1}{3}$.

$x[n] = \left(\frac{1}{3}\right)^n u[n] + 2^n u[-n-1]$	Region of convergence $\frac{1}{3} < |z| < 2$
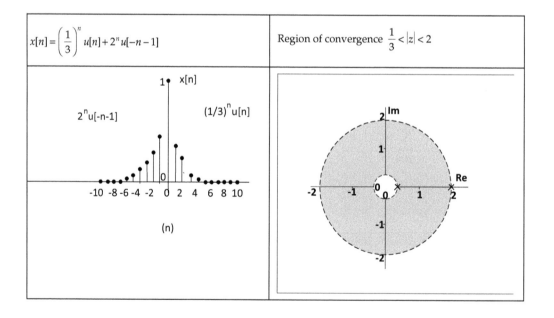	

Based on transform (6), we have $-2^n u[-n-1] \leftrightarrow \dfrac{z}{z-2}$.

From the property of linearity, if $y[n] \leftrightarrow Y(z)$ then $ay[n] \leftrightarrow aY(z)$, hence $x_2[n] = 2^n u[-n-1] \leftrightarrow -\dfrac{z}{z-2}$ with a region of convergence $|z| < 2$.

Consequently, $X(z) = \dfrac{z}{z - \dfrac{1}{3}} + \dfrac{z}{z-2}$

$$= \frac{z^2 - 2z - z^2 + \dfrac{1}{3}z}{\left(z - \dfrac{1}{3}\right)(z-2)} = \frac{-\dfrac{5}{3}z}{\left(z - \dfrac{1}{3}\right)(z-2)} = -5\frac{z}{3z^2 - 7z + 2}$$

The poles of $X(z)$ are $z_1 = 2$ and $z_2 = \dfrac{1}{3}$. Since $x[n]$ is bilateral, the region of convergence will be the ring between the two poles, i.e., it will be defined by the relation $\dfrac{1}{3} < |z| < 2$, which is the intersection of the regions of convergence of $X_1(z)$ and $X_2(z)$.

3.7.7 Compute the z-transform for the function: $x[n] = n \cdot a^n u[n]$.

SOLUTION: According to the transform (5), $x[n] = a^n u[n] \leftrightarrow \dfrac{z}{z-a} = X(z)$. The derivative of $X(z)$, with respect to z, is expressed as:

$$\frac{d}{dz}\left(\frac{z}{z-a}\right) = \frac{(z)'(z-a) - (z-a)' \cdot z}{(z-a)^2} = \frac{1 \cdot (z-a) - 1 \cdot z}{(z-a)^2} = \frac{-a}{(z-a)^2} \tag{1}$$

Based on the property $n \cdot x[n] \rightarrow -z\dfrac{dX(z)}{dz}$, the following is derived:

$$Z\{n \cdot a^n u[n]\} = \frac{az}{(z-a)^2} \tag{2}$$

3.7.8 Compute the z-transform of the function: $x[n] = \left(\dfrac{1}{3}\right)^n u[-n]$

SOLUTION: Based on the transform (6), we have:

$$-\left(\frac{1}{3}\right)^n u[-n-1] \leftrightarrow \frac{z}{z - \dfrac{1}{3}}, \text{when } |z| < \frac{1}{3}. \tag{1}$$

The function is time-shifted by one unit, i.e., n is substituted by $n-1$, so:

$$-\left(\frac{1}{3}\right)^{n-1} u[-(n-1)-1] \leftrightarrow \frac{z}{z - \dfrac{1}{3}}z^{-1} \text{ or}$$

$$-\left(\frac{1}{3}\right)^{-1}\left(\frac{1}{3}\right)^{n}u[-n] \leftrightarrow \frac{1}{z-\dfrac{1}{3}} \quad \text{or} \quad \left(\frac{1}{3}\right)^{n}u[-n] \leftrightarrow -\frac{1}{3}\frac{1}{z-\dfrac{1}{3}} \tag{2}$$

3.7.9 Compute the z-transform of the function: $x[n] = n\left(\dfrac{1}{2}\right)^{n} u[n-2]$

SOLUTION: According to the transform (5), we have $\left(\dfrac{1}{2}\right)^{n} u[n] \leftrightarrow \dfrac{z}{z-\dfrac{1}{2}}$, when $|z| < \dfrac{1}{3}$.
Based on the time-shift property, we have:

$$\left(\frac{1}{2}\right)^{n-2} u[n-2] \leftrightarrow z^{-2}\frac{z}{z-\dfrac{1}{2}} \Rightarrow$$

$$\left(\frac{1}{2}\right)^{2}\left(\frac{1}{2}\right)^{n-2} u[n-2] \leftrightarrow \left(\frac{1}{2}\right)^{2}\frac{z^{-1}}{z-\dfrac{1}{2}} \Rightarrow \left(\frac{1}{2}\right)^{n} u[n-2] \leftrightarrow \left(\frac{1}{4}\right)\frac{z^{-1}}{z-\dfrac{1}{2}}$$

Making use of the differentiation property:

$$n\left(\frac{1}{2}\right)^{n-2} u[n-2] \leftrightarrow -z\frac{d}{dz}\left[\frac{1}{4}\frac{z^{-1}}{z-\dfrac{1}{2}}\right] = -\frac{z}{4}\frac{-z^{-2}\left(z-\dfrac{1}{2}\right)-z^{-1}}{\left(z-\dfrac{1}{2}\right)^{2}}$$

$$\text{or } n\left(\frac{1}{2}\right)^{n-2} u[n-2] \leftrightarrow -\frac{z}{4}\cdot\frac{-z^{-1}+\dfrac{1}{2}z^{-2}-z^{-1}}{\left(z-\dfrac{1}{2}\right)^{2}} \text{ and, after some complicated calculations:}$$

$$n\left(\frac{1}{2}\right)^{n-2} u[n-2] \leftrightarrow \frac{z^{-2}}{2}\cdot\frac{1-\dfrac{1}{4}z^{-1}}{\left(1-\dfrac{1}{2}z^{-1}\right)^{2}}$$

3.7.10 Define the initial and final value of the impulse response of the system when:

$$Y(z) = \frac{2z^2}{(z-1)(z-a)(z-b)}, |a|, |b| < 1$$

SOLUTION: From the initial value theorem:

$$y[0] = \lim_{z \to \infty} Y(z) = \lim_{z \to \infty} \frac{2z^2}{(z-1)(z-a)(z-b)} = 0 \qquad (1)$$

From the final value theorem:

$$y[\infty] = \lim_{z \to 1}(z-1)\, Y(z) = \lim_{z \to 1} \frac{2z^2}{(z-a)(z-b)} = \frac{2}{(1-a)(1-b)} \qquad (2)$$

3.7.11 Find the first four coefficients of the function $x[n]$ when:

$$X_1(z) = \frac{4z^{-1}}{z^{-2} - 2z^{-1} + 2}, \quad X_2(z) = \frac{2 + 2z^{-1} + z^{-2}}{1 + z^{-1}} \quad for\ |z| > 1$$

SOLUTION:
  a. For $X_1(z)$:

$$X_1(z) = \frac{4z^{-1} \cdot z^2}{(z^{-2} - 2z^{-1} + 2) \cdot z^2} = \frac{4z}{2z^2 - 2z + 1} \qquad (1)$$

The polynomial division of the numerator and the denominator of $X_1(z)$, yields:

$4z$	$2z^2 - 2z + 1$
$-4z + 4 - 2z^{-1}$	$-2z^{-1} + 2z^{-2} + z^{-3} + \ldots$
$4 - 2z^{-1}$	
$-4 + 4z^{-1} - 2z^{-2}$	
$2z^{-1} + 2z^{-2}$	
$-2z^{-1} - 2z^{-2} - 2z^{-3}$	
$-2z^{-3}$	

$$\Rightarrow X_1(z) = 2z^{-1} + 2z^{-2} + z^{-3} + \ldots$$

Therefore, the sequence $x_1[n]$ is: $x_1[n] = \left\{ \underset{\uparrow}{0},\ 2,\ 2,\ 1, \ldots \right\}$ $\qquad (2)$

b. For $X_2(z)$, we provide another method for a polynomial division:

$$
\begin{array}{r}
z^{-1} + 0z^{-2} - 3z^{-3} - 4z^{-4} \\
z^3 + 2z + 4 \overline{\smash{\big)}\ z^2 - 1}
\end{array}
$$

$$
\begin{array}{r}
z^2 + 2 + 4z^{-1} \\
\hline
-3 - 4z^{-1} \\
\end{array}
$$

$$
\begin{array}{r}
-3 \qquad -6z^{-2} - 12z^{-3} \\
\hline
-4z^{-1} + 6z^{-2} + 12z^{-3} \\
\end{array}
$$

$$
\begin{array}{r}
-4z^{-1} \qquad -8z^{-3} - 16z^{-4} \\
\hline
6z^{-2} + 20z^{-3} + 16z^{-4}
\end{array}
$$

Thus, the sequence $x_2[n]$ is:

$$
\begin{aligned}
X_2(z) &= z^{-1} + 0z^{-2} - 3z^{-3} - 4z^{-4} + \dots \\
&\Rightarrow x_2[n] = 0\delta[n] + 1\delta[n-1] + 0\delta[n-2] - 3\delta[n-3] + \dots
\end{aligned}
\tag{3}
$$

3.7.12 Compute the inverse z-transform of the functions:

$$
F_1(z) = \frac{z^{-1}}{(1-\alpha z^{-1})^2(1-bz^{-1})} = \frac{z^2}{(z-\alpha)^2(z-b)}
$$

$$
F_2(z) = \frac{2(z^2 - 5z + 6.5)}{(z-2)(z-3)^2}, \; for \quad 2 < |z| < 3
$$

$$
F_3(z) = \frac{2z^2 + 3z}{(z+2)(z-4)}
$$

$$
F_4(z) = \frac{z^3 - 20z}{(z-2)^3(z-4)}
$$

SOLUTION:

a. It is true that : $f_1[n] = \sum residues \left[\dfrac{z^2}{(z-a)^2(z-b)} z^{n-1} \right]$
$\tag{1}$

The residue of the simple pole $z = b$ is:

$$
\left. \frac{z^2 z^{n-1}}{(z-a)^2} \right|_{z=b} = \left. \frac{z^{n+1}}{(z-a)^2} \right|_{z=b} = \frac{b^{n+1}}{(b-a)^2}
\tag{2}
$$

The residue of the double pole $z = a$ is:

$$\frac{d}{dz}\left(\frac{z^2}{z-b}z^{n-1}\right)\bigg|_{z=\alpha} = \frac{[n+1]z^n(z-b)-z^{n+1}}{(z-b)^2}\bigg|_{z=\alpha} = a^n\frac{[n+1](a-b)-a}{(a-b)^2} \tag{3}$$

Hence:

$$f[n] = \frac{a^n\left[n(a-b)-b\right]+b^{n+1}}{(a-b)^2} \tag{4}$$

b. We expand the function $F_2(z)$ in a sum of fractions:

$$F_2(z) = \frac{2(z^2-5z+6.5)}{(z-2)(z-3)^2} = \frac{A}{z-2}+\frac{B}{z-3}+\frac{C}{(z-3)^2} \quad \text{where } A = B = C = 1$$

$$\Rightarrow \quad F_2(z) = \frac{1}{z-2}+\frac{1}{z-3}+\frac{1}{(z-3)^2}$$

$$= \frac{1}{2}\left(1-\frac{2}{z}\right)^{-1} - \frac{1}{3}\left(1-\frac{z}{3}\right)^{-1} + \frac{1}{9}\left(1-\frac{z}{3}\right)^{-2} \quad \text{so that } \frac{2}{z} < 1 \text{ and } \frac{z}{3} < 1$$

$$= \frac{1}{z}\left(1+\frac{2}{z}+\frac{4}{z^2}+\frac{8}{z^3}+\cdots\right) - \frac{1}{3}\left(1+\frac{z}{3}+\frac{z^2}{9}+\frac{z^3}{27}+\cdots\right) + \frac{1}{9}\left(1+\frac{2z}{3}+\frac{3z^2}{9}+\frac{4z^3}{27}+\cdots\right)$$

where $2 < |z| < 3$

$$= \left(\frac{1}{2}+\frac{2}{z^2}+\frac{2^2}{z^3}+\frac{2^3}{z^4}+\cdots\right) - \left(\frac{1}{3}+\frac{z}{3^2}+\frac{z^2}{3^3}+\frac{z^3}{3^4}+\cdots\right) + \left(\frac{1}{3^2}+\frac{2z}{3^3}+\frac{3z^2}{3^4}+\frac{4z^3}{3^5}+\cdots\right)$$

$$= \sum_{n=1}^{\infty} 2^{n-1}z^{-n} - \sum_{n=0}^{\infty}\left(\frac{1}{3}\right)^{n-1}z^n + \sum_{n=0}^{\infty}(n+1)\left(\frac{1}{3}\right)^{n+2}z^n$$

or inversely, we have $u_n = 2^{n-1}$, $n \geq 1$ *and* $u_n = -(n+2)3^{n-2}$, $n \leq 0$.

c. $F_3(z) = \dfrac{2z^2+3z}{(z+2)(z-4)}$ can be written as:

$$\frac{F_3(z)}{z} = \frac{2z+3}{(z+2)(z-4)} = \frac{A}{z+2}+\frac{B}{z-4} \quad \text{when } A = \frac{1}{6} \text{ and } B = \frac{11}{6}$$

Thus,

$$F_3(z) = \frac{1}{6}\frac{z}{z+2}+\frac{11}{6}\frac{z}{z-4}$$

and, by applying the inverse z-transform, we get:

$$f_3[n] = \frac{1}{6}(-2)^{-n} + \frac{11}{6}(4)^n$$

d. $F_4(z) = \dfrac{z^3 - 20z}{(z-2)^3(z-4)}$ can be written as:

$$\frac{F_4(z)}{z} = \frac{z^2 - 20}{(z-2)^3(z-4)} = \frac{A + Bz + Cz^2}{(z-2)^3} + \frac{D}{z-4}$$

Multiplication with $(z-2)^3(z-4)$ yields:

$$z^2 - 20 = (A + Bz + Cz^2)(z-4) + D(z-2)^3.$$

We then compare the coefficients of z in the last equation, which yields:

$$A = 6, B = 0, C = \tfrac{1}{2} \text{ and } D = -\tfrac{1}{2}.$$

Thus, $\dfrac{F_4(z)}{z} = \dfrac{z^2 - 20}{(z-2)^3(z-4)} = \dfrac{6 + 0z + \tfrac{1}{2}z^2}{(z-2)^3} - \dfrac{\tfrac{1}{2}}{z-4}$

$$F_4(z) = \frac{1}{2} \cdot \left\{ \frac{12z + z^3}{(z-2)^3} \right\} - \frac{1}{2}\left(\frac{z}{z-4} \right) = \frac{1}{2}\left\{ \frac{12z + 4z^2 - 4z^2 + z^3}{(z-2)^3} \right\} - \frac{1}{2}\left(\frac{z}{z-4} \right)$$

$$= \frac{1}{2}\left\{ \frac{(8z + 4z) + 4z^2 - 4z^2 + z^3}{(z-2)^3} \right\} - \frac{1}{2}\left(\frac{z}{z-4} \right)$$

$$= \frac{1}{2}\left\{ \frac{\left(z^3 - 4z^2 + 4z\right) + 8z + 4z^2}{(z-2)^3} \right\} - \frac{1}{2}\left(\frac{z}{z-4} \right)$$

$$= \frac{1}{2}\left\{ \frac{z(z-2)^2 + 4z^2 + 8z}{(z-2)^3} \right\} - \frac{1}{2}\left(\frac{z}{z-4} \right) = \frac{1}{2}\left\{ \frac{z}{z-2} + 2\frac{2z^2 + 4z}{(z-2)^3} \right\} - \frac{1}{2}\left(\frac{z}{z-4} \right)$$

The inverse z-transform yields:

$$f_4[n] = \frac{1}{2}\left[(2^n + 2.n^2 2^n) - 4^n \right] = 2^{n-1} + n^2 2^n - 2^{2n-1}$$

3.7.13 Compute the inverse z-transform of the function: $F(z) = \dfrac{z^{-2} + z^{-1} + 1}{0.2z^{-2} + 0.9z^{-1} + 1}$ using

a. the method of the expansion of the sum of partial fractions and

b. the method of complex integration.

SOLUTION: a. Computation of $f[k]$ using the method of partial fraction expansion.

$$F(z) = \frac{z^{-2} + z^{-1} + 1}{0.2z^{-2} + 0.9z^{-1} + 1} = \frac{z^2 + z + 1}{(z + 0.4)(z + 0.5)} \tag{1}$$

$$\frac{F(z)}{z} = \frac{z^2 + z + 1}{z(z + 0.4)(z + 0.5)} = \frac{k_1}{z} + \frac{k_2}{z + 0.4} + \frac{k_3}{z + 0.5} \tag{2}$$

Calculation of k_i

$$\left.\begin{aligned}
k_1 &= \lim_{z \to 0} \frac{F(z)}{z} z = \frac{1}{0.2} = 5 \\[6pt]
k_2 &= \lim_{z \to -0.4} \frac{F(z)}{z}(z + 0.4) = -19 \\[6pt]
k_3 &= \lim_{z \to -0.5} \frac{F(z)}{z}(z + 0.5) = 15
\end{aligned}\right\} \tag{3}$$

$$\text{Hence: } F(z) = 5 - \frac{19z}{z + 0.4} + \frac{15z}{z + 0.5} \tag{4}$$

So, $f[k]$ becomes:

$$f[k] = Z^{-1}\big[F(z)\big] = \big(5\delta[k] - 19(-0.4)^k + 15(-0.5)^k\big)u[k] \tag{5}$$

b. Computation of $f(k)$ using the method of complex integration.

$$f[k] = Z^{-1}\big[F(z)\big] = \sum \text{residues } F(z)z^{k-1} \Rightarrow$$

$$f[k] = \sum \text{residues } \frac{(z^2 + z + 1)z^{k-1}}{(z + 0.4)(z + 0.5)} \tag{6}$$

For k = 0, the poles are: z = 0, z = −0.4, z = −0.5

$$\text{residue } z = 0 \to \left.\frac{z^2 + z + 1}{(z + 0.4)(z + 0.5)}\right|_{z=0} = \frac{1}{0.2} = 5 \quad, k = 0 \tag{7}$$

$$residue\ z=-0.4\rightarrow\frac{(z^2+z+1)z^{k-1}}{z+0.5}\bigg(_{z=0.4}=7.6(0.4)^{k-1}=-19(0.4)^k \tag{8}$$

$$residue\ z=-0.5\rightarrow\frac{(z^2+z+1)z^{k-1}}{z+0.4}\bigg(_{z=-0.5}=-7.5(-0.5)^{k-1}=15(-0.5)^k \tag{9}$$

Therefore:

$$f[k]=\Big(5\delta[k]+19(-0.4)^k+15(-0.5)^k\Big)u[k] \tag{10}$$

3.7.14 A discrete-time system has a transfer function: $H(z)=\dfrac{4z^2}{z^2-\dfrac{1}{4}}$

Calculate the output, $y[n]$, if the input is the ramp function $x[n]=n$.

SOLUTION: Calculation of $y[n]$:

$$y[n]=z^{-1}\{Y(z)\}=z^{-1}\{H(z)X(z)\}\Rightarrow y[n]=z^{-1}\left\{\frac{4z^2}{z^2-\dfrac{1}{4}}\cdot\frac{z}{(z-1)^2}\right\} \tag{1}$$

$$\frac{Y(z)}{z}=\frac{4z^2}{\left(z-\dfrac{1}{2}\right)\cdot\left(z+\dfrac{1}{2}\right)\cdot(z-1)^2}$$

$$=\frac{k_1}{\left(z-\dfrac{1}{2}\right)}+\frac{k_2}{\left(z+\dfrac{1}{2}\right)}+\frac{k_{31}}{(z-1)}+\frac{k_{32}}{(z-1)^2} \tag{2}$$

Calculation of k_i

$$\left.\begin{array}{l}k_1=\lim\limits_{z\to\frac{1}{2}}\dfrac{Y(z)}{z}\left(z-\dfrac{1}{2}\right)=4\\[2mm]k_2=\lim\limits_{z\to-\frac{1}{2}}\dfrac{Y(z)}{z}\left(z+\dfrac{1}{2}\right)=-\dfrac{4}{9}\\[2mm]k_{31}=\lim\limits_{z\to1}\dfrac{d}{dz}\left(\dfrac{Y(z)}{z}(z-1)^2\right)=-\dfrac{32}{9}\\[2mm]k_{32}=\lim\limits_{z\to1}\dfrac{Y(z)}{z}(z-1)^2=\dfrac{16}{3}\end{array}\right\} \tag{3}$$

Thus:

$$Y(z) = 4\frac{z}{z-\frac{1}{2}} - \frac{4}{9}\frac{z}{z+\frac{1}{2}} - \frac{32}{9}\frac{z}{z-1} + \frac{16}{3}\frac{z}{(z-1)^2} \tag{4}$$

By using the inverse z-transform, the output y[n] of the system is computed:

$$y[n] = \left(4\left(\frac{1}{4}\right)^n - \frac{4}{9}\left(-\frac{1}{2}\right)^n - \frac{32}{9} + \frac{16}{3}n\right)u[n] \tag{5}$$

3.7.15 Compute the inverse z-transform of the following functions.

$$F_1(z) = \frac{10z}{(z-1)(z-0.5)}, \quad F_2(z) = \frac{2z^3 + z}{(z-2)^2(z-1)}$$

$$F_3(z) = \frac{z^2 + z + 2}{(z-1)(z^2 - z + 1)}, \quad F_4(z) = \frac{z(z+1)}{(z-1)(z-.25)^2}$$

SOLUTION:

a. $\dfrac{F_1(z)}{z} = \dfrac{k_1}{z-1} + \dfrac{k_2}{z-0.5} = \dfrac{20}{z-1} + \dfrac{20}{z-0.5}$ (1)

Thus,

$$F_1(z) = 20\frac{z}{z-1} - 20\frac{z}{z-0.5} \Rightarrow$$
$$f_1[k] = IZT[F_1(z)] = 20(1 - 0.5^k)u[k] \tag{2}$$

b. First suggested solution: partial fraction expansion

$$\frac{F_2(z)}{z} = \frac{2z^2 + 1}{(z-2)^2(z-1)} = \frac{k_1}{z-1} + \frac{k_{21}}{z-2} + \frac{k_{22}}{(z-2)^2} \tag{3}$$

$$k_1 = \frac{2z^2 + 1}{(z-2)^2(z-1)}(z-1)\Big|_{z=1} = \frac{3}{1} = 3 \tag{4}$$

$$k_{22} = \lim_{z \to 2}\left[\frac{2z^2 + 1}{z-1}\right] = \frac{9}{1} = 9 \tag{5}$$

$$k_{21} = \lim_{z \to 2}\left[\frac{2z^2+1}{z-1}\right]' = \lim_{z \to 2}\left[\frac{4z(z-1)-(2z^2+1)}{(z-1)^2}\right] = \frac{8-(9)}{1} = -1 \tag{6}$$

$$\text{Hence, } \frac{F_2(z)}{z} = \frac{3}{z-1} - \frac{1}{z-2} + \frac{9}{(z-2)^2} \tag{7}$$

$$\text{or } F_2(z) = 3\frac{z}{z-1} - \frac{z}{z-2} + 9\frac{z}{(z-2)^2} \overset{IZT}{\Rightarrow}$$

$$f_2[k] = Z^{-1}[F_2(z)] = (3 - 2^k + 9k2^{k-1})u[k] \tag{8}$$

Second suggested solution: complex integration

$$F_2(z) = \frac{2z^3+z}{(z-2)^2(z-1)} = \frac{z(2z^2+1)}{(z-2)^2(z-1)} \Rightarrow$$

$$F_2(z)z^{k-1} = \frac{z^k(2z^2+1)}{(z-2)^2(z-1)} \tag{9}$$

Using the complex integration method, we get:

$$\text{residue}\Big|_{z=1}\Bigg(\Rightarrow \lim_{z \to 1}\frac{z^k(2z^2+1)}{(z-2)^2} = \frac{3}{1} = 3 \tag{10}$$

$$\text{residue}\Big|_{z=2} \Rightarrow \lim_{z \to 2}\left[\frac{d}{dz}\left(\frac{z^k(2z^2+1)}{z-1}\right)\right] = \lim_{z \to 2}\left[\left(\frac{2z^{k+2}+z^k}{z-1}\right)'\right] =$$

$$\lim_{z \to 2}\left[\frac{(2(k+2)z^{k+1}+kz^{k-1})(z-1)-(2z^{k+2}+z^k)}{(z-1)^2}\right] =$$

$$\frac{(2(k+2)2^{k+1}+k2^{k-1})-(2\cdot 2^{k+2}+2^k)}{1} =$$

$$2k2^{k+1} + 2^2 2^{k+1} + k2^{k-1} - 2 \cdot 2^{k+2} - 2^k$$

$$= 4k2^k + 8 \cdot 2^k + \frac{k}{2}2^k - 8 \cdot 2^k - 2^k = \left(4k + \frac{k}{2}\right)2^k - 2^k = \frac{9k}{2}2^k - 2^k$$

$$\underset{z=2}{residue} = 9k \cdot 2^{k-1} - 2^k \tag{11}$$

$$\text{So, } f_2[k] = (3 + 9k2^{k-1} - 2^k)u[k] \tag{12}$$

c. $$\frac{F_3(z)}{z} = \frac{z^2 + z + 2}{z(z-1)(z^2 - z + 1)} = \frac{k_1}{z} + \frac{k_2}{z-1} + \frac{k_3}{z - e^{j\frac{\pi}{3}}} + \frac{k_4 = \overline{k_3}}{z - e^{j\frac{\pi}{3}}} \tag{13}$$

$$k_1 = \lim_{z \to 0}\left(\frac{F_3(z)}{z}z\right) = \frac{2}{(-1)1} = -2 \tag{14}$$

$$k_2 = \lim_{z \to 1}\left(\frac{F_3(z)}{z}(z-1)\right) = \frac{1+1+2}{(1-1+1)1} = 4 \tag{15}$$

$$k_3 = \lim_{z \to e^{j\frac{\pi}{3}}}\left(\frac{F_3(z)}{z}(z - e^{j\frac{\pi}{3}})\right) = \frac{e^{j\frac{2\pi}{3}} + e^{j\frac{\pi}{3}} + 2}{e^{j\frac{\pi}{3}}\left(e^{j\frac{\pi}{3}} - 1\right)\left(e^{j\frac{\pi}{3}} - e^{-j\frac{\pi}{3}}\right)}$$

$$= \frac{-0.5 + j0.866 + 0.5 + j0.866 + 2}{(0.5 + j0.866 - 1)2 \, j\sin\frac{\pi}{3}e^{j\frac{\pi}{3}}} = \frac{2 + j1.732}{(-0.5 + j0.866)2 \, j0.866 e^{j\frac{\pi}{3}}}$$

$$= \frac{2.645^{\angle 40.9°}}{1^{\angle 60°}1^{\angle 120°}1^{\angle 90°}1.732} = 1.53^{\angle -299.1°}$$

$$\overline{k_3} = k_4 = 1.53^{\angle 229.1°} \tag{16}$$

Thus: $F_3(z) = -2 + 4\dfrac{z}{z-1} + 1.53e^{-j229.1}\dfrac{z}{z - e^{j\frac{\pi}{3}}} + 1.53e^{j229.1}\dfrac{z}{z - e^{-j\frac{\pi}{3}}}$ \hfill (17)

$$f_3[k] = -2\delta[k] + 4u(k) + 1.53e^{-j299.1}\left(e^{j\frac{\pi}{3}}\right)^k + 1.53e^{j229.1}\left(e^{-j\frac{\pi}{3}}\right)^k \Rightarrow$$

$$f_3[k] = -2\delta[k] + 4u(k) + 1.53\left(e^{-j299.1}e^{j\frac{k\pi}{3}} + e^{j229.1}e^{-j\frac{k\pi}{3}}\right)$$

$$= -2\delta[k] + 4u[k] + 1.53\left(e^{j\left(\frac{k\pi}{3} - 229.1\right)} + e^{-j\left(\frac{k\pi}{3} - 229.1\right)}\right)$$

$$f_3[k] = \left(-2\delta[k] + 4u[k] + 3.06\cos\left(\frac{k\pi}{3} - 229.1\right)\right)u[k]$$ \hfill (18)

d. $\dfrac{F(z)}{z} = \dfrac{(z+1)}{(z-1)(z-0.25)^2} = \dfrac{A}{z-1} + \dfrac{B}{(z-0.25)^2} + \dfrac{C}{(z-0.25)}$ \hfill (19)

Subsequently, another calculation method for A, B, C is provided.

$$A = \left.\frac{(z=1)}{(z-0.25)^2}\right|_{z=1} = \frac{2}{(0.75)^2} = 3.56$$ \hfill (20)

To derive B, we proceed to a multiplication with $(z - 0.25)^2$

$$\frac{(z+1)}{z-1} = \frac{A}{z-1}(z-0.25)^2 + B + C(z-0.25)$$

Therefore, at $z = .25$, we have

$$B = \left.\frac{(z+1)}{(z-1)}\right|_{z=0.25} = \frac{1.25}{-0.75} = -\frac{5}{4}\frac{4}{3} = -\frac{5}{3} = -1.67$$ \hfill (21)

Calculation of C

$$\frac{d}{dz}\left[\frac{(z+1)}{z-1}\right] = \frac{d}{dz}\left[\frac{A}{z-1}(z-0.25)^2\right] + C,$$

$$C = \frac{d}{dz}\left[\frac{(z+1)}{(z-1)}\right]_{z=0.25} = \frac{1(z-1)-1(z+1)}{(z-1)^2} \Rightarrow$$

$$C = \frac{-2}{(z-1)^2}\bigg|_{z=0.25} = \frac{-2}{(-0.75)^2} = -3.56$$

(22)

Thus,

$$F(z) = \frac{(z+1)}{(z-1)(z-0.25)^2} = \frac{3.56z}{z-1} + \frac{-1.67z}{(z-0.25)^2} + \frac{-3.56z}{(z-0.25)}$$

(23)

The second term can be written as:

$$-1.67\frac{z}{(z-0.25)^2} = \frac{-1.67}{.25}\frac{0.25z}{(z-0.25)^2} = -6.68\frac{0.25z}{(z-0.25)^2}$$

Thus,

$$f[k] = \left[3.56 - 6.68 \cdot k(.25)^k - 3.56(.25)^k\right]u[k]$$

(24)

3.7.16 Compute the inverse z-transform of:

$$X(z) = \frac{z^2 + 6z}{(z^2 - 2z + 2)(z - 1)}$$

SOLUTION:

$$\frac{X(z)}{z} = \frac{k_1}{z-1} + \frac{k_2}{z-1+j} + \frac{k_3}{z-1-j}$$

(1)

The k_i coefficients are calculated by using Heaviside's formula.

$$k_1 = \lim_{z \to 1} \frac{z^2 + 6z}{(z^2 - 2z + 2)(z - 1)}(z - 1) = 7$$

(2)

$$k_3 = \lim_{z \to 1+j} \frac{z^2 + 6z}{(z^2 - 2z + 2)(z - 1)}(z - 1 - j) = \frac{1+j+6}{(1+j-1)(1+j-1+j)}$$

$$= \frac{7+j}{j \cdot 2j} = \frac{7+j}{2j^2} = -\frac{7+j}{2} = -\frac{7}{2} - \frac{j}{2}$$

(3)

$$k_2 = -\frac{7}{2} + \frac{j}{2} = \bar{k}_3 \tag{4}$$

Hence,

$$X(z) = 7\frac{z}{z-1} + \left(-\frac{7}{2} + j\frac{1}{2}\right)\frac{z}{z-1+j} + \left(-\frac{7}{2} - j\frac{1}{2}\right)\frac{z}{z-1-j} \tag{5}$$

$$(5)\overset{IZT}{\Rightarrow} x[k] = 7 + \left(-\frac{7}{2} + j\frac{1}{2}\right)(\sqrt{2})^k e^{-j\frac{\pi}{4}k} + \left(-\frac{7}{2} - j\frac{1}{2}\right)(\sqrt{2})^k e^{j\frac{\pi}{4}k} \Rightarrow$$

$$x[k] = 7 - \frac{7}{2}(\sqrt{2})^k e^{-j\frac{\pi}{4}k} - \frac{7}{2}(\sqrt{2})^k e^{j\frac{\pi}{4}k}$$
$$+ j\frac{1}{2}(\sqrt{2})^k e^{-j\frac{\pi}{4}k} - j\frac{1}{2}(\sqrt{2})^k e^{j\frac{\pi}{4}k} \Rightarrow$$

$$x[k] = 7 - \frac{7}{2}(\sqrt{2})^k \left[e^{j\frac{\pi}{4}k} + e^{-j\frac{\pi}{4}k}\right] - j\frac{1}{2}(\sqrt{2})^k \left[e^{j\frac{\pi}{4}k} - e^{-j\frac{\pi}{4}k}\right] \Rightarrow$$

$$x[k] = 7 - \frac{7}{2}(\sqrt{2})^k 2\cos\frac{\pi}{4}k - j\frac{1}{2}(\sqrt{2})^k 2j\sin\frac{\pi}{4}k \Rightarrow$$

$$x[k] = \left(7 - 7(\sqrt{2})^k \cos\frac{\pi}{4}k + (\sqrt{2})^k \sin\frac{\pi}{4}k\right)u[k] \tag{5}$$

3.7.17 Compute the inverse z-transform of $X(z) = \dfrac{z^3 + 1}{z^3 - z^2 - z - 2}$

SOLUTION: The denominator of the given function is expressed as:

$$A(z) = z^3 - z^2 - z - 2 = (z-2)(z+0.5+j0.866)(z+0.5-j0.866) \tag{1}$$

So, $\dfrac{X(z)}{z}$ is analyzed into a partial fraction sum:

$$\frac{X(z)}{z} = \frac{c_0}{z} + \frac{c_1}{z+0.5+j0.866} + \frac{\bar{c}_1}{z+0.5-j0.866} + \frac{c_3}{z-2} \tag{2}$$

The c_i coefficients are calculated by using Heaviside's formula.

$$\left.\begin{array}{l}
c_0 = \left[\dfrac{X(z)}{z}(z)\right]_{z=0} = \dfrac{1}{-2} = -0.5 \\[4mm]
c_1 = \left[\dfrac{X(z)}{z}(z+0.5+j0.866)\right]_{z=-0.5-j0.866} = 0.429 + j0.0825 \\[4mm]
c_3 = \left[\dfrac{X(z)}{z}(z-2)\right]_{z=2} = 0.643
\end{array}\right\} \qquad (3)$$

Replacing the corresponding values of Eq. (3), $X(z)$ is calculated, and then, by using the inverse z-transform, $x[n]$ is obtained.

$$\begin{aligned}
X(z) &= c_0 + \dfrac{c_1 z}{z+0.5+j0.866} + \dfrac{\bar{c}_1 z}{z+0.5-j0.866} + \dfrac{c_3 z}{z-2} \\[3mm]
&= c_0 + \dfrac{c_1}{1+0.5+j0.866 z^{-1}} + \dfrac{\bar{c}_1}{1+0.5-j0.866 z^{-1}} + \dfrac{c_3}{1-2z^{-1}}
\end{aligned} \qquad (4)$$

$$x[n] = c_0 \delta[n] + c_1(-0.5 - j0.866)^n u[n] + \bar{c}_1(-0.5 + j0.866)^n u[n] + c_3 2^n u[n] \qquad (5)$$

Based on Eq. (6) and substituting in Eq. (5), $x[n]$ is derived:

$$\left.\begin{array}{l}
|p_1| = \sqrt{(0.5)^2 + (0.866)^2} = 1 \\[3mm]
\angle p_1 = \pi + \tan^{-1}\dfrac{0.866}{0.5} = \dfrac{4\pi}{3}\,\text{rad} \\[3mm]
|c_1| = \sqrt{(0.429)^2 + (0.0825)^2} = 0.437 \\[3mm]
\angle c_1 = \tan^{-1}\dfrac{0.0825}{0.429} = 0.19\,\text{rad}\,(10.89°)
\end{array}\right\} \qquad (6)$$

$$\begin{aligned}
x[n] &= c_0 \delta[n] + c_1(-0.5 - j0.866)^n u[n] + \bar{c}_1(-0.5 + j0.866)^n u[n] + c_3 2^n u[n] \\[3mm]
&= c_0 \delta[n] + 2|c_1||p_1|\cos(\angle p_1 n + \angle c_1) + c_3(2)^n u[n] \\[3mm]
&= -0.5\delta[n] + 0.874\cos\left(\dfrac{4\pi}{3} n + 0.19\right) + 0.643(2)^n u[n]
\end{aligned} \qquad (7)$$

This result can be also be verified using MATLAB, by implementing the following code.

```
num = [1 0 0 1];
den = [1 -1 -1 -2 0];
[r, p] = residue(num, den)
r =                     p=
0.6429                  2.0000
0.4286 - 0.825i         -0.5000 + 0.8660i
0.4286 + 0.825i         -0.5000 - 0.8660i
-0.5000     0
```

The first 20 output samples are calculated by the following code:

```
num = [1 0 0 1];
den = [1 -1 -1 -2 0];
x = filter(num, den, [1 zeros(1,19)]);
```

3.7.18 Compute the inverse z-transform of:

$$\text{a. } X(z) = \frac{(z-1)(z+0.8)}{(z+0.5)(z+0.2)}, \qquad \text{b. } X(z) = \frac{(z^2-1)(z+0.8)}{(z-0.5)^2(z+0.2)}$$

SOLUTION:

a. $X(z) = \dfrac{(z-1)(z+0.8)}{(z+0.5)(z+0.2)} \Rightarrow$

$$\frac{X(z)}{z} = \frac{C_1}{z} + \frac{C_2}{z-0.5} + \frac{C_3}{z+0.2} \tag{1}$$

The c_i coefficients are calculated using Heaviside's formula.

$$C_1 = \frac{X(z)}{z} z \Big|_{z=0} = 8, \qquad C_2 = \frac{X(z)}{z}(z-0.5) \Big|_{z=0.5} = -1.857,$$

$$C_3 = \frac{X(z)}{z}(z+0.2) \Big|_{z=-0.2} = -5.143,$$

$$X(z) = 8 - \frac{1.857z}{z-0.5} - \frac{5.143z}{z+0.2} \Rightarrow$$

$$x[n] = IZT\{X(z)\} = 8\delta[n] - 1.857(0.5)^n u[n] - 5.143(-0.2)^n u[n]$$

b. $X(z) = \dfrac{(z^2 - 1)(z + 0.8)}{(z - 0.5)^2(z + 0.2)}$

$\dfrac{X(z)}{z} = \dfrac{C_1}{z} + \dfrac{C_2}{z + 0.2} + \dfrac{C_3}{z - 0.5} + \dfrac{C_4}{(z - 0.5)^2}$

$C_1 = \dfrac{X(z)}{z} z \Big|_{z=0} = -16, \qquad C_2 = \dfrac{X(z)}{z}(z + 0.2) \Big|_{z=-0.2} = 5.88,$

$C_4 = \dfrac{X(z)}{z}(z - 0.5) \Big|_{z=0.5} = -2.79,$

$C_3 = \dfrac{d}{dz}\left(\dfrac{(z^2 - 1)(z + 0.8)}{z(z + 0.2)} \right)\Bigg|_{z=0.5} = 11.12$

$X(z) = -16 - \dfrac{5.88z}{z + 0.2} - \dfrac{11.12z}{z - 0.5} - \dfrac{2.79z}{(z - 0.5)^2} \Rightarrow$

$x[n] = -16\delta[n] + 5.88(-0.2)^n u[n] + 11.12(0.5)^n u[n] - 2.79n(0.5)^n u[n].$

3.7.19 Use the method of analysis in simple fractions and the division method to find the inverse z-transforms of the following signals:

a. $X(z) = \dfrac{z}{(z - 1)^2}$

b. $X(z) = \dfrac{1 - \dfrac{1}{2}z^{-1}}{1 + \dfrac{3}{4}z^{-1} + \dfrac{1}{8}z^{-2}}, \; |z| > \dfrac{1}{2}$

c. $X(z) = \dfrac{1 - \dfrac{1}{2}z^{-1}}{1 - \dfrac{1}{4}z^{-2}}, \; |z| > \dfrac{1}{2}$

SOLUTION:

a. $X(z) = \dfrac{z}{(z - 1)^2} = \dfrac{z}{z^2 + 2z + 1} = \dfrac{z}{z^2\left(1 + \dfrac{2}{z} + \dfrac{1}{z^2}\right)}$

$= \dfrac{z}{z^2\left(1 + \dfrac{1}{z}\right)^2} = \dfrac{1}{z}\left(1 + \dfrac{1}{z}\right)^{-2}$

$= \dfrac{1}{z}\left[1 - 2\left(\dfrac{1}{z}\right) + 3\left(\dfrac{1}{z^2}\right) - \dfrac{1}{4}\left(\dfrac{1}{z^3}\right) + \cdots \right]$

Thus,

$$X(z) = \dfrac{1}{z} - 2\dfrac{1}{z^2} + 3\dfrac{1}{z^3} - 4\dfrac{1}{z^4} + \cdots = \sum_{n=1}^{\infty}(-1)^{n-1} nz^{-n}$$

and $x[n] = (-1)^{n-1} n$

b. Analysis in simple fractions:

$$X(z) = \frac{1 - \frac{1}{2}z^{-1}}{1 + \frac{3}{4}z^{-1} + \frac{1}{8}z^{-2}} = \frac{A}{1 + \frac{1}{4}z^{-1}} + \frac{B}{1 + \frac{1}{2}z^{-1}} \Rightarrow$$

$$A = \frac{1 - \frac{1}{2}z^{-1}}{1 + \frac{1}{2}z^{-1}}\bigg|_{z^{-1} = 4} = -3 \text{ and } B = \frac{1 - \frac{1}{2}z^{-1}}{1 + \frac{1}{2}z^{-1}}\bigg|_{z^{-1} = -2} = 4.$$

Thus,

$$x[n] = \left[-3\left(-\frac{1}{4} \right)^n + 4\left(-\frac{1}{2} \right)^n \right] \cdot u[n]$$

Long division with remainders:

$1 - \frac{1}{2}z^{-1}$	$1 + \frac{3}{4}z^{-1} + \frac{1}{8}\cdot z^{-2}$
$1 - \frac{3}{4}z^{-1} - \frac{1}{8}\cdot z^{-2}$	$1 - \frac{5}{4}z^{-1} - \frac{13}{16}\cdot z^{-2} + \cdots$
$-\frac{5}{4}z^{-1} - \frac{1}{8}\cdot z^{-2}$	
$+\frac{5}{4}z^{-1} - \frac{15}{16}\cdot z^{-2} + \frac{5}{32}z^{-3}$	
$\frac{13}{16}\cdot z^{-2} + \frac{5}{32}z^{-3}$	

For $n = 1$, from the relation $x[n] = \left[-3\left(-\frac{1}{4} \right)^n + 4\left(-\frac{1}{2} \right)^n \right]\cdot u[n]$, we get:

$x[1] = \left[-3\left(-\frac{1}{4} \right) + 4\left(-\frac{1}{2} \right) \right]\cdot u[1] = +\frac{3}{4} - 2 = -\frac{5}{4},$ which is in agreement with the
corresponding division coefficient.

c. Analysis in simple fractions:

$$X(z) = \frac{1 - \frac{1}{2}z^{-1}}{1 - \frac{1}{4}z^{-2}} = \frac{\left(1 - \frac{1}{2}z^{-1} \right)}{\left(1 - \frac{1}{2}z^{-1} \right)\left(1 + \frac{1}{2}z^{-1} \right)} = \frac{1}{1 + \frac{1}{2}z^{-1}} = \frac{z}{z + \frac{1}{2}}, |z| > \frac{1}{2}$$

$$\Rightarrow x[n] = Z^{-1}\left\{\frac{z}{z+\dfrac{1}{2}}\right\} = \left(-\frac{1}{2}\right)^n u[n]$$

Method of long division with remainders:

$1 - \dfrac{1}{2}z^{-1}$	$1 - \dfrac{1}{4}z^{-2}$
$-1 + \dfrac{1}{4}z^{-2}$	$1 - \dfrac{1}{2}z^{-1} + \dfrac{1}{4}z^{-2} - \cdots$
$-\dfrac{1}{2}z^{-1} - \dfrac{1}{4}\cdot z^{-2}$	
$+\dfrac{1}{2}z^{-1} - \dfrac{1}{8}z^{-3}$	
$\dfrac{1}{4}z^{-2} + \dfrac{1}{8}z^{-3}$	

3.7.20 Solve the following difference equation, where the initial values are zero.

$$y[k+2] - 0.3\,y[k+1] + 0.02y[k] = (0.01)(0.3)^k u[k]$$

SOLUTION: We apply the z-transform in the given difference equation:

$$z^2 Y(z) - 0.3zY(z) + 0.02Y(z) = 0.01\frac{z}{z-0.3} \qquad (1)$$

$$(1) \Rightarrow Y(z)(z^2 - 0.3\,z + 0.02) = 0.01\frac{z}{z-0.3} \Rightarrow$$

$$Y(z) = \frac{0.01z}{(z-0.1)(z-0.2)(z-0.3)} \qquad (2)$$

$$\frac{Y(z)}{z} = \frac{A}{(z-0.1)} + \frac{B}{(z-0.2)} + \frac{C}{(z-0.3)} \qquad (3)$$

$$A = \frac{0.01}{(z-0.2)(z-0.3)}\Bigg|_{z=0.1} = \frac{0.01}{(-0.1)(-0.2)} = 0.5$$

$$B = \frac{0.01}{(z-0.1)(z-0.3)}\Bigg|_{z=0.2} = \frac{0.01}{(0.1)(-0.1)} = -1 \Bigg\}$$

$$C = \frac{0.01}{(z-0.1)(z-0.2)}\Bigg|_{z=0.3} = \frac{0.01}{(0.2)(0.1)} = 0.5 \qquad (4)$$

Thus,

$$y[k] = \left(0.5(0.1)^k - (0.2)^k + 0.5(0.3)^k\right)u[k] \qquad (5)$$

We can now check the solution. From the given difference equation, the first non-zero terms are:

$$k = 0 : y[2] - 0.3y[1] + 0.02y[0] = (0.01), \text{ thus, } y[2] = 0.001$$

$$k = 1 : y[3] - 0.3(0.01) = (0.01)(0.3), \text{ thus, } y[3] = 0.006$$

From the analytical solution of Expression (5), we have:

$$y[0] = (0.5 - 1 + 0.5) = 0$$

$$y[1] = \left(0.5(0.1) - (0.2) + 0.5(0.3)\right) = (0.05 - 2 + 0.15) = 0$$

$$y[2] = \left(0.5(0.01) - (0.04) + 0.5(0.09)\right)$$
$$= 0.005 - 0.04 + .045 = 0.001$$

$$y[3] = \left(0.5(0.1)^3 - (0.2)^3 + 0.5(0.3)^3\right)$$
$$= .0005 - .008 + .0135 = 0.006$$

The derived values agree with the previously derived solution; so, the result is correct.

3.7.21 Solve the second-order difference equation $y[k+2] - 5y[k+1] + 6y[k] = 0$ with the initial conditions $y[0] = 0$, $y[1] = 2$.

SOLUTION: We apply the z-transform in the given equation, thus:

$$Z\left[y[k+2] - 5y[k+1] + 6y[k]\right] = 0$$
$$\Rightarrow z^2Y(z) - z^2y[0] - zy[1] - 5\left(z \cdot Y(z) - z \cdot y[0]\right) + 6Y(z) = 0$$
$$\Rightarrow z^2Y(z) - 2z - 5zY(z) + 6Y(z) = 0 \Rightarrow$$

$$Y(z)(z^2 - 5z + 6) = 2z \Rightarrow Y(z) = \frac{2z}{z^2 - 5z + 6} \tag{1}$$

It is:

$$\frac{Y(z)}{z} = \frac{2}{z^2 - 5z + 6} = \frac{2}{(z-2)(z-3)} = \frac{k_1}{z-2} + \frac{k_2}{z-3} \tag{2}$$

$$k_1 = \lim_{z \to 2} \frac{Y(z)}{z}(z-2) = -2 \tag{3}$$

$$k_2 = \lim_{z \to 3} \frac{Y(z)}{z}(z-3) = 2 \tag{4}$$

Thus,

$$Y(z) = z\left[\frac{Y(z)}{z}\right] = -\frac{2z}{z-2} + \frac{2z}{z-3} \Rightarrow$$
$$y[k] = IZT[Y(z)] = \left(-2(2)^k + 2(3)^k\right)u[k] \tag{5}$$

3.7.22 Solve the second-order difference equation: $y[k + 2] + 5y[k + 1] + 6 + y[k] = 5u[k]$ with the initial conditions: $y[0] = -12$, $y[1] = 59$.

SOLUTION: We apply the *z*-transform in the given equation and then calculate $\frac{Y(z)}{z}$:

$$z^2 Y(z) - z^2 y[o] - zy[1] + 5\left[zY(z) - zy[0]\right] + 6Y(z) = 5\frac{z}{z-1}$$

$$z^2 Y(z) - z^2(-12) - z(59) + 5zY(z) - 5z(-12) + 6Y(z) = 5\frac{z}{z-1}$$

$$(z^2 + 5z + 6)Y(z) = 12z^2 + z + 5\frac{z}{z-1}$$

$$Y(z) = \frac{12z^2 + z}{z^2 + 5z + 6} + \frac{1}{z^2 + 5z + 6} \cdot \frac{5z}{z-1}$$

$$\frac{Y(z)}{z} = \frac{(12z+1)(z-1) + 5}{(z+2)(z+3)(z-1)} = \frac{K_1}{z+2} + \frac{K_2}{z+3} + \frac{K_3}{z-1} \tag{1}$$

We calculate the corresponding numerators:

$$K_1 = \frac{(12z+1)(z-1)+5}{(z+2)(z+3)(z-1)}(z+2)\bigg|_{z=-2} = \frac{64}{3}$$

$$K_2 = \frac{(12z+1)(z-1)+5}{(z+2)(z+3)(z-1)}(z+3)\bigg|_{z=-3} = \frac{-135}{4} \qquad (2)$$

$$K_3 = \frac{(12z+1)(z-1)+5}{(z+2)(z+3)(z-1)}(z-1)\bigg|_{z=1} = \frac{5}{12}$$

We substitute the values of K_i into $\dfrac{Y(z)}{z}$, and then we multiply with z, hence:

$$Y(z) = \frac{64/3z}{z+2} + \frac{135/4z}{z+3} + \frac{5/12z}{z-1} \qquad (3)$$

By using the inverse z-transform, we get:

$$y[k] = \left[\frac{64}{3}(-2)^k - \frac{135}{4}(-3)^k + \frac{5}{12}\right]u[k] \qquad (4)$$

3.7.23 Let the following equation describe a discrete-time system:

$$y[k] = 0.3y[n-1] + 0.7x[n] \text{ and } n \geq 0$$

a. Compute the frequency response, $H(e^{j\Omega})$
b. Compute the response of the system when the input is: $x[n] = \sin(0.3\pi n)$ for $n \geq 0$.

SOLUTION: We apply the z-transform to the given equation:

$$y[n] = 0.3y[n-1] + 0.7x[n] \Rightarrow Y(z) = 0.3z^{-1}Y(z) + 0.7X(z)$$
$$\Rightarrow (1 - 0.3z^{-1})Y(z) = 0.7X(z)$$

The transfer function will be:

$$H(z) = \frac{Y(z)}{X(z)} = \frac{0.7}{1 - 0.3z^{-1}} = \frac{0.7z}{z - 0.3} \qquad (1)$$

The frequency response is calculated by substituting $z = e^{j\Omega}$:

$$H(e^{j\Omega}) = H(z)\big|_{z=e^{j\Omega}} = \frac{0.7e^{j\Omega}}{e^{j\Omega} - 0.3} = \frac{0.7e^{j\Omega}}{\cos\Omega - 0.3 + j\sin\Omega} \qquad (2)$$

The magnitude of $H(e^{j\Omega})$ is:

$$\left| H(e^{j\Omega}) \right| = \left| \frac{0.7e^{jw}}{\cos w - 0.3 + j\sin w} \right| = \frac{0.7}{\sqrt{(\cos w - 0.3)^2 + \sin^2 w}} \tag{3}$$

and the phase of $H(e^{j\Omega})$ is:

$$\varphi(\Omega) = Arg \frac{0.7e^{j\Omega}}{\cos\Omega - 0.3 + j\sin\Omega} = \Omega - \tan^{-1}\frac{\sin\Omega}{\cos\Omega - 0.3} \tag{4}$$

b. For an input signal $x[n] = \sin(0.3\,\pi n)$ when $n \geq 0$, it is: $\Omega = 0.3\pi = 0.94rad$

For $\Omega = 0.3\,\pi$, the magnitude of the frequency response is $|H(e^{j\Omega})| = 0.815$, and the phase is $\phi(\Omega) = 0.3\pi - \tan^{-1}(2.81) = -0.29rad$.

The response of the system is $y[n] = e^{j0.94n} \cdot 0.815e^{-j0.29} = 0.815 \cdot e^{j(0.94n - 0,29)}$.

3.7.24 Derive the inverse z-transform of $X(z) = \dfrac{3z}{z^2 - z + 1}$ with ROC $|z| > 1$. Study the stability of the system.

SOLUTION: From the form of the function and the transforms (10) and (13) of the Table 3.1, we choose to use: $(\sin\Omega n)u[n] \leftrightarrow \dfrac{(\sin\Omega)z}{z^2 - (2\cos\Omega)z + 1}$ (1)

Comparing the denominators yields:

$2\cos\Omega = 1 \Rightarrow \cos\Omega = 0,5 \Rightarrow \cos\Omega = 0,5$ or $\Omega = \dfrac{\pi}{3}$. Then, $\sin\Omega = \dfrac{\sqrt{3}}{2}$ and the

nominator can be written as: $3z = \dfrac{3}{\sin\left(\dfrac{\pi}{3}\right)} \cdot \left[\sin\left(\dfrac{\pi}{3}\right)\right]z$

$$X(z) = \frac{3}{\frac{\sqrt{3}}{2}} \frac{\left[\sin\left(\dfrac{\pi}{3}\right)\right]z}{z^2 - \left[2\cos\left(\dfrac{\pi}{3}\right)\right]z + 1} \leftrightarrow 2\sqrt{3}\sin\left(\frac{\pi}{3}n\right)u[n] \tag{2}$$

An LTI system that has a rational transfer function (with regards to z), is casual when the ROC is in the exterior of the circle, defined by the most remote pole and the rank of the numerator is less than or equal to the rank of the denominator.

The system under discussion is casual.

We can decide on the stability of a causal system by studying the position of

the poles in the unit cycle. The poles are $z_1 = e^{j\frac{\pi}{3}}$ and $z_2 = e^{-j\frac{\pi}{3}}$. The magnitude of the poles is equal to 1, so the poles are located on the unit circle, and therefore the system is marginally stable.

3.7.25 Study the stability of the following system:

$$X(z) = \frac{(z^2 - z)}{(z^2 - 0.4z + 0.16)} \text{ with ROC} |z| > 0.4.$$

SOLUTION: Using the criteria of stability of the casual LTI systems referred to in the solution of Problem 3.7.24, we study the position of the poles with regards to the unit circle: The poles of the system are $z_1 = 0.4e^{j\frac{\pi}{3}}$ and $z_2 = 0.4e^{-j\frac{\pi}{3}}$. The magnitude of the poles is 0.4. Consequently, the poles are positioned inside the unit circle, and the system is stable.

The stability of the system could also be decided by the fact that the unit circle belongs to the ROC.

$$z_1 = 0.4e^{j\frac{\pi}{3}} = 0.4\left[\cos\left(\frac{\pi}{3}\right) + j\sin\left(\frac{\pi}{3}\right)\right] = 0.4\left[\frac{1}{2} + j\frac{\sqrt{3}}{2}\right] = 0.2 + j0.348$$

$$z_2 = 0.4e^{j\frac{\pi}{3}} = 0.4\left[\cos\left(\frac{\pi}{3}\right) + j\sin\left(\frac{\pi}{3}\right)\right] = 0.4\left[\frac{1}{2} + j\frac{\sqrt{3}}{2}\right] = 0.2 - j0.348$$

3.7.26 Discuss the stability of the LTI systems described by the following difference equations.

$$y[n] - \frac{1}{2}y[n-1] + \frac{3}{4}y[n-2] = x[n]$$

$$y[n] - \frac{3}{4}y[n-1] + \frac{1}{8}y[n-2] = x[n]$$

SOLUTION: a. By z-transforming the first difference equation, we have:

$$Y(z) - \frac{3}{4}z^{-1}Y(z) + \frac{3}{4}z^{-2}Y(z) = X(z) \Rightarrow$$

$$\left(1 - \frac{1}{2}z^{-1} + \frac{3}{4}z^{-2}\right)Y(z) = X(z) \Rightarrow$$

$$H(z) = \frac{Y(z)}{X(z)} = \frac{1}{1 - \frac{1}{2}z^{-1} + \frac{3}{4}z^{-2}} \Rightarrow$$

$$H(z) = \frac{4z^2}{4z^2 - 2z + 3}$$

$$\Delta = 4 - 4 \cdot 4 \cdot 3 = -44, \; z_{1,2} = \frac{2 \pm \sqrt{44}}{8} = 0.2 \pm j0.8$$

The poles are included in the unit circle; therefore, the system is stable.

b. By z-transforming the first difference equation, we have:

$$Y(z) - \frac{1}{2}z^{-1}Y(z) + \frac{3}{4}z^{-2}Y(z) = X(z) \Rightarrow$$

$$\left(1 - \frac{3}{4}z^{-1} + \frac{1}{8}z^{-2}\right)Y(z) = X(z) \Rightarrow$$

$$H(z) = \frac{Y(z)}{X(z)} = \frac{1}{1 - \frac{3}{4}z^{-1} + \frac{1}{8}z^{-2}} \Rightarrow$$

$$H(z) = \frac{z^2}{z^2 - \frac{3}{4}z + \frac{1}{8}} = \frac{z^2}{\left(z - \frac{1}{2}\right)\left(z - \frac{1}{4}\right)} \quad where \ |z| > \frac{1}{2}$$

The system is stable because it is casual and the poles are included in the unit circle.

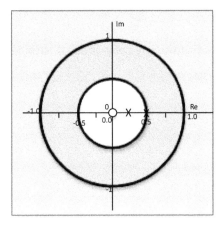

3.7.27 Compute the output, $y(n)$, of a system with the impulse response $h[n] = \left(\frac{2}{3}\right)^n u[n]$, when the input signal is:

a. The complex exponential sequence: $x[n] = Ae^{j\pi n/2}$

b. The complex exponential sequence: $x[n] = Ae^{j\pi n}$

c. $x[n] = 10 - 5\sin\frac{\pi}{2}n + 20\cos\pi n$

SOLUTION: It is well-known that when an input of frequency, ω, is applied to a linear system, the output is of the same frequency but of a different amplitude and phase. The new amplitude and phase are defined by the frequency response of the linear system at that frequency.

$$h[n] = \left(\frac{2}{3}\right)^n u[n] \Leftrightarrow H(z) = \frac{1}{1 - \frac{2}{3}z^{-1}}$$

$$\overset{z=e^{j\omega}}{\Leftrightarrow} H(e^{j\omega}) = \frac{1}{1 - \frac{2}{3}e^{-j\omega}}$$

The frequency response in MATLAB can be calculated as follows:

```
w = [0:1:500]*pi/500;% 501 frequencies in the interval 0 - π
a = 1; % vector of coefficients for numerator of H(z)
b = [1, -2/3]; % vector of coefficients for denominator of H(z)
H = freqz(a,b,w); %calculates frequency response for frequencies
   in vector w
plot(w/pi, abs(H))% abs returns the magnitude of complex value
xlabel('w/pi for 0<w<pi')
ylabel('Magnitude |H(exp(jw)|')
title('Frequency response of H(exp(jw))')
```

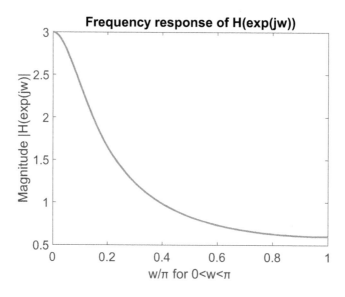

We notice that $|H(0)| = 3$. In the case that the frequency response has to be designed in the interval $-\pi \le \omega \le \pi$, the following change has to be made in the MATLAB code:

$$w = [0:1:500]*2*pi/500;\% \ w \ from \ 0 \ to \ 2*pi$$

a.   Since the input $x[n] = Ae^{j\pi n/2}$ is of the form $Ae^{j\omega n}$ with frequency $\omega = \pi/2$, then:

$$y[n] = H\left(e^{j\pi/2}\right) Ae^{j\pi n/2}$$

$$H(e^{j\pi/2}) = \frac{1}{1-\frac{2}{3}e^{-j\pi/2}} = \frac{1}{1+\frac{2}{3}j} =$$

$$= \frac{1}{\sqrt{1+\left(\frac{2}{3}\right)^2}} e^{-j\tan^{-1}\left(\frac{\frac{2}{3}}{1}\right)} = \frac{3}{\sqrt{5}} e^{-j33.7^{\circ}}$$

Thus,

$$y[n] = \frac{3}{\sqrt{5}} Ae^{j\left(\frac{\pi n}{2}-33.7^{\circ}\right)}$$

b.   Similarly, the second discrete signal is of a frequency π.

So,

$$H(e^{j\pi}) = \frac{1}{1-\frac{2}{3}e^{-j\pi}} = \frac{1}{1+\frac{2}{3}} = \frac{3}{5}$$

Thus,

$$y[n] = \frac{3}{5} Ae^{j\pi n}$$

c.   The third signal has the following frequencies: 0 (constant), $\pi/2$ and π. The response of the last two frequencies has already been calculated. So, for $\omega = 0$:

$$H(e^{j0}) = \frac{1}{1-\frac{2}{3}e^{-j0}} = \frac{1}{1-\frac{2}{3}} = \frac{3}{1} = 3$$

Thus,

$$y[n] = 3\cdot 10 - \frac{3}{5}\cdot 5\cdot\sin\left(\frac{\pi n}{2} - 33.6^{\circ}\right) + \frac{3}{\sqrt{5}}\cdot 20\cdot\cos(\pi n)$$

Validity of the output of the system for $\omega = 0$: $h[n]$ is produced by the difference equation $y[n] = (2/3)\,y[n-1] + x[n]$. Therefore, if we consider the case of a constant input $x[n]$ = 10, n = 0, 1, 2, ... that has ω=0, the output produced by the aforementioned difference equation is:

n=0   y[0] = 10
n=1   y[1] = (2/3)*10 + 10 = 16.667
n=2   y[2] = (2/3)*16.67 + 10 = 11.11 + 10 = 21.11
n=3   y[3] = (2/3)*21.11 + 10 = 14.07 + 10 = 24.07
n=4   y[4] = (2/3)*24.07 + 10 = 16.05 + 10 = 26.05
n=5   y[5] = (2/3)*26.05 + 10 = 17.36 + 10 = 27.36
n=6   y[6] = (2/3)*27.36 + 10 = 18.24 + 10 = 28.24
n=7   y[7] = (2/3)*28.24 + 10 = 18.82 + 10 = 28.82
n=8   y[8] = (2/3)*28.82 + 10 = 19.22 + 10 = 29.22
n=9   y[9] = (2/3)*29.22 + 10 = 19.47 + 10 = 29.47
n=10 y[10] = (2/3)*29.47 + 10 = 19.65 + 10 = 29.65
n=11 y[11] = (2/3)*29.65 + 10 = 19.77 + 10 = 29.

We can see that, after some time (transient time), the output, at its steady state, takes the value 30, just as we previously calculated using the frequency response of the filter.

3.7.28 Let the inverse z-transform:

$$X(z) = \frac{1 + 2z^{-1}}{1 + 0.4z^{-1} - 0.12z^{-2}}$$

Use the **impz.m** and **filter.m** commands in MATLAB to calculate the first 11 samples of the $x[n]$ sequence.

SOLUTION:

    a. MATLAB code using **impz.m** command:

```
L = 11; % Length of output vector
num = [1 2];
den = [1 0.4 -0.12];
[y, t] = impz(num,den,L);
disp('Coefficients of the power series expansion');
disp(y')
```

    b. MATLAB code using **filter.m** command:

```
N = 11;
num = [1 2];
den = [1 0.4 -0.12];
x =[1 zeros(1,N-1)];
y = filter(num,den, x);
disp('Coefficients of the power series expansion');
disp(y)
```

The result for both programs is the same:

$$x[n] = \{1.0000 \quad 1.6000 \quad -0.5200 \quad 0.4000 \quad -0.2224 \; 0.1370$$

$$-0.0815 \quad 0.0490 \quad -0.0294 \quad 0.0176 \; -0.0106\}$$

3.7.29 Compute the *z*-transform of the function $x[n] = [3\,5\,4\,3]$, $0 \leq n \leq 3$ using MATLAB.

SOLUTION: First suggested method

```
syms z
x0=3; x1=5; x2=4; x3=3;
Xz=x0*(z^0)+x1*(z^-1)  +x2*(z^-2)+x3*(z^-3)
pretty(X)
```

$$X(z) = 3 + \frac{5}{z} + \frac{4}{z^2} + \frac{3}{z^3}$$

Second suggested method

```
syms z
x=[3  5  4  3];
n=[0  1  2  3];
X=sum(x.*(z.^-n))
pretty(X)
```

$$X(z) = 3 + \frac{5}{z} + \frac{4}{z^2} + \frac{3}{z^3}$$

3.7.30 Compute the *z*-transform of $f[n] = 2^n$ using MATLAB.

SOLUTION: The code in MATLAB is:

```
syms n z
f= 2^n ;
ztrans(f)
simplify(ans)
```

$$F(z) = z/(z-2)$$

To verify the above result, we calculate the inverse *z*-transform of $F(z) = \dfrac{z}{z-2}$

```
syms n z
F=z/(z-2);
iztrans(F)
ans =2^n
```

3.7.31 Compute the z-transform of the following sequences using MATLAB:

$$\delta[n], u[n], n \cdot u[n], a^n u[n], na^n u[n] \cos(\omega_0 n) u[n], \sin(\omega_0 n) u[n],$$
$$a^n \cos(\omega_0 n) u[n], a^n \sin(\omega_0 n) u[n]$$

SOLUTION: In the following Table 3.5, the main transform relations are provided, while their accuracy is verified with the aid of **ztrans** and **iztrans** *functions*.

It is noteworthy that the **ztrans** *function* in MATLAB calculates the single-sided z-transform ($n \geq 0$), so the unit-step function can be omitted.

3.7.32 Rewrite the following function in a partial fraction expansion formulation.

$$X(z) = \frac{z^2 + 3z + 1}{z^3 + 5z^2 + 2z - 8}$$

TABLE 3.5

z-Transforms and Corresponding MATLAB Functions

Discrete-Time Domain n	z-Domain	Commands – Result	
$x[n]$	$X(z)$	`syms n z a w`	
$\delta[n]$	1	`f=dirac(n);` `ztrans(f,z)`	`ans =` `dirac(0)` `% δ(0) = 1`
$u[n]$	$z/(z-1)$	`f=heaviside(n)` `ztrans(f,z)`	`ans =` `z/(z-1)`
$n \cdot u[n]$	$\dfrac{z}{(z-1)^2}$	`ztrans(n,z)`	`ans =` `z/(z-1)^2`
$a^n u[n]$	$\dfrac{z}{z-a}$	`F=z/(z-a);` `f=iztrans(F,n)`	`f =` `a^n`
$na^n u[n]$	$\dfrac{az}{(z-a)^2}$	`f=n*a^n;` `ztrans(f,z)`	`ans =` `z*a/(-z+a)^2`
$\cos(\omega_0 n)u[n]$	$\dfrac{z^2 - z\cos(\omega_0)}{z^2 - 2z\cos(\omega_0) + 1}$	`f=cos(w*n)` `ztrans(f,z)`	`ans =` `(-z+cos(w))*z /` `(-z^2+2*z *cos(w)-1)`
$\sin(\omega_0 n)u[n]$	$\dfrac{z\sin(\omega_0)}{z^2 - 2z\cos(\omega_0) + 1}$	`f=sin(a*n);` `ztrans(f,z)`	`ans =` `z*sin(a)/` `(z^2-2*z*cos(a)+1)`
$a^n \cos(\omega_0 n)u[n]$	$\dfrac{z^2 - az\cos(\omega_0)}{z^2 - 2az\cos(\omega_0) + a^2}$	`f=(a^n)*cos(a*n)` `ztrans(f,z)` `simplify(ans)`	`ans =` `-(-z+cos(a)*a) *z/` `(z^2-2*z *` `cos(a)*a+a^2)`
$a^n \sin(\omega_0 n)u[n]$	$\dfrac{az\sin(\omega_0)}{z^2 - 2az\cos(\omega_0) + a^2}$	`f=(a^n)*sin(a*n);` `ztrans(f,z)` `simplify(ans)`	`ans =` `z*sin(a)*a/ (z^2-` `2*z * cos(a)*a+a^2)`

SOLUTION:

```
% Calculation of denominator's roots
A=[1 5 2 -8];
riz=roots(A);
% Calculation of numerators of the partial fractions
syms z
X=(z^2+3*z+1)/(z^3+5*z^2+2*z-8);
c1=limit((z-riz(1))*X,z,riz(1))
c2=limit( (z-riz(2))*X,z,riz(2))
c3=limit( (z-riz(3))*X,z,riz(3))
```

Hence, we get:

$$X(z) = \frac{z^2 + 3z + 1}{z^3 + 5z^2 + 2z - 8} = \frac{1/2}{z+4} + \frac{1/6}{z+2} + \frac{1/3}{z-1}$$

3.7.33 Rewrite the following function in a partial fraction expansion formulation.

$$X(z) = \frac{z^2 + 3z + 1}{z^3 - 3z + 2}$$

SOLUTION: First suggested solution

```
% Calculation of denominator's roots
A=[1 0 -3 2];
riz=roots(A)
riz = -2.0000 1.0000 1.0000
```

Observe the existence of a double root at point 1.0000

$X(z)$ can be expressed as:

$$X(z) = \frac{c_1}{z - p_1} + \frac{c_2}{(z - p_1)^2} + .. + \frac{c_r}{(z - p_1)^r} + \frac{c_{r+1}}{z - p_{r+i}} + ... + \frac{c_n}{z - p_n}$$

The coefficients c_1, ..., c_n can be calculated as:

$$c_i = \lim_{z \to \lambda_i} \frac{1}{(r-i)!} \frac{d^{r-1}\left((s - p_i)^r X(z)\right)}{dz^{r-1}} , i = 1,...,r$$

$$c_i = \lim_{z \to \lambda_i} (z - p_i) X(z). \qquad , i = r+1,...,n$$

```
% Calculation of c₁
syms z
X=(z^2+3*z+1)/(z^3-3*z+2);
c1=limit((z-riz(1))*X,z,riz(1))
% Calculation of c₂ (i=1) - 2 common roots (r=2)
r=2
% definition of (z - p_i)^r X(z)
f= ((z-1)^r )*X;
```

% Definition $\dfrac{d^{r-1}\left((z-p_i)^r\,X(z)\right)}{dz^{r-1}}$

```
par=diff(f,z,r-1);
```

% *Calculation* $\dfrac{1}{(r-i)!}$

```
fact=1/factorial(r-1);
% Calculation c₂
c2=limit(fact*par,z,1)
% Calculation c₃ (i=2)
par=diff(f,z,r-2);
fact=1/factorial(r-2);
limit(fact*par,z,1)
```

Therefore:

$$X(z) = \frac{z^2 + 3z + 1}{z^3 - 3z + 2} = \frac{1/9}{z+2} + \frac{10/9}{z-1} + \frac{5/3}{(z-1)^2}$$

Second suggested solution

$X(z)$ can be converted to partial fraction expansion formulation with the use of the **residue** command.

```
% Define the coefficients of numerator and denominator
num=[ 1 3 1];
den=[ 1 0 -3 2]
% Use of residue command
[R,P,K]=residue(num,den)
```

$X(z)$ can now be expressed as a fractional expansion:

$$X(z) = \frac{1/9}{z+2} + \frac{10/9}{z-1} + \frac{5/3}{(z-1)^2},$$

which is the same derived previously.

3.7.34 a. Solve the following difference equations using the *z*-transform: $y[n] - y[n-1] = x[n] + x[n-1]$ where $x[n] = 0.8^n$.

  b. Plot the results for $0 \le n \le 20$.

  c. Verify the accuracy of the solution by replacing the result in the given difference equation.

SOLUTION: a. The code in MATLAB is:

```
syms n z Y
x=0.8^n;
X=ztrans(x,z);  X1=z^(-1)*X;
Y1=z^(-1)*Y;
G=Y-Y1-X-X1;
SOL=solve(G,Y);
y=iztrans(SOL,n)
```

Then $y[n]$ is given by:

$$y = 10 - 9*(4/5)_n$$

  b. The code in MATLAB is:

```
n_s=0:30;
y_s=subs(y,n,n_s);
stem(n_s,y_s);
legend('SOLUTION')
```

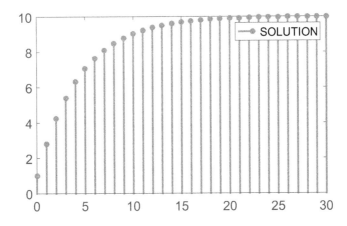

  c. The code in MATLAB is:

```
xn=0.8^(n);
xn_1=.8^(n-1);
```

```
yn=10-9*(4/5)^n;
yn _ 1=10-9*(4/5)^(n-1);
test=yn-yn _ 1-xn-xn _ 1
test =8*(4/5)^(n - 1) - 10*(4/5)^n
simplify(test)
ans =0
```

We would expect to get: test = 0. By using the **simplify** command, we simplify the answer, so we got: ans=0, hence the answer is right.

3.7.35 Let a casual discrete-time system be described by the following difference equation $y[n] = 0.9y[n - 1] + x[n]$. For this system plot using MATLAB:

a. the magnitude of the frequency response, and
b. the zeros and poles

SOLUTION: a. We z-transform the given deference equation:

$$y[n] = 0.9y[n-1] + x[n] \Rightarrow Y(z) = 0.9z^{-1}Y(z) + X(z) \Rightarrow H(z)$$
$$= \frac{Y(z)}{X(z)} = \frac{1}{1 - 0.9z^{-1}} \quad |z| > 0.9$$

The magnitude of the frequency response is: $\left|H(e^{j\omega})\right| = \dfrac{1}{\left|e^{j\omega} - 0.9\right|}$

We use the **freqz** command in MATLAB to design |H| in the frequency interval $\omega = 0 - \pi$ rad:

```
a=1;
b=[1, -0.9];
[H,w]=freqz(a,b,100);
magH=abs(H);
plot(w/pi, magH)
xlabel('Frequency in pi units')
ylabel('Magnitude')
title('Magnitude Response')
```

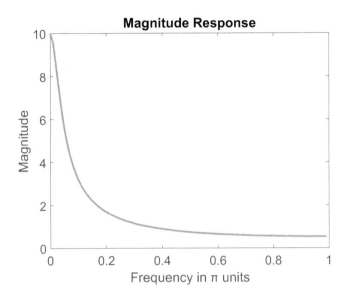

With the following code, we design the frequency response magnitude for the frequency range

ω=0-2π rad, and we notice that the figure is symmetrical.

```
a=1;
b=[1, -0.9];
[H,w]=freqz(a,b,200, 'whole');
magH=abs(H);
plot(w/pi, magH)
xlabel('Frequency in pi units')
ylabel('Magnitude')
title('Magnitude Response')
```

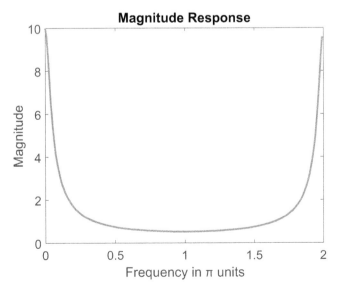

In the code that follows, we design the magnitude of the frequency response for the frequency range ω= -π- π rad.

```
a=1;
b=[1, -0.9];
w=[-pi:pi/200:pi]; % which is w=[-200:1:200]*pi/200;
H=freqz(a,b,w);
plot(w/pi, abs(H))
ylabel('Magnitude')
xlabel('Frequency in pi units')
title('Magnitude Response')
```

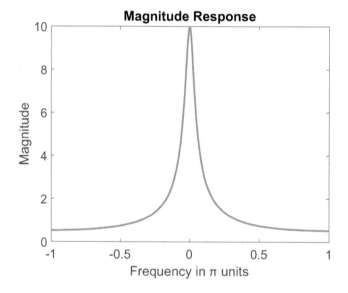

b. We plot the poles and zeros by using the **zplane** command as shown below:

```
a=[1];
b=[1, -0.9];
zplane(a,b)
```

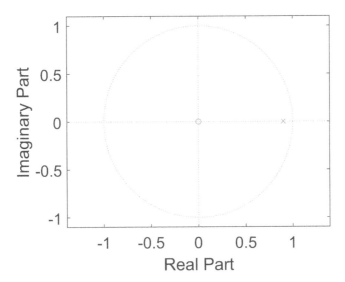

It is easily understood that there is a root in the numerator at $z = 0$ (a zero for $H(z)$) and a root in the denominator at $z = 0.9$ (a pole for $H(z)$).

3.7.36 Plot the magnitude of the frequency response of the system with impulse response

$$h[n] = \left(\frac{2}{3}\right)^n u[n]$$

a. For the range $\omega = 0\text{-}\pi$ and
b. For the range $\omega = \text{-}\pi\text{-}\pi$.

SOLUTION: The solution is:

$$h[n] = \left(\frac{2}{3}\right)^n u[n] \;\Leftrightarrow\; H(z) = \frac{1}{1 - \frac{2}{3}z^{-1}} \overset{z = e^{j\omega}}{\Leftrightarrow} H(e^{j\omega}) = \frac{1}{1 - \frac{2}{3}e^{-j\omega}}$$

a. We write the following MATLAB code:

```
w=[0:1:500]*pi/500; % 501 samples in the range 0 - π
a=1; % coefficients of the numerator of H (z)
b=[1, -2/3]; % coefficients of the denominator H(z)
H=freqz(a,b,w); % Calculate Frequency Response
plot(w/pi, abs(H)) % Plot the magnitude
xlabel('w/pi for 0<w<pi')
ylabel('Magnitude |H(exp(jw)|')
title('Frequency response of H(exp(jw))')
```

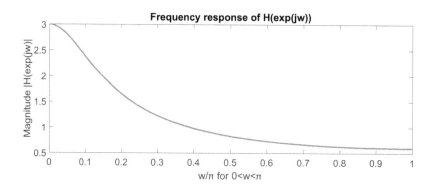

b. In order to plot the frequency response at $-\pi \leq \omega \leq \pi$, we will only change the command to:

w=[0:1:500]*2*pi/500; % w from 0 to 2*pi

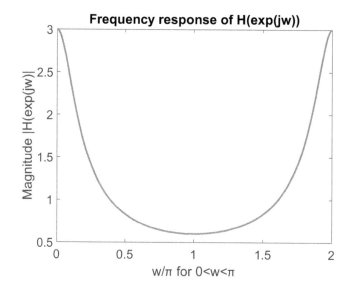

3.7.37 Study the stability of the systems with the following transfer functions.

$$\text{a.} \quad H(z) = \frac{(z-0.5)}{(z+0.75)}, \qquad \text{b.} \quad H(z) = \frac{(z^2+1)}{(z^2-0.25)}$$

$$\text{c.} \quad H(z) = \frac{z(z-1)}{(z^2+0.5z-0.5)}, \qquad \text{d.} \quad H(z) = \frac{(z-0.5)(z+0.5)}{(z^2+z+0.75)}$$

Write the MATLAB code to plot the step responses of the systems.

SOLUTION:

$$\text{a.} \quad H(z) = \frac{(z - 0.5)}{(z + 0.75)}$$

The system is stable because the pole of the system $p = -0.75$ is included in the unit circle.

$$\text{b.} \quad H(z) = \frac{(z^2 + 1)}{(z^2 - 0.25)}$$

The system is stable because the poles of the system $p_1 = -0.5$ and $p_2 = 0.5$ are included in the unit circle. The zero-pole plots for the above two systems are:

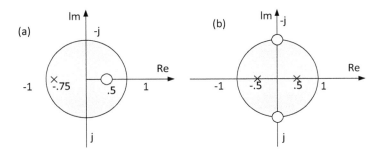

$$\text{c.} \quad H(z) = \frac{z(z - 1)}{(z^2 + 0.5z - 0.5)} = \frac{z(z - 1)}{(z - 1)(z - 0.5)}$$

The system is marginally stable because one of the poles ($p = -1$) is on the circumference of the unit circle.

$$\text{d.} \quad H(z) = \frac{(z - 0.5)(z + 0.5)}{(z^2 + z + 0.75)} = \frac{(z - 0.5)(z + 0.5)}{(z + 0.5 + 07\,j)(z + 0.5 - 07\,j)}$$

The system is stable because its complex conjugate poles $p_1 = 0.86e^{j126^0}$ and $p_2 = 0.86e^{-j126^0}$ are included in the unit circle. The zero-pole plots for the above two systems are:

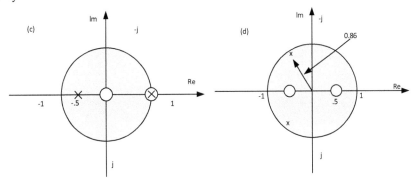

In order to plot the step responses we write the following commands in MATLAB:
step([1, -0.5],[1, 0.75]) % for the system a.

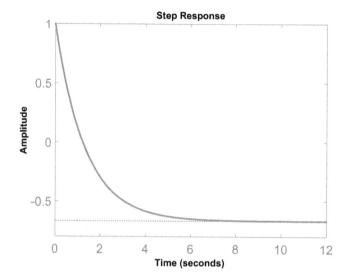

step([1, 0, 1],[1, 0, -0.25]) % for system b

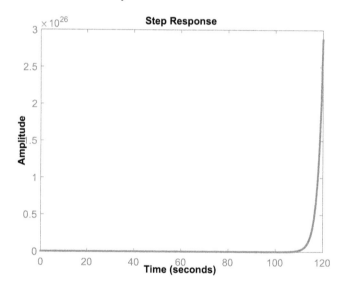

step([1, -1, 0],[1, 0.5, -0.5]) % for the system c

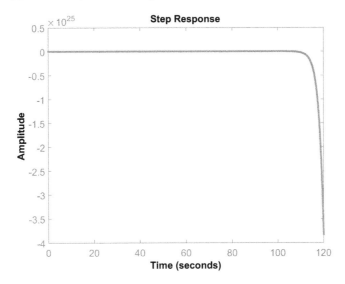

step([1, -0.25],[1, 1, 0.75]) % for the system d.

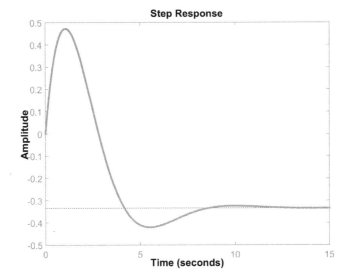

4

Structures for the Realization
of Discrete-Time Systems

4.1 Introduction

The implementation of discrete-time systems can be achieved in two ways:

1. using *software*, i.e., by programming of either a general or special purpose computer system
2. using *hardware*, i.e., in a circuit-like form

The way we choose to describe a system often determines the required memory space, the speed, the quantization error of a digital system, and therefore it constitutes a subject of further study. In this chapter, the schematic way (i.e., using diagrams) of describing the discrete-time systems is discussed that facilitate circuit analysis and implementation.

In general, the implementation of systems requires storage capabilities of previous input-output values of the system and of the intermediate sets of values created (delay elements), as well as the ability to use multipliers and adders. The difference equation of a discrete-time system (filter) could be interpreted as an algorithm for calculating the output sequence, $y[n]$, of the system when the input sequence, $x[n]$, is known. For each difference equation, a *block diagram* can be designed consisting of delay circuits, multipliers and adders. This block diagram is usually described as the *realization structure* of the system.

4.2 Block Diagrams

The block diagram is a graph with *nodes* and *branches* constituting its basic structural components. The block diagrams of the basic components, that are being used to implement the difference equation of a discrete system, are illustrated in Figure 4.1 and are the *adder*, the *multiplier* and the *delay* components.

The symbol z^{-1} represents the delay of one sampling interval (unit delay).

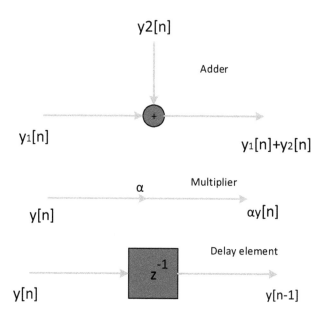

FIGURE 4.1
Basic implementation components.

For example, let the difference equation describing a digital filter (system): $y[n] = 2y[n-1] + 3y[n-2] + 5x[n]$. In Figure 4.2, the block diagram, which depicts the realization of the difference equation, is presented.

z^{-1} corresponds to one unit delay. In digital implementation, each unit delay can be implemented by providing a storage register. If the required number of samples of delay is $D>1$, then the corresponding system function is of z^{-D} degree.

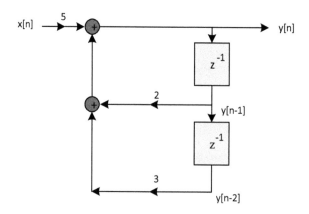

FIGURE 4.2
Realization of difference equation.

4.3 Realization Structures

There are four implementation structures:

1. Direct form I (Figure 4.3)
2. Direct form II – Canonical form (Figure 4.4)
3. Cascade form (Figure 4.5)
4. Parallel form (Figure 4.6)

The different structures of implementation have different characteristics with respect to the required *computational complexity, memory* and due to the *finite word length effects*. The system parameters must necessarily be represented with finite precision. The results derived from the calculation procedure of each output sample of the system must be rounded or truncated at the appropriate finite word length (corresponding accuracy) imposed by the computer or circuit used for the implementation.

Difference equations for discrete-time systems of degree $N>2$ are usually implemented as combinations of first- or second-order structures in series or in parallel to reduce calculation errors in their implementation by digital systems (which already use finite precision to represent coefficients and calculate operations).

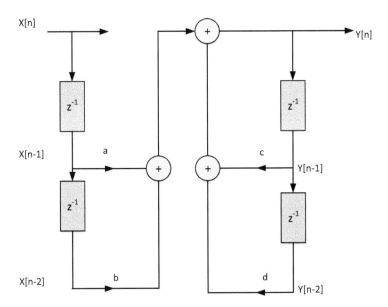

FIGURE 4.3
Direct form I.

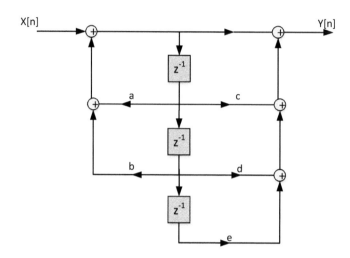

FIGURE 4.4
Direct form II or canonical form.

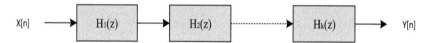

FIGURE 4.5
Cascade form.

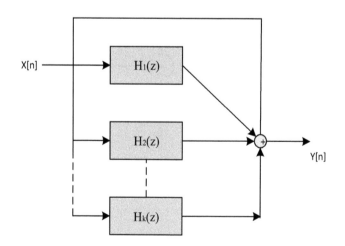

FIGURE 4.6
Parallel form.

The problems that need to be addressed in a series implementation are:

- The way in which the numerators should be combined with the denominators after factorizing the transfer function
- The connection order of the individual blocks
- The reduction of the signal amplitude at intermediate points of the structure so that it is neither too large nor too small.

In the case of parallel form, the interconnection order of the individual structures is not significant. In addition, scaling is easier, since it can be done for each structure independently. However, the zeros of the individual structures are more sensitive to the quantization errors of coefficients.

4.3.1 Implementation Structures of IIR Discrete Systems

Let us consider the case of implementing an IIR discrete system. The general difference equation is:

$$y[n] = -\sum_{k=1}^{N} \frac{b_k}{b_0} y[n-k] + \sum_{m=0}^{M} \frac{\alpha_m}{b_0} x[n-m] \tag{4.1}$$

$$\text{or} \sum_{k=0}^{N} b_k y[n-k] = \sum_{m=0}^{M} \alpha_m x[n-m] \tag{4.2}$$

For $b_0 = 1$ we get:

$$y[n] = \sum_{m=0}^{M} \alpha_m x[n-m] + \sum_{k=1}^{N} b_k y[n-k] \tag{4.3}$$

Based on the above difference equation we can implement the discrete system as shown in Figure 4.7 in direct form I, which is called so since it can be designed directly from the transfer function of the system:

$$H(z) = \frac{Y(z)}{X(z)} = \frac{\displaystyle\sum_{m=0}^{M} \alpha_m z^{-m}}{\displaystyle\sum_{k=0}^{N} b_k z^{-k}} \tag{4.4}$$

$$\text{or} \, Y(z) \sum_{k=0}^{N} b_k z^{-k} = X(z) \sum_{m=0}^{M} \alpha_m z^{-m} \tag{4.5}$$

In Equation 4.5, the terms of the form $Y(z)z^{-k}$ correspond to the inverse z-transform of $y[n-k]$.

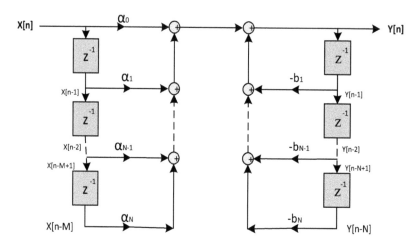

FIGURE 4.7
Implementation of the IIR filter in direct form I.

Figure 4.7 shows the connection of two partial systems in which one implements the zeros (i.e., the roots of the nominator), and the other implements the poles (i.e., the roots of the denominator) of the transfer function of the discrete system. Because in a discrete LTI system, the total transfer function does not change if the order of the individual systems is reversed, we plot the equivalent system of Figure 4.8 (for N = M).

The two branches with the z^{-1} delays have the same input as shown in Figure 4.8. Thus, they are replaced by a single branch as shown in Figure 4.9, which constitutes the implementation as a direct form II or canonical form.

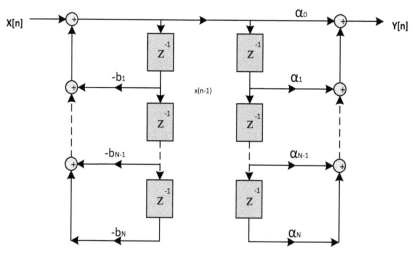

FIGURE 4.8
Implementation of the IIR filter — intermediate stage.

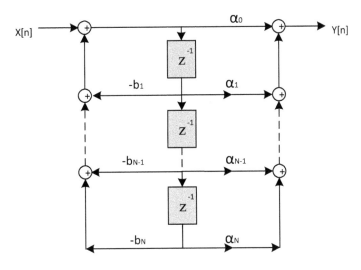

FIGURE 4.9
Implementation of the IIR filter in canonical form.

Another form of IIR filter implementation is the cascade form, which is produced if we express $H(z)$ as presented in Equation 4.6:

$$H(z) = \frac{Y(z)}{X(z)} = \alpha_0 \prod_{i=1}^{k} H_i(z) \tag{4.6}$$

where $H_i(z)$ is a first-degree polynomial with respect to z^{-1}

$$H_i(z) = \frac{1 + \alpha_{1i}z^{-1}}{1 + b_{1i}z^{-1}} \tag{4.7}$$

or a second-degree polynomial

$$H_i(z) = \frac{1 + \alpha_{1i}z^{-1} + \alpha_{2i}z^{-2}}{1 + b_{1i}z^{-1} + b_{2i}z^{-2}} \tag{4.8}$$

with k the integer part of $\left(\frac{N+1}{2}\right)$.

The discrete-time system described by Equation 4.6 is realized using a cascade connection of k individual systems (Figure 4.5). Moreover, each of them is realized using a direct form I or II.

A fourth version of implementation of an IIR system is the parallel form. In this case, $H(z)$ is recorded in the form of:

$$H(z) = C + \sum_{i=1}^{k} H_i(z) \tag{4.9}$$

where $H_i(z)$ is a first-degree polynomial with respect to z^{-1}

$$H_i(z) = \frac{\alpha_{0i}}{1 + b_{1i}z^{-1}} \tag{4.10}$$

or a second-degree polynomial

$$H_i(z) = \frac{\alpha_{0i} + \alpha_{1i}z^{-1}}{1 + b_{1i}z^{-1} + b_{2i}z^{-2}} \tag{4.11}$$

with k the integer part of $\left(\dfrac{N+1}{2}\right)$ and $C = \dfrac{a_N}{b_N}$.

The implementation of the discrete system of Equation 4.9 is accomplished by parallel connection of k individual systems (Figure 4.6). Each of them is implemented using direct form I or II.

4.3.2 Implementation Structures of FIR Discrete Systems

In FIR systems the discrete impulse response is of finite duration and the function $H(z)$ is usually written in the form:

$$H(z) = \sum_{n=0}^{N-1} h[n] \cdot z^{-n} \tag{4.12}$$

The function $h(n)$ consists of N samples. Consequently, $H(z)$ is a polynomial of $N - 1$ degree with respect to z^{-1}. Thus, $H(z)$ has $N - 1$ poles at $z = 0$ and $N - 1$ zeros. FIR systems are characterized by non-recursive difference equations that are in the form of:

$$y[n] = \sum_{k=0}^{N-1} h[k]x[n-k] \tag{4.13}$$

$$\text{or } y[n] = h[0]\,x[n] + h[1]\,x[n-1] + \dots + h[N-1]\,x[N-1-k] \tag{4.14}$$

Figure 4.10 illustrates the realization of Equation 4.14 using direct form I.

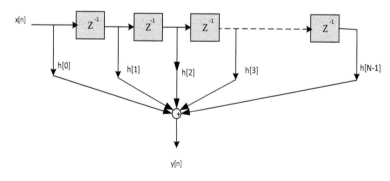

FIGURE 4.10
Implementation of FIR filter in direct form I.

In case of the cascade implementation, $H(z)$ can be written in the form of a product of first- and second-degree terms.

$$H(z) = \prod_{n=1}^{k} H_n(z) \tag{4.15}$$

where the first-degree terms are:

$$H_n(z) = \alpha_{0n} + \alpha_{1n}z^{-1} \tag{4.16}$$

and the second-degree terms are:

$$H_n(z) = \alpha_{0n} + \alpha_{1n}z^{-1} + \alpha_{2n}z^{-2} \tag{4.17}$$

with k the integer part of $\left(\dfrac{N+1}{2}\right)$.

Equation 4.15 is equal to Equation 4.17, where, if N is an even number, one of the coefficient, α_{2n}, will be zero.

$$H(z) = \prod_{n=1}^{\left(N/2\right)} \alpha_{0n} + \alpha_{1n}z^{-1} + \alpha_{2n}z^{-2} \tag{4.18}$$

Figure 4.11 illustrates the realization of the cascade form where each subsystem is in direct form I.

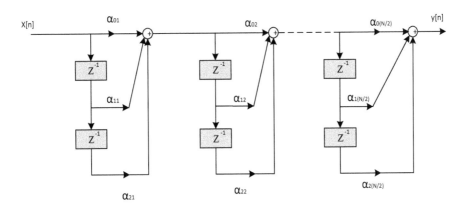

FIGURE 4.11
Block diagram in cascade form for a FIR filter.

4.4 Signal Flow Graphs

Signal flow graphs (SFGs), like block diagrams, provide an alternative method for graphi-cally describing a system. SFG theory was introduced by Samuel J. Mason, and it can be applied to any system without having to simplify the block diagram, which sometimes can be quite a difficult process. An SFG is actually a simplified version of a block diagram. It consists of nodes, branches, and loops. Each node represents a variable (or signal) and belongs to one of the following categories:

a. *Source or input node*: The node from which one or more branches start and no branch ends (Source node - Figure 4.12 (a)).

b. *Sink node*: The node in which one or more branches end and from which no branch starts (Sink node - Figure 4.12 (b)).

c. *Mixed node*: The node that branches can both come in and come out (Mixed node– Figure 4.12 (c)).

A *branch* connects two nodes and is characterized by its direction and gain. The direc-tion shows the signal flow, while the gain is a coefficient (corresponding to a transfer function), which relates the variables x_1 and x_2. For Figure 4.13, the relationship $x_2 = ax_1$ is true.

Path is a succession of branches of the same direction (e.g., x_1, x_2, x_3 and x_4 in Figure 4.14).

Forward path is the path that starts from the input node and ends at the output node (e.g., x_1, x_2, x_3, x_4, x_5 in Figure 4.14).

Loop is a closed path that begins and ends at the same node (e.g., x_2, x_3, x_2 in Figure 4.14). Two loops of a SFG are called non-touching when they have no common node.

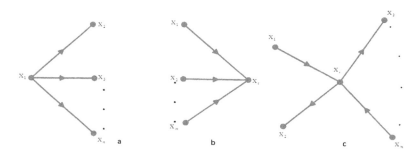

FIGURE 4.12
(a, b, c) Nodes of system flow diagrams.

FIGURE 4.13
Branch of system flow diagrams.

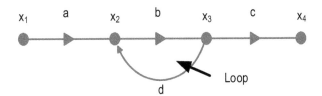

FIGURE 4.14
Loop of systems flow diagrams.

4.4.1 Mason's Gain Formula

Mason's gain formula (1953), or Mason's rule, is a method for deriving the transfer function of a system through its SFG. By applying Mason's rule there is no need to use block diagram reduction. The mathematical formulation is given in Equation 4.19:

$$H(z) = \frac{\sum_{n=1}^{k} T_n D_n}{D} \tag{4.19}$$

where T_n is the gain of the n-th forward path
D is the determinant of the diagram and is given by the following relation:

$$D = 1 - \sum L_1 + \sum L_2 - \sum L_3 + \cdots \tag{4.20}$$

with L_1 the gain of each loop,
L_2 the product of any two non-touching loop gains,
L_3 is the product of the gains of any three pairwise non-touching loops, and so on,
D_n is the subdeterminant of the path T_n and is calculated from Equation 4.20, without taking into account that part of the SFG that is non-touching with the n-th forward path.
For example, let the difference equation of the discrete system:

$$y[n] = 2y[n-1] + x[n] + 5x[n-1].$$

In Figure 4.15, the SFG is illustrated.
In Figure 4.16, the implementation structure of the system as a canonical form is illustrated.

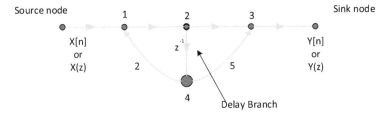

FIGURE 4.15
SFG of the difference equation.

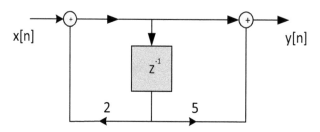

FIGURE 4.16
Canonical form implementation structure.

4.5 Solved Problems

4.5.1 The following systems are connected in series:

$$H_1(z) = \frac{5}{1+3z^{-1}}, H_2(z) = 1+z^{-1}, H_3(z) = \frac{z^{-2}}{(1+2z^{-1})^3}$$

Determine the systems that need to be connected in parallel to lead to the same result as the connection in series.

SOLUTION: $H_1(z)$, $H_2(z)$, H3(z) can be given in the following form: $H_1(z) = \dfrac{5z}{z+3}$, $H_2(z) = \dfrac{1+z}{z}$, $H_3(z) = \dfrac{z}{(z+2)^3}$

Connection of the above systems in series yields:

$$H(z) = H_1(z) \cdot H_2(z) \cdot H_3(z) = \frac{5z \cdot (z+1)}{(z+3)(z+2)^3} \qquad (1)$$

Obviously, the systems to be connected in parallel will be computed by the technique of partial fraction decomposition of $H(z)$.

$$\frac{H(z)}{z} = \frac{5(z+1)}{(z+3)(z+2)^3} = \frac{A}{(z+3)} + \frac{B_1}{z+2} + \frac{B_2}{(z+2)^2} + \frac{B_3}{(z+2)^3} \qquad (2)$$

The coefficients of the fractions are computed next:

$$A = \frac{H(z)}{z}(z+3)\bigg|_{z=-3} = \frac{5(z+1)}{(z+2)^3}\bigg|_{z=-3} = \frac{5(-3+1)}{(-3+2)^3} = \frac{5(-2)}{(-1)^3} = 10$$

$$B_1 = \frac{1}{2}\frac{d^2}{dz^2}\left\{\frac{H(z)}{z}(z+2)^3\right\}\Bigg|_{z=-2} = \frac{1}{2}\frac{d^2}{dz^2}\left\{\frac{5(z+1)}{(z+3)}\right\}\Bigg|_{z=-2}$$

$$= \frac{1}{2}\frac{d}{dz}\left\{\frac{5(z+3)-5(z+1)}{(z+3)^2}\right\}\Bigg|_{z=-2} = \frac{1}{2}\frac{d}{dz}\left\{\frac{10}{(z+3)^2}\right\}\Bigg|_{z=-2}$$

$$= \frac{1}{2}\left\{\frac{-10\cdot2\cdot(z+3)}{(z+3)^4}\right\}\Bigg|_{z=-2} = -\frac{10}{(z+3)^3}\Bigg|_{z=-2} = -\frac{10}{(-2+3)^3} = -10$$

$$B_2 = \frac{d}{dz}\left\{\frac{H(z)}{z}(z+2)^3\right\}\Bigg|_{z=-2} = \frac{d}{dz}\left\{\frac{5(z+1)}{(z+3)}\right\}\Bigg|_{z=-2} = -\frac{10}{(-2+3)^2} = 10$$

$$B_3 = \frac{H(z)}{z}(z+2)^3\Bigg|_{z=-2} = \frac{5(z+1)}{z+3}\Bigg|_{z=-2} = \frac{5(-2+1)}{(-2+3)} = -5$$

Thus,

$$\frac{H(z)}{z} = \underbrace{\frac{10z}{z+3}}_{H_1'(z)} + \underbrace{\frac{(-10)z}{z+2}}_{H_2'(z)} + \underbrace{\frac{10z}{(z+2)^2}}_{H_3'(z)} + \underbrace{\frac{(-5)z}{(z+2)^3}}_{H_4'(z)} \tag{3}$$

From relation (3), it is concluded that the systems that need to be connected in parallel to lead to the same result as the connection in series are:

$$H_1'(z) = 10\cdot\frac{1}{1+3z^{-1}}, H_2'(z) = (-10)\cdot\frac{1}{1+2z^{-1}}$$

$$H_3'(z) = 10\cdot\frac{z^{-1}}{(1+2z^{-1})^2}, H_4'(z) = (-5)\cdot\frac{z^{-2}}{(1+2z^{-1})^3}$$

4.5.2 Let the block diagram of a discrete casual system be described as:

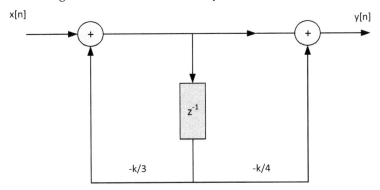

Illustrate the zeros and poles of the transfer function $H(z)$ and define the ROC. For which values of k is the system stable?

SOLUTION: The implementation structure is of direct form II, so the transfer function $H(z)$ will be: $H(z) = \dfrac{Y(z)}{X(z)} = \dfrac{1 - \dfrac{k}{4}z^{-1}}{1 + \dfrac{k}{3}z^{-1}} = \dfrac{z - \dfrac{k}{4}}{z + \dfrac{k}{3}}$

The ROC is: $|z| > \dfrac{k}{3}$. The system is stable for $|k| < 3$

4.5.3 Let the transfer function $H(z)$ of a discrete system:

$$H(z) = \frac{10\left(1 - \dfrac{2}{3}z^{-1}\right)\left(1 - \dfrac{1}{2}z^{-1}\right)(1 - 2z^{-1})}{\left(1 - \dfrac{3}{4}z^{-1}\right)\left(1 - \dfrac{1}{8}z^{-1}\right)\left[1 - \left(\dfrac{1}{2} + j\dfrac{1}{2}\right)\cdot z^{-1}\right]\left[1 - \left(\dfrac{1}{2} - j\dfrac{1}{2}\right)\cdot z^{-1}\right]}$$

Plot the block diagram for the system a) in cascade form and b) in parallel form.

SOLUTION: The transfer function $H(z)$ can be written: $H(z) = 10H_1(z)H_2(z)$

where: $H_1(z) = \dfrac{\left(1 - \dfrac{2}{3}z^{-1}\right)}{\left(1 - \dfrac{3}{4}z^{-1}\right)\left(1 - \dfrac{1}{8}z^{-1}\right)} = \dfrac{\left(1 - \dfrac{2}{3}z^{-1}\right)}{1 - \dfrac{7}{8}z^{-1} + \dfrac{3}{32}z^{-2}}$ (1)

and

$$H_2(z) = \frac{\left(1 - \dfrac{1}{2}z^{-1}\right)(1 - 2z^{-1})}{\left[1 - \left(\dfrac{1}{2} + j\dfrac{1}{2}\right)\cdot z^{-1}\right]\left[1 - \left(\dfrac{1}{2} - j\dfrac{1}{2}\right)\cdot z^{-1}\right]} = \frac{1 + \dfrac{3}{2}z^{-1} - z^{-2}}{1 - z^{-1} + \dfrac{1}{2}z^{-2}}$$ (2)

In the following figure the implementation of the system in a cascade form is illustrated.

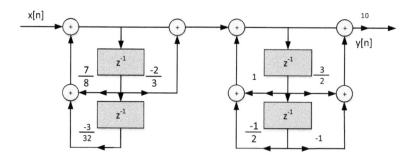

For the implementation in parallel form, $H(z)$ must be written in a form of a partial fractions sum:

$$H(z) = \frac{k_1}{1-\frac{3}{4}z^{-1}} + \frac{k_2}{1-\frac{1}{8}z^{-1}} + \frac{k_3}{1-\left(\frac{1}{2}+j\frac{1}{2}\right)z^{-1}} + \frac{\overline{k_3}}{1-\left(\frac{1}{2}-j\frac{1}{2}\right)z^{-1}} \qquad (3)$$

After some calculations, we get:

$$k_1 = 2.93, k_2 = -17.68, k_3 = 12.25 - j\ 14.57, \overline{k_3} = 12.25 + j\ 14.57$$

Thus,

$$H(z) = \frac{2.93}{1-\frac{3}{4}z^{-1}} + \frac{-17.68}{1-\frac{1}{8}z^{-1}} + \frac{12.25 - j\ 14.57}{1-\left(\frac{1}{2}+j\frac{1}{2}\right)z^{-1}} + \frac{12.25 + j\ 14.57}{1-\left(\frac{1}{2}-j\frac{1}{2}\right)z^{-1}} \Rightarrow$$

$$H(z) = \frac{-14.75 - 12.9\ z^{-1}}{1 - \frac{7}{8}z^{-1} + \frac{3}{32}z^{-2}} + \frac{24.5 - 26.82\ z^{-1}}{1 - z^{-1} + \frac{1}{2}z^{-2}} \qquad (4)$$

The following figure illustrates the implementation of the system in parallel form.

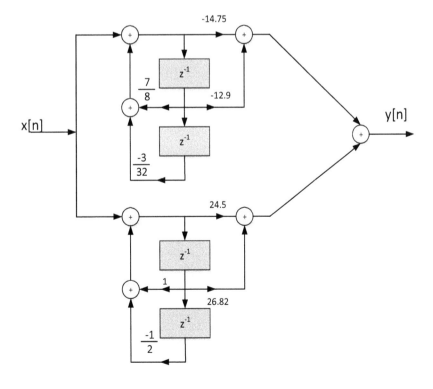

4.5.4 Let the discrete-time system of the following figure:

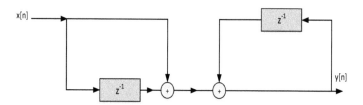

a.   Design the realization structure of the depicted system using only one delay element.

b.   If the input is $x[n] = u[n - 2]$, find the output, $y[n]$, of the system, using z-transform.

SOLUTION:

a.   From the above figure, it is deduced:

$$y[n] = y[n-1] + x[n] + x[n-1] \Rightarrow$$

$$y[n] - y[n-1] = x[n] + x[n-1] \overset{ZT}{\Rightarrow} Y(z)(1 - z^{-1}) = X(z)(1 + z^{-1}) \Rightarrow$$

$$H(z) = \frac{Y(z)}{X(z)} = \frac{1 + z^{-1}}{1 - z^{-1}} \tag{1}$$

With the realization of the system as a direct form II, only one delay element is used as illustrated next:

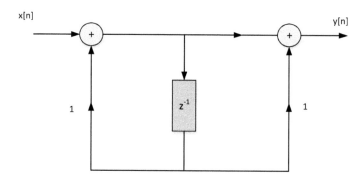

b.   The z-transform of the input $x[n] = u[n-2]$ is: $X(z) = z^{-2}U(z) = z^{-2}\dfrac{z}{z-1} = \dfrac{z^{-1}}{z-1} \tag{2}$

But $Y(z) = H(z)X(z) = \dfrac{z^{-1}(1 + z^{-1})}{(z-1)(1-z^{-1})} = \dfrac{z+1}{z(z-1)^2} \tag{3}$

It is: $\dfrac{Y(z)}{z} = \dfrac{z+1}{z^2(z-1)^2} = \dfrac{k_{11}}{z} + \dfrac{k_{11}}{z^2} + \dfrac{k_{11}}{z-1} + \dfrac{k_{22}}{(z-1)^2} \tag{4}$

We compute the fraction coefficients:

$$k_{11} = 3 \ , \ k_{12} = 1 \ , \ k_{21} = -3 \ , \ k_{22} = 2$$

thus:

$$Y(z) = 3 + \frac{1}{z} - \frac{3z}{z-1} + \frac{2z}{(z-1)^2} \overset{IZT}{\Rightarrow} y[n] = \delta[n] + \delta[n-1] - u[n] + nu[n] \Rightarrow$$

$$y[n] = 3\delta([n] + \delta[n-1] + (2n-3)u[n]$$ (5)

4.5.5 A discrete-time system is described by the input-output relation:

$$y[n] = x[n] + y[n-1] - 0.1y[n-2]$$

a. Develop the system as a parallel connection of the systems H1(z) and H2(z). Then, give its schematic representation.
b. Find and plot the impulse response of the system. Which pole determines the response of the system?

SOLUTION:
a. $y[n] = x[n] + y[n-1] - 0.1y[n-2]$ (1)

From the difference equation (1) of the discrete system, the transfer function is calculated in the form of equation (2), and the implementation diagram in a parallel form is designed.

$$(1) \Rightarrow H(z) = \frac{Y(z)}{X(z)} = \frac{1}{1 - z^{-1} + 0.1z^{-2}} = \frac{z^2}{z^2 - z + 0.1} = \frac{z^2}{(z - 0.887)(z - 0.113)} \Rightarrow$$

$$H(z) = \underbrace{\frac{1.146z}{z - 0.887}}_{H_1(z)} - \underbrace{\frac{0.146z}{z - 0.113}}_{H_2(z)}$$ (2)

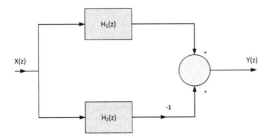

b. The impulse response of the system is calculated:

$$h(n) = IZT\{H(z)\} = \left[1.146(0.887)^n - 0.146(0.113)^n\right]u[n] \tag{3}$$

$$(3) \Rightarrow h[n] = h_1[n] - h_2[n] \tag{4}$$

The impulse response is plotted next:

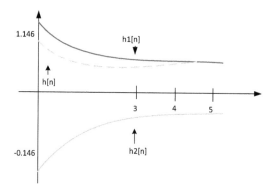

We notice that the term $h_2[n]$ has a much smaller contribution (in terms of values) to $h[n]$ and is practically zero for $n \geq 4$. Thus, the impulse response is approximated by:

$$h[n] \cong h_1[n] = 1.146(0.887)^n u[n] \quad , \quad n \geq 4$$

4.5.6 The discrete-time systems S1 and S2 are described in the following figures.
   a. Calculate the discrete impulse response of the first system.
   b. Compute the coefficients K1, K2, K3 and K4 of the second system, so as to have the same impulse response as the first.
   c. Give other forms of realization and compare each form with the others.
   d. Are the systems stable?

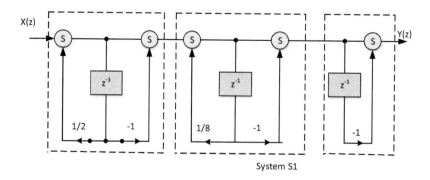

System S1

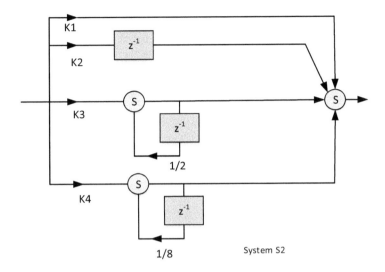

1/8   System S2

SOLUTION:

a. For S1 (implemented in series), the total transfer function is:

$$H(z) = H_1(z)H_2(z)H_3(z) = \frac{1-z^{-1}}{1-\frac{1}{2}z^{-1}} \frac{1-z^{-1}}{1-\frac{1}{8}z^{-1}} \frac{1-z^{-1}}{1} = \frac{(1-z^{-1})^3}{\left(1-\frac{1}{2}z^{-1}\right)\left(1-\frac{1}{8}z^{-1}\right)} \tag{1}$$

$$\frac{H(z)}{z} \overset{(1)}{=} \frac{(z-1)^3}{z^2\left(z-\frac{1}{2}\right)\left(z-\frac{1}{8}\right)} = \frac{A}{z^2} + \frac{B}{z} + \frac{C}{z-\frac{1}{2}} + \frac{D}{z-\frac{1}{8}} \tag{2}$$

$$A = \lim_{z \to 0} \frac{(z-1)^3}{\left(z-\frac{1}{2}\right)\left(z-\frac{1}{8}\right)} = -16, \quad B = \lim_{z \to 0} \frac{d}{dz}\left(\frac{(z-1)^3}{\left(z-\frac{1}{2}\right)\left(z-\frac{1}{8}\right)}\right) = -112$$

$$C = \lim_{z \to \frac{1}{2}} \frac{(z-1)^3}{z^2\left(z-\frac{1}{8}\right)} = -\frac{4}{3}, \quad D = \lim_{z \to \frac{1}{8}} \left(\frac{(z-1)^3}{z^2\left(z-\frac{1}{2}\right)}\right) = \frac{343}{3}$$

Thus,

$$H(z) \overset{(2)}{=} -\frac{16}{z} - 112 - \frac{\frac{4}{3}z}{z-\frac{1}{2}} + \frac{\frac{343}{3}z}{z-\frac{1}{8}} \tag{3}$$

The discrete impulse response of the first system is given by relation (4).

$$h[n] = IZT\left[H(z)\right]^{(3)} = -112\delta[n] - 16\delta([n-1] - \frac{4}{3}\left(\frac{1}{2}\right)^{n}u[n] + \frac{343}{3}\left(\frac{1}{8}\right)^{n}u[n] \qquad (4)$$

b. For S2 (implemented in parallel), we can see that:

$$\frac{Y(z)}{X(z)} = k_1 + k_2 z^{-1} + k_3 \frac{1}{1 - \frac{1}{2}z^{-1}} + k_4 \frac{1}{1 - \frac{1}{8}z^{-1}} \qquad (5)$$

Thus $k_1 = -112$, $k_2 = -16$, $k_3 = -4/3$, $k_3 = -343/3$

c. S1: Implementation in direct form II (Canonical form)

$$H(z) = \frac{1 - 3z^{-1} + 3z^{-2} - z^{-3}}{1 - \frac{5}{8}z^{-1} + \frac{1}{16}z^{-2}} = \frac{Y(z)}{X(z)} \qquad (6)$$

Based on the relation (6), the implementation of the discrete system in canonical form is presented in the following figure:

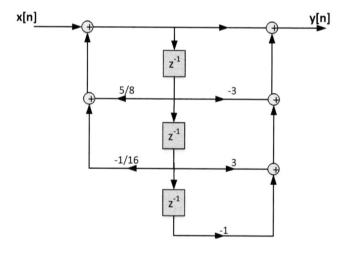

S1- Implementation in direct form I

$$(6) \Rightarrow \left(1 - \frac{5}{8}z^{-1} + \frac{1}{16}z^{-2}\right)Y(z) = (1 - 3z^{-1} + 3z^{-2} - z^{-3})X(z) \Rightarrow$$

$$y[n] = x[n] - 3x[n-1] + 3x[n-2] - x[n-3] + \frac{5}{8}y[n-1] - \frac{1}{16}y[n-2] \qquad (7)$$

Based on the relation (7), the implementation of the discrete system in direct form I is presented in the following figure:

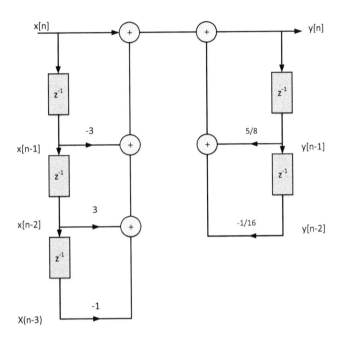

A comparison Table 4.1 follows, which contains the number of memory elements, adders and multipliers for the four different implementations.

d. The poles of the systems are included in the unit circle, so the systems are stable.

TABLE 4.1

Comparison List

Comparison	In Parallel	In Series	Direct Form I	Direct Form II
Memory elements	3	3	5	3
Adders	3	5	5	7
Multipliers	6	5	5	5

4.5.7 The transfer function of a digital controller is given by

$$G(z) = \frac{U(z)}{E(z)} = \frac{5(1+0.25z)}{(1-0.5z)(1-0.1z)}$$

Realize the above transfer function as a direct form, in series and in parallel.

SOLUTION:

$$G(z) = \frac{U(z)}{E(z)} = \frac{5(1+0.2z)}{(1-0.5z)(1-0.1z)} = \frac{5z^{-1}(z^{-1}+0.2)}{(z^{-1}-0.5)(z^{-1}-0.1)} = \frac{5z^{-2}+z^{-1}}{z^{-2}-0.6z^{-1}+0.05} \Rightarrow$$

$$G(z) = \frac{20z^{-1}+100z^{-2}}{1-12z^{-1}+20z^{-2}} \tag{1}$$

$(1) \Rightarrow 0.05U(z) - 0.6z^{-1}U(z) + z^{-2}U(z) = 5z^{-2}E(z) + z^{-1}E(z) \Rightarrow$

$0.05u[n] = 5e[n-2] + e[n-1] + 0.6u[n-1] - u[n-2] \Rightarrow$

$u[n] = 100e[n-2] + 20e[n-1] + 12u[n-1] - 20u[n-2]$ \tag{2}

Based on relation (2), the realization of the discrete system in direct form I is designed. For the design of the realization of the discrete system in cascade form, relation (3) is used, whereas relation (4) is used for the parallel form design.

$$\frac{U(z)}{E(z)} = \frac{z^{-1}}{z^{-1}-0.5} \cdot \frac{1+5z^{-1}}{z^{-1}-0.1} = \frac{2z^{-1}}{1-2z^{-1}} \cdot \frac{10+50z^{-1}}{1-10z^{-1}} \tag{3}$$

$$\frac{U(z)}{E(z)} = \frac{z+5}{(1-0.5z)(1-0.2z)} = \frac{-17.5z^{-1}}{1-2z^{-1}} + \frac{37.5z^{-1}}{1-10z^{-1}} \tag{4}$$

The realization in direct form I will be:

e[n]

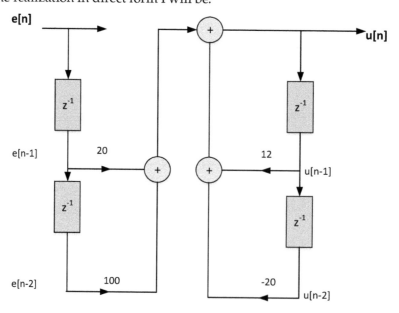

The realization in canonical form will be:

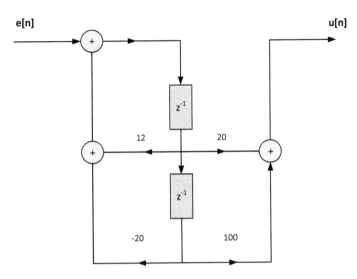

The realization in cascade form will be:

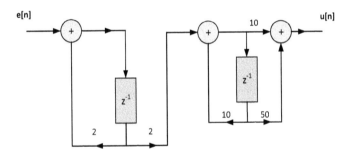

The realization in parallel form will be:

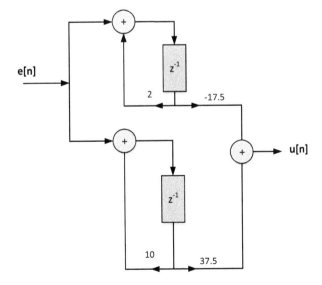

4.5.8 Derive the difference equation for the system described by the following SFG:

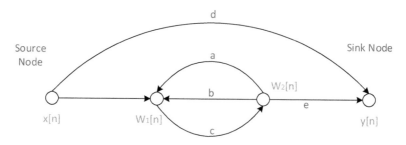

SOLUTION: The SFG results in the following relations:

$$w_1[n] = x[n] + aw_2[n] + bw_2[n] \tag{1}$$

$$w_2[n] = cw_1[n] \tag{2}$$

$$(1),(2) \Rightarrow w_2[n] = cx[n] + acw_2[n] + bcw_2[n] \Rightarrow$$

$$w_2[n] = \frac{cx[n]}{1 - ac - bc} \tag{3}$$

$$\text{But } y[n] = dx[n] + ew_2[n] \tag{4}$$

Hence, the difference equation of the system is given by equation (5).

$$(3),(4) \Rightarrow y[n] = \left(d + \frac{ce}{1 - ac - bc}\right)x[n] \tag{5}$$

4.5.9 Plot the SFG and derive the difference equation of the system described by the following block diagram.

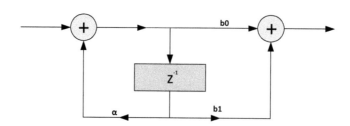

SOLUTION: The SFG is easily plotted using the block diagram of the system.

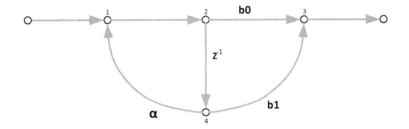

Then, from the SFG, the following relations are derived:

$$w_1[n] = x[n] + aw_4[n] \tag{1}$$

$$w_2[n] = w_1[n] \tag{2}$$

$$w_3[n] = b_0 w_2[n] + b_1 w_4[n] \tag{3}$$

$$w_4[n] = w_2[n-1] \tag{4}$$

$$y[n] = w_3[n] \tag{5}$$

$$(5) \Rightarrow y[n] = w_3[n] \overset{(3)}{=} b_0 w_2[n] + b_1 w_2[n-1] \tag{6}$$

$$(2) \Rightarrow w_2[n] = w_1[n] \overset{(1)}{=} x[n] + aw_2[n-1] \tag{7}$$

In z- plane it is:

$$Y(z) = (b_0 + b_1 z^{-1}) W_2(z) \tag{8}$$

$$W_2(z) = X(z) + az^{-1} W_2(z) = \frac{X(z)}{1 - az^{-1}} \tag{9}$$

$$(8),(9) \Rightarrow Y(z) = \frac{(b_0 + b_1 z^{-1})}{1 - az^{-1}} X(z) \tag{10}$$

Thus, the difference equation is given by equation (11):

$$y[n] = ay[n-1] + b_0 x[n] + b_1 x[n-1] \tag{11}$$

4.5.10 Calculate the transfer function of the discrete-time system described by the following SFG:

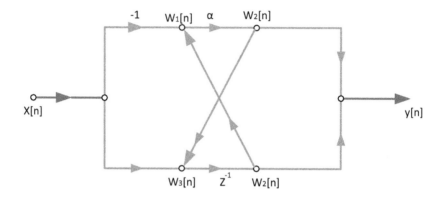

SOLUTION: First suggested solution: Derivation of the mathematical model describing the system.

From the SFG we get:

$$w_1[n] = w_4[n] - x[n] \tag{1}$$

$$w_2[n] = \alpha w_1[n] \tag{2}$$

$$w_3[n] = w_2[n] + x[n] \tag{3}$$

$$w_4[n] = w_3[n-1] \tag{4}$$

$$y[n] = w_2[n] + w_4[n] \tag{5}$$

Assuming zero initial conditions we z-transform the mathematical model:

$$W_1(z) = W_4(z) - X(z) \tag{6}$$

$$W_2(z) = \alpha W_1(z) \tag{7}$$

$$W_3(z) = W_2(z) + X(z) \tag{8}$$

$$W_4(z) = W_3(z)z^{-1} \tag{9}$$

$$Y(z) = W_2(z) + W_4(z) \tag{10}$$

$$W_2(z) \overset{(6)}{=} \alpha W_4(z) - \alpha X(z) \tag{11}$$

$$W_4(z) \overset{(8)}{=} z^{-1}W_2(z) + z^{-1}X(z) \tag{12}$$

$$W_2(z) \overset{(12)}{=} \alpha z^{-1}W_2(z) + \alpha z^{-1}X(z) - \alpha X(z) \Rightarrow$$
$$W_2(z) = \frac{\alpha z^{-1}X(z) - \alpha X(z)}{1 - \alpha z^{-1}} \tag{13}$$

$$W_4(z) = \alpha z^{-1}W_4(z) - \alpha z^{-1}X(z) + z^{-1}X(z) \Rightarrow$$
$$W_4(z) = \frac{z^{-1}X(z) - \alpha z^{-1}X(z)}{1 - \alpha z^{-1}} \tag{14}$$

Therefore, from relation (14) the transfer function of the discrete system is derived:

$$H(z) = \frac{Y(z)}{X(z)} = \frac{z^{-1} - \alpha}{1 - \alpha z^{-1}} \tag{15}$$

Then, the difference equation that describes the system is:

$$h[n] = \alpha^{n-1}u[n-1] - \alpha^{n+1}u[n] \tag{16}$$

Second suggested solution: Using Mason's gain formula

It is true that: $H(z) = \dfrac{\displaystyle\sum_{n=1}^{k} T_n D_n}{D} = \dfrac{T_1 D_1 + T_2 D_2 + T_3 D_3 + T_4 D_4}{D} \tag{17}$

where: T_n is the gain of the n-th forward path that connects the input with the output. In the SFG of this problem, there are four paths.

$$T_1 = -1 \cdot \alpha, \ T_2 = z^{-1}, \ T_3 = -1 \cdot \alpha \cdot z^{-1}, \ T_4 = \alpha \cdot z^{-1} \tag{18}$$

D is the determinant of the diagram and is given by

$$D = 1 - \sum L_1 + \sum L_2 - \sum L_3 = 1 - az^{-1} \tag{19}$$

D_n is the subdeterminant of T_n path and is calculated from (19), not taking into account the loops touching the n-th forward path. Obviously $D_n = 1$.

Thus, the transfer function of the discrete system using Mason's formula would be:

$$(17)\overset{(18)}{\underset{(19)}{\Longrightarrow}} H(z) = \frac{-1 \cdot \alpha + z^{-1} + (-1 \cdot \alpha \cdot z^{-1}) + \alpha \cdot z^{-1}}{1 - az^{-1}} = \frac{-\alpha + z^{-1}}{1 - az^{-1}} \tag{20}$$

The result is the same with the one derived at the first suggested solution.

4.5.11 a. Implement the causal filter with transfer function $H_1(z) = \dfrac{1 - z^{-1}}{1 - 0.4z^{-3}}$ with both IIR and FIR digital filter.

b.   Implement the causal filter with transfer function

$$H_3(z) = H_1(z) \cdot H_2(z) = \frac{(1 - z^{-1})(1 - z^{-1} - 1.6z^{-2})}{(1 - 0.4z^{-3})(1 - 0.8z^{-1})}$$

SOLUTION:

a.   Realization with IIR digital filter: $H_1(z) = \dfrac{Y_1(z)}{X_1(z)} = \dfrac{1 - z^{-1}}{1 - 0.4z^{-3}}$ $\tag{1}$

The above filter is stable as all of its poles ($\alpha = 0.4$) are included in the unit circle. Thus, from equation (1), the difference equation is obtained as follows. We first convert it to: Y1(z)·(1 − 0.4z–3) = X1(z)·(1–z–1) ⇒Y1(z) − 0.4Y1(z)z–3 = X1(z) − X1(z)z–1

Since the terms of the form $Y(z)z^{-i}$ correspond to the inverse z-transform of the form $y[n - 1]$, we can write:

$$y_1[n] - 0.4y_1[n - 3] = x_1[n] - x_1[n - 1] \Rightarrow y_1[n] = 0.4y_1[n - 3] + x_1[n] - x_1[n - 1] \tag{2}$$

From equation (2), the filter desired can be implemented in canonical form:

Direct form II-Canonical form

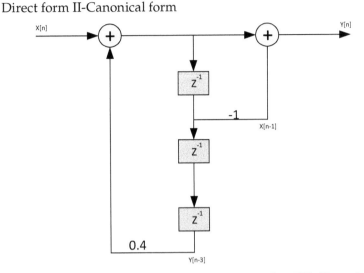

Realization with FIR digital filter: i.e., realization with a FIR filter of the form:

$$y[n] = \sum_{i=0}^{9} h[i]x[n-i]$$

We only consider the ten most important terms. We write the transfer function of the filter to the following form:

$$H_1(z) = \frac{1-z^{-1}}{1-0.4z^{-3}} = \frac{z^3 - z^2}{z^3 - 0.4} \tag{3}$$

By dividing (using only the first ten most important terms) the polynomial of the numerator and the polynomial of the denominator, the following is derived:

$$H_1(z) = 1 - z^{-1} + 0.4z^{-3} - 0.4z^{-4} + 0.16z^{-6} - 0.16z^{-7} + 0.064z^{-9} \Rightarrow$$

$$H_1(z) = \frac{Y_1(z)}{X_1(z)} = 1 - z^{-1} + 0.4z^{-3} - 0.4z^{-4} + 0.16z^{-6} - 0.16z^{-7} + 0.064z^{-9} \Rightarrow$$

$$Y_1(z) = X_1(z) - z^{-1}X_1(z) + 0.4z^{-3}X_1(z) - 0.4z^{-4}X_1(z) + 0.16z^{-6}X_1(z)$$
$$-0.16z^{-7}X_1(z) + 0.064z^{-9}X_1(z)$$

Since the terms of the form $Y(z)z^{-i}$ correspond to an inverse z-transform of the form $y[n-i]$, we obtain the corresponding difference equation:

$$y_1[n] = x_1[n] - x_1[n-1] + 0.4x_1[n-3] - 0.4x_1[n-4] + 0.16x_1[n-6]$$

$$-0.16x_1[n-7] + 0.064x_1[n-9]$$

The realization of the filter is illustrated in the following figure. By comparing
the two different ways of implementation of the filter (IIR and FIR), one can easily
notice that the IIR filter is much simpler (and consequently cheaper) than the FIR
filter.

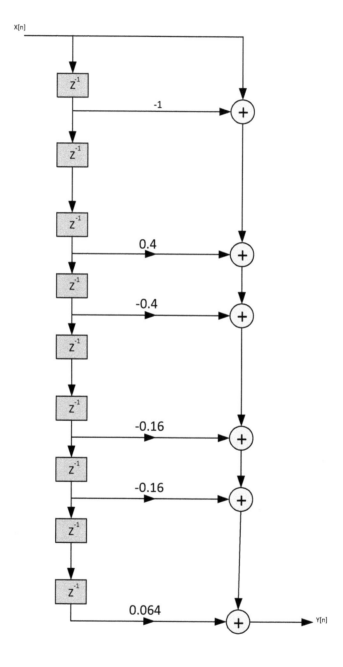

b. $H_3(z) = H_1(z) \cdot H_2(z) = \dfrac{(1-z^{-1})(1-z^{-1}-1.6z^{-2})}{(1-0.4z^{-3})(1-0.8z^{-1})} = \dfrac{1-2z^{-1}-0.6z^{-2}+1.6z^{-3}}{1-0.8z^{-1}-0.4z^{-3}+0.32z^{-4}}$

Since, $H_3(z) = \dfrac{Y_3(z)}{X_3(z)} \Rightarrow$

$Y_3(z)(1-0.8z^{-1}-0.4z^{-3}+0.32z^{-4}) = X_3(z)(1-2z^{-1}-0.6z^{-2}+1.6z^{-3}) \Rightarrow$

(and after applying inverse z-transform)

$$y_3[n] = x_3[n] - 2 \cdot x_3[n-1] - 0.6x_3[n-2] + 1.6x_3[n-3] + 0.8 \cdot y_3[n-1]$$
$$+ 0.4y_3[n-3] - 0.32y_3[n-4]$$

The realization is illustrated in the next figure:

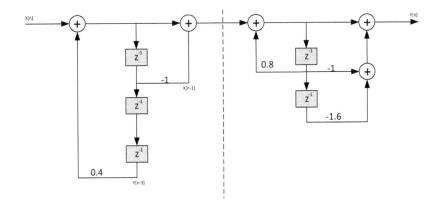

5

Frequency Domain Analysis

5.1 Introduction

Fourier analysis is an important tool for extracting significant information of the frequency domain of a signal that is not apparent in its representation in the time domain. It is used for signal analysis, describing the spectral behavior of the signal. Since the spectral information is given within the frequency domain, the Fourier transform domain is called the frequency domain.

From the signal and the noise spectrum, it is possible to draw useful conclusions, such as:

1. The frequency range which the sinusoidal waveform components, composing the signal on the whole or most of it, are located in.

2. The frequency range which the sinusoidal waveform components, composing the noise of the signal are located in.

3. The type and source of noise.

4. The frequency range that can be cut off to reduce the presence of noise to the minimum possible, without significant distortion of the signal itself.

5. The minimum possible sampling frequency that does not introduce artifacts, according to the Nyquist theorem, in cases when sampling is implemented.

Each discrete-time signal can be represented as a sum of properly shifted impulse functions. In this section, an alternative representation of discrete-time signals will be studied as a combination of complex exponential sequences of the form $e^{-j\omega n}$.

The following are going to be analyzed:

- The Discrete-Time Fourier Transform (DTFT), which is applied to non-periodic discrete-time signals.

- The Discrete Fourier Series (DFS), which is applied to periodic sequences.

- The Discrete Fourier Transform (DFT), which is applied to non-periodic discrete-time signals of finite length.

- The Fast Fourier Transform (FFT), which is an algorithm for calculating the DFT.
- The Discrete Cosine Transform (DCT).
- The Discrete Wavelets Transform (DWT).

5.2 Discrete-Time Fourier Transform (DTFT)

The discrete-time Fourier transform of a discrete-time signal, $x(n)$, is the representation of this signal as a combination of complex exponential sequences of the form $e^{-j\omega n}$, where ω is the angular frequency in rad/sec. The original signal can be computed when its DTFT is given, by means of the inverse discrete-time Fourier transform.

The DTFT of a sequence $x[n]$ is defined as:

$$X\left(e^{j\omega}\right) = F\{x[n]\} = \sum_{n=-\infty}^{\infty} x[n]e^{-j\omega n} \tag{5.1}$$

whereas the inverse DTFT is defined as

$$x[n] = F^{-1}\left\{X\left(e^{j\omega}\right)\right\} = \frac{1}{2\pi}\int_{-\pi}^{\pi} X\left(e^{j\omega}\right)e^{j\omega n}\,d\omega \tag{5.2}$$

Equations 5.1 and 5.2 form the pair of discrete-time Fourier transforms. Equation 5.1 is called an analysis equation and expresses the analysis of the discrete-time signal, $x[n]$, in exponential signals $e^{-j\omega n}$ that extend into a continuous range of angular frequencies ω limited in the interval $0 \leq \omega < 2\pi$, while Equation 5.2 is called a synthesis equation. The sufficient condition for the existence of DTFT is for $x[n]$ is an absolutely summable sequence, i.e., $\sum_{n=-\infty}^{\infty} |x[n]| < \infty$.

Table 5.1 includes the DTFT of basic signals, while Table 5.2 includes some of the properties of the DTFT, which are used to analyze discrete-time signals and systems.

Parseval's theorem refers to the conservation of energy during the transition from the time domain to the frequency domain. The quantity $|X(e^{j\omega})|^2$ is called energy-density spectrum of the discrete signal, $x[n]$, and describes the distribution of the energy of the discrete signal, $x[n]$, over a frequency range.

TABLE 5.1

DTFT of Fundamental Signals

Signal	DTFT
$\delta[n]$	$1, -\infty < \omega < \infty$
$\delta[n-N]$	$e^{-j\omega N}, N = \pm 1, \pm 2, \dots$
$u[n]$	$\dfrac{1}{1-e^{j\omega}} + \pi \displaystyle\sum_{k=-\infty}^{\infty} \delta(\omega - 2\pi k)$
$\alpha^n u[n],\ \lvert \alpha \rvert < 1$	$\dfrac{1}{1-ae^{-j\omega}}$
$\sin(\omega_0 n)$	$\dfrac{\pi}{j} \displaystyle\sum_{k=-\infty}^{\infty} \left[\delta(\omega - \omega_0 + 2\pi k) - \delta(\omega + \omega_0 + 2\pi k)\right]$
$\sin(\omega_0 n + \theta)$	$\displaystyle\sum_{k=-\infty}^{\infty} j\pi \left[e^{-j\theta}\delta(\omega + \omega_0 - 2\pi k) - e^{j\theta}\delta(\omega - \omega_0 - 2\pi k)\right]$
$\cos(\omega_0 n)$	$\pi \displaystyle\sum_{k=-\infty}^{\infty} \left[\delta(\omega - \omega_0 + 2\pi k) + \delta(\omega + \omega_0 + 2\pi k)\right]$
$\cos(\omega_0 n + \theta)$	$\displaystyle\sum_{k=-\infty}^{\infty} \pi \left[e^{-j\theta}\delta(\omega + \omega_0 - 2\pi k) + e^{j\theta}\delta(\omega - \omega_0 - 2\pi k)\right]$
$u[n] - u[n-M]$	$\dfrac{\sin(\omega M/2)}{\sin(\omega/2)} e^{-j(M-1)/2}$
$\text{sgn}[n] = \begin{cases} 1 & n = 0,1,2,\dots \\ 1 & n = -1,-2,\dots \end{cases}$	$\dfrac{2}{1-e^{-j\Omega}}$
$\dfrac{B}{\pi}\,\text{sinc}\left(Bn/\pi\right)$	$\displaystyle\sum_{k=-\infty}^{\infty} p_{2B}(\omega + 2\pi k)$
	Where $p_L[n] = p_{2q+1}[n] \leftrightarrow \dfrac{\sin\left[\left(q+\frac{1}{2}\right)\omega\right]}{\sin\left(\omega/2\right)}$
	(Rect. Pulse)

TABLE 5.2

Properties of DTFT

Property	Non Periodic Signal	DTFT
	$\left.\begin{array}{c} x_1[n] \\ x_2[n] \end{array}\right\}$	$\left.\begin{array}{c} x_1(e^{j\omega}) \\ x_2(e^{j\omega}) \end{array}\right\}$ periodic functions with period of 2π
Linearity	$ax[n] + by[n]$	$aX(e^{j\omega}) + bY(e^{j\omega})$
Time shifting	$x[n - n_0]$	$e^{-j\omega n 0}\,X(e^{j\omega})$
Frequency shifting	$e^{j\omega 0 n}x[n]$	$X(e^{j\omega - \omega 0})$
Conjugate sequence	$x^*[n]$	$X^*(e^{-j\omega})$
Time reversal	$x[-n]$	$X(e^{j\omega})$
Time expansion	$x(n/k)$ for n multiple of k	$X(e^{jk\omega})$
Convolution	$x_1[n] * x_2[n]$	$X_1(e^{j\omega})\,X_2(e^{j\omega})$
Multiplication	$x_1[n]x_2[n]$	$\dfrac{1}{2\pi} \displaystyle\int_{2\pi} X_1(e^{j\theta})X_2(e^{j(\omega-\theta)})d\theta$
Derivative in frequency	$nx[n]$	$j\dfrac{dX(e^{j\omega})}{d\omega}$
Conjugate symmetry for real signals	$x[n]$ real	$\left\{\begin{array}{l} X(e^{j\omega}) = X^*(e^{-j\omega}) \\ \mathrm{Re}\left\{X(e^{j\omega})\right\} = \mathrm{Re}\left\{X(e^{-j\omega})\right\} \\ \mathrm{Im}\left\{X(e^{j\omega})\right\} = \mathrm{Im}\left\{X(e^{-j\omega})\right\} \\ \left|X(e^{j\omega})\right| = \left|X(e^{-j\omega})\right| \\ \angle X(e^{j\omega}) = -\angle X(e^{-j\omega}) \end{array}\right.$
Symmetry for real and even signals	$x[n]$ real and even	$X(e^{j\omega})$ real and even
Symmetry for real and odd signals	$x[n]$ real and odd	$X(e^{j\omega})$ imaginary and odd
Parseval's Theorem		$\displaystyle\sum_{n=-\infty}^{\infty} |x(n)|^2 = \dfrac{1}{2\pi}\int_{2\pi} |X(e^{j\omega})|^2\,d\omega$

5.3 Discrete Fourier Series (DFS)

Let the periodic sequence $\tilde{x}[n]$ with period N. Obviously it is true that $\tilde{x}[n] = \tilde{x}[n + rN]$ (where n, r are integers). As in the case of continuous-time signals, a periodic sequence can be represented by a Fourier series, that is, by a sum of complex exponential terms whose frequency is an integer multiple of the fundamental frequency, $2\pi/N$, as shown in the following relations:

$$\textit{Synthesis Equation}: \tilde{x}[n] = \frac{1}{N}\sum_{k=0}^{N-1} \tilde{X}[k]e^{j2\pi\frac{k}{N}n} \tag{5.4}$$

Unlike the case of continuous time signals, the Fourier series representation does not require an infinite number of harmonics (multiples of the fundamental frequency), but only N complex exponential terms. The reason is that the complex exponential terms are periodic to N:

$$e_{k+lN}\{n\} = e^{j2\pi\frac{k+lN}{N}n} = e^{j2\pi\frac{k}{N}n}e^{j2\pi\frac{lN}{N}n} = e^{j2\pi\frac{k}{N}n}e^{j2\pi ln} = e^{j2\pi\frac{k}{N}n} = e_k\{n\}$$

where l is an integer.

The coefficients of the Discrete Fourier series are given by:

$$Analysis\ Equation: \tilde{X}[k] = \sum_{n=0}^{N-1}\tilde{x}[n]e^{j2\pi\frac{k}{N}n} \tag{5.5}$$

The pair of Equations 5.4 and 5.5 define the Discrete Time Fourier Series of the discrete-time periodic signal, $\tilde{x}[n]$. The coefficients $\tilde{X}[k]$ are called Fourier coefficients or spectral lines.

By designating

$$W_N = e^{-j\left(\frac{2\pi}{N}\right)} \tag{5.6}$$

the following relations are obtained:

$$\tilde{x}[n] = \frac{1}{N}\sum_{k=0}^{N-1}\tilde{X}[k]\ W_N^{-nk} \quad n = 0,1,\dots,N-1 \tag{5.7}$$

$$\tilde{X}[k] = \sum_{n=0}^{N-1}\tilde{x}[n]\ W_N^{nk} \quad k = 0,1,\dots,N-1 \tag{5.8}$$

5.3.1 Periodic Convolution

Let the periodic sequences $\tilde{x}_1[n]$ and $\tilde{x}_2[n]$ with a period, N, and with discrete Fourier series components $\tilde{X}_1[k]$ and $\tilde{X}_2[k]$, respectively.

The sequence:

$$\tilde{x}[n] = \sum_{m=0}^{N-1}\tilde{x}_1[m]\tilde{x}_2[n-m] \tag{5.9}$$

is the periodic convolution of $\tilde{x}_1[n]$ and $\tilde{x}_2[n]$, and the components of its discrete Fourier series are given by the following relation:

$$\tilde{X}[k] = \tilde{X}_1[k]\tilde{X}_2[k] \tag{5.10}$$

5.3.2 The Relation of the DFS Components and the DTFT over a Period

Consider the periodic sequence, $\tilde{x}[n]$, with the components of the Discrete Fourier Series represented as $\tilde{X}[k]$ and the Discrete Time Fourier Transform over a period of the sequence $\tilde{x}[n]$, represented as $X(e^{j\omega})$. In this case, the following relation is true:

$$\tilde{X}[k] = X(e^{j\omega})\Big|_{\omega=2\pi\frac{k}{N}} \tag{5.11}$$

5.4 Discrete Fourier Transform

The Discrete Fourier Transform (DFT) is used for the frequency representation of finite-time signals. In fact, what is sought is the calculation of the DTFT, but what can actually be calculated is the Discrete Fourier Transform (since DTFT is a continuous function of the discrete frequency ω). Since the DFT is obtained by sampling in the DTFT frequency domain for finite-length signals, the sampling effects must be made completely clear. The Discrete Fourier Transform, apart from its importance for the representation of discrete-time sequences in the frequency domain, is the basis of many applications of signal processing, such as spectrum analysis and data encoding (JPEG, MP3, MPEG video).

Additionally, in many cases (such as in spectrum analysis and filtering applications), the signals are not of finite lengths, and the DFT implementation requires the use of a time window, the processing of frames, or the use of an alternative DFT form suitable for time variable signals, which is called *Time Dependent Discrete Fourier Transform*.

The Discrete Fourier Transform (DFT) is an alternative form of the Fourier transform in which the finite length sequences are transformed into other sequences, which represent samples in the frequency domain and are defined by the relations:

$$X[k] = \sum_{n=0}^{N-1} x[n] e^{-j2\pi\frac{k}{N}n} = \sum_{n=0}^{N-1} x[n] W_N^{kn} \tag{5.12}$$

$$x[n] = \frac{1}{N}\sum_{k=0}^{N-1} X[k] e^{j2\pi\frac{k}{N}n} = \frac{1}{N}\sum_{n=0}^{N-1} x[n] W_N^{-kn} \tag{5.13}$$

where $W_N = e^{-j\frac{2\pi}{N}}$. The factors W_N^{kn} are called twiddle factors and are the complex sequences on which the DFT is based. Note that the following is true:

$$W_N^N = e^{-j\frac{2\pi}{N}N} = e^{-j2\pi} = 1 = W_N^0 \tag{5.14}$$

Generally, it is true:

$$W_N^{N+A} = W_N^A \tag{5.15}$$

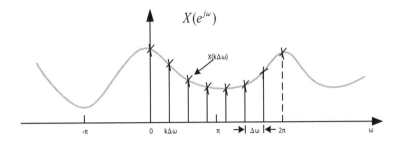

FIGURE 5.1
Sampling at the DTFT frequency.

The procedure of sampling at the DTFT frequency is illustrated in Figure 5.1. The DFT of Equation 5.12 in the form of a system of algebraic equations is:

$$\left.\begin{aligned} X[0] &= x[0]W_N^0 + x[1]W_N^0 + x[2]W_N^0 + ... \\ X[1] &= x[0]W_N^0 + x[1]W_N^1 + x[2]W_N^2 + ... \\ X[2] &= x[0]W_N^0 + x[1]W_N^2 + x[2]W_N^4 + ... \end{aligned}\right\} \tag{5.16}$$

For example, in the case of $N = 3$ samples the DFT matrix is:

$$\begin{bmatrix} X[0] \\ X[1] \\ X[2] \end{bmatrix} = \begin{bmatrix} W_3^0 & W_3^0 & W_3^0 \\ W_3^0 & W_3^1 & W_3^2 \\ W_3^0 & W_3^2 & W_3^4 \end{bmatrix} \begin{bmatrix} x[0] \\ x[1] \\ x[2] \end{bmatrix} \tag{5.17}$$

where $W_3^4 = W_3^{3+1} = W_3^1$. Substituting in Equation 5.17, the following is obtained:

$$\begin{bmatrix} X[0] \\ X[1] \\ X[2] \end{bmatrix} = \begin{bmatrix} 1 & 1 & 1 \\ 1 & 0.5 - j0.867 & -0.5 + j0.867 \\ 1 & -0.5 + j0.867 & 0.5 - j0.867 \end{bmatrix} \begin{bmatrix} x[0] \\ x[1] \\ x[2] \end{bmatrix} \tag{5.18}$$

A method of calculating the *N-point DFT* is based on factorizing Equation 5.12 to reduce complex multiplications.

$$\begin{aligned} X[k] &= x[0] + x[1]W_N^k + x[2]W_N^{2k} + x[3]W_N^{3k} + ... + x[N-1]W_N^{k(N-1)} \\ &= x[0] + W_N^k \left[x[1] + x[2]W_N^k + x[3]W_N^{2k} + ... + x[N-1]W_N^{k(N-2)} \right] \\ \Rightarrow X[k] &= x[0] + W_N^k \left[x[1] + W_N^k \left[x[2] + W_N^k \left[x[3] + ...W_N^k x[N-1] \right] \right] \right] \end{aligned} \tag{5.19}$$

The sequences $x[n]$ and $X[k]$ are periodic with a period N, that is,

$$x[n+N] = x[n] \forall n \tag{5.20}$$

$$X[k+N] = X[k] \forall k \tag{5.21}$$

5.4.1 Properties of the DFT

Some of the properties of the Discrete Fourier Transform are proportional to the corresponding properties of the Discrete-Time Fourier Transform, but there are other different properties, as well, due to the finite length of the sequences and their Discrete Transform.

5.4.1.1 Linearity

$$DFT\left[\alpha x_1[n] + b x_2[n]\right] = \alpha DFT\left[x_1[n]\right] + b DFT\left[x_2[n]\right] = \alpha X_1[k] + b X_2[k] \tag{5.22}$$

5.4.1.2 Circular Shift

To understand the circular shift, consider the finite sequence $x[n]$ ($n = 0, 1, ..., N-1$), which is set on the circumference of a circle clockwise and at equal intervals. Then, $x[n-k]$ is the clockwise rotation of the circle for k interval.

A circular shift of the sequence $x[n] = \{x[0], x[1], x[2], ..., x[N-1]\}$ to the right is mathematically stated as $x([n-1] \bmod N)$, and it denotes:

$$x([n-1]\bmod N) = \left\{x[N-1], x[0], x[1], ..., x[N-2]\right\} \tag{5.23}$$

Two circular shifts to the right, i.e., $x[n-2]$, create the sequence:

$$x([n-2]\bmod N) = \left\{x[N-2], x[N-1], x[0], x[1], ..., x[N-3]\right\} \tag{5.24}$$

N right shifts recover the original sequence, $x[n]$, thus,

$$x([n-N]\bmod N) = x[n] = \left\{x[0], x[1], ..., x[N-1]\right\} \tag{5.25}$$

DFT of the circular shift in time domain:

It is true that:

$$DFT\left[x[n-m] \bmod N\right] = W_N^{km} X[k] \tag{5.26}$$

Equation 5.26 indicates that the DFT of a circularly shifted sequence is of the same magnitude but of a different phase with the corresponding DFT of the original sequence.

DFT of circular shift in frequency domain:

It is true that:

$$DFT\left[W_N^{-\alpha n} x[n]\right] = X[k-\alpha]\bmod N \tag{5.27}$$

Equation 5.27 indicates that the multiplication of the sequence $x[n]$ with the exponential sequence $W_N^{-\alpha n}$ is equivalent to a circular shift by α units in the frequency domain.

5.4.1.3 Circular Convolution

Consider two sequences $x[n]$ and $h[n]$ of length N. $x[n] \otimes h[n]$ denotes their circular convolution and it is defined as follows:

$$x[n] \otimes h[n] = \sum_{k=0}^{N-1} x\big([n-k]\bmod N\big)h[k] = \sum_{k=0}^{N-1} x[k]h\big([n-k]\bmod N\big) \qquad (5.28)$$

When the two sequences are of different lengths, N_1 and N_2, respectively, then the smaller of the two is filled in with zeros so that both get to the same length, N, where $N = \max(N_1, N_2)$. The circular convolution is of length $M = \max (N1, N2)$.

The steps for the calculation of the circular convolution of two sequences are:

1. Circular reflection of one sequence
2. Circular shift of the reflected sequence
3. Element-by-element multiplication of the reflected and shifted sequence with the other sequence and summation of the products.

DFT of the circular convolution
The DFT of a circular convolution of two sequences of finite length is given by:

$$DFT\big[x[n] \otimes h[n]\big] = DFT\big[x[n]\big]DFT\big[h[n]\big] = X[k]H[k] = Y[k] \qquad (5.29)$$

Thus, the response of the discrete system will be:

$$y[n] = IDFT\big[X[k]\,H[k]\big] \qquad (5.30)$$

Equation 5.30 shows that the DFT can be used to calculate the linear convolution by increasing the length of the sequences, $x[n]$ and $h[n]$, to N (by adding zero elements to the end of each one) and performing circular convolution of the resulting sequences. The result is the same as the one obtained by the linear convolution of the original sequences. The illustration of the procedure of linear convolution with the use of DFT is shown in Figure 5.2.

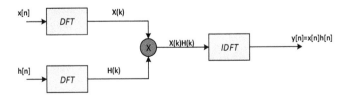

FIGURE 5.2
Linear convolution using DFT.

5.4.1.4 Multiplication of Sequences

$$DFT\left[x_1[n] \cdot x_2[n]\right] = \frac{1}{N} X_1[k] \otimes X_2[k] \tag{5.31}$$

Based on Equation 5.31, multiplication of two sequences in the time domain corresponds to the circular convolution of their DFTs.

5.4.1.5 Parseval's Theorem

It is true that:

$$\sum_{n=0}^{N-1} \left|x(n)\right|^2 = \frac{1}{N} \sum_{k=0}^{N-1} \left|X(k)\right|^2 \tag{5.32}$$

Parseval's theorem expresses the conservation of energy in the transition from the time domain to the frequency domain. That is, the sum of the squared values of the samples is equal to the average value of the squares of the spectral lines. The $|X(k)|^2$ representation is called power spectrum of $x[n]$ and depends only on the spectrum magnitude and not on its phase.

In Table 5.3, all properties of the Discrete Fourier Transform are listed.

TABLE 5.3

DFT Properties

Property	Sequence of Length N	N Point DFT
	$x_1[n]$	$X_1[k]$
	$x_2[n]$	$X_2[k]$
Linearity	$a_1 x_1[n] + a_2 x_2[n]$	$a_1 X_1[k] + a_2 X_2[k]$
Circular shift in time	$x[<n - n_0>_N]$	$W^{kn}_0 X[k]$
Circular shift in frequency	$W^{-k}_0{}^n x[n]$	$X[<k - k_0>_N]$
Conjugate sequence	$x^*[n]$	$X^*[<-k>_N]$
Reflection in time	$x[<-n>_N]$	$X^*[k]$
Circular Convolution	$x_1[n] \otimes x_2[n]$	$X_1[k] X_2[k]$
Multiplication	$x_1[n] x_2[n]$	$\dfrac{1}{N} \displaystyle\sum_{m=0}^{N-1} X_1[m] X_2[<k-m>_N]$
Conjugate symmetry for real signals	$x[n]$ real	$\begin{cases} X[k] = X^*[<-k>_N] \\ \mathrm{Re}\{X[k]\} = \mathrm{Re}\{X[<-k>_N]\} \\ \mathrm{Im}\{X[k]\} = -\mathrm{Im}\{X[<-k>_N]\} \\ \|X[k]\| = \|X[<-k>_N]\| \\ \angle X[k] = -\angle X[<-k>_N] \end{cases}$
Parseval's Theorem	$\displaystyle\sum_{n=0}^{N-1} \|x[n]\|^2 = \frac{1}{N} \sum_{k=0}^{N-1} \|X[k]\|^2$	

5.5 Fast Fourier Transform

The Fast Fourier Transform (FFT) is an efficient algorithm for calculating DFT. In practice, there is not only one algorithm but a variety of algorithms that can achieve the same result. The differences among them are mainly in the number and type of operations, as well as in the size of the required computer memory.

When the sequence $x[n]$ is known, then the DFT $X[k]$ can be derived by using Equation 5.12 or the inverse by using Equation 5.13. In order to calculate the sum for a k value in Equation 5.12, N complex multiplications and $(N-1)$ complex additions are required. For N values of k, M complex operations are required where: $M = N[N + (N-1)] \approx N^2$ (that is, N^2 multiplications and $N \cdot (N-1)$ additions). The same is true for the IDFT calculation. For large N, there is a huge amount of complex operations, which creates many difficulties in the calculation of the DFT or IDFT of the $x[n]$ and $X[k]$ sequences, respectively.

These difficulties are diminished with the implementation of the Fast Fourier Transform and the Inverse Fast Fourier Transform (FFT and IFFT, respectively). The method to be presented next was introduced by Couley and Tukey (1965) and is based on the technique of breaking up the original DFT into smaller segments, which, in the end, are reunited to create the DFT.

The basic FFT method reduces the number of multiplications from N^2 to $\dfrac{N}{2}log_2N = \dfrac{N \cdot v}{2}$, where $N = 2^n$. For example, if $N = 1024 = 2^{10}$, then for the calculation of the DFT $N^2 = 2^{20} = 1.048.576$ complex multiplications are required, while the FFT requires only $\dfrac{N \cdot v}{2} = \dfrac{1024.10}{2} = 5120$ multiplications. When N is a power of 2, then the algorithm is called *radix - 2 FFT*.

5.5.1 FFT Equations

Let the sequence $x[n]$ have a length of N. It is divided into two segments of $\dfrac{N}{2}$ length each, containing even-indexed and odd-indexed samples, respectively, i.e.,

$$x[n] = x_e[n] + x_o[n] \tag{5.33}$$

where:

$$\left. \begin{array}{l} x_e[n] = \left\{ x[0], x[2], x[4], \ldots, x[N-2] \right\} \\ \text{or} \\ x_e[n] = x[2n] \quad , \quad n = 0,1,2,\ldots, \dfrac{N}{2} - 1 \end{array} \right\} \tag{5.34}$$

and

$$\left. \begin{array}{l} x_o[n] = \left\{ x[1], x[3], x[5], \ldots, x[N-1] \right\} \\ x_o[n] = x[2n+1] \quad , \quad n = 0,1,2,\ldots, \dfrac{N}{2} - 1 \end{array} \right\} \tag{5.35}$$

The DFT of $x[n]$ is: $DFT[x(n)] = \sum_{n=0}^{N-1} x[n]W_N^{kn}$, $k = 0,1,2,\ldots,N-1$, therefore, due to Equations 5.33, 5.34 and 5.35, it is:

$$X[k] = \sum_{n=0}^{\frac{N}{2}-1} x[2n]W_N^{2nk} + \sum_{n=0}^{\frac{N}{2}-1} x[2n+1]W_N^{(2n+1)k} \tag{5.36}$$

But, $W_N^2 = e^{-j\frac{2\pi}{N}2} = e^{-j\frac{4\pi}{N}}$ and $W_{\frac{N}{2}}^1 = e^{-j\frac{2\pi}{\frac{N}{2}}} = e^{-j\frac{4\pi}{N}}$ thus,

$$W_N^2 = W_{\frac{N}{2}}^1 \tag{5.37}$$

Hence, Equation 5.36 can be written as:

$$X[k] = \sum_{n=0}^{\frac{N}{2}-1} x_e[n]W_{\frac{N}{2}}^{nk} + W_N^k \sum_{n=0}^{\frac{N}{2}-1} x_o[n]W_{\frac{N}{2}}^{nk}, k = 0,1,2,\ldots,\frac{N}{2}-1 \tag{5.38}$$

Equation 5.38 shows that the DFT of $x[n]$ is equal to the sum of the DFT of length $N/2$ of the even-indexed elements of $x[n]$ and the DFT of length $N/2$ of the odd-indexed elements of $x[n]$. That is:

$$X[k] = X_e[k] + W_N^k X_o[k] \, , \, k = 0,1,2,\ldots,\frac{N}{2}-1 \tag{5.39}$$

where $X_e[k] = DFT[x_e[n]]$ and $X_o[k] = DFT[x_o[n]]$ of length $N/2$.

$X_e[k]$ and $X_o[k]$ are periodic functions with a period $\frac{N}{2}$, that is:

$$\left. \begin{array}{c} X_e[k] = X_e\left[k \pm \frac{N}{2}\right] \\ \text{and } X_o[k] = X_o\left[k \pm \frac{N}{2}\right] \end{array} \right\} \tag{5.40}$$

For example, $X_e\left[k + \frac{N}{2}\right] = \sum_{n=0}^{\frac{N}{2}-1} X_e[k]W_{\frac{N}{2}}^{n(k+\frac{N}{2})} = \sum_{n=0}^{\frac{N}{2}-1} X_e[k]W_{\frac{N}{2}}^{nk}W_{\frac{N}{2}}^{n\frac{N}{2}}$

But $W_{\frac{N}{2}}^{n\frac{N}{2}} = e^{-j\frac{2\pi n}{\frac{N}{2}}\cdot\frac{N}{2}} = e^{-j2\pi n} = 1 \Rightarrow X_e\left[k + \frac{N}{2}\right] = \sum_{n=0}^{\frac{N}{2}-1} X_e[k]W_{\frac{N}{2}}^{nk} = X_e[k]$

Hence, $X\left[k + \frac{N}{2}\right] = X_e[k] + W_N^{k+\frac{N}{2}} X_o[k], k = 0,1,2,\ldots,\frac{N}{2}-1$ (5.41)

But since $W_N^{k+\frac{N}{2}} = e^{\left(-j\frac{2\pi}{N}\right)\left(k+\frac{N}{2}\right)} = e^{-j\frac{2\pi k}{N}} e^{-j\pi} = -e^{-j\frac{2\pi k}{N}}$

$\Rightarrow W_N^{k+\frac{N}{2}} = -e^{-j\frac{2\pi}{N}k} = -W_N^k$ (5.42)

By combining Equations 5.41 and 5.42, the following relation is derived:

$$X\left[k + \frac{N}{2}\right] = X_e[k] - W_N^k X_o[k], \ k = 0,1,2,\ldots,\frac{N}{2}-1 \tag{5.43}$$

Equations 5.39 and 5.43 constitute *the fundamental equations of the radix-2 FFT algorithm.* The number of complex multiplications required for the DFT of $x[n]$ (by implementing only one stage of analysis) is: $\left(\frac{N}{2}\right)^2 + \left(\frac{N}{2}\right)^2 + \left(\frac{N}{2}\right) = \frac{N(N+1)}{2}$.

The process continues by breaking up the sequences $x_e[n]$ and $x_0[n]$ of length $\frac{N}{2}$ in sequences with odd-indexed and even-indexed samples as follows: $x_{ee}[n]$, $x_{eo}[n]$, $x_{oe}[n]$ and $x_{oo}[n]$ with $n = 0,1,\ldots,\frac{N}{4}-1$. In this second stage of segmentation, and based on Equations 5.41 and 5.42, the following relations are derived:

$$\left.\begin{array}{ll}X_e[k] = X_{ee}[k] + W_N^{2k} X_{eo}[k] & , \quad k = 0,1,2,\ldots,\frac{N}{4}-1 \\[2mm] X_o[k] = X_{oe}[k] + W_N^{2k} X_{oo}[k] & , \quad k = 0,1,2,\ldots,\frac{N}{4}-1\end{array}\right\} \tag{5.44}$$

and, respectively

$$\left.\begin{array}{ll}X_e\left[k + \frac{N}{4}\right] = X_{ee}[k] - W_N^{2k} X_{eo}[k] & , \quad k = 0,1,2,\ldots,\frac{N}{4}-1 \\[2mm] X_o\left[k + \frac{N}{4}\right] = X_{oe}[k] - W_N^{2k} X_{oo}[k] & , \quad k = 0,1,2,\ldots,\frac{N}{4}-1\end{array}\right\} \tag{5.45}$$

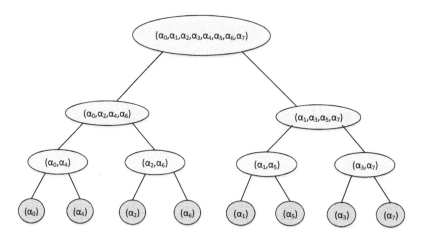

FIGURE 5.3
Iterative FFT algorithm.

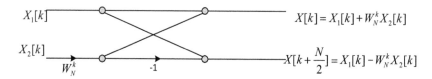

FIGURE 5.4
Two-point butterfly.

In this way, the segmentation process continues until the sequence reaches to the length of one sample (Figure 5.3). Since $N = 2^v$, the partition stages are v, and v stages require $v \cdot \dfrac{N}{2}$ multiplications.

This process can be illustrated by means of the so-called 'butterfly' graph of Figure 5.4. It is the depiction of the calculation of a two-point DFT.

For the case where the number of samples is N=8, the depiction of the DFT is given in Figure 5.5. The calculation of the 8-point DFT has changed to the calculation of two DFTs of $N/2 = 4$ elements each and the final combination of their results. The analysis procedure previously followed may continue for both new DFTs of N/2 points. So, we continue the analysis into even- and odd-indexed elements, and following the same technique we obtain the diagram of Figure 5.6. The process continues until it results in a 2-point DFT calculation (Figure 5.7). This will be done after m stages, where $m = \log2N$ At each stage, there are $N/2$ butterflies, that is, $(N/2) \log2N$ butterflies on the whole.

The flow diagram of an 8-point FFT is shown in Figure 5.8, which suggests that one complex multiplication and two complex additions are required to calculate each butterfly. Thus, the N-point FFT requires $(N/2) \log2N$ complex multiplications and $N \log2N$ complex additions. The arrangement of the output samples is 'proper,' i.e., X[0], X[1], ... , X[7], whereas the arrangement of the input samples, x[0], x[4], x[2], x[6], x[1], x[5], x[3], x[7], is not in the right turn. This kind of arrangement of the input samples is the result of their gradual decomposition applied during the development of the algorithm. This decomposition process is called *decimation in time* (DIT). The arrangement of the input elements is not

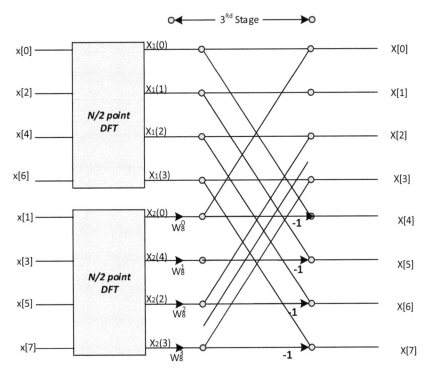

FIGURE 5.5
Third stage of development of the 8-point FFT algorithm.

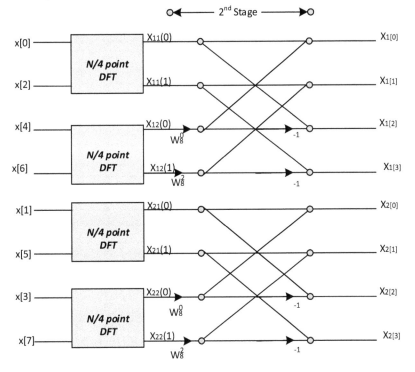

FIGURE 5.6
Second stage of development of the 8-point FFT algorithm.

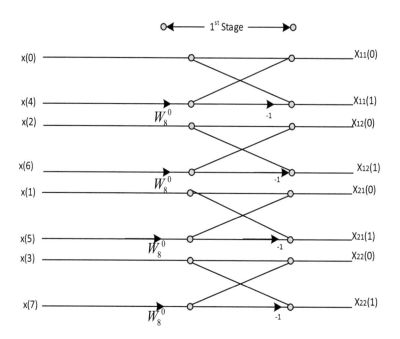

FIGURE 5.7
First stage of development of the 8-point FFT algorithm.

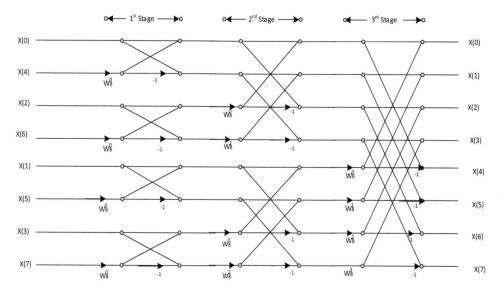

FIGURE 5.8
Final flow graph of 8-point FFT.

random, but results from the 'proper' arrangement of samples with bit reversal. The full match of 'proper' and reversed indices for $N = 8$ is given in Table 5.4.

The FFT algorithm is a very fast and efficient algorithm, and when applied, the gain in processing time is undeniable, as demonstrated for various numbers of samples in Table 5.5.

TABLE 5.4

Match of 'Proper' and Reversed Indices for $N = 8$

'Proper' Arrangement		Arrangement Resulting from Bit Reversal	
Decimal	Binary	Binary	Decimal
0	000	000	0
1	001	100	4
2	010	010	2
3	011	110	6
4	100	001	1
5	101	101	5
6	110	011	3
7	111	111	7

TABLE 5.5

Complex Multiplications for DFT and FFT Calculations

N	DFT Number of Complex Multiplications	FFT Number of Complex Multiplications	Ratio of Complex Multiplications (DFT/FFT)
2	4	1	4
4	16	4	4
8	64	12	5,3
16	256	32	8,0
32	1 024	80	12,8
64	4 096	192	21,3
128	16 348	448	36,6
256	65 536	1 024	64,0
512	262 144	2 304	113,8
1024	1 048 576	5 120	204,8
2048	4 194 304	11 264	372,4
4096	16 777 216	24 576	682,7
8192	67 108 864	53 248	1 260,3

The FFT algorithm, which has just been developed, is applied only to a number of samples that are power of 2 (radix-2) and not for a random number of samples, as for the DFT. There are algorithms that are applied only to a number of samples that are power of 4 (radix-4), or algorithms only for real-value inputs (real-valued algorithms), or algorithms for numbers of samples that are prime numbers (prime length algorithms).

The previous analysis was based on the decimation in time (DIT). In a similar way, there is the decimation in frequency (DIF) technique resulting in an FFT algorithm (Gentleman & Sande, 1966), which has the same computational complexity as the corresponding DIT algorithm. A characteristic of the DIF algorithm is that the multiplications in the 'butterflies' are made after adding the input values and that the input samples are in 'proper' arrangement, while the output samples are bit-reversed.

5.5.2 Computation of the IDFT Using FFT

For the computation of IDFT, which is obtained by $x[n] = IDFT\left[X[k]\right] = \dfrac{1}{N}\sum\limits_{k=0}^{N-1} X[k]W_N^{-kn}$,

where $W_N = e^{-j\frac{2\pi}{N}}$, the FFT algorithm can be applied. At first, $W_N^k = e^{j\frac{2\pi k}{N}}$ and the output sequence $\dfrac{X[k]}{N}$ are defined where $k = 0, 1, \dots, N-1$. Then, the output of the FFT algorithm will be the sequence, $x[n]$.

5.5.3 Fast Convolution

The result of the convolution of two sequences $h[n]$ and $x[n]$ is calculated by:

$$y[n] = x[n] \otimes h[n] = IFFT\left[FFT[x[n]] \cdot FFT[h[n]]\right] = IFFT\left[X[k] \times H[k]\right]$$

Let N_1 represent the length of the sequence $x[n]$, and N_2 represent the length of the sequence $h[n]$. After the linear convolution of $h[n]$ and $x[n]$, the length of $y[n]$ is $N_1 + N_2 - 1$. The addition of zeros at the end of the sequence increases the length of $h[n]$ and $x[n]$ to M, where $M \geq (N_1 + N_2 - 1)$ and $M = 2^v$ in order to make use of radix-2FFT. Obviously, for the calculation of $y[n]$ the M-Point FFT is required three times (two times for the calculation of $X[k]$ and $H[k]$ and once for the calculation $IFFT[X[k] \times H[k]]$).

In the case that one of the sequences, $h[n]$ or $x[n]$, is much longer than the other, the algorithm becomes slower due to the increment of the shorter sequence with zero samples. To avoid this problem, the following methods are applied:

a. *Overlap and add*
b. *Overlap and save*

5.5.3.1 Overlap and Add Method

Let sequence $h[n]$ be of length N, and sequence $x[n]$ be of a much longer one (which could be practically infinite). $x[n]$ is divided into segments of length, $M - N + 1$ (where $M \geq N$ and $M = 2^v$), and, then, the $N - 1$ zero samples are added at the end of each subsequence so that its length becomes equal to M. In Table 5.6, the values of Musually chosen are given as a function of N.

TABLE 5.6

Values of M Given as a Function of N

N	$M = 2^v$
≤10	32
11–19	64
20–29	128
30–49	256
50–59	512
100–199	1024

The length of the sequence $h[n]$ is increased to M by adding $M - N$ zeros at the end. The circular convolution of $h[n]$ with the segment of $x[n]$ is calculated by an M-point FFT, and its length is M. It is: $M = (M - N + 1) + (N - 1)$, where $M - N + 1$ the first samples of the circular convolution, and $N - 1$, the last samples of the circular convolution to be added to the next $(N - 1)$ samples in order to create the linear convolution.

5.5.3.2 Overlap and Save Method

In this method, the length of the sequence $h[n]$ is increased to M by adding $M - N$ zero samples at the end. $x[n]$ is divided in segments in such a way that the length of each segment is equal to M ($M \leq N$ and $M = 2^v$), and its first $N - 1$ samples are coming from the immediately preceding segment, i.e., they are overlapping it by $N - 1$ the samples.

The linear convolution, $x[n] * h[n]$, is derived by the successive circular convolutions of the segments of $x[n]$ with $h[n]$, ignoring, however, the first $N - 1$ elements, which results in the formation of a sequence consisting of their $M - N + 1$ remaining samples.

5.6 Estimation of Fourier Transform through FFT

Let the Fourier Transform (FT) of a continuous-time signal $x(t)$, that is, $X(\Omega)$. The process is as follows:

a. We sample the signal $x(t)$ with a sampling period T, resulting in the discrete-time signal $x[nT]$, $n = 0, 1, \ldots, N - 1$.

b. We calculate the DFT X_k (using FFT) of the discrete-time signal, $x[nT]$.

c. We can calculate the FT $X(\Omega)$ at frequencies $\Omega_k = \dfrac{2\pi k}{NT}$, $k = 0, 1, \ldots, N - 1$, according to the relation:

$$X(\Omega_k) = NT \frac{1 - e^{-j\frac{2\pi k}{N}}}{j2\pi k} X_k, k = 0, 1, \ldots . N - 1 \tag{5.46}$$

The larger the N (i.e. the number of samples we get), and the lower the sampling frequency, the better we can approach the MF $X(\Omega)$ of the continuous-time signal $x(t)$ from the sequence $X(\Omega_k)$ of Equation 5.46.

5.7 Discrete Cosine Transform

The discrete cosine transform is a method that is effectively implemented in digital image compression. With the DCT transform, the information contained in an image can be transferred from the time domain to the frequency domain (which is an abstract domain), where its description occupies significantly less memory space. The first frequencies in the set are of the greatest importance, and the last of less importance. Decisions on compression of a part of these last frequencies (resulting in the loss of them), affect directly the tolerance standards set for image quality. The algorithm is applied in various MPEG and JPEG encodings.

The DCT transform is defined as follows: For each pixel (x, y)

$$DCT(i,j) = \frac{1}{\sqrt{2N}} C(i)C(j) \sum_{x=0}^{N-1} \sum_{y=0}^{N-1} pixel(x,y) \cos\left[\frac{(2x+1)i\pi}{2N}\right] \cos\left[\frac{(2y+1)j\pi}{2N}\right] \quad (5.47)$$

where $C(x) = 0.7071$, $x = 0$ and $C(x) = 1$, $x > 0$

Procedure: The DCT value (i, j) is obtained, that is the value of the transform coefficients in the frequency domain. Hence, pixel values correspond to coefficient values. Each of these coefficients contains a piece of the original information (corresponding to the part of the spectrum it describes). It is known that the human eye perceives phenomena associated with low frequencies (such as specific colors) much better than other involving high-frequency areas of the signal (e.g., edges of the image). This is why the transform coefficients corresponding to low frequencies have greater weighting factors than those describing the high frequencies, and why the first ones are described with maximum accuracy. During reproduction the inverse process is applied using the *Inverse Discrete Cosine Transform* (IDCT), which is given by the following equation:

$$pixel(x,y) = \sum_{i=0}^{N-1} \sum_{j=0}^{N-1} C(i)DCT(j) \cos\left[\frac{(2x+1)i\pi}{2N}\right] \cos\left[\frac{(2y+1)j\pi}{2N}\right] \quad (5.48)$$

The result is that the original information is taken back almost intact (except some unavoidable rounding errors).

Thus, in the DCT, the information that the image contains can be transferred from the time domain to the frequency domain where its description can be achieved with a significantly smaller number of bits. After applying the transform to the image, the information of the result consists of numbers indicating the weighting factors of each frequency at the reconstruction procedure of the image.

The DTC can be computed using the DFT, as follows: Let a sequence, $x[n]$, of N length, i.e., $0 \le n \le N-1$. At first, $x[n]$ is extended to $2N$ length by adding zero elements:

$$x_e[n] = \begin{cases} x[n], & 0 \le n \le N-1 \\ 0, & N \le n \le 2N-1 \end{cases}.$$

Then, a new sequence, $y[n]$, of length $2N$ is created from $x_e[n]$ using $y[n] = x_e[n] + x_e[2N-1-n]$, $0 \le n \le 2N-1$, and its DFT $Y[k]$, of length $2N$, is calculated by:

$$Y[k] = \sum_{n=0}^{2N-1} y[n]W_{2N}^{nk}, \quad 0 \le k \le 2N-1$$

FIGURE 5.9
Block Diagram of a JPEG encoder.

Then, the DCT Cx[k] of $x[n]$ is obtained by using:

$$C_x[k] = \begin{cases} W_{2N}^{k/2}Y[k], & 0 \le k \le N-1 \\ 0, & elsewhere \end{cases} \tag{5.49}$$

The DCT transform applies mostly in compression of video and audio signals. This transform converts the discrete signal into a series of simple integer values (the transform coefficients) that constitute the amplitudes of the frequencies of the original image. It is then easy to make the smaller coefficients zero and achieve significant compression.

JPEG belongs to the category of transform coding techniques, i.e., techniques that compress the transform of the signal and not the signal, itself. The most widely used technique is the DCT. The DCT's energy-compression capability has as a result that only a few of the transform coefficients have significant values so that almost all of the energy is contained in these particular components.

The JPEG encoder is illustrated in Figure 5.9. It consists of three blocks, the gradient-DCT descriptor, the quantizer and the entropy coder.

An uncompressed monochrome digital image (color depth of 8-bit) is a discrete signal, i.e., it is a series of values ranging from 0–255. Consider the image of $N \times M$ pixels arranged in a matrix that is divided into 8x8 pixel submatrices. The coefficients of this matrix are the 8-bit values (0–255) that represent the intensity of the pixels and constitute the input data in the DCT transform. By calculating the DCT of each submatrix, 64 DCT coefficients are generated for each one, which indicate the weighting factor of each frequency that contributes to the synthesis of the discrete image. The purpose of the quantizer is to achieve greater compression by representing each coefficient with no more digits than needed. That is, the quantizer tries to eliminate any information that is not perceived by human vision. After quantization, many of the DCT coefficients have been reset. For compressing these data, JPEG is usually combined with an entropy coding algorithm. Entropy coding refers to techniques that do not take into account the type of information to be compressed. In other words, these techniques treat information as a simple sequence of bits. That is the reason why entropy coding can be applied independently of the type of information. In addition, entropy coding techniques offer coding without losses.

5.8 Wavelet Transform

If a time-series is stationary, it is reasonable to assume that its harmonic components can be detected by means of Fourier analysis, e.g., using DFT. Nevertheless, in various practical applications, it became evident that many time-series are not stationary (i.e., their mean statistical properties change in time). The waves of infinite support that form the harmonic components are not adequate in such a case, whereas waves localized not only in frequency but in time are required, as well. These waves are generally known as *wavelets* and allow a time-scale decomposition of a signal. The Discrete Wavelet Transform (DWT) is used in a variety of signal processing applications, such as video compression, Internet communications compression, and numerical analysis. It can efficiently represent non-stationary signals.

As already mentioned, the wavelet transform is discrete in time and scale. In other words, the DWT coefficients may have real values; yet, under the condition of integer-only time and scale values used to index these coefficients. This subsection discusses the fundamental principles of wavelet transform and shows how the wavelet breaks a signal down into detail signals and an approximation.

Define a certain level of resolution as *octave*. Then, each octave can be encountered as a pair of FIR filters. One filter of the analysis (wavelet transform) pair is a LPF, while the other is a HPF, as illustrated in Figure 5.10. Each filter is followed by a down-sampler so as to enhance the transform efficiency. Figure 5.11 shows the synthesis (inverse wavelet transform) pair, consisting of an inverse low-pass filter (ILPF) and an inverse high-pass filter (IHPF), each preceded by an up-sampler. The low-pass and high-pass filters derive the average and detail signal, respectively. For example, a simple low-pass filter may have coefficients {1/2, 1/2}, producing an output $(x[n] + x[n-1])/2$, which is clearly the average of two samples. A corresponding simple high-pass filter would have coefficients {1/2, −1/2}, producing an output $(x[n] - x[n-1])/2$, which is half the difference of the given samples. While the average signal would be similar to the original one, the details are required so as to make the reconstructed signal match the original one. Multi-resolution analysis feeds the average signal into another set of filters, which produces the average and detail signals at the next octave. Since only the detail signals are kept as the higher octave averages can

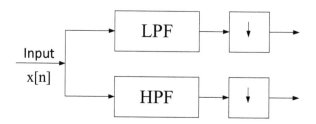

FIGURE 5.10
Analysis filters.

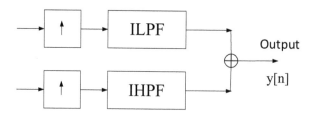

FIGURE 5.11
Synthesis filters.

be recomputed during the synthesis, each of the octave's outputs have only half the input's amount of data. Thus, the wavelet representation is almost the same size as the original.

Using the discrete wavelet transform on 1D data can be accomplished via, for example, the MATLAB commands **dwt** and **idwt**. Furthermore, DWT on images (or other 2D data) can be accomplished with the **dwt2** and **idwt2** *functions* in MATLAB. 2D transform can be accomplished either by applying the low- and high-pass filters along the rows of the data and then applying these filters along the columns of the previous results, or by applying four matrix convolutions—one for each low-pass/high-pass, horizontal/vertical combination. Separability means that the 2D transform is simply an application of the 1D DWT in the horizontal and vertical directions. The non-separable 2D transform works in a slightly different manner, since it computes the transform based on a 2D sample of the input convolved with a matrix, but the results are identical. This can be considered as two ways of reaching the same answer, since we can combine two filters into one, we can therefore combine vertical and horizontal operations into a matrix multiplication.

The main difference between the 1D and higher-order dimension DWT is that a pair of filters operates on each channel for its corresponding dimension.

5.8.1 Wavelet Transform Theory

The discrete wavelet transform convolves the input by the shifts and scales of the wavelet. Below are variables commonly used in wavelet literature:

- g represents the high-pass (wavelet) filter
- h represents the low-pass (scaling) filter
- J is the total number of octaves
- j is the current octave (used as an index $1 \leq j \leq J$)
- N is the total number of inputs
- n is the current input (used as an index $1 \leq n \leq N$)
- L is the width of the filter (number of taps)
- k is the current wavelet coefficient
- $W_f(a,b)$ represents the continuous wavelet transform (CWT) of function f
- $W_h(j,n)$ represents the discrete wavelet transform of function f
- $W(j,n)$ represents the discrete scaling function of f, except $W(0,n)$, which is the input signal.

The continuous wavelet transform is represented by:

$$W_f(a,b) = \int f(t)\psi(at+b)dt$$

where $f(t)$ is the function to analyze, ψ is the wavelet, and $\psi(at + b)$ is the shifted and scaled version of the wavelet at time, b, and scale, a. An alternate form of the equation is:

$$W_f(s,u) = \int\limits_{-\infty}^{+\infty} f(t)\sqrt{s}\psi(s(t-u))dt$$

again, where ψ is the wavelet, while the wavelet family is shown above as $\sqrt{s}\psi(s(t-u))$, shifted by u and scaled by s. We can rewrite the wavelet transform as an inner product:

$$W_f(s,u) = \langle f(t), \sqrt{s}\psi(s(t-u)) \rangle.$$

This inner product is essentially computed by the involved filters.

Thus far, we focused on how to get the wavelet transform given a wavelet, but how does one get the wavelet coefficients and implement the transform with filters? The relationship between wavelets and the filter banks that implement the wavelet transform is as follows. The scaling function, $\varphi(t)$, is determined by recursively applying the filter coefficients, since multi-resolution recursively convolutes the input vector after shifting and scaling. All the required information about the scaling and wavelet functions is obtained by the coefficients of the scaling function and of the wavelet function, respectively. The scaling function is given by:

$$\varphi(t) = \sqrt{2}\sum_k h[k]\varphi(2t-k).$$

The wavelet function is expressed as:

$$\psi(t) = \sqrt{2}\sum_k g[k]\varphi(2t-k).$$

There is a finite set of coefficients, $h[k]$. Once these coefficients are found, allowing us to design the low-pass filter, then the high-pass filter coefficients can be extracted quite easily. The low-pass (scaling) coefficients are:

$$h[0], h[1], h[2], h[3] = \frac{1+\sqrt{3}}{4\sqrt{2}}, \frac{3+\sqrt{3}}{4\sqrt{2}}, \frac{3-\sqrt{3}}{4\sqrt{2}}, \frac{1-\sqrt{3}}{4\sqrt{2}}.$$

These produce a space, V_j, which is invariant to shift and scale and which is required for multi-resolution analysis. Now that we have coefficients for the low-pass filter, the corresponding high-pass (wavelet) filter coefficients can be calculated, as follows:

$$g[0], g[1], g[2], g[3] = \frac{1-\sqrt{3}}{4\sqrt{2}}, \frac{-3+\sqrt{3}}{4\sqrt{2}}, \frac{3+\sqrt{3}}{4\sqrt{2}}, \frac{-1-\sqrt{3}}{4\sqrt{2}}.$$

Notice that $g[0] = h[3]$, $g[1] = -h[2]$, $g[2] = h[1]$, and $g[3] = -h[0]$. This pattern is explicitly shown in Figure 5.12.

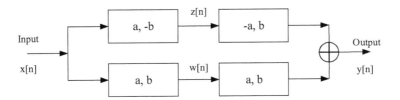

FIGURE 5.12
A quadrature mirror filter.

The significance of the above coefficients is that they are used in a filter bank to generate the wavelet transform. Writing the wavelet equation with regard to the discrete sampled sequence $x[n]$ we get:

$$W_f[a,b] = \frac{1}{\sqrt{a}} \sum_{n=b}^{aL+b-1} x[n]g\left(\frac{n-b}{a}\right)$$

where the wavelet is replaced by function g, which is obtained by sampling the continuous wavelet function. For the discrete case, we let $a = 2k$ such that the parameters (a, b) and k take integer-only values. Particularly, we replace a and b with j and n. By using the discrete case, redundancy is not required in the transformed signal. To reduce redundancy, the wavelet equation must satisfy certain conditions. First, we must introduce an orthogonal scaling function, which allows us to have an approximation at the last level of resolution (the initial state is simply denoted by the input signal). Hence, the coefficients for the low- and high-pass filter are intimately related. The functions that produce these coefficients are also dependent on each other. Second, the representation of the signal at octave j must have all the information of the signal at octave $j + 1$, that is: the input signal has more information than the first approximation of this signal. The function $x[n]$ is thus changed to $W[j − 1, m]$, which is the scaling function's decomposition from the previous level of resolution, with m as an index. Third, after J levels of resolution, the result of the scaling function on the signal will be 0. After repeatedly viewing a signal in successive approximation rounds, eventually the scaling function will not produce any useful information. Fourth, the scaling function allows us to approximate any given signal with a variable amount of precision. The scaling function, h, provides an approximation of the signal via the following equation. This is also known as the low-pass output:

$$W[j,n] = \sum_{m=0}^{N-1} W[j-1,m]h[2n-m].$$

The wavelet function gives us the detail signal, namely, the high-pass output:

$$W_h[j,n] = \sum_{m=0}^{N-1} W[j-1,m]g[2n-m].$$

The n term gives us the shift, the starting points for the wavelet calculations. The index $2n - m$ incorporates the scaling, resulting in half of the outputs for octave j compared to the previous octave $j - 1$.

5.9 SOLVED PROBLEMS

5.9.1 Let the discrete sequence $x[n] = \begin{cases} 1, & 0 \le n \le 3 \\ 0, & elsewhere \end{cases}$. Find the magnitude and phase of its DTFT.

SOLUTION: The DTFT of $x[n]$ is derived by:

$$X(\omega) = \sum_0^3 x(n)e^{-j\omega n} = 1 + e^{-j\omega} + e^{-j2\omega} + e^{-j3\omega}$$

$$= \frac{1 - e^{-j4\omega}}{1 - e^{-j\omega}} = \frac{\sin(2\omega)}{\sin(\omega/2)} e^{-j3\omega/2}$$

Thus, the magnitude of DTFT is $|X(\omega)| = \left|\frac{\sin(2\omega)}{\sin(\omega/2)}\right|$, and the phase:

$$\angle X(\omega) = \begin{cases} -\dfrac{3\omega}{2}, & when & \dfrac{\sin(2\omega)}{\sin(\omega/2)} > 0 \\ -\dfrac{3\omega}{2}, & when & \dfrac{\sin(2\omega)}{\sin(\omega/2)} < 0 \end{cases}.$$

5.9.2 Let the discrete sequence $x[n] = \cos w_0 n$ *where* $w_0 = \dfrac{2\pi}{5}$. Find and plot the DTFT of the signal.

SOLUTION: From the property $X(e^{jw}) = \sum_{l=-\infty}^{\infty} 2\pi\delta(w - w_0 - 2\pi l)$, we obtain:

$$X(e^{jw}) = \sum_{l=-\infty}^{\infty} \pi\delta\left(w - \frac{2\pi}{5} - 2\pi l\right) + \sum_{l=-\infty}^{\infty} \pi\delta\left(w + \frac{2\pi}{5} - 2\pi l\right).$$

Thus,

$$X(e^{jw}) = \pi\delta\left(w - \frac{2\pi}{5}\right) + \pi\delta\left(w + \frac{2\pi}{5}\right) \quad -\pi \le w < \pi$$

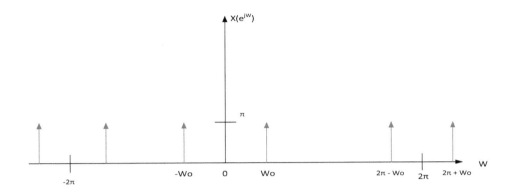

The figure above shows that $X(e^{jW})$ is repeated with a period 2π.

5.9.3 Consider an LTI system. It is given that:

$$h[n] = a^n u[n]$$
$$x[n] = b^n u[n]$$

where $h[n]$ its impulse response and $x[n]$, the input, respectively, and $|a| < 1$, $|b| < 1$. Find the response of the system using DTFT.

SOLUTION: The DTFTs of $h[n]$ and $x[n]$ are:

$$X(e^{jw}) = \frac{1}{1 - be^{-jw}}, \qquad H(e^{jw}) = \frac{1}{1 - \alpha e^{-jw}}$$

Thus, the system response will be: $y[n] = IDTFT[DTFT\,(h[n])DTFT(x[n])]$.

$$Y(e^{jw}) = X(e^{jw}) \cdot H(e^{jw}) = \frac{1}{\left(1 - \alpha e^{-jw}\right)\left(1 - be^{-jw}\right)}$$

$$Y(e^{jw}) = \frac{A}{\left(1 - \alpha e^{-jw}\right)} + \frac{B}{\left(1 - be^{-jw}\right)}$$

From the equations above we calculate A and B:

$$A = \frac{\alpha}{\alpha - b}, B = -\frac{b}{\alpha - b} \Rightarrow Y(e^{jw}) = \frac{\dfrac{\alpha}{\alpha - b}}{\left(1 - \alpha e^{-jw}\right)} + \frac{-\dfrac{b}{\alpha - b}}{\left(1 - be^{-jw}\right)}$$

Thus, $y[n] = \dfrac{1}{\alpha - b}\left[\alpha^{n+1}u[n] - b^{n+1}u[n]\right]$ *for* $\alpha \neq b$.

5.9.4 Let the periodic sequence be $\tilde{x}[n]$, with a period $N=2$. Find the Fourier coefficients of the sequence.

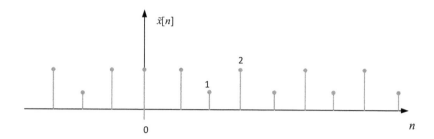

SOLUTION: The Fourier coefficients of the periodic sequence, $\tilde{x}[n]$, are:

$$\tilde{X}[k] = \sum_{n=0}^{1} \tilde{x}[n]e^{-j\frac{2\pi}{2}nk} = \tilde{x}[0] + \tilde{x}[1](-1)^k = 1 + 2 \cdot (-1)^k, \ k = 0,1$$

Thus, $\tilde{X}[0] = 3, \quad \tilde{X}[1] = -1$

and the sequence, $\tilde{x}[n]$, can be written as:

$$\tilde{x}[n] = IDFS\{\tilde{X}[k]\} = \frac{1}{2}\sum_{k=0}^{1} \tilde{X}[k]e^{j\frac{2\pi}{2}kn} = 1.5 - 0.5 \cdot (-1)^n$$

5.9.5 Let the periodic sequence be $\tilde{x}[n]$ with a period $N = 10$.
    a. Find the Fourier coefficients $\tilde{X}[k]$ of the sequence
    b. Plot the magnitude $\left|\tilde{X}[k]\right|$ and the phase $Arg(\tilde{X}[k])$.

$x[n]$

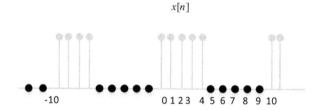

SOLUTION:

    a. $\tilde{x}[n] = \left\{ \overbrace{...1,1,1,1,1,0,0,0,0,0,1,1...}^{N=10} \right\}$ (1)

The Fourier coefficients of the sequence $\tilde{x}[n]$ are:

$$\tilde{X}[k] = \sum_{n=0}^{9} \tilde{x}[n]W_N^{kn} \quad k = 0,1,2,\ldots,N-1$$

$$\Rightarrow \tilde{X}[k] = \sum_{n=0}^{4} W_{10}^{kn} = \sum_{n=0}^{4} e^{-j\left(\frac{2\pi}{10}\right)nk} = \sum_{n=0}^{4}\left(e^{-j\frac{2\pi k}{10}}\right)^n = \frac{1-\left(e^{-j\frac{2\pi k}{10}}\right)^5}{1-e^{-j\frac{2\pi k}{10}}}$$

$$= \frac{1-e^{-j\pi k}}{1-e^{-j\frac{2\pi k}{10}}} = \frac{1-e^{-j\left(\frac{\pi k}{2}+\frac{\pi k}{2}\right)}}{1-e^{-j\left(\frac{\pi k}{10}+\frac{\pi k}{10}\right)}} \Rightarrow \tilde{X}[k] = \frac{e^{-j\frac{\pi k}{2}}e^{j\frac{\pi k}{2}}-e^{-j\frac{\pi k}{2}}e^{-j\frac{\pi k}{2}}}{e^{-j\frac{\pi k}{10}}e^{j\frac{\pi k}{10}}-e^{-j\frac{\pi k}{10}}e^{-j\frac{\pi k}{10}}} \quad (2)$$

$$= \frac{e^{-j\frac{\pi k}{2}}\left(e^{j\frac{\pi k}{2}}-e^{-j\frac{\pi k}{2}}\right)}{e^{-j\frac{\pi k}{10}}\left(e^{j\frac{\pi k}{10}}-e^{-j\frac{\pi k}{10}}\right)} \Rightarrow \tilde{X}[k] = e^{-j\frac{4\pi k}{10}}\frac{2j\sin\frac{\pi k}{2}}{2j\sin\frac{\pi k}{10}} = e^{-j\frac{4\pi k}{10}}\frac{\sin\frac{\pi k}{2}}{\sin\frac{\pi k}{10}}$$

b. The magnitude of $\tilde{X}[k]$ is:

$$\left|\tilde{X}[k]\right| \overset{(3)}{=} \left|e^{-j\frac{4\pi k}{10}}\frac{\sin\frac{\pi k}{2}}{\sin\frac{\pi k}{10}}\right| = \left|\frac{\sin\frac{\pi k}{2}}{\sin\frac{\pi k}{10}}\right| \quad (3)$$

For $k=0 \Rightarrow \lim_{k\to0}\left|\tilde{X}[k]\right| = \lim_{k\to0}\left|\frac{\sin\frac{\pi k}{2}}{\sin\frac{\pi k}{10}}\right| = \lim_{k\to0}\left|\frac{\frac{\pi}{2}\cos\frac{\pi k}{2}}{\frac{\pi}{10}\cos\frac{\pi k}{10}}\right| = 5$

For $k=1 \Rightarrow \lim_{k\to1}\left|\tilde{X}[k]\right| = 3.24$

For $k=2 \Rightarrow \lim_{k\to2}\left|\tilde{X}[k]\right| = 0$

For $k=3 \Rightarrow \lim_{k\to3}\left|\tilde{X}[k]\right| = 1.24$

For $k=4 \Rightarrow \lim_{k\to4}\left|\tilde{X}[k]\right| = 0$

For $k=5 \Rightarrow \lim_{k\to5}\left|\tilde{X}(k)\right| = 1$

The phase of $\tilde{X}[k]$ is: $Arg\left[\tilde{X}[k]\right] = Arg\left(e^{-j\frac{4\pi k}{10}}\frac{\sin\frac{\pi k}{2}}{\sin\frac{\pi k}{10}}\right) = -\frac{4\pi k}{10} \quad (4)$

The magnitude and phase of the sequence, $\tilde{x}[n]$, are illustrated in the following figure.

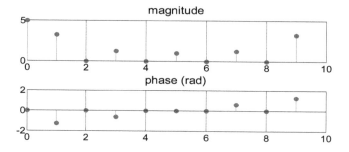

5.9.6 Compute and plot the magnitude spectra of signals

$$x[n] = a^n u[n], \ |a| < 1 \text{ and } h[n] = \begin{cases} 0, & n < -N_1 \\ 1, & -N_1 \le n \le N_1 \\ 0, & n > N_1 \end{cases}$$

SOLUTION:

a. The magnitude spectrum of the signal $x[n] = a^n u[n]$, $|a| < 1$ is computed as follows:

$$X\left(e^{j\omega}\right) = \sum_{n=-\infty}^{\infty} x[n]e^{-j\omega n} = \sum_{n=0}^{\infty} a^n e^{-j\omega n} = \sum_{n=0}^{\infty} \left(ae^{-j\omega}\right)^n = \frac{1}{1 - ae^{-j\omega}}$$

$$\Rightarrow \left|X\left(e^{j\omega}\right)\right| = \left|\frac{1}{1 - ae^{-j\omega}}\right| = \left|\frac{1}{(1 - a\cos(\omega)) - ja\sin(\omega)}\right| \qquad (1)$$

$$= \frac{1}{\sqrt{(1 - a\cos(\omega))^2 + (a\sin(\omega))^2}}$$

NOTE THAT:

$$(1 - a\cos(\omega))^2 + (a\sin(\omega))^2 = 1 - 2a\cos(\omega) + a^2\cos^2(\omega) + a^2\sin^2(\omega)$$
$$= 1 - 2a\cos(\omega) + a^2 \qquad (2)$$

Consequently,

$$\left|X\left(e^{j\omega}\right)\right| = \frac{1}{\sqrt{1 - 2a\cos(\omega) + a^2}} \qquad (3)$$

$$\left| X\left(e^{j\omega}\right) \right|_{\omega=0} = \frac{1}{\sqrt{1-2a\cos(0)+a^2}} = \frac{1}{\sqrt{(1-2a+a^2)}}$$

$$= \frac{1}{\sqrt{(1-a)^2}} = \frac{1}{1-a} \tag{4}$$

$$\left| X\left(e^{j\omega}\right) \right|_{\omega=\pi} = \frac{1}{\sqrt{1-2a\cos(\pi)+a^2}} = \frac{1}{\sqrt{(1+2a+a^2)}}$$

$$= \frac{1}{\sqrt{(1+a)^2}} = \frac{1}{1+a} \tag{5}$$

The magnitude spectrum of the signal is plotted in the figures below. Note that for $a > 0$, we have a low-pass filter (LPF), while for $a < 0$, we have a high-pass filter (HPF).

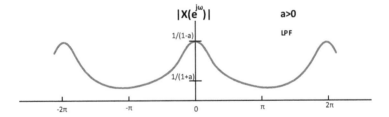

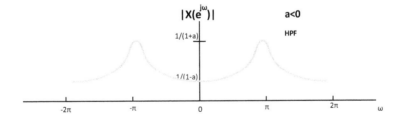

b. The magnitude spectrum of the signal $h[n] = \begin{cases} 0, & n < -N_1 \\ 1, & -N_1 \le n \le N_1 \\ 0, & n > N_1 \end{cases}$ is computed as follows:

$$H\left(e^{j\omega}\right) = \sum_{n=-N_1}^{N_1} e^{-j\omega n} = \sum_{n=-N_1}^{N_1} \left(e^{-j\omega}\right)^n = \frac{\sin\left(\omega(N_1+1/2)\right)}{\sin(\omega/2)} \tag{6}$$

The sequence $h[n]$ for $N_1 = 2$ is plotted as follows

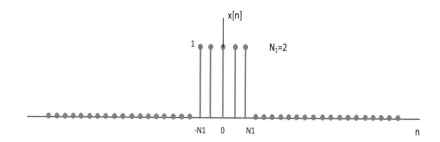

The function $H(e^{j\omega})$ is very similar to $\sin c = \sin(x)/x$ and is periodic with a period 2π.

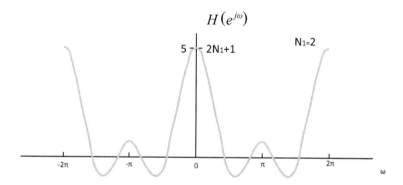

5.9.7 Plot the energy spectral density of the signal $x[n] = a^n u[n]$, $|a| < 1$ for $a = 0.7$ and $a = -0.7$.

SOLUTION: The signal spectrum is

$$X\left(e^{j\omega}\right) = \sum_{n=-\infty}^{\infty} x[n]e^{-j\omega n} = \sum_{n=0}^{\infty} a^n e^{-j\omega n} = \sum_{n=0}^{\infty} \left(ae^{-j\omega}\right)^n = \frac{1}{1 - ae^{-j\omega}}$$

$$\Rightarrow X\left(f\right) = \frac{1}{1 - ae^{-j2\pi f}}$$

Thus, the energy spectral density is

$$\left|X(f)\right|^2 = X(f)X^*(f) = \frac{1}{1 - ae^{-j2\pi f}} \cdot \frac{1}{1 - ae^{j2\pi f}} = \frac{1}{1 - 2a\cos(2\pi f) + a^2}$$

In the figures below the energy spectral density for $a = 0.7$ and for $a = -0.7$ is illustrated. Note that for $a = 0.7$, the signal has stronger low frequencies, while for $a = -0.7$, the signal has stronger high frequencies. This can be explained as follows:

The signal $x[n] = a^n u[n]$ converges exponentially to zero as $n \to \infty$. However, if α is negative, for example $a = -0.7$, this convergence occurs with alternate positive and negative values, which means that the signal exhibits sharp changes and, consequently, a strong high frequency spectral content.

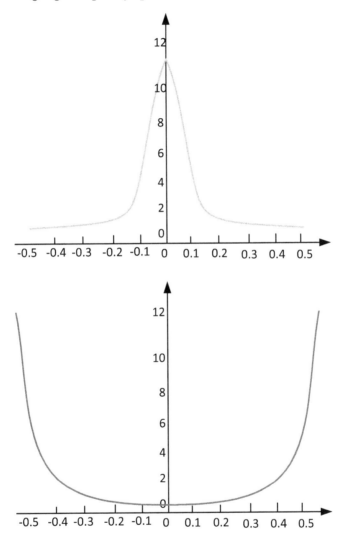

5.9.8 Let the discrete-time signal:

$$x[n] = \begin{cases} 1, |n| \le k \\ 0, |n| > k \end{cases}$$

a. Compute the DFT of N length of the signal $x[n]$, where $N > 2k + 1$.
b. Plot the magnitude and phase of the DFT when $k = 2$ and $N = 12$.
c. Estimate the energy of the signal contained between $f_1 = 1kHz$ and $f_2 = 3kHz$, if the sampling frequency f_s is $8kHz$.

SOLUTION:

a. From the signal, it is derived:

$$x[n] = \begin{cases} 1, & |n| \le k \\ 0, & |n| > k \end{cases} \Rightarrow x[n-k] = \begin{cases} 1, & 0 \le n \le 2k \\ 0, & elsewhere \end{cases}$$

It is true that:

$$x[n-k] \xrightarrow{F_{(DFT)}} w^{+km} X[m], \ w \triangleq e^{-j\frac{2\pi}{N}}$$

$$\Rightarrow X[m] \leftarrow w^{-km} F_{DFT}\{x[n-k]\} \Rightarrow DFT\{x[n-k]\} = \sum_{n=0}^{N-1} w^{mn} x[n-k] = \sum_{n=0}^{2k} w^{mn}$$

$$\Rightarrow DFT\{x[n-k]\} = \frac{w^{(2k+1)m} - 1}{w^m - 1} \Rightarrow X[m] = w^{-km} \frac{w^{\left(k+\frac{1}{2}\right)m}\left[w^{\left(k+\frac{1}{2}\right)m} - w^{-\left(k+\frac{1}{2}\right)m}\right]}{w^{\frac{m}{2}}\left[w^{\frac{m}{2}} - w^{-\frac{m}{2}}\right]}$$

$$\Rightarrow X[m] = \frac{2j \cdot \sin\left[(2k+1)\frac{\pi m}{N}\right]}{2j \cdot \sin\left(\frac{\pi m}{N}\right)} = \frac{\sin\left[(2k+1)\frac{\pi m}{N}\right]}{\sin\left(\frac{\pi m}{N}\right)}, m = 0,1,2,..,(N-1)$$

Another way of calculating $X[m]$ is:

$$X[m] = \sum_{n=-k}^{+k} w^{mn} x[n] = \sum_{n=-k}^{k} w^{mn} = w^{-mk} + .. + 1 + .. + w^{mk}$$

$$= w^{-mk}(1 + .. + w^{2mk}) \Rightarrow X[m] = w^{-mk} \frac{w^{(2k+1)m} - 1}{w^m - 1}$$

b. For k = 2 και N = 12, we have:

$$X[m] = \frac{\sin\frac{5\pi m}{12}}{\sin\left(\frac{\pi m}{12}\right)}, m = 0,1,2,..,(N-1).$$

It is true that: $|X[0] = X[12]|$, $|X[1] = X[11]|$, $|X[2] = X[10]|$ ..., etc.
Generally, it is: $|X[m] = X[N-m]|$.
After some calculations, the following is obtained:

$X[0] = 5, X[1] = 3.73, X[2] = 1, X[3] = -1, X[4] = -1, X[5] = 0.268,$ and $X[6] = 1.$

Knowing that the magnitude of the DFT will have an even symmetry with respect to the point $\dfrac{N}{2} = 6$, and the phase will have an odd symmetry with respect to the same point, we plot the magnitude and phase spectra of the DFT of the signal below.

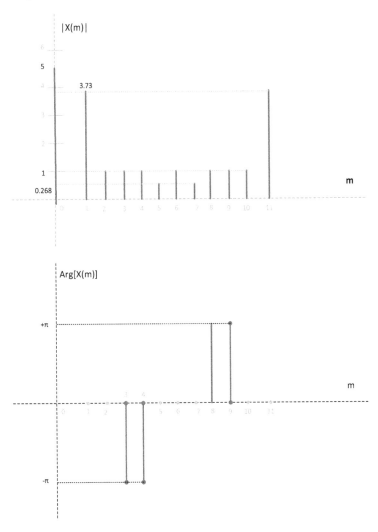

c.   It is $F_1 = \dfrac{f_1}{f_S} = \dfrac{1kHz}{8kHz} = \dfrac{1}{8}$, $F_2 = \dfrac{f_2}{f_S} = \dfrac{3kHz}{8kHz} = \dfrac{3}{8}$.

Thus:

$$\Omega_1 = 2\pi F_1 = 2\pi \frac{1}{8} = \frac{\pi}{4} \text{ and } \Omega_2 = 2\pi F_2 = 2\pi \frac{3}{8} = \frac{3\pi}{4}.$$

From the relations above, it is: $\Omega_1 \leq \Omega \leq \Omega_2 \Rightarrow 2 \leq m \leq 4$, hence, the energy of the signal between the frequencies f_1 1kHz and f_2 3kHz is:

$$\Rightarrow E = \frac{1}{N} 2 \sum_{m=2}^{4} |X(m)|^2 = \frac{1}{6}(1^2 + 1^2 + 1^2) = \frac{1}{2}.$$

5.9.9 a. Show that the DFT of the unit pulse, $\delta[n]$, is a constant, which is independent of the number of zeros that follow the pulse within the analysis window of length N.

   b. Show that the DFT of a constant signal contains only one frequency within the analysis window (i.e., $x_2[n] = A$ for $0 \leq n \leq N - 1$).

SOLUTION:

   a. After applying the definition of the DFT, we get:

$$x_1[n] = \delta[n] \Rightarrow X_1[m] = DFT[\delta[n]] = \sum_{n=0}^{N-1} \delta[n] \cdot e^{-j\frac{2\pi n \cdot m}{N}} = 1 \cdot e^{-j\frac{2\pi m}{N}0} = 1.$$

   which is true for $\forall m \in [0, N - 1]$ and $\forall N > 1$.

   b. For the constant signal, $x_2[n] = A$ for $0 \leq n \leq N-1$, it is $x_2[n] = A[u[n] - u[n - N]]$. The DFT of the signal is derived as follows:

$$X_2[m] = DFT\left[A(u[n] - u[n - N])\right] = A\sum_{n=0}^{N-1} 1 \cdot e^{-j\frac{2\pi n \cdot m}{N}} = A \cdot N \cdot \delta[m].$$

   The equation above derives from:

$$\sum_{n=0}^{N-1} e^{-j\frac{2\pi m \cdot n}{N}} = \begin{cases} 0, & \forall m \neq 0 \\ N, & m = 0 \end{cases}.$$

The duality in the above cases is easy to explain from a physical point of view if we consider the corresponding analog signals. Indeed, the signal, $\delta(t)$, contains all frequencies, with constant amplitude and, respectively, the signal $\delta[n]$ gives a constant signal for all m values in the analysis window, $[0, N - 1]$. Respectively, the continuous (constant) signal contains a single frequency: zero. Similarly, the constant signal in the analysis window, $[0, N - 1]$, gives a non-zero value only for $m = 0$.

5.9.10 Compute the DFT of the sequence $x[n] = \{1, 0, 0, 1\}$. The IDFT of the sequence $X(k)$ will be evaluated in order to verify the result.

SOLUTION: $x[0] = 1, x[T] = 0, x[2T] = 0, x[3T] = 1, N = 4$

$$k = 0 \Rightarrow X[0] = \sum_{n=0}^{3} x[nT]e^{j0} = \sum_{n=0}^{3} x[nT] = x[0] + x[T] + x[2T] + x[3T]$$

$$= 1 + 0 + 0 + 1 = 2$$

$$k = 1 \Rightarrow X[1] = \sum_{n=0}^{3} x[nT]e^{-j\Omega nT}$$

$$= \sum_{n=0}^{3} x[nT]e^{-j2\pi n/N} = 1 + 0 + 0 + 1e^{-j\frac{6\pi}{4}} = 1 + j$$

$$k = 2 \Rightarrow X[2] = \sum_{n=0}^{3} x[nT]e^{-j2\pi n2/N} = 1 + 0 + 0 + 1e^{-j3\pi} = 1 - 1 = 0$$

$$k = 3 \Rightarrow X[3] = \sum_{n=0}^{3} x[nT]e^{-j2\pi n3/N} = 1 + 0 + 0 + 1e^{-j\frac{9\pi}{2}} = 1 - j$$

Thus,

$$X[k] = \left\{2, (1+j), 0, (1-j)\right\}$$

We can study the result be computing $x[nT]$

$$n = 0 \Rightarrow x[0] = \frac{1}{N}\sum_{k=0}^{N-1} X[k] = \frac{1}{4}\left[X[0] + X[1] + X[2] + X[3]\right]$$

$$= \frac{1}{4}\left[2 + 1 + j + 0 + 1 - j\right] = 1$$

$$n = 1 \Rightarrow x[T] = 0$$

$$n = 2 \Rightarrow x[2T] = 0$$

$$n = 3 \Rightarrow x[3] = \frac{1}{N}\sum_{k=0}^{N-1} X[k]e^{jk3\pi/2} = \frac{1}{4}\left[2 + (1+j)e^{j\frac{3\pi}{2}} + (1-j)e^{j9\pi/2}\right]$$

$$= \frac{1}{4}\left[2 + (1+j)(-j) + (1-j)(j)\right] = 1$$

Thus,

$$x[nT] = \left\{1 \quad 0 \quad 0 \quad 1\right\}$$

5.9.11 Compute the DFT of length, N, of the signal $x[n] = \cos\left(\dfrac{2\pi kn}{N}\right), 0 \le n \le N-1$, where k is an integer in the interval 0 to N–1.

SOLUTION: We will use Euler's formula:

$$x[n] = \cos\left(\frac{2\pi kn}{N}\right) = \frac{1}{2}e^{j\frac{2\pi kn}{N}} + \frac{1}{2}e^{-j\frac{2\pi kn}{N}} = \frac{1}{2}e^{j\frac{2\pi kn}{N}} + \frac{1}{2}e^{j\frac{2\pi(N-k)n}{N}}$$

$$\text{or: } x[n] = \frac{1}{N}\left[\frac{N}{2}e^{j\frac{2\pi kn}{N}} + \frac{N}{2}e^{j\frac{2\pi(N-k)n}{N}}\right].$$

From the equation above and the definition of the inverse DFT, $x[n] = \dfrac{1}{N}\sum_{m=0}^{N-1}X[m]e^{j\frac{2\pi mn}{N}}$, the following is obtained:

For $k = 0$ or $k = N/2$ (if N is even): $X[m] = \begin{cases} N, & m = k \\ 0, & m \ne k \end{cases} = N\delta(m-k)$.

Otherwise: $X[m] = \begin{cases} \dfrac{N}{2}, & m = k \\ \dfrac{N}{2}, & m = N-k \\ 0, & elsewhere \end{cases}$

5.9.12 The DFT $X[k]$ of length, N, of the signal $x[n]$, $0 \le n \le N-1$ is considered known. Compute the DFTs of length N of the signals $x_c[n] = x[n]\cos\dfrac{2\pi kn}{N}$ and $x_s[n] = x[n]\sin\dfrac{2\pi kn}{N}, 0 \le n \le N-1$ where k is an integer in the interval 0 to N-1.

SOLUTION: This is a case of amplitude modulation (AM) leading in a frequency shift.

According to Euler's formula:

$$x_c[n] = x[n]\frac{e^{j\frac{2\pi kn}{N}} + e^{-j\frac{2\pi kn}{N}}}{2} = \frac{1}{2}x[n]e^{j\frac{2\pi kn}{N}} + \frac{1}{2}x[n]e^{-j\frac{2\pi kn}{N}}.$$

Making use of the linearity and shifting properties of the DFT in the frequency domain we have:

$$X_c[m] = \frac{1}{2}X[[m-k]]_N + \frac{1}{2}X[[m+k]]_N, \ 0 \le m \le N-1,$$

where $((\cdot))_N$ denotes modulo N. In the same way, we find:

$$X_s[m] = \frac{1}{2j} X[[m-k]]_N - \frac{1}{2j} X[[m+k]]_N, \ 0 \le m \le N-1.$$

5.9.13 Let the sequence $x[n] = \left\{\underset{\uparrow}{1},\ 2,\ 3,\ 4\right\}$. Illustrate the sequences $x[(n-1)\bmod 4]$, $x[(n+1)\bmod 4]$ and $x[(-n)\bmod 4]$.

SOLUTION: The sequence $x[n]$ is illustrated in the following figure.

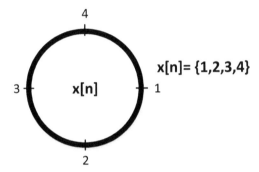

The sequence $x[(n-1)\bmod 4]$, i.e., a circular clockwise shift of the sequence $x[n]$ is illustrated below:

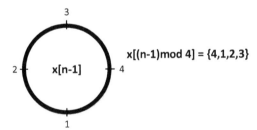

The sequence $x[(n+1)\bmod 4]$, i.e., a circular anticlockwise shift of the sequence $x[n]$ is illustrated below:

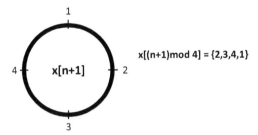

Let's assume that the sequence $x[n]$ is periodic with a period, N (here $N = 4$), that is, $x[n] = x[n \pm N]$, and it is illustrated in the following figure.

Making use of $x[n] = x[n \pm N]$, we have:

$$x[0] = 1$$
$$x[-1] = x[-1+4] = x[3] = 4$$
$$x[-2] = x[-2+4] = x[2] = 3$$
$$x[-3] = x[-3+4] = x[1] = 2$$

Thus,

$$x[-n] = \left\{x[0], x[-1], x[-2], x[-3]\right\} \Rightarrow x[-n] = \left\{\underset{\uparrow}{1}, 4, 3, 2\right\}$$

The sequence $x[(-n)\bmod 4]$ is illustrated below:

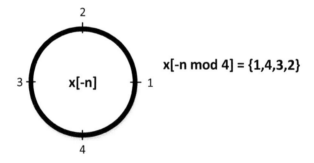

5.9.14 Compute the 4-point circular convolution for the signals $x_1[n] = \{1, 2, 2, 0\}$ and $x_2[n] = \{1, 2, 3, 4\}$.

SOLUTION:

$$x_1[n] \otimes_N x_2[n] = \sum_{m=0}^{3} x_1[m]x_2[[n-m]]_4 \quad 0 \le n \le 3$$

$$x_1[n] = \{1, 2, 2, 0\} \text{ and } x_2[n] = \{1, 2, 3, 4\}$$

For n = 0

$$\sum_{m=0}^{3} x_1[m]x_2[[0-m]]_4 = \sum_{m=0}^{3}\left[\{1,2,2,0\}\cdot\{1,4,3,2\}\right]\sum_{m=0}^{3}\{1,8,6,0\} = 15$$

For n = 1

$$\sum_{m=0}^{3} x_1[m]x_2[[0-m]]_4 = \sum_{m=0}^{3}\left[\{1,2,2,0\}\cdot\{2,1,4,3\}\right]\sum_{m=0}^{3}\{2,2,8,0\} = 12$$

For n = 2

$$\sum_{m=0}^{3} x_1[m]x_2[[0-m]]_4 = \sum_{m=0}^{3}\left[\{1,2,2,0\}\cdot\{3,2,1,4\}\right]\sum_{m=0}^{3}\{3,4,2,0\} = 9$$

For n = 3

$$\sum_{m=0}^{3} x_1[m]x_2[[0-m]]_4 = \sum_{m=0}^{3}\left[\{1,2,2,0\}\cdot\{4,3,2,1\}\right]\sum_{m=0}^{3}\{4,6,4,0\} = 14$$

Thus, $x_1[n] \otimes_4 x_2[n] = \{15, 12, 9, 14\}$

We can easily verify that

DFT (length 4) of $x_1[n] = \{5, -1 - j2, 1, -1 + j2\}$

DFT (length 4) of $x_2[n] = \{10, -2 + j2, 2, -2 - j2\}$

and DFT $[x_1[n]]\cdot$DFT $[x_2[n]] = \{50, 6+ j2, -2, 6 - j2\}$

Finally: IDFT $\{50, 6+ j2, -2, 6 - j2\} = \{15, 12, 9, 14\}$

For the previous example, if we get a 3 + 4-1 = 6-point circular convolution (length (x1) = 3), we will have $x_1[n] \otimes_6 x_2[n] = \{1, 4, 9, 14, 14, 8\}$, which is the same as the result of the linear convolution: conv(x1, x2) = 1 4 9 14 14 8.

5.9.15 Let the sequences $x[n] = \left\{\underset{\uparrow}{1}, 2, 3\right\}$ and $h[n] = \left\{\underset{\uparrow}{1}, 2\right\}$.
  a. Compute the circular convolution of the sequences.
  b. Compute the linear convolution of the sequences using the circular convolution.
  c. Compute the linear convolution of the sequences using the DFT.

SOLUTION:
  a. The circular convolution of $x[n]$ and $h[n]$ is:

$$f[n] = x[n] \otimes h[n] = \sum_{k=0}^{2} x\big([n-k] \bmod 3\big)\, h[k] \tag{1}$$

We will implement relation (1) by placing the samples of the sequences $h[k] = \{1, 2, 0\}$ and $x[-k] = \{1, 3, 2\}$ in two concentric circles, as shown in the figures below.

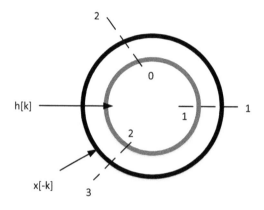

It is: $f[0] = (1, 2, 0) \times (1, 3, 2)^T = 1 \cdot 1 + 2 \cdot 3 = 7$

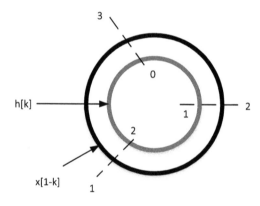

It is: $f[1] = (1, 2, 0) \times (2, 1, 3)^T = 1 \cdot 2 + 2 \cdot 1 = 4$

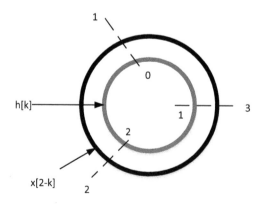

It is: $f[2] = (1, 2, 0) \times (3, 2, 1)^T = 1 \cdot 3 + 2 \cdot 2 = 7$

Consequently, the circular convolution of $x[n]$ and $h[n]$ is $f[n] = \left\{\underset{\uparrow}{7}, 4, 7\right\}$ $\qquad$ (2)

b. The length of the sequence $x[n] = \left\{\underset{\uparrow}{1}, 2, 3\right\}$ is $N_1 = 3$, whereas the length of the sequence $h[n]$ = {1, 2} is $N_2 = 2$. The linear convolution of $x[n]$ and $h[n]$ creates a sequence of length $N_1 + N_2 - 1 = 3 + 2 - 1 = 4$. Consequently, we must add one and two zeros to $x[n]$ and $h[n]$ sequences, respectively, i.e., $x[n] = \left\{\underset{\uparrow}{1}, 2, 3, 0\right\}$ and $h[n] = \left\{\underset{\uparrow}{1}, 2, 0, 0\right\}$.

The circular convolution, $x[n] \otimes h[n]$, will be:

$$y[n] = x[n] \otimes h[n] = \sum_{k=0}^{3} h[k]x[(n-k) \bmod 4] \qquad (3)$$

Thus,

$$\begin{aligned}
y[0] &= (1,2,0,0) \times (1,0,3,2)^T = 1 \cdot 1 = 1 \\
y[1] &= (1,2,0,0) \times (2,1,0,3)^T = 1 \cdot 2 + 2 \cdot 1 = 4 \\
y[2] &= (1,2,0,0) \times (3,2,1,0)^T = 1 \cdot 3 + 2 \cdot 2 = 7 \\
y[3] &= (1,2,0,0) \times (0,3,2,1)^T = 2 \cdot 3 = 6
\end{aligned} \qquad (4)$$

Using the circular convolution, the linear convolution of the sequences is $y[n] = \left\{\underset{\uparrow}{1}, 4, 7, 6\right\}$. The result is the same with that of the linear convolution of the sequences $x[n]$ and $h[n]$.

c. In order to calculate the linear convolution of $x[n]$ and $h[n]$, first we will add the appropriate number of zeros to the sequences so that their length is $N = N_1 + N_2 - 1 = 4$, that is $x[n]$ = {1, 2, 3, 0} and $y[n]$ = {1, 2, 0, 0}.

The DFT of $x[n]$ is: $X[k] = DFT\left[x[n]\right] = \sum_{n=0}^{3} x[n]W_4^{kn}$ $\qquad$ (5)

Consequently, the DFT matrix for $x[n]$ will be:

$$\begin{bmatrix} X[0] \\ X[1] \\ X[2] \\ X[3] \end{bmatrix} = \begin{bmatrix} W_4^0 & W_4^0 & W_4^0 & W_4^0 \\ W_4^0 & W_4^1 & W_4^2 & W_4^3 \\ W_4^0 & W_4^2 & W_4^0 & W_4^2 \\ W_4^0 & W_4^3 & W_4^2 & W_4^1 \end{bmatrix} \begin{bmatrix} x[0] \\ x[1] \\ x[2] \\ x[3] \end{bmatrix} \qquad (6)$$

where:

$$W_4^0 = \left(e^{-j\frac{2\pi}{4}}\right)^0 = 1$$

$$W_4^1 = e^{-j\frac{2\pi}{4}} = e^{-j\frac{\pi}{2}} = \cos\frac{\pi}{2} - j\sin\frac{\pi}{2} = -j$$

$$W_4^2 = e^{-j\frac{2\pi}{4}2} = e^{-j\pi} = \cos\pi - j\sin\pi = -1$$
$$\eta' \ W_4^2 = \left(W_4^1\right)^2 = (-j)^2 = j^2 = -1$$
$$W_4^3 = W_4^2 \cdot W_4 = (-1)(-j) = j$$
$$W_4^4 = \left(W_4^2\right)^2 = (-1)^2 = 1 = W_4^0$$

Thus,

$$
\begin{bmatrix} X[0] \\ X[1] \\ X[2] \\ X[3] \end{bmatrix} =
\begin{bmatrix}
1 & 1 & 1 & 1 \\
1 & -j & -1 & j \\
1 & -1 & 1 & -1 \\
1 & j & -1 & -j
\end{bmatrix}
\begin{bmatrix} 1 \\ 2 \\ 3 \\ 0 \end{bmatrix} =
\begin{bmatrix} 6 \\ -2-2j \\ 2 \\ -2+2j \end{bmatrix}
\tag{7}
$$

That is, $X[k] = DFT[x[n]] = \{6, -2-2j, -2+2j\}$ \hfill (8)

Likewise, for $H[k]$:

$$
\begin{bmatrix} H[0] \\ H[1] \\ H[2] \\ H[3] \end{bmatrix} =
\begin{bmatrix}
1 & 1 & 1 & 1 \\
1 & -j & -1 & j \\
1 & -1 & 1 & -1 \\
1 & j & -1 & -j
\end{bmatrix}
\begin{bmatrix} 1 \\ 2 \\ 0 \\ 0 \end{bmatrix} =
\begin{bmatrix} 3 \\ 1-2j \\ -1 \\ 1+2j \end{bmatrix}
\tag{9}
$$

That is, $H[k] = DFT[h[n]] = \{3, \ 1-2j, \ -1, \ 1+2j\}$ \hfill (10)

It is true that:

$$Y[k] = DFT[y[n]] = X[k]H[k] \Rightarrow$$

$$Y(k) = \{6, \ -2-2j, \ 2, \ -2+2j\} \times \{3, \ 1-2j, \ -1, \ 1+2j\} \Rightarrow$$

$$Y[k] = \{18, \ -6+2j, \ -2, \ -6-2j\} \tag{11}$$

But $y[n] = IDFT\left[Y[k]\right] = \dfrac{1}{4}\displaystyle\sum_{k=0}^{3} Y[k]W_4^{-kn}, \ \ 0 \le n \le 3$ \hfill (12)

So, the IDFT matrix for $y[n]$ will be:

$$
\begin{bmatrix} y[0] \\ y[1] \\ y[2] \\ y[3] \end{bmatrix} = \frac{1}{4}
\begin{bmatrix}
W_4^0 & W_4^0 & W_4^0 & W_4^0 \\
W_4^0 & W_4^{-1} & W_4^{-2} & W_4^{-3} \\
W_4^0 & W_4^{-2} & W_4^0 & W_4^{-2} \\
W_4^0 & W_4^{-3} & W_4^{-2} & W_4^{-1}
\end{bmatrix}
\begin{bmatrix} Y[0] \\ Y[1] \\ Y[2] \\ Y[3] \end{bmatrix}
\tag{13}
$$

where

$$W_4^0 = 1$$

$$W_4^{-1} = \left(W_4^1\right)^{-1} = (-j)^{-1} = \frac{1}{-j} = \frac{j}{-j^2} = j$$

$$W_4^{-2} = \left(W_4^2\right)^{-1} = \frac{1}{-1} = -1 \qquad (14)$$

$$W_4^{-3} = \left(W_4^3\right)^{-1} = \frac{1}{j} = \frac{j}{j^2} = -j$$

$$(13), (14) \Rightarrow \begin{bmatrix} y[0] \\ y[1] \\ y[2] \\ y[3] \end{bmatrix} = \frac{1}{4} \begin{bmatrix} 1 & 1 & 1 & 1 \\ 1 & j & -1 & -j \\ 1 & -1 & 1 & -1 \\ 1 & -j & -1 & j \end{bmatrix} \begin{bmatrix} 18 \\ -6+2j \\ -2 \\ -6-2j \end{bmatrix} = \begin{bmatrix} 1 \\ 4 \\ 7 \\ 6 \end{bmatrix}$$

Thus, $y[n] = \{1, 4, 7, 6\}$. The result is the same with that of the linear convolution of $x[n]$ and $h[n]$.

5.9.16 Let $\{g[n]\} = \{g[n]\} = \{5, 2, 4, -1, 0, 0\}$ and $h[n] = \{-3, 4, 0, 2, -1, 2\}$, two sequences of finite length.
   a. Compute the circular convolution $y_C[n] = g_e[n] *_6 h[n]$.
   b. Compute the circular convolution using the DFT.

SOLUTION: The circular convolution $y_C[n] = g_e[n]*_6 h[n]$ is: $y_C[n] = \sum_{k=0}^{5} g_e[k]h[((n-k))_6]$, where $\{g_e[n]\} = \{5, 2, 4, -1, 0, 0\}$.
   Thus,

$$y_C[0] = g_e[0]h[0] + g_e[1]h[5] + g_e[2]h[4] + g_e[3]h[3] + g_e[4]h[2]$$
$$+ g_e[5]h[1] = g[0]h[0] + g[1]h[5] + g[2]h[4] + g[3]h[3]$$
$$= 5 \times (-3) + 2 \times 2 + 4 \times (-1) + (-1) \times 2 = -17$$

$$y_C[1] = g_e[0]h[1] + g_e[1]h[0] + g_e[2]h[5] + g_e[3]h[4] + g_e[4]h[3]$$
$$+ g_e[5]h[2] = g[0]h[1] + g[1]h[0] + g[2]h[5] + g[3]h[4]$$
$$= 5 \times 4 + 2 \times (-3) + 4 \times 2 + (-1) \times (-1) = 23$$

$$y_C[2] = g_e[0]h[2] + g_e[1]h[1] + g_e[2]h[0] + g_e[3]h[5] + g_e[4]h[4]$$
$$+ g_e[5]h[3] = g[0]h[2] + g[1]h[1] + g[2]h[0] + g[3]h[5]$$
$$= 5 \times 0 + 2 \times 4 + 4 \times (-3) + (-1) \times 2 = -6$$

$$y_C[3] = g_e[0]h[3] + g_e[1]h[2] + g_e[2]h[1] + g_e[3]h[0] + g_e[4]h[5]$$
$$+ g_e[5]h[4] = g[0]h[3] + g[1]h[2] + g[2]h[1] + g[3]h[0]$$
$$= 5 \times 2 + 2 \times 0 + 4 \times 4 + (-1) \times (-3) = 29$$

$$y_C[4] = g_e[0]h[4] + g_e[1]h[3] + g_e[2]h[2] + g_e[3]h[1] + g_e[4]h[0]$$
$$+ g_e[5]h[5] = g[0]h[4] + g[1]h[3] + g[2]h[2] + g[3]h[1]$$
$$= 5 \times (-1) + 2 \times 2 + 4 \times 0 + (-1) \times 4 = -5$$

$$y_C[5] = g_e[0]h[5] + g_e[1]h[4] + g_e[2]h[3] + g_e[3]h[2] + g_e[4]h[1]$$
$$+ g_e[5]h[0] = g[0]h[5] + g[1]h[4] + g[2]h[3] + g[3]h[2]$$
$$= 5 \times 2 + 2 \times (-1) + 4 \times 2 + (-1) \times 0 = 16$$

b. Next, the two DFTs (length 6) are derived:

$$G_e[k] = \sum_{n=0}^{5} g_e[n]e^{-j\frac{2\pi kn}{6}} \text{ and } H[k] = \sum_{n=0}^{5} h[n]e^{-j\frac{2\pi kn}{6}}, \text{ resulting to:}$$

$$\{G_e[k]\} = \{10, 5 - j5.1962, 1 + - j1.7322, 8, 1 - j1.7322, 5 + j5.1962\} \text{ and}$$
$$\{H[k]\} = \{4, -1.5 - j2.5981, -3.5 - j0.866, -12, -3.5 + j0.866, -1.5 + j2.5981\}$$

Thus,

$$\{G_e[k]H[k]\} = \{40, -21 - j5.1962, -2 - j6.9282, -96, -2 +$$
$$+ j6.9282, -21 + j5.1962\}$$

The circular convolution can be computed by the inverse DFT of the product above, i.e., $y_C[n] = \frac{1}{6}\sum_{k=0}^{5} G_e[k]H[k]e^{j\frac{2\pi kn}{6}}$, providing the following result:

$\{y_C[n]\} = \{-17, 23, -6, 29, -5, 16\}$, which is the same with the one estimated in (a).

5.9.17 Derive the DFT of a 4-point sequence using the radix-2 FFT algorithm when:
a. $x[n] = \{x[0], x[1], x[2], x[3]\}$ and
b. $x[n] = \{1, ,1\ 1, 1\}$.

SOLUTION:
a. $N = 4 = 2^2 \Rightarrow v = 2$. Thus, there are two segmentation stages. The DFT is:

$$X[k] = DFT\left[x[n]\right] = \sum_{n=0}^{3} x[n]W_4^{kn}, \quad k = 0,1,2,3 \Rightarrow$$

$$X[k] = x[0]W_4^0 + x[1]W_4^k + x[2]W_4^{2k} + x[3]W_4^{3k}, \quad k = 0, 1, 2, 3 \tag{1}$$

Then, from Eq. (1) we can conclude that in order to calculate $X[k]$, $4^2 = 16$ complex multiplications are needed.

By partition of $x[n]$, we get:
$$\begin{aligned} x_e[n] &= \{x[0], \ x[2]\} \\ x_0[n] &= \{x[1], \ x[3]\} \end{aligned} \tag{2}$$

The DFT of $x_e[n]$ and $x_o[n]$ is:

$$X_e[k] = DFT\left[x_e[n]\right] = \sum_{n=0}^{1} x_e[n]W_2^{kn} = x_e[0]W_2^0 + x_e[1]W_2^k \ \Rightarrow \tag{3}$$

$$X_e[k] = x[0]W_2^0 + x[1]W_2^k, \quad k = 0, 1$$

Likewise, $X_o[k] = DFT\left[x_o[n]\right] = x[1]W_2^0 + x[3]W_2^k, \quad k = 0, 1 \tag{4}$

By writing equations (3) and (4) in a general form, we obtain:

$$X_e[k] = x[0]W_{\frac{N}{2}}^0 + x[2]W_{\frac{N}{2}}^k = x[0]W_N^0 + x[2]W_N^{2k}, \quad k = 0, 1, 2, \dots, \frac{N}{2} - 1 \tag{5}$$

and $X_o[k] = x[1]W_{\frac{N}{2}}^0 + x[3]W_{\frac{N}{2}}^k = x[1]W_N^0 + x[3]W_N^{2k}, \quad k = 0, 1, 2, \dots, \frac{N}{2} - 1 \tag{6}$

From (1), obviously it is:

$$X[k] = X_e[k] + W_N^k X_o[k], \quad k = 0, 1, 2, \dots, N - 1 \tag{7}$$

And for the remaining part of the DFT, it is:

$$X\left[k + \frac{N}{2}\right] = X[k+2] = X_e[k] - W_N^k X_o[k], \quad k = 0, 1 \tag{8}$$

where

$$X_e[0] = x[0] + x[2] \tag{9}$$

$$\begin{aligned} X_e[1] &= x[0] + x[2]W_2 = x[0] + x[2] \, e^{-j\frac{2\pi}{2}} \ \Rightarrow \\ X_e[1] &= x[0] + x[2] \, e^{-j\pi} = x[0] - x[2] \end{aligned} \tag{10}$$

$$X_o[0] = x[1] + x[3] \tag{11}$$

$$X_o[0] = x[1] - x[3] \tag{12}$$

b. $x[n] = \{1, 1, 1, 1\}, N = 4$

  In this case, it is: $x[0] = x[1] = x[2] = x[3] = 1$

$$
\begin{aligned}
(9) &\Rightarrow X_e[0] = 2 \\
(10) &\Rightarrow X_e[1] = 0 \\
(11) &\Rightarrow X_o[0] = 2 \\
(12) &\Rightarrow X_o[1] = 0
\end{aligned} \tag{13}
$$

$\overset{k=0}{(7)} \Rightarrow X[0] = X_e[0] + X_o[0] = 4$

$\overset{k=1}{(7)} \Rightarrow X[1] = X_e[1] + W_4^1 X_o[1] \Rightarrow$

  $X[1] = X_e[1] + e^{-j\frac{2\pi}{4}} X_o[1] = X_e[1] + e^{-j\frac{\pi}{2}} X_o[1] \Rightarrow$
  $X[1] = X_e[0] + jX_o[1] = 0$

$\overset{k=0}{(8)} \Rightarrow X[2] = X_e[0] + W_4^0 X_o[0] = X_e[0] - X_o[0] = 0$

$\overset{k=1}{(8)} \Rightarrow X[3] = X_e[1] + W_4^1 X_o[1] = X_e[1] - jX_o[1] = 0$

Hence, the DFT of $x[n]$, using radix-2 FFT, is $X\{k\} = \{4, 0, 0, 0\}$

5.9.18 Let the real signal $x[n]$, $n = 0, 1, 2, \ldots, 7$:

$$x[n] = \{2, 0, 1, 0, -1, 1, -2, 1\}$$

Compute the DFT of the signal using the FFT. How many complex multiplications are required for the computation of the DFT and of the FFT, respectively?

SOLUTION: $x[n] = \{2, 0, 1, 0, -1, 1, -2, 1\}$, $N = 8 \Rightarrow v = 3$, so three stages of segmentation are required.

1st stage   2nd stage

$x[n] = x_e[n] + x_o[n]$
$x_e[n] = \{2, 1, -1, -2\}$ $x_{ee}[n] = \{2, -1\}$ $x_{oe}[n] = \{0, 1\}$
$x_o[n] = \{0, 0, 1, 1\}$ $x_{eo}[n] = \{1, -2\}$ $x_{oo}[n] = \{0, 1\}$ etc.

For the first segmentation stage, the FFT equations are:

$$X[k] = X_e[k] + W_N^k X_o[k], \qquad k = 0,1,2,\ldots,\frac{N}{2}-1 \quad (1)$$

$$X[k+\frac{N}{2}] = X_e[k] - W_N^k X_o[k], \qquad k = 0,1,2,\ldots,\frac{N}{2}-1 \quad (2)$$

$$(1) \Rightarrow X[k] = X_e[k] + W_8^k X_o[k], \qquad k = 0,1,2,3 \quad (1)'$$

$$(2) \Rightarrow X[k+4] = X_e[k] - W_8^k X_o[k], \qquad k = 0,1,2,3 \quad (2)'$$

$$(1)' \Rightarrow X[0]=X_e[0] + X_o[0]$$

$$X[1]=X_e[1]+e^{-j\frac{2\pi}{8}} X_o[1]$$

$$X[2]=X_e[2]+e^{-j\frac{2\pi}{8}2} X_o[2]$$

$$X[3]=X_e[3]+e^{-j\frac{2\pi}{8}3} X_o[3]$$

The DFT of $x_e[n]$ is:

$$X_e[k] = \sum_{n=0}^{3} x_e[n]W_4^{kn} = x_e[0] + x_e[1]W_4^k + x_e[2]W_4^{2k} + x_e[3]W_4^{3k} \text{ or}$$

$$\begin{bmatrix} X_e[0] \\ X_e[1] \\ X_e[2] \\ X_e[3] \end{bmatrix} = \begin{bmatrix} 1 & 1 & 1 & 1 \\ 1 & W_4^1 & W_4^3 & W_4^3 \\ 1 & W_4^2 & W_4^0 & W_4^2 \\ 1 & W_4^3 & W_4^2 & W_4^1 \end{bmatrix} \cdot \begin{bmatrix} x_e[0] \\ x_e[1] \\ x_e[2] \\ x_e[3] \end{bmatrix} \Rightarrow$$

$$\begin{bmatrix} X_e[0] \\ X_e[1] \\ X_e[2] \\ X_e[3] \end{bmatrix} = \begin{bmatrix} 1 & 1 & 1 & 1 \\ 1 & -j & -1 & j \\ 1 & -1 & 1 & -1 \\ 1 & j & -1 & -j \end{bmatrix} \cdot \begin{bmatrix} 2 \\ 1 \\ -1 \\ -2 \end{bmatrix} = \begin{bmatrix} 0 \\ 3-3j \\ 2 \\ 3+3j \end{bmatrix}$$

Likewise, for $X_o(k)$, it is:

$$\begin{bmatrix} X_o[0] \\ X_o[1] \\ X_o[2] \\ X_o[3] \end{bmatrix} = \begin{bmatrix} 1 & 1 & 1 & 1 \\ 1 & -j & -1 & j \\ 1 & -1 & 1 & -1 \\ 1 & j & -1 & -j \end{bmatrix} \cdot \begin{bmatrix} 0 \\ 0 \\ 1 \\ 1 \end{bmatrix} = \begin{bmatrix} 2 \\ -1+j \\ 0 \\ -1-j \end{bmatrix}$$

$$(3) \Rightarrow X[0] = 2$$

$$X[1] = 3 - 3j + e^{-j\frac{\pi}{4}}(-1+j) = 3 - j1.586$$

$$X[2] = 2 + e^{-j\frac{\pi}{2}} \cdot 0 = 2$$

$$X[3] = 3 + 3j + e^{-j\frac{3\pi}{4}}(-1-j) = 3 + j4.414$$

For the remaining part of the DFT, it is:

$$(2)' \Rightarrow X[4] = X_e[0] - X_o[0] = -2$$

$$X[5] = X_e[1] - W_8^1 X_o[1] = 3 - 3j + e^{-j\frac{2\pi}{8}}(-1+j) = 3 - j4.414$$
$$X[6] = X_e[2] - W_8^2 X_o[2] = 2$$
$$X[7] = X_e[3] - W_8^3 X_o[3] = 3 + j1.585$$

Note that if we used the second segmentation stage, the FFT equations would be:

$$X_e[k] = X_{ee}[k] + W_N^{2k} X_{eo}[k]$$
$$X_e\left[k + \frac{N}{4}\right] = X_{ee}[k] - W_N^{2k} X_{eo}[k]$$
$$X_o[k] = X_{oe}[k] + W_N^{2k} X_{oo}[k] \qquad k = 0,1,2,\ldots,\frac{N}{4} - 1$$
$$X_o\left[k + \frac{N}{4}\right] = X_{oe}[k] - W_N^{2k} X_{oo}[k]$$

The computation of the DFT requires $8^2=64$ complex multiplications, whereas the FFT requires $\dfrac{N \cdot v}{2} = \dfrac{8 \cdot 3}{2} = 12$ complex multiplications.

5.9.19 Calculate the DFT of the sequence $x[n] = \{\alpha, 2\alpha, 4\alpha, 8\alpha\}$ with radix-2 FFT and describe the process rigorously. How many multiplications are needed for the DFT and the FFT, respectively?

SOLUTION: $x[n] = \{\alpha, 2\alpha, 4\alpha, 8\alpha\}$, $N = 4 = 2^2 \Rightarrow v = 2$, so two stages of analysis are required.

$$x[n] = x_e[n] + x_0[n] \begin{cases} x_e[n] = \{\alpha, 4\alpha\} \\ x_0[n] = \{2\alpha, 8\alpha\} \end{cases}$$

FFT Equations

$$X[k] = X_e[k] + W_N^k X_0[k] \qquad k = 0, 1 \quad (1)$$
$$X[k+2] = X_e[k] - W_N^k X_0[k] \qquad k = 0, 1 \quad (2)$$

$$(1) \Rightarrow X[0] = X_e[0] + X_0[0]$$

$$X_e[k] = \begin{bmatrix} W_2^0 & W_2^0 \\ W_2^0 & W_2^1 \end{bmatrix} \begin{bmatrix} x_e[0] \\ x_e[1] \end{bmatrix} = \begin{bmatrix} 1 & 1 \\ 1 & -1 \end{bmatrix} \begin{bmatrix} \alpha \\ 4\alpha \end{bmatrix} = \begin{bmatrix} 5\alpha \\ -3\alpha \end{bmatrix}$$

$$X_0[k] = \begin{bmatrix} W_2^0 & W_2^0 \\ W_2^0 & W_2^1 \end{bmatrix} \begin{bmatrix} x_0[0] \\ x_0[1] \end{bmatrix} = \begin{bmatrix} 1 & 1 \\ 1 & -1 \end{bmatrix} \begin{bmatrix} 2\alpha \\ 8\alpha \end{bmatrix} = \begin{bmatrix} 10\alpha \\ -6\alpha \end{bmatrix}$$

$$(1) \Rightarrow \begin{cases} X[0] = 5\alpha + 10\alpha = 15\alpha \\ X[1] = X_e[1] + W_4^1 X_0[1] = -3\alpha + j6\alpha \end{cases}$$

$$(2) \Rightarrow \begin{cases} X[2] = X_e[0] - W_4^0 X_0[0] = 5\alpha - 10\alpha = -5\alpha \\ X[3] = X_e[1] - W_4^1 X_0[1] = X_e[1] + jX_0[1] = -3\alpha - j6\alpha \end{cases}$$

Thus, $X[k] = \{15\alpha, -3\alpha + j6\alpha, -5\alpha, -3\alpha - j6\alpha\}$.

The same result can be obtained if we divide again $X_e[k]$ and $X_0[k]$ The DFT requires 16 multiplications, whereas the FFT requires 4 multiplications.

5.9.20 Compute the linear convolution of $h[n]$ and $x[n]$ using the *overlap and add* method, when $h[n] = \{1, 1, 1\}$ and $x[n] = \{0, 1, 2, 3,...\}$

SOLUTION: The length of $h[n]$ is $N=3$, whereas $x[n]$ is much longer. By choosing $M=2^3=8$, $h[n]$ expands to a sequence of length 8 (by adding 5 (= M-N) zero samples at its end), i.e.,

$$h[n] = \{1, 1, 1, 0, 0, 0, 0, 0,\} \qquad (1)$$

The sequence $x(n)$ is divided in segments of length M-N+1=8-3+1=6, and then, N-1=3-1=2 zeros are added at the end of each subsequence so that each of them becomes of length 8. Hence:

$$
\left.\begin{array}{l}
x_0[n] = \{0,1,2,3,4,5,0,0\} \\
x_1[n] = \{6,7,8,9,10,11,0,0\} \\
x_2[n] = \{12,13,14,15,16,17,0,0\} \\
\vdots
\end{array}\right\} \tag{2}
$$

The circular convolution of each of the subsequences with $h[n]$ is:

$$y_0[n] = x_0[n] \otimes h[n] = \{0,1,2,3,4,5,0,0\} \otimes \{1,1,1,0,0,0,0,0\} \Rightarrow$$

$$y_0[n] = \{1,1,1,0,0,0,0,0\} \otimes \{0,1,2,3,4,5,0,0\}$$

$$x_0[-n] = \{0,0,0,5,4,3,2,1\}$$

thus:

$$y_0[0] = (1,1,1,0,0,0,0,0) \times (0,0,0,5,4,3,2,1)^T = 0$$

$$y_0[1] = (1,1,1,0,0,0,0,0) \times (1,0,0,0,5,4,3,2)^T = 1 \times 1 = 1$$

$$y_0[2] = (1,1,1,0,0,0,0,0) \times (2,1,0,0,0,5,4,3)^T = 1 \times 2 + 1 \times 1 = 3$$

$$y_0[3] = (1,1,1,0,0,0,0,0) \times (3,2,1,0,0,0,5,4)^T = 1 \times 3 + 1 \times 2 + 1 \times 1 = 6$$

$$y_0[4] = (1,1,1,0,0,0,0,0) \times (4,3,2,1,0,0,0,5)^T = 1 \times 4 + 1 \times 3 + 1 \times 2 = 9$$

$$y_0[5] = (1,1,1,0,0,0,0,0) \times (5,4,3,2,1,0,0,0)^T = 1 \times 5 + 1 \times 4 + 1 \times 3 = 12$$

$$y_0[6] = (1,1,1,0,0,0,0,0) \times (0,5,4,3,2,1,0,0)^T = 1 \times 0 + 1 \times 5 + 1 \times 4 = 9$$

$$y_0[7] = (1,1,1,0,0,0,0,0) \times (0,0,5,4,3,2,1,0)^T = 1 \times 0 + 1 \times 0 + 1 \times 5$$

$$y_0[n] = \left\{0,1,3,6,9,12,\underbrace{9,5}_{\substack{N-1=2 \\ \text{last samples}}}\right\} \tag{3}$$

Likewise, $y_1[n] = x_1[n] \otimes h[n] = h[n] \otimes x_1[n] \Rightarrow$

$$y_1[n] = \{1,1,1,0,0,0,0,0\} \otimes \{6,7,8,9,10,11,0,0\}$$

$$x_2[-n] = \{6,0,0,11,10,,9,8,7\}$$

Likewise, $y_1[n] = \left\{6,13,21,24,27,30,\underbrace{21,11}_{N-1}\right\}$ last samples (4)

and $y_2[n] = x_2[n] \otimes h[n] = \{1, 1, 1, 0, 0, 0, 0, 0,\} \otimes \{12, 13, 14, 15, 16, 17, 0, 0\} \Rightarrow$
$y_2[n] = \left\{12,25,39,42,45,48,\underbrace{33,17}_{N-1}\right\}$ last samples (5) etc.

The sequence of the linear convolution, $y[n]$, consists of the above subsequences, $y_0[n]$, $y_1[n]$, $y_2[n]$, etc., with the difference that the last N-1=3-1=2 samples of each segment must be added to the next N-1=2 first samples of the next segment. Therefore:

$$y[n] = \{0,1,3,6,9,12,9+6,5+13,21,24,27,30,21+12,11+25,39,42,45,...\} \Rightarrow$$
$$y[n] = \{0,1,3,6,9,12,15,18,21,24,27,30,33,36,39,42,45,...\}.$$ (6)

5.9.21 Compute the linear convolution of $h[n]$ and $x[n]$ using the *overlap and save* method, when: $h[n] = \{1,1,1\}$ and $x[n] = \{0,1,2,3,...\}$

SOLUTION: Since, now, N=3 and M=2^3=8, $h[n]$ is extended to a sequence of length 8 by adding M-N=5 zero samples at the end, i.e.,

$$h[n] = \{1,1,1,0,0,0,0,0\}$$ (1)

$x[n]$ is divided in segments of length 8 (=M) overlapped by N-1=3-1=2 samples, i.e.,

$$x_0[n] = \{0,0,0,1,2,3,4,5\}$$
$$x_1[n] = \{4,5,6,7,8,9,10,11\}$$ (2)
$$x_2[n] = \{10,11,12,13,14,15,16,17\}$$
$$\vdots$$

The circular convolution of each segment of $x[n]$ with $h[n]$ is:

$$y_0[n] = x_0[n] \otimes h[n] = h[n] \otimes x_0[n] \Rightarrow$$
$$y_0[n] = \{1,1,1,0,0,0,0,0\} \otimes \{0,0,0,1,2,3,,4,5\}$$
$$x_0[-n] = \{0,5,4,3,2,1,0,0\}$$

$$\Rightarrow y_0[0] = (1,1,1,0,0,0,0,0) \times (0,5,4,3,2,1,0,0)^T = 1 \times 5 + 1 \times 4 = 9$$

$$y_0[1] = (1,1,1,0,0,0,0,0) \times (0,0,5,4,3,2,1,0)^T = 1 \times 5 = 5$$

$$y_0[2] = (1,1,1,0,0,0,0,0) \times (0,0,0,5,4,3,2,1)^T = 0$$

$$y_0[3] = (1,1,1,0,0,0,0,0) \times (1,0,0,0,5,4,3,2)^T = 1 \times 1 = 1$$

$$y_0[4] = (1,1,1,0,0,0,0,0) \times (2,1,0,0,0,5,4,3)^T = 1 \times 2 + 1 \times 1 = 3$$

$$y_0[5] = (1,1,1,0,0,0,0,0) \times (3,2,1,0,0,0,5,4)^T = 1 \times 3 + 1 \times 2 + 1 \times 1 = 6$$

$$y_0[6] = (1,1,1,0,0,0,0,0) \times (4,3,2,1,0,0,0,5)^T = 1 \times 4 + 1 \times 3 + 1 \times 2 = 9$$

$$y_0[7] = (1,1,1,0,0,0,0,0) \times (5,4,3,2,1,0,0,0)^T = 1 \times 5 + 1 \times 4 + 1 \times 3 = 12$$

Thus,

$$y_0[n] = \{9,5,0,1,3,6,9,12\} \tag{3}$$

Likewise, $y_1[n] = x_1[n] \otimes h[n] = \{25,12,15,18,21,24,27,30\}$ (4)

and $y_2(n) = x_2(n) \otimes h(n) = \{43,38,33,36,39,42,45,48\}$ (5)

The linear convolution sequence required is derived from all the above sequences provided that previously the first $N-1=2$ elements of each segment are eliminated, i.e.,

$$y[n] = \{0,1,3,6,9,12,15,18,21,24,27,30,33,36,39,42,45,48,\dots\} \tag{6}$$

We notice that the result of $y[n]$ is the same with what we found by applying the *overlap and add* method.

5.9.22 Let the signal $cos(2\pi20t)$ be defined for $n = 0{:}30$ and sampled with a sampling frequency, $f_s = 200Hz$. Plot its DTFT from $-f_s$ to f_s. Do the same for the signal $x[n] = (-0.9)^n$ defined for $n=-5{:}5$ and for 100 frequencies in the interval $0 - \pi$, but in total from -2π to 2π.

SOLUTION:

a. The discrete-time signal is:

$$\cos(2\pi20nT_s) = \cos\left(2\pi\frac{20}{f_s}n\right) = \cos(2\pi0.1n)$$

The MATLAB code for the computation of the DTFT of the signal is:
clear;

```
M=100; % M frequencies in the interval 0 - 2pi
Fs = 200;
n=0:30;
k=-100:100; % (k=100 is equal to w=2*pi)
x=cos(0.2.*pi.*n); % vector with 31 values, since vector n
has 31 points
X = x * (exp(-j*2*pi/M)).^( n'*k);
F= (2*(Fs/2)/M)*k; % since fs/2 corresponds to pi, the freq.
step size 2(Fs/2)/M
% is a vector of length equal to the length of k !
plot(F, abs(X))
axis([-200 200 0 18])
xlabel('Frequency (Hz)')
ylabel('Magnitude of X, |X|')
```

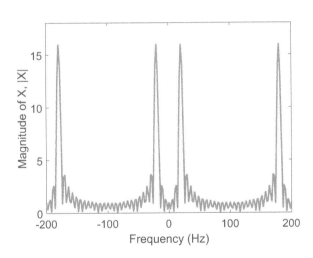

We can easily observe that the signal has two components at 20 Hz and 20 Hz, as expected. Also, the spectrum is periodic with frequency *fs* = 200 *Hz*.

b. The MATLAB code for the computation of the DTFT of the signal is:
clear;

```
M=100;
```

```
n=-5:5;
k=-200:200; % (k=200 is equal to w=2*pi since k=100 is w=pi)
x=(-0.9).^n;
X = x * (exp(-j*pi/M)).^( n'*k);
W= (pi/M)*k;
plot(W/pi, abs(X)) % (We avoid π and its multiples at the
horizontal axis!)
axis([-2 2 0 15])
```

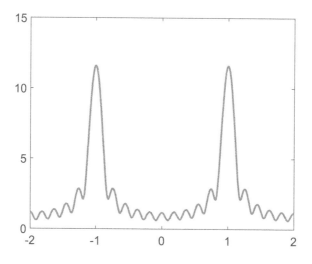

5.9.23 Plot the DTFT of the signal $x[n] = 0.8^n$, $0 \leq n \leq 20$ at the frequency range $-5\pi \leq \Omega \leq 5\pi$.

SOLUTION: Since the DTFT of a signal is a complex function, we shall plot its magnitude and phase. The MATLAB code for the computation of the DTFT of the signal is:

```
% Generation of the signal x[n] = 0.8ⁿ, 0 ≤ n ≤ 20.
syms w
n= 0:20;
x=0.8.^n;
% Computation of the DTFT of x[n]
X=sum(x.*exp(-j*w*n));
ezplot(abs(X), [-5*pi 5*pi]);
```

legend('Magnitude of the DTFT at 5 periods')

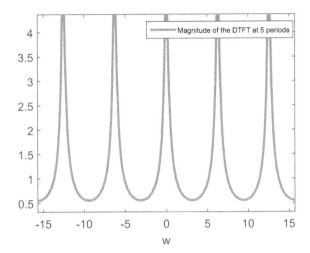

It is evident that the DTFT is periodic (at the frequency domain) with a period of 2π.

```
>>% Plot of the phase of the DTFT at -5π ≤ Ω ≤ 5π
w1=-5*pi:.01:5*pi;
XX=subs(X,w,w1);
plot(w1,angle(XX));
xlim([-5*pi 5*pi])
legend('Phase of the DTFT at 5 periods')
```

5.9.24 Compute the DFT of the sequence $x[n] = [1, 2, 2, 1]$, $0 \le n \le 3$ Plot the magnitude, the phase, the real and the imaginary part of the DFT X_k of the signal.

SOLUTION: We compute the DFT of the sequence $x[n]$ according to the following

relation $X_k = \displaystyle\sum_{n=0}^{N-1} x(n)e^{-j\frac{2\pi nk}{N}}, k = 0,1,...,N-1.$

```
x=[1 2 2 1];
N=length(x);
for k=0:N-1
for n=0:N-1
X(n+1)=x(n+1)*exp(-j*2*pi*k*n/N);
end
Xk(k+1)=sum(X);
end
Xk
```

The DFT of the sequence $x[n] = [1, 2, 2, 1]$, $0 \leq n \, 3$ is: $X_k = \{6, -1 - i, 0, -1 + i\}$.

```
Plot of the magnitude |X_k| of the DFT X_k
% magnitude=abs(Xk);
stem(0:N-1,magnitude);
legend ('Magnitude of Xk')
xlim([-.5 3.5]);
ylim([-.5 6.5]);
```

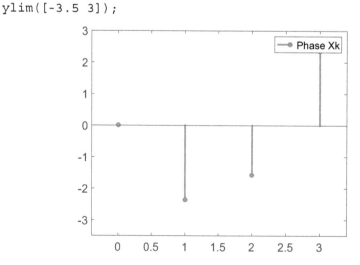

```
%Plot of the phase ∠X_k of the DFT X_k
Phase=angle(Xk);
stem(0:N-1,Phase);
legend ('Phase Xk')
xlim([-.4 3.4]);
ylim([-3.5 3]);
```

```
% Plot of the real part of Xk
realpart=real (Xk);
stem(0:N-1,realpart);
xlim([-.4 3.4]);
ylim([-1.5 6.5]);
legend ('real Xk')
```

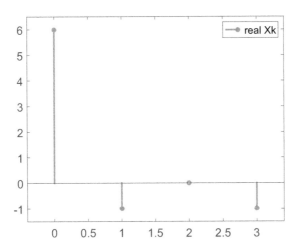

```
% Plot of the imaginary part of Xk
impart=imag (Xk);
stem(0:N-1,impart);
xlim([-.4 3.4]);
ylim([-1.5 1.5]);
legend ('imag Xk')
```

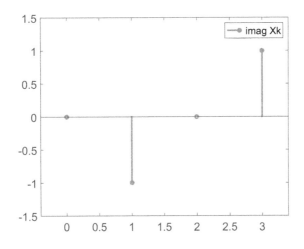

5.9.25 Create the *function* in MATLAB which computes the DFT of a sequence, $x[n]$ Apply this code to the sequence $x[n] = [1, -2, 2, 1], 0 \le n \le 3$.

SOLUTION: We will present two different methods for implementing this *function*. The first way is based on a double *for*-loop, and the second and more complex way is more efficient since it uses only one loop.

```
function Xk = dft(x);

N=length(x);

for k=0:N-1

for n=0:N-1

X(n+1)=x(n+1)*exp(-j*2*pi*k*n/N);

end

Xk(k+1)=sum(X);

end
```

First suggested solution:

We run the *function* in MATLAB.

```
x=[1 -2 2 1];

Xk=dft(x)
```

The DFT of $x[n]$ is: $X[k] = \{2, -1 + 3i, 4, -1 - 3i\}$

Second suggested solution: The '*if*' in the *function* ensures that x is a column vector. So, the multiplication in the loop is, in fact, an inner product, resulting in a one-dimensional element.

```
function Xk =dft2(x);

[N,M]=size(x);

if M~=1

x=x';

N=M;

end

Xk=zeros(N,1);

n=0:N-1;

for k=0:N-1

Xk(k+1)=exp(-j*2*pi*k*n/N)*x;

end
```

We run the *function* in MATLAB.

```
x=[1 -2 2 1];
Xk=dft2(x)
```

The DFT of $x[n]$ is: $X[k] = \{2, -1 + 3i, 4, -1 - 3i\}$

5.9.26 Let the analog signal $x_a(t) = e^{-1000|t|}$. Plot the DFT of the signal obtained after sampling the analog signal with sampling frequency $F_s = 1000$ *samples*/sec.

SOLUTION: Using the following code, we plot the analog signal and the Continuous Time Fourier Transform.

```
%Analog signal
Dt=0.00005;
t=-0.005:Dt:0.005;
xa=exp(-1000*abs(t));
% Continues Time Fourier Transform
Wmax=2*pi*2000;
K=500;k=0:1:K;
W=k*Wmax/K;
Xa=xa*exp(-j*t'*W) * Dt;
Xa=real(Xa);
W=[-fliplr(W),W(2:501)];%Frequency from -Wmax to Wmax
Xa=[fliplr(Xa),Xa(2:501)];
subplot(1,1,1)
subplot(2,1,1);plot(t*1000,xa);
xlabel('t in msec.');ylabel('xa(t)')
title('Analog Signal')
subplot(2,1,2);plot(W/(2*pi*1000),Xa*1000);
xlabel('Frequency in KHz');ylabel('Xa(jW)*1000')
title(' Continues Time Fourier Transform ')
```

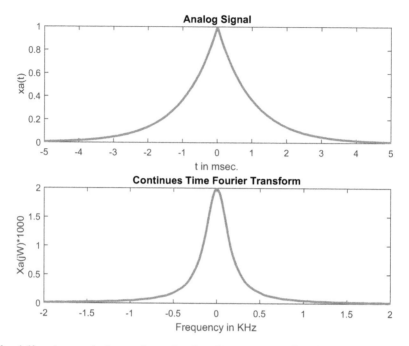

The following code is used to plot the discrete signal and its DFT.

```
%Analog Signal
Dt=0.00005;
t=-0.005:Dt:0.005;
xa=exp(-1000*abs(t));
%Discrete Signal
Ts=0.001;n=-5:1:5;
x=exp(-1000*abs(n*Ts));
%DFT
K=500;k=0:1:K;
w=pi*k/K;
X=x*exp(-j*n'*w);
X=real(X);
w=[-fliplr(w),w(2:K+1)];
X=[fliplr(X),X(2:K+1)];
subplot(1,1,1)
subplot(2,1,1);plot(t*1000,xa);
xlabel('t in msec.');ylabel('xa(t)')
```

```
title('Discrete signal');hold on
stem(n*Ts*1000,x);
gtext('Ts=1 msec');
hold off
subplot(2,1,2);plot(w/pi,X);
xlabel('Frequency in KHz');ylabel('X(w)')
title('DFT')
```

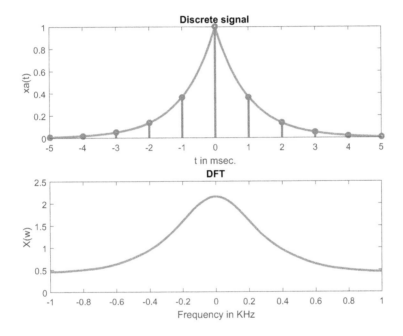

5.9.27 Compute the IDFT of the sequence $X_k = \{6, -1-j, 0, -1+j\}, 0 \le k \le 3$.

SOLUTION: We compute the IDFT of X_k according to the relation:

$$x(n) = \frac{1}{N}\sum_{k=0}^{N-1} X_k e^{j\frac{2\pi nk}{N}}, \ n = 0, 1, \ldots, N-1.$$

```
Xk=[6, -1-j,0,-1+j];
N=length(Xk);
for n=0:N-1
for k=0:N-1
xn(k+1)=Xk(k+1)*exp(j*2*pi*n*k/N);
```

```
end

    x(n+1)=sum(xn);

end

x=(1/N)*x
```

The result is: $x[n] = [1, 2, 2, 1], 0 \leq n \leq 3.$

5.9.28 Write the *function* in MATLAB that computes the IDFT of a sequence, X_k.

SOLUTION: As in problem **5.9.21**, we will develop two *functions*, one implemented by using a double *for*-loop and one implemented by using an inner vector product. We will apply them to $X_k = \{6, -1 - j, 0, -1 + j\}, 0 \leq k \leq 3.$

First *function* for the computation of the IDFT of a sequence, X_k.

```
function x=idft(Xk)

N=length(Xk);

for n=0:N-1

        for k=0:N-1

        n(k+1)=Xk(k+1)*exp(j*2*pi*n*k/N);

        end

        x(n+1)=sum(xn);

end

x=(1/N)*x;
```

We run the *function* in MATLAB as follows:

```
Xk=[6, -1-j,0,-1+j];

x=idft(Xk)
```

The IDFT of X_k is: $x[n] = \{1, 2, 2, 1\}.$

Second *function* for the computation of the IDFT of a sequence, X_k. The *if* of the *function* ensures that the X_k is a column vector. The dot before the apostrophe ensures that we do not get the conjugate complex of X_k Then, the multiplication in the loop is an inner product that produces a one-dimensional element.

```
function x=idft2(Xk);

[N,M]=size(Xk);

if M~=1

        Xk=Xk.';
```

```
        N=M;
   end
   x=zeros(N,1);
   k=0:N-1;
   for n=0:N-1
           x(n+1)=exp(j*2*pi*k*n/N)*Xk;
   end
   x=(1/N)*x;
```

We run the *function* in MATLAB as follows:

```
Xk=[6, -1-j,0,-1+j];
x=idft2(Xk)
```

The IDFT of X_k is: $x[n] = \{1, 2, 2, 1\}$. That is the same result as the one obtained by the first *function*.

5.9.29 Find the circular shift, $x_{c,2}[n] = x[(n - 2)\bmod N]$, of the sequence $x[n] = [0.1, 0.2, 0.3, 0.4, 0.5, 0.6, 0.7, 0.8]$, $0 \le n \le 7$.

SOLUTION: First suggested way:

```
% We generate x[n], N and M = 2
x=[.1,.2,.3,.4,.5,.6,.7,.8]
N=8;
m=2;
x'
%We compute (n - m)mod N for the first element (n=0).
n=0;
p=mod((n-m),N)
% The 1st term of the circularly shifted sequence xc, 2[n],
i.e., xc, 2(0) is the 7th term of the sequence x[n], i.e.,
x[6]
xc(n+1)=x(p+1);
xcm=xc'
% The 2nd term of the circularly shifted sequence xc, 2[n],
i.e., xc, 2(1) is the 8th term of the sequence x[n], i.e.,x[7]
```

```
n=1;
p=mod((n-m),N)
xc(n+1)=x(p+1);
xcm=xc'
```

% The 3rd term of the circularly shifted sequence $x_{c,\,2}[n]$, i.e., $x_{c,\,2}(2)$ is the 1st term of the sequence $x[n]$, i.e., $x[0]$

```
n=2;
p=mod((n-m),N)
xc(n+1)=x(p+1);
xcm=xc'
```

% The 4th term of the circularly shifted sequence $x_{c,\,2}[n]$, i.e., $x_{c,\,2}(3)$ is the 2nd term of the sequence $x[n]$, i.e., $x[1]$

```
n=3;
p=mod((n-m),N)
xc(n+1)=x(p+1);
xcm=xc'
```

% The 5th term of the circularly shifted sequence $x_{c,\,2}[n]$, i.e., $x_{c,\,2}(4)$ is the 3rd term of the sequence $x[n]$, i.e., $x[2]$

```
n=4;
p=mod((n-m),N)
xc(n+1)=x(p+1);
xcm=xc'
```

% By calculating the remaining 3 terms of $x_{c,\,2}[n]$ in the same way, we obtain the final value of the circularly shifted sequence.

```
for n=5:N-1
p=mod((n-m),N);
xc(n+1)=x(p+1);
end
xcm=xc'
```

The circular shift, $x_{c,2}[n] = x[(n - 2)\bmod N]$, of the sequence, $x[n]$, is $x_{c,2}[n]=\{0.7, 0.8, 0.1, 0.2, 0.3, 0.4, 0.5, 0.6\}$.

Second suggested way:

A simple way to apply a circular shift to a sequence is by using the **circshift** *function*. The *function* is compiled as **xcm = circshift (x′, m)**, i.e., the necessary condition for the *function* to run is that the input vector is a column vector. The *m* expresses the shift.

```
m=2;
xcm=circshift(x',m)
```

5.9.30 Let the sequences $x_1[n] = [1, 0, 2.5, 1.5]$, $0 \le n \le 3$ and $x_2[n] = [1, 2, 0.5, 2]$, $0 \le n \le 3$. Use MATLAB to prove the property: $DFT\{x_1[n] \otimes x_2[n]\} = X_1[k] \cdot X_2[k]$.

SOLUTION: We define the signal $y[n] = x_1[n] \otimes x_2[n]$ and compute its DFT (the left part of $DFT\{x_1[n] \otimes x_2[n]\} = X_1[k] \cdot X_2[k]$).

y = [3.75 6.75 6 6];

```
Ar=dft(y);

Ar.'
```

The following result is obtained: $Ar = \{22.5, -2.25 -0.75i, -3, -2.25 + 0.75i\}$

We define $x_1(n)$, $x_2(n)$ and compute their DFT, $X_1[k]$ and $X_2[k]$. Next, we multiply $X_1[k]$ and $X_2[k]$ computing in this way the right part of: $DFT\{x_1[n] \otimes x_2[n]\} = X_1[k] \cdot X_2[k]$.

```
x1=[1 0 2.5 1.5];

x2=[1 1 .5 2];

X1=dft(x1);

X2=dft(x2);

De=X1.*X2;

De.'
```

The following result is obtained: $De = \{22.5, -2.25 - 0.75i, -3, -2.25 + 0.75i\}$

Since the two results are the same, the property of the DFT of the circular convolution has been proved.

5.9.31 Let the sequences $x_1[n] = [1, 2, 3, 4]$, $0 \le n \le 3$ and $x_2[n] = [3, 2, 5, 1]$, $0 \le n \le 3$. Compute:

a. The circular convolution of $N = 4$ points.

b. The circular convolution of $M = 2N - 1 = 7$ points of the sequences $x_{11}[n] = [1, 2, 3, 4, 0, 0, 0]$, $0 \le n \le 6$ and $x_{22}[n] = [3, 2, 5, 1, 0, 0, 0]$, $0 \le n \le 6$, which are filled in with zeros.

c. The linear convolution of the sequences $x_1[n]$ and $x_2[n]$.

SOLUTION:

a. The MATLAB code for the computation of the circular convolution of $N=4$ points is:

```
% We define x₁[n], x₂[n] and N
x1=[1,2,3,4];
x2=[3,2,5,1];
N=length(x1);
% We compute the sequence which is the circular reflection
of x₂[n]
for m=0:N-1
p(m+1)=mod(-m,N);
end
% We compute the circular convolution of 4 points of x₁[n]
and x₂[n]
for m=0:N-1
x2s(1+m)=x2(1+p(m+1));
end
for n=0:N-1
x2sn=circshift (x2s',n);
y1(n+1)=x1*x2sn;
end
y1
y1 =
28 31 22 29
```

b. The MATLAB code for the computation of the circular convolution of $M=2N-1=7$ points of the sequences filled in with zeros is:

```
% Definition of x₁₁[n] , x₂₂[n] and of the new N.
x11=[ x1 0 0 0];
x22=[x2 0 0 0];
N=length(x11);
% We compute the sequence which is the circular reflection
of x₂₂[n]
for m=0:N-1
```

```
p(m+1)=mod(-m,N);

end

for m=0:N-1

x22s(1+m)=x22(1+p(m+1));

end
```

% We compute the circular convolution of 7 points of $x_{11}[n]$ and $x_{22}[n]$

```
for n=0:N-1

x22sn=circshift(x22s',n);

y2(n+1)=x11*x22sn;

end

y2

y2 =

3  8  18  29  25  23  4
```

c. The MATLAB code for the computation of the linear convolution of the sequences $x_1[n]$ and $x_2[n]$ is:

% We compute the linear convolution of $x_1[n]$ and $x_2[n]$ and

% we note that is equal to the circular convolution of the

% of the sequences filled in with zeros

```
y3=conv(x1,x2)

y3 =

3  8  18  29  25  23  4
```

5.9.32 Let the sequence $x[n] = 0.7^n$, $0 \le n \le 19$. Plot both the DTFT of $x[n]$ for $0 \le \omega \le 2\pi$ and the DFT of $x[n]$ for $\omega = \dfrac{2\pi k}{N}$, $k = 0,1,......N-1$ in in the same figure.

SOLUTION: The MATLAB code is:

```
n=0:19;

x=0.7.^n;

syms w

Xdtft=sum(x.*exp(-j*w*n));

Xdft=dft(x);

N=length(Xdft);

k=0:N-1;
```

```
wk=2*pi*k/N;
ezplot(abs(Xdtft), [0 2*pi]);
hold on
plot(wk,abs(Xdft),'o')
legend('Xdtft', 'Xdft')
```

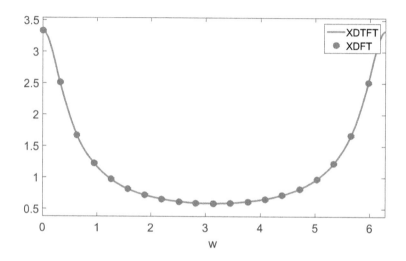

5.9.33 Let the discrete signal $x[n] = \cos\dfrac{7\pi n}{8}$. Find the DTFT of the signal via its DFT.

SOLUTION: It is $X(k) = X(e^{j\omega})\big|_{\omega=2\pi k/N}$.

We plot the DTFT of the signal using the following MATLAB code:

```
n=0:15;
N=length(n);
x=cos(2*pi*n*7/16);
xx=fft(x);
k=-(N/2):(N/2)-1;
NN=512;
xe=fft(x,NN);
kk=-(NN/2):(NN/2)-1;
plot(2*pi*kk/NN,abs(xe));
hold on; axis tight; grid on;
stem(2*pi*k/N,abs(xx),'or');
xlabel('Normalized angular frequency');
ylabel('Magnitude');
```

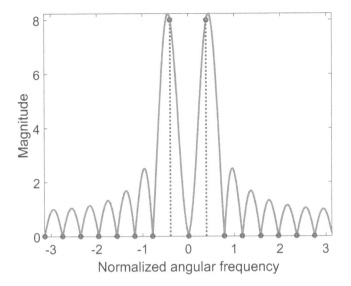

5.9.34   a.   Create a *function* in MATLAB that computes the circular convolution of two sequences (which do not necessarily have the same number of samples).

b.   Create a *function* in MATLAB that computes the circular convolution of two sequences by using the *functions* fft and ifft.

c.   Compute, using the result of exercise (a), the circular convolution of the sequences $x_1[n]$ = [1, 2, 3, 4] and $x_2[n]$ = [5, 4, 3, 2, 1].

SOLUTION:

a.   The *function* in MATLAB that computes the circular convolution of two sequences:

```
function y=circonv2(x,h)

N1=length(x);

N2=length(h);

N=max(N1,N2);

if N>length(x),

        x=[x zeros(1,N-length(x))];

end

if N>length(h),

        h=[h zeros(1,N-length(h))];

end

for m=0:N-1

p(m+1)=mod(-m,N);
```

```
hs(1+m)=h(1+p(m+1));

end

for n=0:N-1

        hsn=circshift(hs',n);

        y(n+1)=x*hsn;

end
```

b. The *function* in MATLAB that computes the circular convolution of two sequences by using the *functions* **fft** and **ifft**:

```
function y=circonv3(x,h);

N1=length(x);

N2=length(h);

N=max(N1,N2);

X=fft(x,N);

H=fft(h,N);

y=ifft(X.*H);
```

c. Computation of the circular convolution of the sequences $x_1[n] = [1, 2, 3, 4]$ and $x_2[n] = [5, 4, 3, 2, 1]$

```
x1=[1,2,3,4];

x2=[5,4,3,2,1];

y=circonv2(x1,x2)

y =

25 25 30 40 30

y=circonv3(x1,x2)

y =

25 25 30 40 30
```

5.9.35 Compute and plot the FT $X(\Omega)$ of the continuous-time signal $x(t) = t - 1, 0 \leq t \leq 2$. Then, compute and plot at the same figure the approximation $X(\Omega_k)$ for $N=128$ and $T=0.1$.

SOLUTION: Description of the procedure:
- We define the sampled signal $x[nT]$, n = 0, 1,.... N − 1, for $N=128$ and $T = 0.1$.
- We compute the DFT X_k of $x[nT]$

- We compute the frequencies, Ω_k, and the approximation, $X(\Omega_k)$, according to

  the relation $X(\Omega_k) = NT \dfrac{1 - e^{j\frac{2\pi k}{N}}}{j2\pi k} X_k, k = 0, 1, \ldots, N-1.$

- We express $x(t)$ as a symbolic function and compute its FT $X(\Omega)$

We plot $X(\Omega)$ and $X(\Omega_k)$ at the frequency range $\Omega_0 \leq \Omega \leq \Omega_{N-1}.$

```
T=0.1;
N=128;
t=0:1/(N*T):2;
x=t-1;
Xk=fft(x,N);
k=0:N-1;
wk=2*pi*k/(N*T);
Xwk=(N*T*(1-exp(-j*2*pi*k/N))./ (j*2*pi*k) ).*Xk;
syms t w
x=(t-1) *(heaviside(t)-heaviside(t-2));
X=fourier(x,w);
ezplot(abs(X), [0 wk(N)]);
hold on
plot(wk,abs(Xwk),'o')
hold off
```

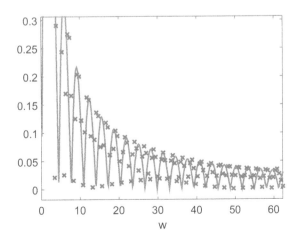

5.9.36 a. Create a *function* in MATLAB that computes the linear convolution of two sequences, using the *functions* **fft** and **ifft**.

   b. Compute, using the *function* of exercise (a), the linear convolution of the sequences $x_1[n] = [1, 2, 3, 4]\ 0 \le n \le 3$ and $x_2[n] = [7, 6, 5, 4, 3, 2, 1],\ 0 \le n \le 6$.

   c. Verify the result using the command **conv**.

SOLUTION:

a. We create the *function*:

```
function y=linconv(x,h);
N1=length(x);
N2=length(h);
N= N1 + N2-1 ;
X=fft(x,N);
H=fft(h,N);
y=ifft(X.*H);
```

b. We compute the linear convolution of the sequences:

```
x1=1:4;
x2=7:-1:1;
y=linconv(x1,x2)
y=7.0000 20.0000 38.0000 60.0000 50.0000 40.0000 30.0000
20.0000 11.0000 4.0000
```

c. We compute, using the command **conv**, the linear convolution of the sequences:

```
y=conv(x1,x2)
y =
7 20 38 60
50 40 30 20
11 4
```

5.9.37 Let the sequence $x(n) = \cos(0.48\pi n) + \cos(0.52\pi n)$. Compute the 10-point DFT of $x(n)$ and plot its amplitude with respect to the frequency.

SOLUTION:

```
n = [0:1:99]; x = cos(0.48*pi*n)+cos(0.52*pi*n);
n1 = [0:1:9]; y1 = x(1:1:10);
```

```
subplot(2, 1, 1); stem(n1, y1);
title('signal x(n), 0 <= n <= 9');
xlabel('n');
Y1 = dft(y1, 10); magY1 = abs(Y1(1:1:6));
k1 = 0:1:5; w1 = 2*pi/10*k1;
subplot(2, 1, 2); plot(w1/pi, magY1);
title('Samples of DFT Magnitude');
xlabel('Frequency in pi units');
```

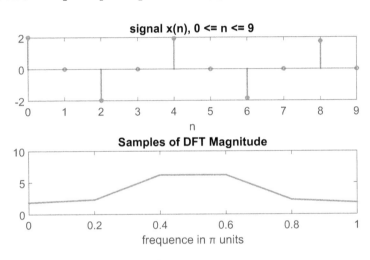

NOTE: The *function* **dft.m** that was used is given next:

```
function [Xk] = dft(xn, N)
% Computes Discrete Fourier Transform
% ----------------------------------------------
% [Xk] = dft(xn, N)
% Xk = DFT coeff. Array over 0 <= k <= N-1
% xn = N-point finite-duration sequence
% N = Length of DFT
n = [0:1:N-1];     % row vector for n
k = [0:1:N-1];     % row vector for k
WN = exp(-j*2*pi/N);    % Wn factor
nk = n'*k;  % creates a NxN matrix of nk values
WNnk = WN .^ nk;  % DFT matrix
```

```
Xk = xn * WNnk;    % row vector for DFT coefficients
% End of function
function [xn] = idft(Xk, N)
% Computes Inverse Discrete Fourier Transform
% ---------------------------------------------
% [xn] = idft(Xk, N)
% xn = N-point sequence over 0 <= n <= N-1
% Xk = DFT coeff. Array over 0 <= n <= N-1
% N = Length of DFT
%
n = [0:1:N-1];    % row vector for n
k = [0:1:N-1];    % row vector for k
WN = exp(-j*2*pi/N);    % Wn factor
nk = n'*k;  % creates a NxN matrix of nk values
WNnk = WN .^ (-nk);     % IDFT matrix
xn = (Xk*WNnk)/N; % row vector of IDFT values
% End of function
```

6

Design of Digital Filters

6.1 Introduction

A filter is the system that processes a signal in order to detect and extract desired components from unwanted parts, like noise or other signals that distort it. There are two main types of filters, the *analog* and *digital* filters. They are completely different both in their physical structure and in the way they work.

- An analog filter uses electronic circuits produce filtering effect. These filter circuits are widely applied in video signal amplification, graphic equalizer in hi-fi systems, noise reduction, etc.
- A digital filter uses a digital processor to perform numerical calculations on sampled signal values. The processor may be a PC, or a Digital Signal Processor (DSP).

In Figure 6.1, the process of filtering a signal using digital filters is illustrated.

The analog input signal must first be sampled and digitized using an analog-to-digital (ADC) converter. The resulting binary numbers, which represent successive values of the input signal obtained by the sampling procedure, are transferred to the processor, which, in turn, performs the appropriate numerical operations. These calculations typically contain multiplications of input values with certain constant numbers and summation of the products. In cases where this is necessary, the results of these calculations, which represent sampled values of the filtered signal, are driven to a digital-to-analog converter (DAC) in order to return the signal back into the analog form.

The most important advantages of digital filters with respect to analog are the following:

1. A digital filter is programmable. Its operation is determined by a program in the processor memory, which can be easily changed, whereas an analog filter can only be changed by redesigning the filter circuit.
2. A digital filter can be easily designed and implemented on a general purpose computer.
3. The characteristics of the circuit designs of an analog filter (especially those containing active components) are voltage-dependent and temperature-based. Digital filters are not affected by such problems and are therefore particularly stable in terms of time and temperature.
4. Unlike analog filters, digital filters can handle low-frequency signals with precision. As the DSP technology continues to develop, digital filters get to be

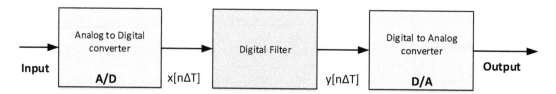

FIGURE 6.1
Signal filtering process using digital filters.

implemented even to high frequency signals in the radio frequency (RF) domain, which was formerly an exclusive area of the analog technology.

5. Digital filters are much more adaptable in their ability to produce signals in a variety of ways: this includes the ability of some types of digital filters to adjust easily to changes of the input signal characteristics.

6. Complex combinations of filters (for example in parallel or in series) can easily be implemented by fast DSP processors, making the hardware requirements relatively simple compared to the corresponding analog circuit design.

6.2 Types of Digital Filters

Linear digital filters are divided into two categories according to the length of their impulse response: The *Infinite Impulse Response* (IIR) filters and the *Finite Impulse Response* (FIR) filters.

Linear digital FIR filters have a finite response to the unit impulse function input with the great advantage of ensuring stability and an absolutely linear phase (as long as the impulse response is symmetrical or anti-symmetrical to its origin)—properties that easily explain their wide application in telecommunication systems. One disadvantage can be the need for a large filter length (therefore, increased computational power and large group delay—equal to half the length of the filter) for a sufficient approximation of the amplitude response.

Linear digital IIR filters have an infinite time response to the unit impulse function input. For the calculation of the output, we usually use input as well as output values of previous instances, so we refer to them as recursive filters. These filters have less computational complexity than the corresponding FIR filters, but they cannot ensure stability or linear phase, resulting in distortions. Digital IIR filters are widely applied in Digital Signal Processing, since all types of digital filter can be implemented as IIR filters.

6.3 Digital Filter Design Specifications

A digital filter is primarily used by its behavior with respect to frequency. This behavior is described by the frequency response of the filter both in relation to its effect on the range of the harmonic amplitudes of the signal (amplitude response) and in the way it affects the harmonic phases (phase response). Since the frequency response of digital systems is periodic, it is sufficient to determine the behavior of the digital filter in the range from 0 to π.

The steps for designing a digital filter are:

a. Determine the properties required for the filter

b. Optimally approximate the desired system by a casual time-invariant discrete system

c. Implement the system of step (b).

In Figure 6.2, the four basic types of filters are illustrated:

a. Low-pass filter

b. High-pass filter

c. Band-pass filter

d. Band-stop filter

The desired magnitude and phase response of the digital filter constitute the required specifications for the filter design. In order to understand the specifications of a digital filter, let us consider the case of Figure 6.3, where the frequency response of the filter is illustrated (dashed line).

In Figure 6.3, three bands are present:

a. Pass band: The area $[0,\omega_p]$ where the magnitude $\left|H\left(e^{j\omega}\right)\right|$ is within $1\pm\delta_1$, where $\pm\delta_1$ the deviation of the pass band (the required tolerance of the pass band).

$$1-\delta_1 \leq \left|H\left(e^{j\omega}\right)\right| \leq 1+\delta_1 \quad ,\forall\omega \leq \omega_p \tag{6.1}$$

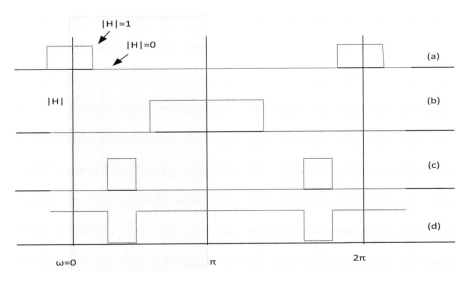

FIGURE 6.2
Filter types.

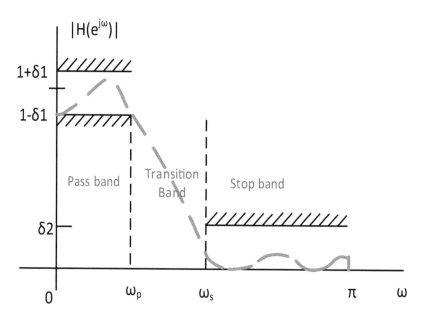

FIGURE 6.3
Frequency response of an analog low-pass filter.

  b. Transition band: The area between ω_p and ω_s, i.e., the pass band and stop band cut-off frequencies, where δ_2 is the deviation of the stop band (ripple) and there are no constraints on the response magnitude.

$$\delta_2 < \left|H\left(e^{j\omega}\right)\right| < 1 - \delta_1 \tag{6.2}$$

  c. Stop band: The area that the magnitude $\left|H\left(e^{j\omega}\right)\right|$ is almost zero with a deviation δ_2, i.e.,

$$\left|H\left(e^{j\omega}\right)\right| \le \delta_2 \ , \ \omega_s \le |\omega| \le \pi \tag{6.3}$$

ω_s and ω_p are defined as fractions of π, whereas the deviations δ_1 and δ_2 are in decibels (dB).

6.4 Design of Digital IIR Filters

In this section, we study IIR filters on the condition that they are realizable and stable. Therefore, the discrete impulse response must be:

$$h[n] = 0 \text{ for } n < 0 \quad and \quad \sum_{n=0}^{\infty} \left|h[n]\right| < \infty \tag{6.4}$$

The general form of the transfer function of an IIR filter is:

$$H(z) = \sum_{n=0}^{\infty} h[n]z^{-n} = \frac{\sum_{m=0}^{M} \alpha_m z^{-m}}{1 + \sum_{k=1}^{N} b_k z^{-k}} \tag{6.5}$$

The filter of the Equation 6.5 has M zeros and N poles, which must be within the unit circle to satisfy the stability condition. Usually, it is $M \leq N$, and the filter is known as being of N order.

The design methods of IIR filters are classified into the following categories:

a. *Indirect design methods*, whereby an analog filter is, at first, designed as a standard (Butterworth low-pass, normalized Chebychev low-pass, elliptical low-pass, Bessel low-pass), where its transfer function, $H(s)$, is determined, and it is of the form:

$$H(s) = \frac{\sum_{m=0}^{M} \alpha_m s^m}{\sum_{k=0}^{N} b_n s^n} = \frac{\prod_{m=0}^{M}(s + z_m)}{\prod_{k=1}^{N}(s + p_k)} \tag{6.6}$$

with differential Equation 6.7:

$$\sum_{k=0}^{N} b_k \frac{d^k y(t)}{dt^k} = \sum_{m=0}^{M} \alpha_m \frac{d^m x(t)}{dt^m} \tag{6.7}$$

The indirect design methods are:
1. The impulse invariant method
2. The step invariant method
3. The backward and forward difference methods
4. The bilinear or Tustin method
5. The matched pole-zero method

b. *Direct design methods*, whereby the digital filter is directly designed without having an analog standard filter designed, at first. The direct methods are:
1. The method of $|H(e^{j\omega})|^2$ design
2. The method of $h[n]$ design

c. *Optimization methods*, whereby the coefficients of the filters are calculated using optimization criteria. The optimization methods are:
1. The least squares method of frequency response
2. The linear programming method
3. The use of an all-pass filter

6.5 Indirect Methods of IIR Filter Design

The most important points of comparison of the indirect methods of filter design to be studied next are:

- Ease of use
- Conservation of stability
- Conservation of impulse response
- Conservation of harmonic (frequency) response

6.5.1 The Impulse Invariant Method

In this method, the impulse response of the digital filter is obtained by the impulse response sampling of the analog filter. This method has the advantage of retaining the form of the impulse response of the analog filter to the corresponding digital filter and is usually applied to design low-pass and band-pass filters.

For a system in which the transfer function is $H(s) = \dfrac{Y(s)}{X(s)}$, its impulse response in the z-plane will be: $H(z) = \dfrac{Y(z)}{X(z)} = Z\big[H(s)\big]$. The equivalent discrete-time filter of the impulse invariant method is that whose impulse response is identical to the impulse response of the continuous time filter at kT time points, $k = 0,1,2,3 \dots, T$, where T is the sampling period.

The impulse response in the z-plane is the inverse z-transform of the transfer function $H(z)$, whereas in the s-plane, the impulse response is the inverse Laplace transform of the transfer function $H(s)$. The digital transfer function is obtained from the analog transfer function as follows:

- Applying an inverse z-transform to $H(s)$ yields $h(t)$ in the time domain
- By sampling $h(t)$, $h[kT]$ is obtained where T is the sampling period.
- z-transforming $h[kT]$ results in $H(z)$ in the z-plane.

Thus,

$$H(s) \xrightarrow{\ L^{-1}\ } h(t) \xrightarrow{\ t=kT\ } h[kT] \xrightarrow{\ ZT\ } H(z) \tag{6.8}$$

or

$$H(z) = Z\big[h[kT]\big], \quad where\ h[kT] = \big[L^{-1}H(s)\big]_{t=kT} \tag{6.9}$$

Since the z-transform always correlates a stable pole of the s-plane to a stable pole in the z-plane, we conclude that the digital filter will be stable as long as the original analog filter is also stable. This method results in non-conservation of the frequency response (frequency distortion is observed due to overlapping). The step response is also not conserved.

6.5.2 Step Invariant Method (or *z*-Transform Method with Sample and Hold)

The purpose of this method is to construct a digital filter, $H(z)$, whose step response will consist of samples of the continuous system step response at time points kT, $k = 0,1,2,3...$ where T is the sampling period.

Consider the system illustrated in Figure 6.4.

The digital transfer function $H(z)$ is derived from the analog transfer function using Equation 6.10 or 6.11.

$$H(z) = Z\left[G_h(s) \cdot H(s)\right] = Z\left[\frac{1-e^{Ts}}{s} \cdot H(s)\right] \tag{6.10}$$

$$H(z) = (1-z^{-1})Z\left(\frac{H(s)}{s}\right) \tag{6.11}$$

If the original analog filter $H(s)$ is stable, then the equivalent discrete $H(z)$, resulting from the step invariant method, will be stable. This method maintains neither the frequency response (harmonic response) nor the impulse response.

Here, the zero-order hold filter (ZOH) has been used with transfer function $G_h(s)$. A DAC is a sample and hold circuit whose output is a partly continuous function. In Figure 6.5, the

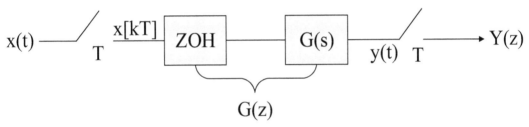

FIGURE 6.4
System of sampled data using a ZOH (zero-order hold).

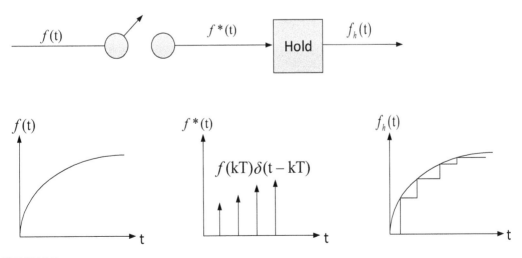

FIGURE 6.5
The operation of zero-order hold (ZOH).

operation of the zero-order hold filter (with the objective to hold the last sampled value of the $f(t)$ signal) is illustrated.

Consider the system $h_0(t)$, with the impulse response of Figure 6.6. The reconstructed signal is obtained by convolving $h_0(t)$ with $g^*(t)$, where $g^*(t)$ is the signal after sampling (i.e., $g(t) = g^*(t)^*h_0(t)$), and is illustrated in Figure 6.7.

Let $f_h(t)$, the impulse response of the hold filter, be a gate function (rectangle function), i.e., $f_h(t) = u(t) - u(t - T)$ or

$$F_h(s) = L\{u(t) - u(t - T)\} = L\{u(t)\} - L\{u(t - T)\}$$

$$\Rightarrow F_h(s) = \frac{1}{s} - \frac{1}{s}e^{-Ts} = \frac{1 - e^{-Ts}}{s} \qquad (6.12)$$

Obviously, since the input is the impulse function, the transfer function of the hold circuit is given by the following relation:

$$G_h(s) = \frac{1 - e^{-Ts}}{s} \qquad (6.13)$$

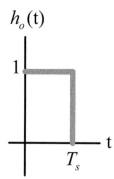

FIGURE 6.6
The impulse response $h_0(t)$.

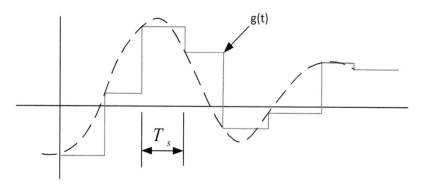

FIGURE 6.7
Reconstruction of the signal with ZOH.

6.5.3 Backward Difference Method

One way of estimating the derivative $\overset{\circ}{h}(t)$ at $t = kT$ is to calculate the difference of the current sample from the previous divided by the sampling period, as shown in Equation 6.14:

$$\overset{\circ}{h}(t) = \frac{dh(t)}{dt} = \frac{h[k] - h[k-1]}{T} \tag{6.14}$$

The derivative of $\overset{\circ}{h}(t)$ in the s- plane is $sH(s)$, whereas in the z-plane, it is:

$$z\left[\overset{\circ}{h}(t)\right] = z\left[\frac{h[k] - h[k-1]}{T}\right] = \frac{1 - z^{-1}}{T} \cdot H(z)$$

Comparing the two derivatives, it can be deduced that the conversion of n in the z-plane is given by

$$H(z) = H(s)\Big|_{s = \frac{1 - z^{-1}}{T}} \tag{6.15}$$

Relations $s = \dfrac{1 - z^{-1}}{T}$ and $z = \dfrac{1}{1 - sT}$ represent the mapping of the s-plane to the z-plane, as illustrated in Figure 6.8. By substituting $s = j\omega_s$ in $z = \dfrac{1}{1 - sT}$, we obtain:

$$z = \frac{1}{1 - sT}\Big|_{s = j\omega_s} = \frac{1}{1 - j\omega_s T} = \frac{1 + j\omega_s T}{1 + (\omega_s T)^2} = x + jy$$

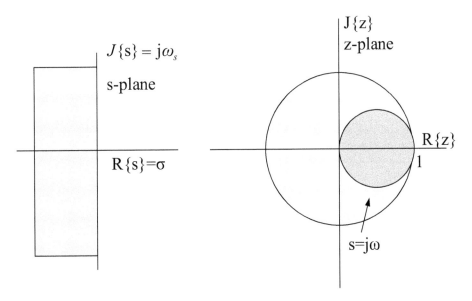

FIGURE 6.8
Mapping of the s-plane to the z-plane with the backward difference method.

$$x^2 + y^2 = \left(\frac{1}{1+(\omega_s T)^2}\right)^2 + \left(\frac{\omega_s T}{1+(\omega_s T)^2}\right)^2 = \frac{1+(\omega_s T)^2}{\left(1+(\omega_s T)^2\right)^2} \Rightarrow$$

$$x^2 + y^2 = \frac{1}{1+(\omega_s T)^2} \Rightarrow x^2 + y^2 = x \Rightarrow x^2 - x + y^2 = 0 \Rightarrow$$

$$x^2 - x + \frac{1}{4} - \frac{1}{4} + y^2 = 0 \Rightarrow \left(x - \frac{1}{2}\right)^2 + y^2 = \frac{1}{4} \qquad (6.16)$$

Equation 6.16 is an equation of a circle with a radius of $\left(\frac{1}{2}\right)$ and a center of $\left(\frac{1}{2}\right)$.

Substituting $s = \sigma + j\omega_s$ in $z = \frac{1}{1 - sT}$, the same equation is derived, and the values of σ define points in the circle $\left(\frac{1}{2}, \frac{1}{2}\right)$.

The backward difference method conserves the stability but not the harmonic response. The forward difference method is not used due to instability problems.

6.5.4 Forward Difference Method

Another way of obtaining the derivative $\overset{\circ}{h}(t)$ at $t = kT$ is by computing the difference of the next sample from the present divided by the sampling period, as shown in Equation 6.17.

$$\overset{\circ}{h}(t) = \frac{dh(t)}{dt} = \frac{h[k+1] - h[k]}{T} \qquad (6.17)$$

The derivative of $\overset{\circ}{h}(t)$ in the s-plane is $sH(s)$, whereas in the z- plane, it is:

$$z\left[\overset{\circ}{h}(t)\right] = z\left[\frac{h[k+1] - h[k]}{T}\right] = \frac{z-1}{T} \cdot H(z).$$

By comparing the two derivatives, it follows that the transform of H(s) to the z-plane is given by the Equation 6.18.

$$H(z) = H(s)\Big|_{s = \frac{z-1}{T}} \qquad (6.18)$$

The forward difference method does not always conserve the stability or the harmonic response, and therefore its application is usually avoided.

6.5.5 Bilinear or Tustin Method

This method is based on the integration of the differential equation and then on the numerical approach of the integral. With this method, the transform of the function $G(s)$ to the z-plane is given by Equation 6.19.

$$H(z) = H(s)\Big|_{s=\frac{2}{T}\frac{z-1}{z+1}} \tag{6.19}$$

The Tustin transform is defined by:

$$s = \frac{2}{T}\frac{1-z^{-1}}{1+z^{-1}} \quad \text{and} \quad z = \frac{1+s\dfrac{T}{2}}{1-s\dfrac{T}{2}} \tag{6.20}$$

It is one of the most popular methods because the stability of the analog system is conserved while the frequency response can also be conserved considering the non-linear relation between the analog and the digital frequency.

If we substitute $s = j\omega_s$ in Equation 6.20 (in order to map n-axis to the z-plane), we obtain:

$$z = \frac{1+s\dfrac{T}{2}}{1-s\dfrac{T}{2}}\Bigg|_{s=j\omega_s} = \frac{1+j\dfrac{\omega_s T}{2}}{1-j\dfrac{\omega_s T}{2}} = \frac{1-\left(\dfrac{\omega_s T}{2}\right)^2 + j\omega_s T}{1+\left(\dfrac{\omega_s T}{2}\right)^2} \Rightarrow$$

$$z = x + jy \qquad = \frac{1-\left(\dfrac{\omega_s T}{2}\right)^2}{1+\left(\dfrac{\omega_s T}{2}\right)^2} + j\frac{\omega_s T}{1+\left(\dfrac{\omega_s T}{2}\right)^2}$$

But

$$x^2 + y^2 = \left(\frac{1-\left(\dfrac{\omega_s T}{2}\right)^2}{1+\left(\dfrac{\omega_s T}{2}\right)^2}\right)^2 + \left(\frac{\omega_s T}{1+\left(\dfrac{\omega_s T}{2}\right)^2}\right)^2 = \frac{1+\left(\dfrac{\omega_s T}{2}\right)^2}{1+\left(\dfrac{\omega_s T}{2}\right)^2} = 1 \tag{6.21}$$

Equation 6.21, i.e., $x^2 + y^2 = 1$, is the equation of a circle with a radius of 1 and a center of 0, as illustrated in Figure 6.9.

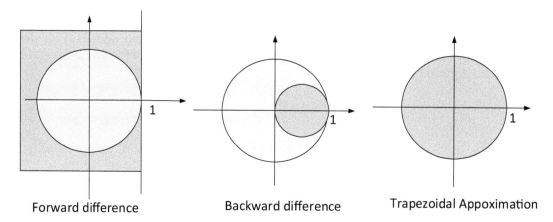

FIGURE 6.9
Stability regions for different methods.

According to Figure 6.9, the left half of the s-plane is mapped by the Tustin method within the unit circle in the z-plane, therefore the discrete system will always be stable. The disadvantage of the Tustin method is the introduction of distortion on the axis of the frequency ω due to the non-linear relation between the frequency, ω_s (in the s-plane) and ω (in the z-plane). Substituting $s = j\omega_s$ and $z = e^{j\omega}$ in Equation 6.20 yields:

$$j\omega_s = \frac{2}{T}\frac{1-e^{-j\omega}}{1+e^{-j\omega}} = \frac{2}{T}\frac{e^{-j\frac{\omega}{2}}\left(e^{j\frac{\omega}{2}} - e^{-j\frac{\omega}{2}}\right)}{e^{-j\frac{\omega}{2}}\left(e^{j\frac{\omega}{2}} + e^{-j\frac{\omega}{2}}\right)} = \frac{2}{T}\frac{2j\sin\frac{\omega}{2}}{2\cos\frac{\omega}{2}} = \frac{2}{T}j\tan\frac{\omega}{2} \Rightarrow$$

$$\omega = 2\tan^{-1}\left(\frac{\omega_s T}{2}\right) \quad \text{and} \quad \omega_s = \frac{2}{T}\tan\frac{\omega}{2} \tag{6.22}$$

The non-linear relation between ω_s and ω, referred to as frequency warping is illustrated in Figure 6.10. When the specifications of the digital filter are known, the significant analog frequencies are deliberately distorted (frequency pre-warping), so that the digital filter has the desired specifications. The limitation in the use of the bilinear transform is that the magnitude $\left|H(\omega_s)\right|$ of the analog filter must be, in parts, linear function of ω_s.

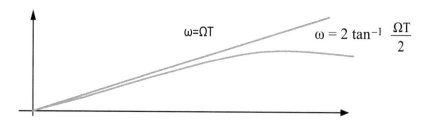

FIGURE 6.10
The non-linear relation between ω_s (in s-plane) and ω (in z-plane).

6.5.6 Matched Pole-Zero Method

A simple but, at the same time, very efficient way to find the equivalent discrete form of a continuous-time transfer function is the pole-zero matching method. If we calculate the z-transform of the $e[kT]$ samples of a continuous signal, $e(t)$, then the poles of the discrete signal $E(z)$ are associated to $E(s)$ poles according to $z = e^{sT}$. The pole-zero matching method is based on the fact that we can use z-transform and the relation $z = e^{sT}$ in order to locate $E(z)$ zeros.

According to this method, we consider separately the numerator and the denominator of the transfer function $H(s)$ of the analog filter and depict $H(s)$ poles at the poles of the discrete-time transfer function, $H(z)$, and $H(s)$ zeros at $H(z)$ zeros. An important point in this method is the way in which the zeros and the poles in infinity in the s-plane ($s = \infty$) are depicted in the z-plane. The $j\omega$-axis from $\omega = 0$ to $\omega = \dfrac{\pi}{T}\left(= \dfrac{\omega_s}{2}\ \text{Nyquist frequency}\right)$ in the s-plane is depicted using $z = e^{sT}$ to the unit circle in the z-plane, from $z = e^{j0} = 1$ to $z = e^{j\frac{\pi T}{T}} = -1$. So, by choosing ω_s, the angular frequency that satisfies the sampling theorem, we can then consider $\omega = \dfrac{\omega_s}{2}$, and not $\omega = \infty$, as the maximum possible frequency.

We can therefore assume that the frequency response $G(j\omega)$ tends to zero as ω tends to $\dfrac{\pi}{T}$. Similarly, we can say that $H(z)$ (the discrete form of $H(s)$) tends to zero as z tends to –1. Consequently, point $z = -1$ in the z-plane depicts the maximum possible frequency, $\omega = \dfrac{\pi}{T}$.

The pole-zero matching is realized as follows:

1. All finite poles of $H(s)$ are depicted in the z-plane according to $z = e^{sT}$.
2. All finite zeros of $H(s)$ are depicted in the z-plane according to $z = e^{sT}$.
3. (i) The zeros at infinity of $H(s)$, that is, the zeros in $s = \infty$ are located at $z = -1$ in $H(z)$. Therefore, for every zero at infinity in $H(s)$, we will have a component $z + 1$ in the numerator of $H(s)$. Similarly, the poles at infinity of $H(s)$, if any, are depicted in the $z = -1$ point. So, for each pole in $s = \infty$, we have a component $z + 1$ in the denominator of $H(z)$.

   (ii) If the numerator of $H(z)$ is of a smaller degree than the denominator, then we add powers of $z + 1$ to the numerator until the numerator and the denominator are of the same degree.
4. We adjust the $H(z)$ gain.

For low-frequency filters, it must be: $H(z)|_{z=1} \equiv H(s)|_{s=0}$
For high-frequency filters, it must be: $H(z)|_{z=-1} \equiv H(s)|_{s=\infty}$
If we calculate $H(z)$ by the pole-zero matching method, difference equations are derived in which the calculation of the $y[kT + T]$ value depends on the $u[kT + T]$ value, i.e., the output at the time $kT + T$ depends on the input value at the same time. However, if a time delay is permissible, i.e., if the output value calculation $y[kT + T]$ needs the input value $u[kT]$, then the numerator of $H(z) = \dfrac{Y(z)}{U(z)}$ should be of one degree smaller than its denominator. In this case, we use the modified pole-zero matching method. The modified pole-zero matching method results from the pole-zero matching method described, by excluding step 3(ii).

Both methods conserve the stability of the original system because, through the depiction $z = e^{sT}$, all points with $s < 0$ are depicted at z points with $z < 1$, since $0 < e^{sT} < e^{0T} < 1$.

Let the analog transfer function be of the form

$$H(s) = K_s \frac{(s+\mu_1)(s+\mu_2)\ldots\ldots(s+\mu_m)}{(s+\pi_1)(s+\pi_2)\ldots\ldots(s+\pi_n)}, m \leq n \tag{6.23}$$

Then, based on the pole-zero matching method, $H(z)$ will be of the form:

$$H(z) = K_z \frac{(z+1)^{n-m}(z+z_1)(z+z_2)\ldots(z+z_m)}{(z+p_1)(z+p_2)\ldots(z+p_n)} \tag{6.24}$$

The z_i and p_i are matched with the corresponding μ_i and μ_i according to relation (6.25):

$$z_i = -e^{-\mu_i T} \quad \text{and} \quad p_i = -e^{-\pi_i T} \tag{6.25}$$

The $n - m$ multiple zeros $((z + 1)^{n-m})$ that appear in $H(z)$ represent the numerator and denominator polynomial degree difference.

The constant K_z (dc gain factor) is chosen so that $H(s)$ and $H(z)$ are equal for $s = 0$ and $z = 1$ (for systems of interest in their low-frequency behavior), i.e.,:

$$H(z)(\text{for } z = 1) = H(s)(\text{for } s = 0) \tag{6.26}$$

So, we can easily compute the constant K_z from Relation 6.26:

$$K_z \cdot z^{n-m} \frac{(1+z_1)(1+z_2)\ldots(1+z_m)}{(1+p_1)(1+p_2)\ldots(1+p_m)} = K_s \frac{\mu_1\mu_2\ldots\ldots\mu_m}{\pi_1\pi_2\ldots\ldots\pi_n} \tag{6.27}$$

Alternatively, constant K_z can easily be computed by

$$K_z = \frac{\lim\limits_{t\to\infty} y(t)}{\lim\limits_{n\to\infty} y(k)} = \frac{\lim\limits_{s\to 0} s\left(\frac{1}{s}\right) \cdot H(s)}{\lim\limits_{z\to 1}(z-1)\frac{z}{(z-1)}H(z)} \tag{6.28}$$

6.6 Direct Methods of IIR Filter Design

6.6.1 Design of $\left|H(e^{j\omega})\right|^2$ Method

This method is used when the digital filter is described by a given $\left|H(e^{j\omega})\right|^2$ and is implemented by the relation (6.29).

$$\left|H(e^{j\omega})\right|^2 = \frac{1}{1+A_n^2(\omega)} \tag{6.29}$$

where $A_n(\omega)$ is a trigonometric polynomial of n-order. $A_n(\omega)$ must be first calculated to approximate $\left| H(e^{j\omega}) \right|^2$ for which it is:

$$\left| H(e^{j\omega}) \right|^2 = H(z)H(z^{-1}) \Big|_{z=e^{j\omega}} \tag{6.30}$$

6.6.2 The Method of Calculating $h[n]$

Based on

$$\frac{\displaystyle\sum_{i=0}^{m-1} a_i z^{-i}}{\displaystyle\sum_{i=0}^{n-1} b_i z^{-i}} = H(z) = \sum_{i=0}^{\infty} h[k]z^{-k} \tag{6.31}$$

we conclude that if we want to design a digital filter with a specific impulse response $h_\varepsilon[k]$, $k = 0,1, \ldots, h_\varepsilon[k]$, $k = 0,1,\ldots,\ell-1$, we should approach $h[k]$ based on the given $h_\varepsilon[k]$.

By setting the least squares error criterion, i.e.,

$$\langle \varepsilon \rangle = \sum_{k=0}^{\ell-1} \big[h_\varepsilon[k] - h[k] \big]^2 \omega[k] \tag{6.32}$$

we should minimize it by calculating the parameters of the filter.

6.7 IIR Filter Frequency Transformations

To design different types of digital filters (low-pass, high-pass, band-pass, band-reject), the next steps are followed:

1. Design a standard analog low-pass filter and then transform it, using Equations 6.33 to 6.36, into the corresponding selective analog filter, which, in turn, is transformed into a digital filter using the methods developed in Section 6.6.

$$low\ pass \rightarrow low\ pass \quad s \rightarrow \frac{s}{\omega_n} \tag{6.33}$$

$$low\ pass \rightarrow high\ pass \quad s \rightarrow \frac{\omega_u}{\omega_n} \tag{6.34}$$

$$low\ pass \rightarrow band\ pass \quad s \rightarrow \frac{s^2 + \omega_1\omega_u}{s(\omega_u - \omega_1)} \tag{6.35}$$

$$low\ pass \rightarrow band\ reject \quad s \rightarrow \frac{s(\omega_u - \omega_1)}{s^2 + \omega_1 \omega_u} \tag{6.36}$$

where: ω_1 = low cut-off frequency,

$\quad\quad \omega_u$ = high cut-off frequency

2. We design a low-pass analog filter as a reference, which we transform into a digital low-pass filter, and finally, we transform the latter, using the Equations 6.37 to 6.40, into the desired digital filter.

a. *low-pass* → *low-pass*

$$\left. \begin{array}{l} z^{-1} \rightarrow \dfrac{z^{-1} - \alpha}{1 - \alpha z^{-1}} \\[2em] \alpha = \dfrac{\sin\left[\left[(\omega_c - \omega_u)/2\right]T\right]}{\sin\left[\left[(\omega_c + \omega_u)/2\right]T\right]} \end{array} \right\} \tag{6.37}$$

where: ω_c = cut-off frequency of the initial filter,

$\quad\quad \omega_u$ = cut-off frequency of the final filter

b. *low-pass* → *high-pass*

$$\left. \begin{array}{l} z^{-1} \rightarrow -\dfrac{z^{-1} + \alpha}{1 + \alpha z^{-1}} \\[2em] \alpha = -\dfrac{\cos\left[\left[(\omega_c + \omega_u)/2\right]T\right]}{\cos\left[\left[(\omega_c - \omega_u)/2\right]T\right]} \end{array} \right\} \tag{6.38}$$

where: ω_c = cut-off frequency of the low-pass filter,

$\quad\quad \omega_u$ = cut-off frequency of the high-pass filter

c. *low-pass* → *band-pass*

$$\left. \begin{array}{l} z^{-1} \rightarrow -\dfrac{\left[z^{-2} - \left[2\alpha k/(k+1)\right]z^{-1} + (k-1)/(k+1)\right]}{\left[\left[(k-1)/(k+1)\right]z^{-2} - \left[2\alpha k/(k+1)\right]z^{-1} + 1\right]} \\[2em] \alpha = \cos(\omega_0 T) = \dfrac{\cos\left[\left[(\omega_u + \omega_1)/2\right]T\right]}{\cos\left[\left[(\omega_u - \omega_1)/2\right]T\right]} \\[2em] k = \cot\left[\left(\dfrac{\omega_u - \omega_1}{2}\right)T\right]\tan\left[\dfrac{\omega_c T}{2}\right] \end{array} \right\} \tag{6.39}$$

where: ω_c = cut-off frequency of the low-pass filter,

$\quad\quad \omega_0$ = central frequency of the band-pass filter

ω_1 = low cut-off frequency

ω_u = high cut-off frequency

d. *low-pass → band-reject*

$$\left.\begin{aligned} z^{-1} \rightarrow -\frac{\left[z^{-2} - \left[2\alpha/(1+k)\right]z^{-1} + (1-k)/(1+k)\right]}{\left[\left[(1-k)/(1+k)\right]z^{-2} - \left[2\alpha/(1+k)\right]z^{-1} + 1\right]} \\ \alpha = \cos(\omega_0 T) = \frac{\cos\left[\left[(\omega_u + \omega_1)/2\right]T\right]}{\cos\left[\left[(\omega_u - \omega_1)/2\right]T\right]} \\ k = \tan\left[\left(\frac{\omega_u - \omega_1}{2}\right)T\right]\tan\left[\frac{\omega_c T}{2}\right] \end{aligned}\right\} \qquad (6.40)$$

6.8 FIR Filters

There are critical advantages of FIR (Finite Impulse Response) filters over IIR filters, such as their easy implementation due to FIR filter approximation algorithms, as well as their stability. Their disadvantages are that they require a higher order filter than IIR filters. For the implementation of steep-slope FIR filters, a large number of samples is required, resulting in a delay due to the convolution process, which is computationally heavy.

If the input signal in the FIR filter is the unit impulse sequence $\delta[n]$, then its output (i.e., the impulse response of the filter) will be successively equal to each of the coefficients $h[n]$. Thus, for FIR filters, we have a finite impulse response. It has already been mentioned that the difference equation of a FIR filter is non-recursive, i.e., it is of the form:

$$y[n] = \sum_{k=0}^{n} b_k x[n-k] \qquad (6.41)$$

or

$$y[n] = h[n] * x[n] = \sum_{k=0}^{n} h[k]x[n-k] \qquad (6.42)$$

The order N of the filter is characterized by the number of impulse response terms (number of $h[n]$ coefficients). The transfer function of an N order filter will be:

$$H(z) = \sum_{n=0}^{N-1} h[n]z^{-n} = h[0] + h[1]z^{-1} + \dots + h[N-1]z^{-(N-1)} \qquad (6.43)$$

By substituting $z = e^{j\omega}$, the frequency response $H(e^{j\omega})$ is computed:

$$H\left(e^{j\omega}\right) = H(z)\Big|_{z=e^{j\omega}} = \sum_{k=0}^{N-1} h[k] e^{-j\omega k} \overset{Period\ 2\pi}{\Longrightarrow}$$

$$H\left(e^{j\omega}\right) = \sum_{k=0}^{N-1} h[k]\left(\cos \omega k - j \sin \omega k\right) \tag{6.44}$$

The magnitude $\left|H(e^{j\omega})\right|$, then, will be:

$$\left|H\left(e^{j\omega}\right)\right| = \sqrt{\left(\sum_{k=0}^{N-1}\left(h[k]\cos \omega k\right)\right)^2 + \left(\sum_{k=0}^{N-1}\left(h[k]\sin \omega k\right)\right)^2} \tag{6.45}$$

The phase $\Phi(\omega)$ of $H(e^{j\omega})$ will be:

$$\Phi(\omega) = Arg H(e^{j\omega}) = \tan^{-1}\frac{-\displaystyle\sum_{k=0}^{N-1} h[k]\sin \omega k}{\displaystyle\sum_{k=0}^{N-1} h[k]\cos \omega k} \tag{6.46}$$

For $T \neq 1$, we will have:

$$H(e^{j\omega T}) = \sum_{k=0}^{N-1} h[kT] e^{-j\omega kT} \quad , \quad time\ period\ \frac{2\pi}{T} \tag{6.47}$$

6.9 FIR Linear Phase Filters

A filter has a linear phase response when the phase difference $\Phi(\omega)$ between the input and output for a signal of angular frequency ω is given by:

$$\Phi(\omega) = -m\omega \quad or \quad \Phi(\omega) = b - m\omega \tag{6.48}$$

where m and b constants that depend on the characteristics of the filter. When the harmonic components of a signal pass through a system with a linear phase response according to Equation 6.48, all of them are subjected to the same time delay in *msec*. This has as a result that the form of the signal is maintained, i.e., the input signal is not distorted when it passes through the filter.

FIR filters can be designed to have a linear phase. A FIR filter has the linear phase response property if its impulse response, $h[n]$, exhibits some kind of symmetry.

From Equations 6.46 and 6.48, it is derived:

$$\tan^{-1}\frac{-\sum_{k=0}^{N-1}h[k]\sin\omega k}{\sum_{k=0}^{N-1}h[k]\cos\omega k}=-m\omega \Rightarrow \frac{-\sum_{k=0}^{N-1}h[k]\sin\omega k}{\sum_{k=0}^{N-1}h[k]\cos\omega k}=\tan(-m\omega)\Rightarrow$$

$$\frac{\sum_{k=0}^{N-1}h[k]\sin\omega k}{\sum_{k=0}^{N-1}h[k]\cos\omega k}=\frac{\sin m\omega}{\cos m\omega}\Rightarrow \sum_{k=0}^{N-1}h[k]\sin\omega k\cos m\omega-$$

$$-\sum_{k=0}^{N-1}h[k]\cos\omega k\sin m\omega=0\Rightarrow \sum_{k=0}^{N-1}h[k]\left(\sin\omega k\cos m\omega-\cos\omega k\sin m\omega\right)=0\Rightarrow$$

$$\sum_{k=0}^{N-1}h[k]\sin(m\omega-k\omega)=0 \tag{6.49}$$

Expanding the sum, we have:

$$h[0]\sin m\omega+h[1]\sin(m\omega-\omega)+h[2]\sin(m\omega-2\omega)+...$$
$$+h[N-1]\sin\left(m\omega-(N-1)\omega\right)=0\Rightarrow$$

$$h[0]\sin m\omega+h[1]\sin(m-1)\omega+h[2]\sin(m-2)\omega+...$$
$$...+h(N-1)\sin(m-N+1)\omega)=0 \tag{6.50}$$

Choosing the constant $m=\dfrac{N-1}{2}$ yields:

$$\overset{(6.50)}{\Rightarrow}h[0]\sin\left(\frac{N-1}{2}\omega\right)+h[1]\sin\left(\frac{N-3}{2}\omega\right)+...$$
$$......+h[N-1]\sin\left(-\frac{N-1}{2}\omega\right)=0 \tag{6.51}$$

From Equation 6.51, we conclude that, by choosing a FIR filter with symmetric coefficients* of $h[k]$ with respect to the middle (i.e., $\frac{N-1}{2}$), the filter will have a linear phase with respect to the frequency and will generate a delay of the input signal by $\frac{N-1}{2} T$ sec.

$$^* \text{i.e., it should be true that}: h[k] = h[N-1-k], \quad 0 \le k \le N-1 \tag{6.52}$$

$$\left(h[0] = h[N-1], \ h[1] = h[N-2], ... \right)$$

In conclusion, a FIR filter with a symmetric impulse response and an odd number of terms has a linear phase, and consequently, it introduces a time delay in all signal harmonics by $\frac{N-1}{2}$ samples. There are four different types of linear phase FIR filters, depending on whether the number N of their coefficients of $h[n]$ is even or odd and whether $h[n]$ is symmetric or anti-symmetric.

Type I: The impulse response $h[n]$ is symmetric, and M is odd.

$$h[n] = h[M-1-n]$$

$$H(\omega) = e^{-j\omega(M-1)/2} \left(h\left(\frac{M-1}{2}\right) + 2 \sum_{k=1}^{(M-3)/2} h\left(\frac{M-1}{2} - k\right) \cos(\omega k) \right)$$

Type II: The impulse response $h[n]$ is symmetric, and M is even.

$$h[n] = h[M-1-n]$$

$$H(\omega) = e^{-j\omega(M-1)/2} \cdot 2 \sum_{k=1}^{(M-3)/2} h\left(\frac{M}{2} - k\right) \cos\left(\omega\left(k - \frac{1}{2}\right)\right)$$

Type III: The impulse response $h[n]$ is anti-symmetric, and M is odd.

$$h[n] = -h[M-1-n]$$

$$h[n] = -h[M-1-n]$$

$$H(\omega) = e^{-j[\omega(M-1)/2 - \pi/2]} \cdot 2 \sum_{k=1}^{(M-1)/2} h\left(\frac{M-1}{2} - k\right) \sin(\omega k)$$

Type IV: The impulse response $h[n]$ is anti-symmetric, and M is even.

$$h[n] = -h[M-1-n]$$

$$H(\omega) = e^{-j[\omega(M-1)/2 - \pi/2]} \cdot 2 \sum_{k=1}^{(M-1)/2} h\left(\frac{M}{2} - k\right) \sin\left(\omega(k-1/2)\right)$$

6.10 Stability of FIR Filters

In FIR filters, the output depends only on the input values $x[n-k]$ with $k = 0,..., N-1$ and delayed output values $y[n-k]$ are not involved in its calculation. Thus, the impulse response is finite, and the transfer function $H(z)$ of the filter given by Equation 6.43 has no denominator; consequently, it has no poles to create instability. All of this results in filter stability, which means that for any finite-length input signal, the output is also of a finite length. The position of zeros can be defined such that rejection of specific harmonic components of the signal can be achieved by the filter.

6.11 Design of FIR Filters

The procedure for designing FIR filters includes defining filter specifications, calculating filter coefficients by one of the available methods, determining the implementation structure, error analysis due to the finite length of the coefficients, and choosing between implementation using either software or hardware. An important step in FIR filter design is the calculation of the coefficients $h[n]$ so that the filter satisfies the desired magnitude and phase specifications in frequency.

The methods that are widely used are:

- The windowing method
- The optimal designing method,
- The frequency sampling method.

6.12 The Moving Average Filters

In the specific case where all coefficients of the FIR filter are equal to one another and equal to $1/(M+1)$, where M is the filter order, the output is going to be the average of the $(M+1)$ most recent samples of the input. This is a 'moving' average of input samples (Moving

Average Filter), where the value of each point is replaced by the average of its neighboring point values. The moving average filters, despite their simplicity, are very effective in the reduction of the random noise of a signal.

The difference equation of the filter is given by:

$$y[n] = \frac{x[n] + x[n-1] + \ldots \ldots x[n-N+1]}{N} \tag{6.53}$$

The filter coefficients are:

$$h[n] = \frac{1}{N} \quad (\ldots), \quad n = 0, 1, \ldots N-1 \tag{6.54}$$

The order N of the filter is given by:

$$N = 2M + 1 \tag{6.55}$$

The frequency response (DTFT) is:

$$H(\omega) = \frac{1}{N} \sum_{k=0}^{N-1} e^{-jk\omega} = \frac{\sin\left(N\frac{\omega}{2}\right)}{N\sin\left(\frac{\omega}{2}\right)} e^{-j(N-1)\frac{\omega}{2}} \tag{6.56}$$

or, for odd N:

$$H(\omega) = \frac{1}{2M+1}\{1 + 2\cos\omega + 2\cos 2\omega + \ldots + 2\cos M\omega\} \tag{6.57}$$

The moving average filter uses a sliding window of odd length (e.g., 3 points, etc.), replacing the value of the center point of the window with their average value.

For example, a filter with $M=1$ and $N = 2M+1=3$ (filter length) will have $h[n] = [h[1] \ h[2] \ h[3]] = (1/3) [111]$ coefficients and a time response that is derived by the sliding counterbalance of the values of the input samples, i.e.,

$$y[n] = \{y[1], y[2], y[3], \ldots\} = \left\{\frac{x[1]+x[2]+x[3]}{3}, \frac{x[2]+x[3]+x[4]}{3}, \frac{x[3]+x[4]+x[5]}{3}, \ldots\right\}, \quad \text{as}$$

illustrated in Figure 6.11.

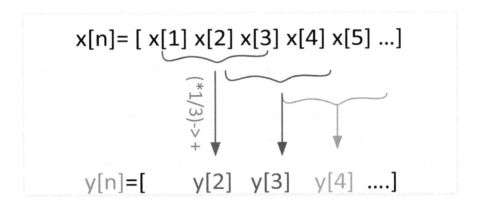

FIGURE 6.11
The output of a 3-point moving average filter.

6.13 FIR Filter Design Using the Frequency Sampling Method

For the design of FIR filters using the frequency sampling method, the following are required:

a. N samples ($H[k]$) of the desired spectrum using DFT
b. Applying IDFT to $H[k]$ for calculation of the discrete impulse response $h[n]$ of the FIR filter.

Taking into account that the coefficients $h[n]$ must be real, as well as the symmetry of $h[n]$ terms with respect to the $\dfrac{N-1}{2}$ point, to satisfy the linear phase requirement, Equations 6.58 and 6.59 are obtained for N as an even and as an odd number, respectively.

For N as an even number,

$$h[n] = \frac{H(0)}{N} + \frac{1}{N}\sum_{k=1}^{\frac{N}{2}-1} |H(k)| 2(-1)^k \cos\frac{(\pi k + 2\pi kn)}{N}, \quad 0 \le n \le N-1 \tag{6.58}$$

For N as an odd number,

$$h[n] = \frac{H(0)}{N} + \frac{1}{N}\sum_{k=1}^{\frac{N-1}{2}} 2(-1)^k |H(k)| \cos\frac{(\pi k + 2\pi kn)}{N}, \quad 0 \le n \le N-1 \tag{6.59}$$

where

$$H[k] = H\left(e^{j2\frac{\pi}{N}k}\right) = \sum_{n=0}^{N-1} h[n]e^{-j\frac{2\pi}{N}kn}, \quad -0 \le k \le N-1 \tag{6.60}$$

$$H[k] = |H[k]|e^{-j\Phi(k)} \tag{6.61}$$

For N as an even number, the linear phase condition is

$$\Phi(k) = \begin{cases} -\dfrac{\pi k(N-1)}{N} , & 0 \le k \le \dfrac{N}{2} \\[2mm] \pi - \dfrac{\pi k(N-1)}{N} , & \dfrac{N}{2} \le k \le N-1 \end{cases} \tag{6.62}$$

And, for N as an odd number, the linear phase condition is

$$\Phi(k) = \begin{cases} -\dfrac{\pi k(N-1)}{N} , & 0 \le k \le \dfrac{N-1}{2} \\[2mm] \pi - \dfrac{\pi k(N-1)}{N} , & \dfrac{N-1}{2} \le k \le N-1 \end{cases} \tag{6.63}$$

Likewise, for the magnitude $|H[k]|$, it will be:
For N as an even number,

$$|H[k]| = \begin{cases} |H[N-k]| , & 1 \le k \le \dfrac{N}{2}-1 \\[2mm] \textit{Magnitude symmetry with respect to } \dfrac{N}{2} \\[2mm] 0 , & k = \dfrac{N}{2} \\[2mm] \textit{Zero magnitude at } \dfrac{N}{2} \end{cases} \tag{6.64}$$

For N as an odd number,

$$|H[k]| = |H[N-k]| , \quad 1 \le k \le \dfrac{N-1}{2}$$

$$\textit{Magnitude symmetry with respect to } \dfrac{N-1}{2} \tag{6.65}$$

If the conditions of Equations 6.62 and 6.64 are true for N=even, and also of Equations 6.63 and 6.65 for N=odd, then Equations 6.58 and 6.59 can be applied for the design of FIR filters of a linear phase with real and symmetric coefficients of the impulse response, $h[n]$.

The algorithm for designing a FIR filter using spectrum sampling is briefly presented in the following steps:

1. Depending on the sampling frequency f_s and the number N of the samples, the distance of the samples is calculated by $Df = \dfrac{f_s}{N}$.

2. The desired magnitude of $H(k)$ is determined, according to the specifications for $k = 0$ to $(N-1)/2+1$ for an odd N, or $N/2$ for an even N. (The magnitude is defined according to the specifications and taking into account the distance between the samples.)

3. Calculation of the coefficients $h(i)$, depending on whether N is even or odd, for $i = 0,1,2..N - 1$.

6.14 FIR Filter Design Using the Window Method

The window method is a relatively simple FIR filter design process. The ideal filter coefficients $h_i[n]$ are calculated by the inverse Fourier transform of the frequency response $H_i(e^{j\omega})$.

The method starts with the implementation of an ideal low-pass filter. In the following Figures 6.12 and 6.13, an ideal low-pass filter is illustrated firstly with respect to the digital frequency, ω, and then with respect to the normalized frequency, F.

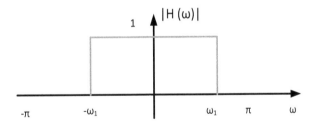

FIGURE 6.12

Desired $H(\omega)$–Ideal low-pass filter $\omega = \Omega T = 2\pi F = 2\pi \dfrac{f}{f_s}$.

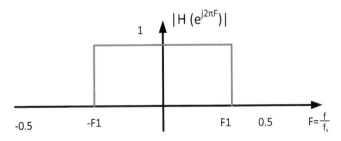

FIGURE 6.13
Desired $H(F)$–Ideal low-pass filter with respect to the normalized frequency F.

This method is recommended for the cut-off process of the central part of N length of the infinite impulse response. If $W[n]$ is the window sequence, then it is true that:

$$h[n] = h_i[n]W[n] \tag{6.66}$$

where $h_i[n]$ is the impulse response (of infinite length) of the ideal filter, i.e.,

$$h_i[n] = \frac{1}{2\pi} \int_{-\pi}^{\pi} H_i\left(e^{j\omega}\right) e^{j\omega n} \, d\omega \tag{6.67}$$

$$\Rightarrow h[n] = \frac{1}{2\pi} \int_{-\pi}^{\pi} H(\omega) e^{jn\omega} \, d\omega = \frac{1}{2\pi} \int_{-\omega_1}^{\omega_1} 1 . e^{jn\omega} \, d\omega = \frac{1}{2\pi} \left[\frac{e^{jn\omega}}{jn} \right]_{-\omega_1}^{\omega_1} = \frac{\sin(n\omega_1)}{n\pi}$$

$$h[n] = \frac{\sin\left(n\omega_1\right)}{n\pi} \quad \text{or} \quad h[n] = \frac{\sin\left(n2\pi F_1\right)}{n\pi} \tag{6.68}$$

and

$$W[n] = \begin{cases} \neq 0 & , \quad -\dfrac{N}{2} \leq n \leq \dfrac{N}{2} \\ 0 & , \quad elsewhere \end{cases} \tag{6.69}$$

We can see that the filter described by Equation 6.68 has an infinite impulse response, so it is an IIR filter. We can then create an FIR filter by choosing the suitable window. In Table 6.1, the formulas for the calculation of $h[n]$ coefficients for all types of filters are given.

TABLE 6.1

Ideal Impulse Responses

Filter Type	h(n) Coefficients	Width
Low pass	$h[k] = \dfrac{\sin(2\pi F_1 k)}{\pi k}$	$2F_1$
High pass	$h[k] = -\dfrac{\sin(2\pi F_1 k)}{\pi k}$	$1 - 2F_1$
Band pass	$h[k] = \dfrac{\sin(2\pi F_2 k)}{\pi k} - \dfrac{\sin(2\pi F_1 k)}{\pi k}$	$2(F_2 - F_1)$
Band stop	$h[k] = \dfrac{\sin(2\pi F_1 k)}{\pi k} - \dfrac{\sin(2\pi F_2 k)}{\pi k}$	$1 - 2(F_2 - F_1)$

Function $W[e^{j\omega}]$, which is the spectrum of window $w[n]$, consists of a main lobe containing most of the spectral energy of the window, as well as side lobes that generally have rapidly decreasing amplitudes. The width of the transition band depends on the width of the main lobe, $W[e^{j\omega}]$. The side lobes also influence the form of the final $W[e^{j\omega}]$, introducing small ripples at all frequencies. Thus, the spectrum of a window function must have a narrow main lobe that contains as much spectral energy as possible, with the side lobes rapidly decaying.

The main window functions are the following:

1. *Rectangular* (Figure 6.14): It corresponds to the simple cut-off process of the sequence $h_i[n]$. The spectrum of the window function is given by:

$$W[n] = 1 \quad , \quad 0 \leq n \leq N \tag{6.70}$$

2. *Bartlett* (Triangular) (Figure 6.15):

$$W[n] = 1 - \frac{2\left|n - \dfrac{N}{2}\right|}{N} \quad , \quad 0 \leq n \leq N \tag{6.71}$$

3. *Hanning* (Figure 6.16):

$$W[n] = 0.5 - 0.5\cos\left(\frac{2\pi n}{N}\right) \quad , \quad 0 \leq n \leq N \tag{6.72}$$

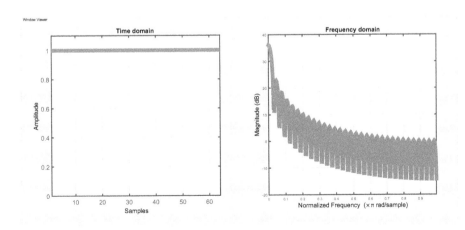

FIGURE 6.14
64-point Rectangular Window.

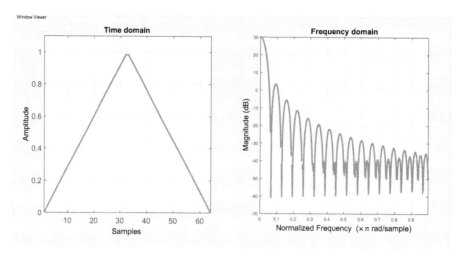

FIGURE 6.15
64-point Bartlett Window.

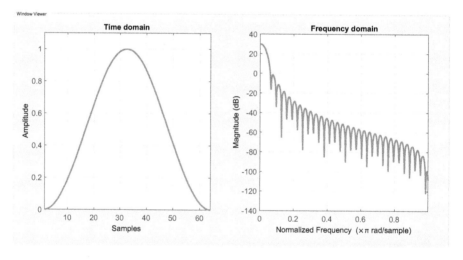

FIGURE 6.16
64-point Hanning Window.

4. *Hamming* (Figure 6.17): The main lobe of $W[e^{j\omega}]$ contains 99.96% of the total spectral energy, while -52 dB attenuation is achieved in the cut-off zone of the filters being designed.

$$W[n] = 0.54 - 0.46\cos\left(\frac{2\pi n}{N}\right) \quad , \quad 0 \le n \le N \tag{6.73}$$

5. *Blackman* (Figure 6.18):

$$W[n] = 0.42 - 0.5\cos\left(\frac{2\pi n}{N}\right) + 0.08\cos\left(\frac{4\pi n}{N}\right), \quad 0 \le n \le N \tag{6.74}$$

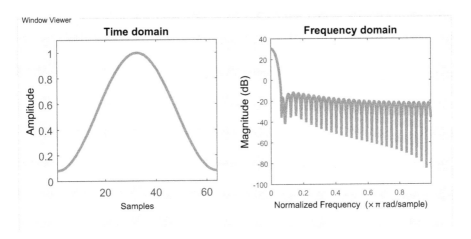

FIGURE 6.17
64-point Hamming Window.

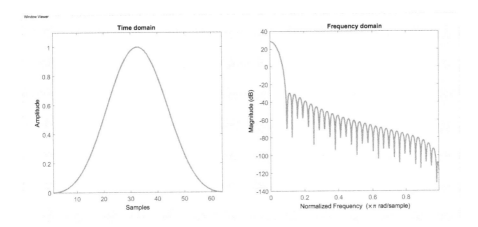

FIGURE 6.18
64-point Blackman Window.

    6. *Kaiser* (Figure 6.19): It has most of the energy in the main lobe for a specific width of the side lobes.

$$W[n] = \frac{I_0 \left[\beta \sqrt{1 - \left[(n - \alpha)/\alpha \right]^2} \right]}{I_0 \left[\beta \right]}, \quad 0 \leq n \leq N \tag{6.75}$$

where: $\alpha = N/2$ and $I_0(\cdot)$ is a zero order modified Bessel function of the first kind, which can be generated by the power series:

$$I_0(x) = 1 + \sum_{k=1}^{\infty} \left[\frac{(x/2)^k}{k!} \right]^2 \tag{6.76}$$

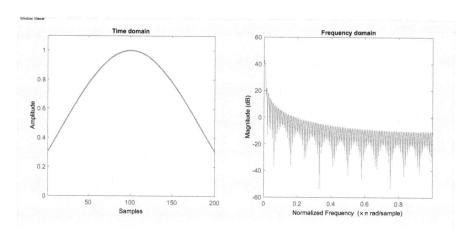

FIGURE 6.19
200-point Kaiser Window with a beta β = 2.5.

The parameter β (beta) defines the shape of the window and, thus, controls the changes between the width of the main lobe and the width of the side lobes.

There are two empirical equations, the first of which associates the ripple of the stop band of a low-pass filter ($a_s = -20 \log \delta_2$) with the parameter, β (beta), while the second associates N with the transition band width of Δf, and the attenuation, a_s, at the stop band.

$$
\begin{aligned}
\beta &= 0.1102(a_s - 8.7), \qquad a_s > 50 \\
\beta &= 0.5842(a_s - 21)^{0.4} + 0.07886(a_s - 21), \qquad 21 \le a_s \le 50 \\
\beta &= 0, \qquad a_s < 21
\end{aligned}
\tag{6.77}
$$

$$
N = \frac{a_s - 7.95}{14.36 \Delta f}
\tag{6.78}
$$

The method of designing FIR filters with the window method, although simple, does not eliminate the main disadvantage of FIR filters, which require many coefficients, thereby increasing the hardware requirements for their implementation. However, the proper use of the windows can contribute to the reduction of the hardware requirements.

In Table 6.1, the ideal impulse responses of various types of filters are given.

The filters in Table 6.2 have an impulse response of infinite length and are, therefore, IIR. Table 6.2 associates the choice of the window function with the desired signal attenuation in the stop band.

The choice of the length, N, depends on K and ΔF, where K (see Table 6.2) depends on the choice of window, while $\Delta F = \dfrac{f_2 - f_1}{f_s}$ is the transition band of the digital frequency. The following should be true:

$$
N \ge \frac{K}{\Delta F}
\tag{6.79}
$$

TABLE 6.2

How the Choice of the Window Function Associates with the Desired Signal Attenuation in the Stop Band

a/a	Window Type	Width of Transition Band (Hz)	δ_1 (db)	Maximum Signal Attenuation in the Stop Band (dB)	K
1	RECTANGULAR	0.9/N	0.7416	21	0.9
2	HANNING	3.1/N	0.0546	44	3.1
3	HAMMING	3.3/N	0.0194	53	3.3
4	BLACKMAN	5.5/N	0.0017	75	5.5

The steps followed to design a FIR filter using the window method are:

1. Based on the pass band specifications of the filter to be designed, a window is chosen, and the number of filter coefficients is calculated using the correct relation between the width of the filter and the width of the transition band.

2. Calculation of the N order of the filter.

3. Calculation of the coefficients of the ideal impulse response of the filter.

4. Shift of the ideal $h[n]$ to the right by $(N - 1)/2$ according to:

$$h_{shifted} = h_{ideal}\left(n - \frac{N-1}{2}\right), \quad n = 0,1,2,......,(N-1) \text{ (for } N \text{ odd, otherwise by } N/2).$$

5. Calculation of the finite length sequence $h[n]$ of the FIR filter by multiplying the ideal $h[n]$ with the window $h[n]$,

$$h[n] = h_{shifted}w[n], \quad 0 \le n \le (N-1) \tag{6.80}$$

6.15 Optimal Equiripple FIR Filter Design

The term *optimal FIR filter* means achieving the best possible specifications of the frequency response with the smallest number of coefficients. The method of designing optimal FIR filters was developed in the early 1970s and is encoded into a computer program, making it easy to apply. With this method, the approximate error between the ideal and the actual FIR filter spreads evenly over the entire frequency range, resulting in the approximation of the FIR filter with a lower order filter than the one that approximates the ideal filter at one frequency, while at other frequencies, its approximation is not good.

Consider a Type 1 FIR filter with respect to N, where $M = N/2$ and $h[n]$ are the impulse response coefficients. The frequency response will be:

$$H(\omega) = \sum_{n=0}^{M} b_n \cos \omega n \tag{6.81}$$

$$b_0 = h\left(\frac{N}{2}\right)$$
$$b_n = 2h\left(\frac{N}{2} - n\right) \quad , n = 1, 2, 3 \tag{6.82}$$

Figures 6.20 and 6.21 illustrate the desired and the actual frequency response, respectively. With this method, the error is $E(\omega) = D(\omega) - H(\omega)$ and $W(\omega)$ in the pass band is equal to $\frac{\delta_2}{\delta_1}$, while in the stop band, is equal to 1. The equalized error is $E_{eq}(\omega) = W(\omega)[D(\omega) - H(\omega)]$ or

$$E_{eq}(\omega) = W(\omega)\left[D(\omega) - \sum_{n=0}^{M} b_n \cos(n\omega) \right] \tag{6.83}$$

To achieve equiripple, it is sufficient to calculate b_n so that the magnitude of the maximum $E_{eq}(\omega)$ is minimized for each ω (of all frequency bands).

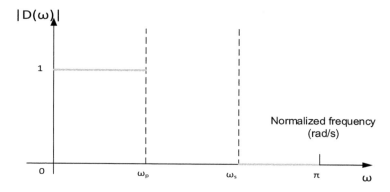

FIGURE 6.20
Desired frequency response.

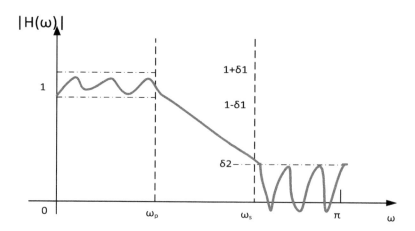

FIGURE 6.21
Actual frequency response.

The method of optimal equiripple filter design, also known as Parks-McClellan, leads to optimal filters in the sense of the mean square error criterion from the desired response. The resultant filters present equiripples in both the pass band and the stop band. The length of the two-zone Parks-McClellan filters (e.g., low-pass or high-pass) is determined by the following empirical equation:

$$L = \frac{-10\log_{10}(\delta_1\delta_2) - 13}{2.324} \frac{}{\omega} + 1, \quad \Delta\omega = 2\pi\left(f_s - f_p\right) \tag{6.84}$$

where δ_1 and δ_2 represent the relative ripple magnitudes in the two zones. The Parks-McClellan algorithm keeps the quantities N, ω_p, ω_s constant in order to control the boundaries of the bands and allows δ_1 and δ_2 variation. The number of coefficients required are known in advance.

The steps followed to design a FIR filter using the optimal method are:

1. Identification of the boundaries of the bands, of the ripple magnitude of pass band (δ_1) and the stop band (δ_2) and of the sampling frequency (f_s).
2. Normalization of the band frequencies and of the width of the pass band (Δf), i.e., division by f_s.
3. Calculation of the filter size, N, by using δ_1, δ_2 and Δf.
4. Calculation of the ratio δ_1/δ_2 as the weighting coefficients.
5. Use of algorithm to calculate filter coefficients.
6. Check if δs and δp satisfy the specifications.
7. If no, increase N and repeat steps (5) and (6) until step (6) is satisfied.
8. Check of fulfillment of the frequency response specifications.

6.16 Comparison of the FIR Filter Design Methods

A comparison of the FIR filter design methods is given next in Table 6.3.

TABLE 6.3

Comparison of the FIR Filter Design Methods

Design Method	Advantages	Disadvantages
WINDOW	Simple to understand. Suitable when rippling in the bands is the same.	It does not allow absolute control of band boundaries values.
OPTIMAL	Simple and flexible. There is absolute control of the filter specifications.	Necessity of algorithm use.
FREQUENCY SAMPLING	Applies to recursive and non-recursive filters. Filter design is accomplished with any magnitudes-phase responses.	Lack of frequency boundaries control.

6.17 Solved Problems

6.17.1 Find the equation of a digital filter resulting from the transform of the analog filter with the transfer function

$$H(s) = \frac{s+1}{(s+1)^2 + 4}$$

applying the pole-zero match method for $T=1$sec.

SOLUTION: $H(s)$ can be written as:

$$H(s) = \frac{s+1}{(s+1+2j)(s+1-2j)} \tag{1}$$

Because the denominator degree is $n = 2$ and the numerator degree is $m = 1$, it is assumed to have $n-m = 2\text{-}1 = 1$ zero at infinity. Note that the $H(s)$ zeros that tend to infinity are substituted by $z = -1$ in the z-plane. So, $H(z)$ will be:

$$H(z) = \frac{(1 - z^{-1}e^{-j\pi})^*(1 - z^{-1}e^{-T})}{1 - 2z^{-1}e^{-T}\cos(2T) + e^{-2T}z^{-2}} \tag{2}$$

Explanation of the term explanation of the term $1-z^{-1}e^{-jp}$: The frequency of the s-plane, $\omega_s \to \infty$ (when zero gets to infinity) is equivalent to the frequency of the z-field, so that the match of the zero at infinity with the zero in the z-plane is:

$$1 - z^{-1}e^{j\omega_s T} = 1 - z^{-1}e^{j\frac{\omega}{T}T} = 1 - z^{-1}e^{j\pi} = 1 - z^{-1}(-1) = 1 + z^{-1} \tag{3}$$

$$\overset{(2),(3)}{\Rightarrow} H(z) = \frac{(1 + z^{-1})(1 - z^{-1}e^{-T})}{1 - 2z^{-1}e^{-T}\cos(2T) + e^{-2T}z^{-2}} \Rightarrow H(z) = \frac{(z+1)(z - e^{-T})}{z^2 - 2ze^{-T}\cos(2T) + e^{-2T}} \tag{4}$$

For $T = 1$sec, it is:

$$(4) \Rightarrow H(z) = \frac{(z+1)(z - 0.3678)}{z^2 + 0.306z + 0.135} \tag{5}$$

but,

$$H(z) = \frac{Y(z)}{X(z)} \overset{(5)}{=} \frac{z^2 + 0.6322z - 0.3678}{z^2 + 0.306z + 0.135} \tag{6}$$

Equation (6) must be multiplied by the dc gain correction factor, i.e.,

$$H(z) = \frac{Y(z)}{X(z)} = k_{dc} \cdot \frac{z^2 + 0.6322z - 0.3678}{z^2 + 0.306z + 0.135} \tag{7}$$

where k_{dc} is calculated as follows:

$$K = \frac{\lim_{t \to \infty} y(t)}{\lim_{k \to \infty} y(k)} \tag{8}$$

$$\lim_{t \to \infty} y(t) \overset{L.T}{=} \lim_{s \to 0} s\left(\frac{1}{s}\right) H(s) = \lim_{s \to 0} H(s) = 0.2$$

$$\lim_{k \to \infty} y(k) \overset{Z.T}{=} \lim_{z \to 1}(z-1)\frac{z}{z-1} H(z) = \lim_{z \to 1} zH(z) = \frac{1.2644}{1.441} = 0.877$$

$$(8) \Rightarrow k = \frac{0.2}{0.877} = 0.23$$

$$(7) \Rightarrow H(z) = \frac{Y(z)}{X(z)} = 0.23\frac{1 + 0.6322z^{-1} - 0.3678z^{-2}}{1 + 0.306z^{-1} + 0.135z^{-2}} \Rightarrow$$

$$\left(1 + 0.306z^{-1} + 0.135z^{-2}\right)Y(z) = 0.23\left(1 + 0.6322z^{-1} - 0.3678z^{-2}\right)X(z) \Rightarrow$$

$$Y(z) = -0.135z^{-2}Y(z) - 0.306z^{-1}Y(z) + 0.23X(z) + 0.1454z^{-1}X(z) - 0.084z^{-2}X(z)$$

Therefore, the digital filter of the difference equation will be:

$$y[k] = -0.135y[k-2] - 0.306y[k-1]$$
$$+0.23x[k] + 0.1454x[k-1] - 0.0846x[k-2] \tag{9}$$

6.17.2 Convert the analog band-pass filter with the transfer function

$$H(s) = \frac{1}{(s+0.1)^2 + 9} \tag{1}$$

to a digital IIR filter using:
a. The Backward Difference method
b. The Impulse Invariant method

SOLUTION:

    a. Using the Backward Difference method, it is:

$$H(z) = H(s)\Big|_{s=\frac{1-z^{-1}}{T}} = \frac{1}{\left(\dfrac{1-z^{-1}}{T}+0.1\right)^{2}+9} \Rightarrow \tag{2}$$

$$\Rightarrow H(z) = \frac{T^{2}\big/\left(1+0.2T+9.01T^{2}\right)}{1-\dfrac{2(1+0.1T)}{1+0.2T+9.01T^{2}}z^{-1}+\dfrac{1}{1+0.2T+9.01T^{2}}z^{-2}} \tag{3}$$

       For T = 0.1sec, the poles are: $P_{1,2} = 0.91 \pm j0.27 = 0.949e^{\pm j16.5°}$

       That is, they are located close to the unit circle (the same happens if we choose T≤0.1).

    b. H(s) is expressed as a sum of fractions:

$$H(s) = \frac{1\big/2}{s+0.1-j3} + \frac{1\big/2}{s+0.1+j3}$$

       Thus: $H(z) = Z\big[H(s)\big] \overset{(4)}{=} Z\left[\dfrac{1\big/2}{s+0.1-j3}\right] + Z\left[\dfrac{1\big/2}{s+0.1+j3}\right] \Rightarrow$

$$\Rightarrow H(z) = \frac{1\big/2}{1-e^{-0.1T}e^{j3T}z^{-1}} + \frac{1\big/2}{1-e^{-0.1T}e^{-j3T}z^{-1}} \Rightarrow$$

$$\Rightarrow H(z) = \frac{1-\left(e^{-0.1T}\cos 3T\right)z^{-1}}{1-\left(2e^{-0.1T}\cos 3T\right)z^{-1}+e^{-0.2T}z^{-1}}$$

       The magnitude of the frequency response for this filter is illustrated in the following figure for T = 0.1sec and T = 0.5sec.

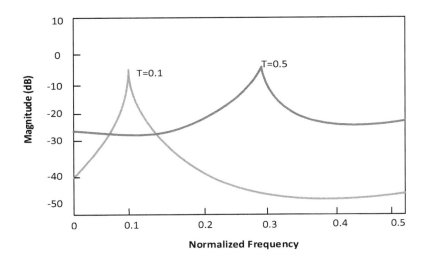

6.17.3 The transfer function of an analog Chebyshev, low-pass filter of order $n = 4$ and cut-off frequency $\omega_u = 0.2\pi$ is given next. Design a digital high-pass filter with the cut-off frequency $\omega_c = 0.6\pi$ and $\Delta T = 1sec$ using the Tustin method.

$$H_\alpha(s) = \frac{0.038286}{(s^2 + 0.4233s + 0.1103)(s^2 + 0.1753s + 0.3894)}$$

SOLUTION: We apply the Tustin method:

$$H(z) = H_\alpha(s)\Big|_{s = \frac{2}{T}\frac{1-z^{-1}}{1+z^{-1}}} \Rightarrow$$

$$H(z) = \frac{0.001836(1 + z^{-1})^4}{(1 - 1.5548z^{-1} + 0.6493z^{-2})(1 - 1.4996z^{-1} + 0.8482z^{-2})} \qquad (1)$$

To design the requested high-pass digital filter, we use Relation 6.38, so:

$$\alpha = -\frac{\cos\left[\left[(\omega_c + \omega_u)/2\right]T\right]}{\cos\left[\left[(\omega_c - \omega_u)/2\right]T\right]} = -\frac{0.3090169}{0.8090169} = -0.38197 \qquad (2)$$

The transfer function of a high-pass filter will be:

$$H_\alpha(z) = H(z)\Big|_{z^{-1} \to -\frac{z^{-1}+\alpha}{1+\alpha z^{-1}} = -\frac{z^{-1}-0.38197}{1-0.38197z^{-1}}} \Rightarrow$$

$$H_\alpha(z) = \frac{0.02426(1-z^{-1})^4}{(1-1.0416z^{-1}+0.4019z^{-2})(1-0.5561z^{-1}+0.7647z^{-2})} \tag{3}$$

The difference equation describing the system is derived from:

$$H_\alpha(z) = \frac{Y(z)}{X(z)} \Rightarrow Y(z) = H_\alpha(z)X(z) \Rightarrow y[n] = z^{-1}\big[H_\alpha(z)X(z)\big]$$

and is implemented using the structures already discussed, i.e., Direct form I, Canonic form, Cascade form or Parallel form.

6.17.4   Let the transfer function of a low-pass Chebyshev filter of order $n = 3$

$$H_\alpha(s) = \frac{0.4431}{(s+0.4606)(s+0.2303-j0.9534)(s+0.2303+j0.9534)}$$

Find:

a. The frequency response of $H_\alpha(s)$ (magnitude-phase).
b. The equivalent to the impulse invariant method digital filter (magnitude–phase of $H(z)$)
c. The impulse response $h_\alpha(t)$ of $H_\alpha(s)$.
d. The impulse response $h[nT]$ of the digital filter.
e. The filter difference equation
f. The realization of the digital filter, using the Parallel form, for $T = 1$sec.

SOLUTION:

a. To find the frequency response of $H_\alpha(s)$, we substitute $s = j\omega$, which yields:

$$H_\alpha(j\omega) = \frac{0.4431}{(j\omega+0.4606)(j\omega+0.2303-j0.9534)(j\omega+0.2303+j0.9534)} \tag{1}$$

The magnitude $|H_\alpha(j\omega)|$ is:

$$|H_\alpha(j\omega)| = \frac{0.4431}{|j\omega+0.4606||0.2303+j(\omega-0.9534)||0.2303+j(\omega+0.9534)|} \Rightarrow$$

$$|H_\alpha(j\omega)| = \frac{0.4431}{\sqrt{(0.4606)^2+\omega^2}\sqrt{(0.2303)^2+(\omega-0.9534)^2}\sqrt{(0.2303)^2+(\omega+0.9534)^2}} \Rightarrow \tag{2}$$

$$|H_\alpha(j\omega)| = \frac{0.4431}{\sqrt{0.2122+\omega^2}\sqrt{0.05303+(\omega-0.9534)^2}\sqrt{0.05303+(\omega+0.9534)^2}}$$

The magnitude $|H_\alpha(j\omega)|$ in dB is:

$$20\log\left|H_\alpha(j\omega)\right| = -7.07 - 20\log\sqrt{0.2122 + \omega^2} - 20\log\sqrt{0.05303 + (\omega - 0.9534)^2}$$
$$- 20\log\sqrt{0.05303 + (\omega + 0.9534)^2} \tag{3}$$

and $Arg(H_\alpha(j\omega))$ is:

$$Arg\left(H_\alpha(j\omega)\right) = -\tan^{-1}\frac{\omega}{0.4606} - \tan^{-1}\frac{\omega - 0.9534}{0.2303} - \tan^{-1}\frac{\omega + 0.9534}{0.2303} \tag{4}$$

NOTE: Another way of the derivation of magnitude and phase of $H_\alpha(j\omega)$ would be:

$$H_\alpha(j\omega) = \frac{0.4431}{\left(0.4431 - 0.9212\omega^2\right) + j\left(1.1742\omega - \omega^3\right)} \Rightarrow$$

$$\left.\begin{array}{l} \left|H_\alpha(j\omega)\right| = \dfrac{0.4431}{\sqrt{\left(0.4431 - 0.9212\omega^2\right)^2 + \left(1.1742\omega - \omega^3\right)^2}} \\[3ex] Arg\left(H_\alpha(j\omega)\right) = -\tan^{-1}\dfrac{1.1742\omega - \omega^3}{0.4431 - 0.9212\omega^2} \end{array}\right\} \tag{5}$$

The magnitude and phase diagram of $H_\alpha(s)$ are illustrated in the following figures:

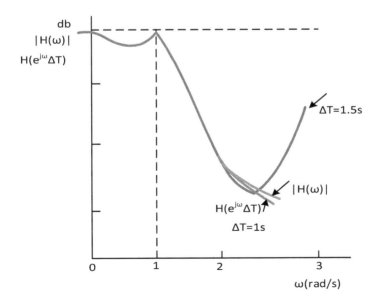

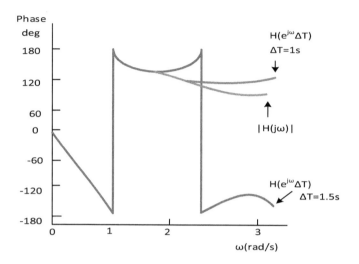

$$H_\alpha(s) = \frac{0.4431}{(s+0.4606)(s+0.2303 - j0.9534)(s+0.2303 + j0.9534)}$$

$$= \frac{0.4431}{(s+0.4606)\left[(s+0.2303)^2 + (0.9534)^2\right]} \Rightarrow$$

$$H_\alpha(s) = \frac{k_1}{s+0.4606} + \frac{k_2 s + k_3}{(s+0.2303)^2 + (0.9534)^2}$$

where $k_1 = \lim\limits_{s \to -0.4606} H_\alpha(s)(s+0.4606) = 0.4606$ and

$$k_1\left[(s+0.2303)^2 + (0.9534)^2 + \left(k_2 s + k_3\right)\left(s+0.4606\right)\right] \equiv 0.4431 \Rightarrow$$
$$k_1 + k_2 = 0 \Rightarrow k_2 = -k_1 = -0.4606$$
$$0.4606(k_1 + k_2) + k_3 = 0 \Rightarrow k_3 = 0$$

Thus:

$$H_\alpha(s) = \frac{0.4606}{s+0.4606} - \frac{0.4606s}{(s+0.2303)^2 + (0.9534)^2} \tag{6}$$

The transfer function of the digital filter of the impulse invariant method is:

$$H(z) = Z\left[H_\alpha(s)\right] \overset{(6)}{=} Z\left[\frac{0.4606}{s+0.4606} - \frac{0.4606s}{(s+0.2303)^2 + (0.9534)^2}\right] \Rightarrow$$

$$H(z) = Z\left[\frac{0.4606}{s + 0.4606}\right] - Z\left[\frac{0.4606s}{(s + 0.2303)^2 + (0.9534)^2}\right] \qquad (7)$$

But $Z\left[\dfrac{0.4606}{s + 0.4606}\right] = \dfrac{0.4606}{1 - z^{-1}e^{-0.4606T}} = \dfrac{z \cdot 0.4606}{z - e^{-0.4606T}}$ and

$$Z\left[\frac{0.4606s}{(s + 0.2303)^2 + (0.9534)^2}\right] = Z\left[\frac{0.4606\left[s + 0.2303 - 0.2303\right]}{(s + 0.2303)^2 + (0.9534)^2}\right]$$

$$= 0.4606Z\left[\frac{s + 0.2303}{(s + 0.2303)^2 + (0.9534)^2}\right] - 0.11128Z\left[\frac{0.9534}{(s + 0.2303)^2 + (0.9534)^2}\right]$$

$$= 0.4606\frac{z^2 - ze^{-0.2303T}\cos(0.9534T)}{z^2 - 2ze^{-0.2303T}\cos(0.9534T) + e^{-0.4606T}}$$

$$- 0.11128\frac{ze^{-0.2303T}\sin(0.9534T)}{z^2 - 2ze^{-0.2303T}\cos(0.9534T) + e^{-0.4606T}}$$

Thus,

$$\overset{(7)}{\Rightarrow} H(z) = 0.4606\left[\frac{z}{z - e^{-0.4606T}} - \frac{z^2 - ze^{-0.2303T}\cos(0.9534T)}{z^2 - 2ze^{-0.2303T}\cos(0.9534T) + e^{-0.4606T}}\right.$$

$$\left. + 0.2416\frac{ze^{-0.2303T}\sin(0.9534T)}{z^2 - 2ze^{-0.2303T}\cos(0.9534T) + e^{-0.4606T}}\right] \qquad (8)$$

The requested frequency response of the digital filter is obtained by substituting $Z = e^{j\omega T}$

$$\overset{(8)}{\Rightarrow} H\left(e^{j\omega}\right) = 0.4606\left[\frac{e^{j\omega T}}{e^{j\omega T} - e^{-0.4606T}} - \frac{e^{j2\omega T} - e^{j\omega T}e^{-0.2303T}\cos(0.9534T)}{e^{j2\omega T} - 2e^{j\omega T}e^{-0.2303T}\cos(0.9534T) + e^{-0.4606T}}\right.$$

$$\left. + 0.2416\frac{e^{j\omega T}e^{-0.2303T}\sin(0.9534T)}{e^{j2\omega T} - 2e^{j\omega T}e^{-0.2303T}\cos(0.9534T) + e^{-0.4606T}}\right] \qquad (9)$$

c. The impulse response $h_\alpha(t)$ of $H_\alpha(s)$ is:

$$h_\alpha(t) = L^{-1}\left\{H_\alpha(s)\right\} \overset{(6)}{=} L^{-1}\left\{\frac{0.4606}{s + 0.4606} - \frac{0.4606s}{(s + 0.2303)^2 + (0.9534)^2}\right\}$$

$$= L^{-1}\left\{ \frac{0.4606}{s+0.4606} - 0.4606\frac{s+0.2303}{(s+0.2303)^2 + (0.9534)^2} \right.$$

$$\left. +0.11128\frac{0.9534}{(s+0.2303)^2 + (0.9534)^2} \right\} \Rightarrow$$

$$h_\alpha(t) = 0.4606\left[e^{-0.4606t} - e^{-0.2303t}\cos(0.9534t) + 0.2416e^{-0.2303t}\sin(0.9534t) \right] \quad (10)$$

d. The impulse response of the digital filter is:

$$h[nT] = Z\left[h_\alpha(t) \right]^{(10)} = 0.4606\left[e^{-0.4606nT} - e^{-0.2303nT}\cos(0.9534nT) \right.$$
$$\left. + e^{-0.2303nT}\sin(0.9534nT) \right] \quad (11)$$

$h[nT]$ can also be derived from the inverse z-transform of $H(z)$ (Eq. 8), that is $h[nT] = Z^{-1}[H(z)]$.

e. The difference equation of the filter is derived as follows:

$$H(z) = \frac{Y(z)}{X(z)} \stackrel{(8)}{=} 0.4606\left[\frac{1}{1-e^{-0.4606T}z^{-1}} - \frac{1-z^{-1}e^{-0.2303T}\cos(0.9534T)}{1-2z^{-1}e^{-0.2303T}\cos(0.9534T)+z^{-2}e^{-0.4606T}} \right.$$

$$\left. + 0.2416\frac{z^{-1}e^{-0.2303T}\cos(0.9534T)}{1-2z^{-1}e^{-0.2303T}\cos(0.9534T)+z^{-2}e^{-0.4606T}} \right]$$

$$\Rightarrow H(z) = 0.4606\left[H_1(z)+H_2(z)+H_3(z) \right]$$
$$= 0.4606\left[\frac{Y_1(z)}{X(z)} + \frac{Y_2(z)}{X(z)} + \frac{Y_3(z)}{X(z)} \right] = \frac{Y(z)}{X(z)} \quad (12)$$

where

$$Y(z) = (Y_1(z)+Y_2(z)+Y_3(z))\cdot 0.4606 \quad (13)$$

We can see that the digital filter consists of three subsystems connected in parallel. Each of them has a difference equation:

$$\frac{Y_1(z)}{X(z)} = \frac{1}{1-e^{-0.4606T}z^{-1}} \Rightarrow (1-e^{-0.4606T}z^{-1})Y_1(z) = X(z) \Rightarrow$$

$$\stackrel{I.Z.T.}{Y_1(z) = X(z)+e^{-0.4606T}z^{-1}Y_1(z)} \Rightarrow$$

$$y_1[nT] = x[nT] + e^{-0.4606T} y_1[(n-1)T] \tag{14}$$

$$\frac{Y_2(z)}{X(z)} = -\frac{1 - z^{-1}e^{-0.2303T}\cos(0.9534T)}{1 - 2z^{-1}e^{-0.2303T}\cos(0.9534T) + z^{-2}e^{-0.4606T}} \Rightarrow$$

$$Y_2(z) = -X(z) + e^{-0.2303T}\cos(0.9534T)z^{-1}X(z) + 2e^{-0.2303T}\cos(0.9534T)z^{-1}Y_2(z)$$
$$\quad\quad\quad\quad\quad\quad\quad\quad\quad {\scriptstyle I.Z.T.}$$
$$- e^{-0.4606T}z^{-2}Y_2(z) \Rightarrow \tag{15}$$
$$y_2[nT] = -x[nT] + e^{-0.2303T}\cos(0.9534T)x[(n-1)T] + 2e^{-0.2303T}\cos(0.9534T)y_2[(n-1)T]$$
$$- e^{-0.4606T}y_2[(n-2)T]$$

$$\frac{Y_3(z)}{X(z)} = 0.2416\frac{z^{-1}e^{-0.2303T}\sin(0.9534T)}{1 - 2z^{-1}e^{-0.2303T}\cos(0.9534T) + z^{-2}e^{-0.4606T}} \Rightarrow$$
$$Y_3(z) = 0.2416e^{-0.2303T}z^{-1}X(z)\cdot\sin(0.9534T) + 2e^{-0.2303T}\cos(0.9534T)z^{-1}Y_3(z)$$
$$\quad\quad\quad\quad\quad\quad\quad\quad\quad {\scriptstyle I.Z.T.}$$
$$- e^{-0.4606T}z^{-2}Y_3(z) \Rightarrow \tag{16}$$
$$y_3[nT] = 0.2416e^{-0.2303T}x[(n-1)T]\sin(0.9534T) + 2e^{-0.2303T}\cos(0.9534T)y_3[(n-1)T]$$
$$- e^{-0.4606T}y_2[(n-2)T]$$

f.   Relations (14), (15), (16) for $T = 1$sec yield:

$$(14) \Rightarrow y_1[n] = x[n] + 0.6309y_1[n-1] \Rightarrow$$

$$\frac{Y_1(z)}{X(z)} = \frac{1}{1 - 0.6309z^{-1}} \tag{17}$$

$$(15) \Rightarrow y_2(n) = -x(n) + 0.4598x(n-1) + 0.9196y_2(n-1) - 0.6309y_2(n-2)$$
$$\frac{Y_2(z)}{X(z)} = \frac{-(1 - 0.4598z^{-1})}{1 - 0.9196z^{-1} + 0.6309z^{-2}} \tag{18}$$

$$(16) \Rightarrow y_3(n) = 0.1565x(n-1) + 0.9196y_3(n-1) - 0.6309y_3(n-2)$$
$$\frac{Y_3(z)}{X(z)} = \frac{0.1565z^{-1}}{1 - 0.9196z^{-1} + 0.6309z^{-2}} \tag{19}$$

The requested parallel realization of the digital filter is illustrated in the following figures.

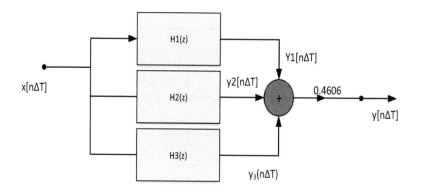

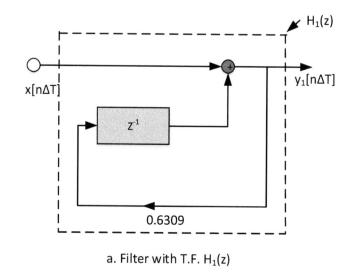

a. Filter with T.F. $H_1(z)$

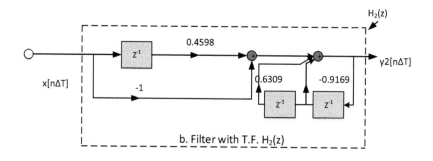

b. Filter with T.F. $H_2(z)$

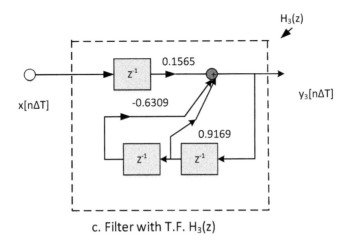

c. Filter with T.F. $H_3(z)$

6.17.5   Calculate the discrete transfer function, $H(z)$, if the corresponding analog is

$$H(s) = \frac{s+0.1}{(s+0.1)^2 + 9}, \quad T = 0.1\,\text{sec},$$ using the impulse invariance method. Plot approximately the frequency response of $H(z)$ and compare it to the frequency response of the analog system.

SOLUTION:

$$H(s) = \frac{s+0.1}{(s+0.1)^2 + 9} \tag{1}$$

Using the impulse invariant method, it is:

$$H(z) = Z\big(H(s)\big) = Z\left[\frac{s+0.1}{(s+0.1)^2 + 9}\right] = \frac{z^2 - e^{-0.1T}\cos 3Tz}{z^2 - 2e^{-0.1T}\cos 3Tz + e^{-0.2T}} \tag{2}$$

$$(1) \Rightarrow |H(j\omega_s)| = \frac{|j\omega_{s+0.1}|}{|(j\omega_{s+0.1})^2 + 9|} = \frac{\sqrt{0.01 + \omega_s^2}}{\sqrt{(9.01 - \omega_s^2)^2 + 0.04\omega_s^2}} \tag{1'}$$

$$(2) \Rightarrow |H(e^{j\omega})| = \frac{|e^{j2\omega} - e^{-0.01}\cos 0.3e^{j\omega}|}{|e^{j2\omega} - 2e^{-0.01}\cos 0.3e^{j\omega} + e^{-0.02}|} \Rightarrow$$

$$\left|H(e^{j\omega})\right| = \frac{\left|\cos 2\omega + j\sin 2\omega - 0.946(\cos\omega + j\sin\omega)\right|}{\left|\cos 2\omega + j\sin 2\omega - 1.8916(\cos\omega + j\sin\omega) + 9802\right|}$$

$$\Rightarrow \left|H(e^{j\omega})\right| = \frac{\sqrt{(\cos 2\omega - 0.946\cos\omega)^2 + (\sin 2\omega - 0.946\sin\omega)^2}}{\sqrt{(0.9802 + \cos 2\omega - 1.8916\cos\omega)^2 + (\sin 2\omega - 1.8916\sin\omega)^2}} \quad (2)'$$

Based on (1)′ and (2)′, we plot the frequency responses of the analog and digital filter as illustrated in the following figures. We notice that for $T = 0.1$sec, the frequency response is closer to that of the analog system (in the second figure, the plot for $T = 0.1$sec and $T = 0.5$sec is illustrated to show the necessity of choosing a short sampling period).

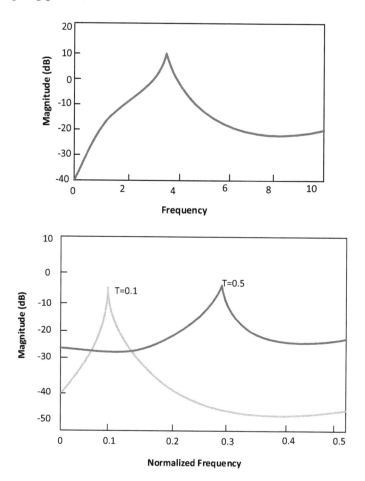

6.17.6 Compute the discrete TF $H(z)$ if the corresponding analog is $H(s) = \dfrac{2}{(s+1)(s+2)}$ with the backward difference method. Plot the frequency response of $H(z)$ and compare it to the frequency response of the analog system.

SOLUTION: Using the backward difference method, the transfer function of the digital system is:

$$H(z) = H(s)\Big|_{s=\frac{1-z^{-1}}{T}} = \frac{2}{(2-z^{-1})(3-z^{-1})} = \frac{2z^2}{(2z-1)(3z-1)} \Rightarrow$$

$$H(e^{j\omega}) = H(z)\Big|_{z=e^{j\omega}} = \frac{2e^{j2\omega}}{(2e^{j\omega}-1)(3e^{j\omega}-1)} \tag{1}$$

The magnitude and phase of the digital filter are computed as follows:

$$\left.\begin{aligned} \left|H(e^{j\omega})\right| &= \frac{2}{\sqrt{(2\cos\omega-1)^2+4\sin^2\omega}\sqrt{(3\cos\omega-1)^2+9\sin^2\omega}} \\ \phi(\omega) &= 2\omega - \tan^{-1}\frac{2\sin\omega}{2\cos\omega-1} - \tan^{-1}\frac{3\sin\omega}{3\cos\omega-1} \end{aligned}\right\} \tag{2}$$

For the analog system, it is true that:

$$H(j\omega) = H(s)\Big|_{s=j\omega} = \frac{2}{(j\omega+1)(j\omega+2)} \tag{3}$$

The magnitude and phase of the analog filter are computed as follows:

$$\left.\begin{aligned} \left|H(j\omega)\right| &= \frac{2}{\sqrt{1+\omega^2}\sqrt{4+\omega^2}} \\ \theta(\omega) &= -\tan^{-1}\omega - \tan^{-1}\frac{\omega}{2} \end{aligned}\right\} \tag{4}$$

The frequency responses of the analog and the digital filters are designed in the following figure. Note that, with the backward difference method, a quite accurate reproduction of the analog system to z-plane is obtained.

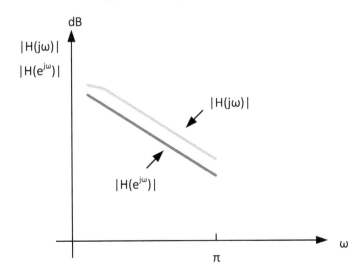

6.17.7   Design an FIR low-pass filter with cut-off frequency $\omega_p = 0.4\pi$ and linear phase. (Choose N=15.)

SOLUTION: According to the specifications, the magnitude and phase of the low-pass filter are:

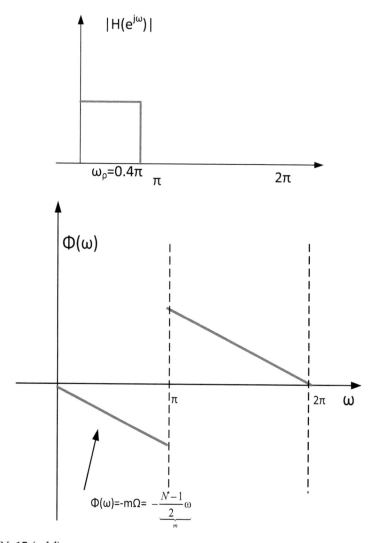

For N=15 (odd),

$$\left|H(k)\right| = \left|H(15-k)\right| \quad , \quad 1 \le k \le 7 \tag{1}$$

$$\Phi(k) = \begin{cases} -\dfrac{\pi k 14}{15} & , \quad 0 \le k \le 7 \\[2mm] \pi - \dfrac{\pi k 14}{15} = \dfrac{\pi k}{15} & , \quad 7 \le k \le 14 \end{cases} \tag{2}$$

The sampling intervals of $H(e^{j\omega})$ are $\dfrac{2\pi}{N} = \dfrac{2\pi}{15}$. The following figure illustrates the ideal magnitude and phase specifications for $N=15$.

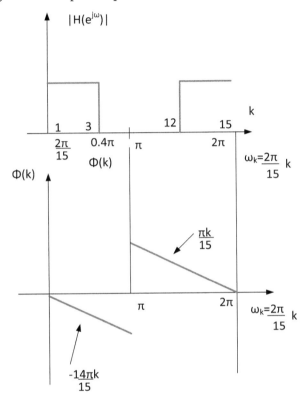

According to Equation 6.54, the impulse response $h[n]$ of the FIR filter is:

$$h[n] = \frac{H(0)}{15} + \frac{1}{15}\sum_{k=1}^{7} 2(-1)^k \left|H(k)\right| \cos\frac{(\pi k + 2\pi k n)}{15} \quad , \quad 0 \le n \le 14 \tag{3}$$

$\begin{aligned} H(0) &= H(1) = H(2) = H(3) = 1 \\ H(4) &= H(5) = H(6) = H(7) = 0 \end{aligned}$. So, Equation (3) becomes:

$$h[n] = \frac{1}{15} + \frac{1}{15}\sum_{k=1}^{3} 2(-1)^k \left|H(k)\right| \cos\frac{(\pi k + 2\pi n)}{15} \quad , \quad 0 \le n \le 14 \tag{4}$$

$$h[0] = \frac{1}{15} + \frac{1}{15}\left(2(-1)\cos\frac{\pi}{15} + 2(-1)^2 \cos\frac{2\pi}{15} + 2(-1)^3 \cos\frac{3\pi}{15} \right) \Rightarrow$$

$$h[0] = \frac{1}{15} + \frac{1}{15}(-1.95629 + 1.827091 - 1.618034) = -0.0498155 \tag{5}$$

Likewise,

$h[1] = 0.0412022$, $h[2] = 0.0666667$, $h[3] = -0.0364878$, $h[4] = -0.107868$,

$h[5] = 0.034078$, $h[6] = 0.318892$, $h[7] = 0.466667$, $h[8] = 0.318892$,

$h[9] = 0.0340779$, $h[10] = -0.10787$, $h[11] = -0.036488$, $h[12] = 0.066667$,

$h[13] = 0.0412022$, $h[14] = -0.0498159$

6.17.8 Design the digital Butterworth filter based on the following specifications: $20\log_{10}\left|H\left(e^{j\omega_p}\right)\right| \geq -0.5$ dB, $20\log_{10}\left|H\left(e^{j\omega_s}\right)\right| \leq -15$ dB, where $\omega_p = 0.25\pi$ and $\omega_s = 0.55\pi$.

Use:

a. the impulse invariant method, and

b. the bilinear method.

SOLUTION:

a. The impulse invariant method

We choose as a sampling period $T = 1s$, since it does not affect the design of the filter. The corresponding analog frequencies ($\Omega = \omega/T$) will be $\Omega_p = 0.25\pi$ and $\Omega_s = 0.55\pi$. Therefore, the given specifications for the continuous-time filter are written as follows:

$$20\log_{10}\left|H_a\left(j\Omega_p\right)\right| \geq -0.5 \text{ dB}, 20\log_{10}\left|H_a\left(j\Omega_s\right)\right| \leq -15 \text{ dB},$$

or:

$$\left|H_a\left(j\Omega_p\right)\right|^2 \geq 10^{-0.05}, \ \left|H_a\left(j\Omega_s\right)\right|^2 \leq 10^{-1.5}.$$

The squared magnitude of the Butterworth filter of order N is:

$$\left|H_\alpha\left(j\Omega\right)\right|^2 = \frac{1}{1+\left(\dfrac{j\Omega}{j\Omega_c}\right)^{2N}}.$$

Now, considering only the equalities to the above equations and the form of $|H_a(j\Omega)|^2$, we obtain the following system of equations with respect to N and Ω_c:

$$1+\left(\frac{j\Omega_p}{j\Omega_c}\right)^{2N} = 10^{0.05} \quad \text{and} \quad 1+\left(\frac{j\Omega_s}{j\Omega_c}\right)^{2N} = 10^{1.5}.$$

We rewrite them in the form

$$\left(\frac{j\Omega_p}{j\Omega_c}\right)^{2N} = 10^{0.05} - 1 \quad \text{and} \quad \left(\frac{j\Omega_s}{j\Omega_c}\right)^{2N} = 10^{1.5} - 1$$

or $\left(\dfrac{\Omega_p}{\Omega_s}\right)^{2N} = \dfrac{10^{0.05} - 1}{10^{1.5} - 1}$

resulting to the following value for N:

$$N = \frac{\log\left(\dfrac{10^{0.05} - 1}{10^{1.5} - 1}\right)}{2\log\left(\dfrac{\Omega_p}{\Omega_s}\right)} = 3.5039.$$

Since N must be an integer, we choose the order of the filter by rounding the above result to the nearest integer, i.e., $N=4$. By substituting it to the first of the two equations, the following is obtained: $\left(\dfrac{\Omega_p}{\Omega_c}\right)^8 = 10^{0.05} - 1$ and solving for Ω_c:

$$\Omega_c = \frac{\Omega_p}{(10^{0.05} - 1)^{1/8}} = \frac{0.25\pi}{0.7688} = 1.0216.$$

This value satisfies the specification for the stop band. It is:

$$|H_a(j\Omega_s)|^2 = 0.0147 < 0.0316 = 10^{-1.5}.$$

The poles of $H_a(s)H_a(-s)$, which are located in the left half of the complex plane (leading to a stable filter) are:
$s_0 = -0.3909 + j\,0.9438, s_1 = -0.9438 + 0.3909\,j$ as well as their conjugates.
Thus, the following transfer function is derived:

$$H_a(s) = \frac{\Omega_c^N}{(s - s_0)(s - s_0^*)(s - s_1)(s - s_1^*)}$$

$$= \frac{\Omega_c^N}{(s^2 - 2\,\mathrm{Re}(s_0)s + |\,s_0\,|^2)(s^2 - 2\,\mathrm{Re}(s_1)s + |\,s_1\,|^2)}$$

$$= \frac{1.0892}{(s^2 + 0.7818s + 1.0436)(s^2 + 1.8876s + 1.0436)}.$$

Rewriting the above function as a sum of partial fractions, we obtain:

$$H_a(s) = \frac{a}{s - s_0} + \frac{b}{s - s_0^*} + \frac{c}{s - s_1} + \frac{d}{s - s_1^*}$$

where $a = -0.4719 + j\,0.1955, b = a^*, c = 0.4719 - j1.1393, d = c^*$. The transfer function is then:

$$
\begin{aligned}
H(z) &= \frac{a}{1 - e^{s_0 T} z^{-1}} + \frac{a^*}{1 - e^{s_0^* T} z^{-1}} + \frac{c}{1 - e^{s_1 T} z^{-1}} + \frac{c^*}{1 - e^{s_1^* T} z^{-1}} \\[2mm]
&= \frac{2\,\mathrm{Re}(a) - 2\,\mathrm{Re}(a^* e^{s_0 T})z^{-1}}{1 - 2\,\mathrm{Re}(e^{s_0 T})z^{-1} + e^{2\,\mathrm{Re}(s_0)T} z^{-2}} + \frac{2\,\mathrm{Re}(c) - 2\,\mathrm{Re}(c^* e^{s_1 T})z^{-1}}{1 - 2\,\mathrm{Re}(e^{s_1 T})z^{-1} + e^{2\,\mathrm{Re}(s_1)T} z^{-2}} \\[2mm]
&= \frac{-0.9438 + 0.1604 z^{-1}}{1 - 0.7937 z^{-1} + 0.4576 z^{-2}} + \frac{0.9438 - 0.0017 z^{-1}}{1 - 0.7196 z^{-1} + 0.1514 z^{-2}} \\[2mm]
&= \frac{0.0888 z^{-1} + 0.1749 z^{-2} + 0.0235 z^{-3}}{1.0000 - 1.5133 z^{-1} + 1.1801 z^{-2} - 0.4495 z^{-3} + 0.0693 z^{-4}}
\end{aligned}
$$

The magnitude of the frequency response of the digital filter is plotted (in dB) in the following figure:

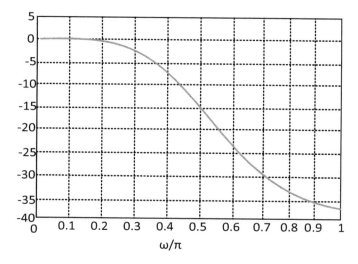

We observe that the specifications are satisfied. The overlapping problem (as a result of the use of the impulse invariant method) is not apparent, here, as the analog filter is of limited bandwidth.

b. The bilinear method

The analog frequencies are computed as follows:

$$\Omega_p = 2\tan\left(\frac{\omega_p}{2}\right) = 0.8284, \ \Omega_s = 2\tan\left(\frac{\omega_s}{2}\right) = 2.3417.$$

For N, it is true, as before:

$$N = \frac{\log\left(\dfrac{10^{0.05} - 1}{10^{1.5} - 1}\right)}{2\log\left(\dfrac{\Omega_p}{\Omega_s}\right)} = 2.6586.$$

Therefore, the required order for the Butterworth filter is $N = 3$ and is less than that required in the impulse invariant method. The frequency, Ω_c, is derived as follows:

$$\Omega_c = \frac{\Omega_p}{(10^{0.05} - 1)^{1/6}} = \frac{0.8284}{0.7043} = 1.1762.$$

and it satisfies the second condition:

$$|H_a(j\Omega_s)|^2 = 0.0158 < 0.0316 = 10^{-1.5}$$

The poles located in the left half plane are: $s_0 = -0.5881 + j1.0186, s_0^*, s_1 = -1.1762.$ The transfer function of the analog filter is:

$$H_a(s) = \frac{\Omega_c^N}{(s - s_0)\left(s - s_0^*\right)(s - s_1)} = \frac{1.6272}{\left(s^2 - 2\operatorname{Re}(s_0)s + |s_0|^2\right)(s - s_1)}$$

$$= \frac{1.6272}{(s^2 + 1.1762s + 1.3834)(s + 1.1762)}$$

Finally, we apply the bilinear method:

$$H(z) = H_a(s)_{s=2(1-z^{-1})/(1+z^{-1})} = \frac{0.0662 + 0.1987z^{-1} + 0.1987z^{-2} + 0.0662z^{-3}}{1 - 0.9359z^{-1} + 0.5673z^{-2} - 0.1016z^{-3}}.$$

The magnitude of the frequency response of the resulting digital filter is illustrated in the figure below. We notice that the specifications are satisfied. In addition, although of lower order, the resulting filter has a response much closer to the ideal than the filter we designed using the impulse invariant method.

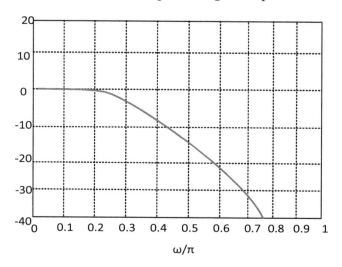

6.17.9   Design a digital differentiator using the Hamming window.

SOLUTION:

Step 1: The desired frequency response of the analog differentiator, $H(r)$, is defined:

$$H(s) = s \Rightarrow H(r) = j\omega \big|_{\substack{\omega = 2\pi f \\ f = rf_s}} = j2\pi f_s r$$

Step 2: $h_i[nT]$ is computed.

$$h_i[nT] = \int_{-1/2}^{1/2} H(r)e^{j2\pi \, nr} \, dr = \int_{-1/2}^{1/2} (j2\pi \, f_s r)e^{j2\pi \, nr} \, dr$$

$$\int_{-1/2}^{1/2} H(r)e^{j2\pi \, nr} \, dr = \int_{-1/2}^{1/2} (j2\pi \, f_s r)e^{j2\pi \, nr} \, dr = \frac{f_s}{n} \int_{-1/2}^{1/2} j2\pi \, nr e^{j2\pi \, nr} \, dr$$

$$= \frac{f_s}{n} \int_{-1/2}^{1/2} r \, de^{j2\pi \, nr} \, r = \frac{f_s}{n} \left[re^{j2\pi \, nr} \Big|_{r=-1/2}^{r=1/2} - \int_{-1/2}^{1/2} e^{j2\pi \, nr} d(r) \right]$$

$$= \frac{f_s}{n} \left[re^{j2\pi \, nr} - \frac{1}{j2\pi n} e^{j2\pi \, nr} \right]_{r=-1/2}^{r=1/2}$$

$$= \frac{f_s}{n} \left[\frac{1}{2} - \frac{1}{j2\pi n} \right] e^{j\pi \, n} - \frac{f_s}{n} \left[-\frac{1}{2} - \frac{1}{j2\pi n} \right] e^{-j\pi \, n}$$

$$= \frac{f_s}{2n} [e^{j\pi \, n} + e^{-j\pi \, n}] - \frac{f_s}{j2\pi \, n^2} [e^{j\pi \, n} - e^{-j\pi \, n}]$$

$$= \frac{f_s}{2n} 2\cos \pi \, n - \frac{f_s}{j2\pi n^2} [2j \sin \pi \, n]$$

Thus,

$$
h_i[nT] = \begin{cases} \dfrac{f_s}{n}(-1)^n - \dfrac{f_s 2j}{j2\pi n^2}\sin \pi n = \dfrac{f_s}{n}(-1)^n & n \neq 0 \\[4mm] \dfrac{f_s}{n} - \dfrac{f_s}{\pi n^2}\sin \pi n = \dfrac{f_s[\pi n - \sin \pi n]}{\pi n^2} \end{cases}
$$

$$
\Rightarrow f_s \lim_{n \to 0} \frac{\dfrac{d}{dn}[\pi n - \sin \pi n]}{\dfrac{d\pi n^2}{dn}}
$$

$$
= \frac{f_s}{2}\lim_{n \to 0}\frac{\pi - \pi \cos \pi n}{\pi n}
$$

$$
= \frac{f_s}{2}\lim_{n \to 0}\frac{d(\pi - \pi \cos \pi n)/dn}{d(\pi n)/dn}
$$

$$
= \frac{f_s}{2}\lim_{n \to 0}\frac{\pi^2 \sin \pi n}{\pi} = 0 \qquad n = 0
$$

$$
h_i[nT] = \begin{cases} \dfrac{f_s}{n}(-1)^n & n \neq 1 \\[4mm] 0 & n = 0 \end{cases}
$$

Step 3: Design of the filter using the Hamming window (M=7).

$$
\text{Hamming window}: w_h[n] = \begin{cases} 0.54 + 0.46\cos\dfrac{n\pi}{M} & |n| \leq M \\[4mm] 0 & |n| > M \end{cases}
$$

$$
H(z) = \sum_{n=-7}^{n=-1}\frac{f_s}{n}(-1)^n w_h[n]z^{-n} + \sum_{n=1}^{n=7}\frac{f_s}{n}(-1)^n w_h[n]z^{-n}
$$

$$
H_c(z) = z^{-7}H(z)
$$

The frequency response of the digital differentiator is illustrated in the following figure.

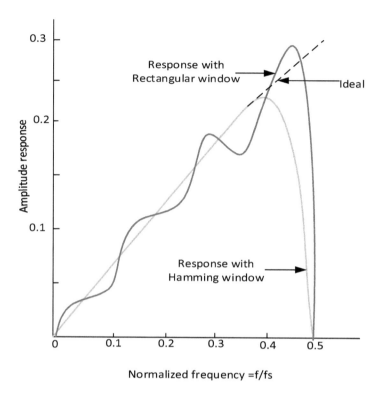

Normalized frequency =f/fs

6.17.10 Design, using the window method, an FIR filter of linear phase and of $N = 24$ order that approximates the following magnitude of the ideal frequency response:

$$\left|H_i(e^{j\omega})\right| = \begin{cases} 1 & |\omega| \leq 0.2\pi \\ 0 & 0.2\pi < |\omega| \leq \pi \end{cases}$$

SOLUTION: We want to approach a low-pass filter with a cut-off frequency $\omega_p = 0.2\pi$. As $N = 24$, the frequency response of the filter to be designed is of the form

$$H(e^{j\omega}) = \sum_{n=0}^{24} h[n]e^{-jn\omega}.$$

Therefore, the delay of $h[n]$ is $N/2 = 12$, and the ideal impulse response is

$$h_i[n] = \frac{\sin\left(0.2\pi(n-12)\right)}{\pi(n-12)}.$$

To complete the design, we have to choose the window function. With a fixed window length, we 'move', during the design, between the width of the transition band and the ripple-width in the pass band and stop band.

By using a rectangular window, which results to the smallest width of the transition band, we get:

$$\Delta\omega = 2\pi \cdot \frac{0.9}{24} = 0.075\pi$$

and the filter is

$$h[n] = \frac{\sin(0.2\pi(n-12))}{\pi(n-12)} \quad \text{for } 0 \le n \le 24 \quad \text{and } h[n] = 0 \text{ elsewhere}$$

The stop band attenuation is only 21dB, resulting in a ripple of $\delta_1 = 0.089$. Using the Hamming window, the filter obtained is:

$$h[n] = \left[0.54 - 0.46\cos\left(\frac{2\pi n}{24}\right) \right] \cdot \frac{\sin(0.2\pi(n-12))}{\pi(n-12)} \quad \text{for } 0 \le n \le 24$$

The attenuation of the stop band is 53dB, resulting in a ripple of $\delta_1 = 0.0022$.

The width of the transition band is consequently quite large: $\omega = 2\pi \cdot \frac{3.3}{24} = 0.275\pi$

6.17.11 Design a FIR high-pass filter with the minimum possible filter order using the window method and having the following characteristics: stop band cut-off frequency of $\omega_s = 0.22\pi$, pass band cut-off frequency of $\omega_p = 0.28\pi$ and a stop band ripple of $\delta_2 = 0.003$.

SOLUTION: The stop band ripple $\delta_2 = 0.003$ corresponds to the stop band attenuation $a_s = -20 \log \delta_2 = 50.46$.

For a filter of the minimum order, we use the Kaiser window with parameter, β, that is equal to $\beta = 0.1102(a_s - 8.7) = 4.6$. Since the transition band width is $\Delta\omega = 0.06\pi = 0.03$, the required window width $N = \frac{a_s - 7.95}{14.36\Delta f} = 98.67$.

By choosing $N = 99$, a linear phase Type II filter is derived, the transfer function of which will have a zero at $z = -1$. However, in this way, the frequency response gets to zero for $\omega = \pi$, so the value is not acceptable. We, therefore, increase the order of the filter to $N = 100$ to produce a linear phase Type I filter.

To have a transition band that extends from $\omega_s = 0.22\pi$ to $\omega_p = 0.28\pi$, we set the cut-off frequency of the ideal high-pass filter equal to the average value of these two frequencies, i.e., $\omega_c = \frac{\omega_s + \omega_p}{2} = 0.25\pi$.

The impulse response of an ideal high-pass filter with zero phase and the cut-off frequency $\omega_c = 0.25\pi$, is $h_{hp}[n] = \delta[n] - \frac{\sin(0.25\pi n)}{\pi n}$, where the second term corresponds to a low-pass filter with a cut-off frequency $\omega_c = 0.25\pi$. In introducing a delay in $h_{hp}[n]$ equal to $N/2 = 50$, we obtain:

$$h_i[n] = \delta[n-50] - \frac{\sin(0.25\pi(n-50))}{\pi(n-50)}$$

The high-pass filter derived is $h[n] = h_i[n] \cdot w[n]$, where $w[n]$ represents the Kaiser window with $N = 100$ and $\beta = 4.6$.

6.17.12 Design a digital, of linear phase, FIR filter using the window method that satisfies the following specifications:

$$0.99 \le \left|H(e^{j\omega})\right| \le 1.01, \qquad 0 \le |\omega| \le 0.3\pi$$

$$\left|H(e^{j\omega})\right| \le 0.01, \qquad 0.35 \le |\omega| \le \pi$$

SOLUTION: Since the pass band ripple and the stop band ripple are the same, we can focus only on the requirement of the stop band ripple. A stop band ripple $\delta_2 = 0.01$ corresponds to a stop band attenuation of -40dB. Therefore, based on Table 6.2, we can use the Hanning window for which an attenuation of approximately -44dB is obtained. The transition band, according to the specifications, is $\Delta\omega = 0.05\pi \Rightarrow \Delta f = 0.025$.

The required filter order is $N = \dfrac{3.1}{f} = 12.4$, and it is $w[n] = 0.5 - 0.5\cos\left(\dfrac{2\pi n}{124}\right)$ for $0 \le n \le 124$.

For an ideal low-pass filter with a cut-off frequency defined by the point corresponding to the middle of the transition band, i.e., $\omega_c = 0.325\pi$, and such a delay that $h_i[n]$ is symmetrically placed in the interval $0 \le n \le 124$, that is $N/2 = 62$, it is:

$$h_i[n] = \frac{\sin\big(0.325\pi(n-62)\big)}{\pi(n-62)}$$

Thus, the filter corresponds to

$$h[n] = \left[0.5 - 0.5\cos\frac{2\pi n}{124}\right] \cdot \frac{\sin\big(0.325\pi(n-62)\big)}{\pi(n-62)}, \qquad 0 \le n \le 124$$

If, instead of the Hanning window, we used the Hamming or Blackman window, we would exceed the threshold of the ripple requirements in the pass band and the stop band, as a larger filter order would be required. For example, using the Blackman window, the order of the filter, which satisfies the transition band specification would be $N = \dfrac{5.5}{0.025} = 220$.

6.17.13 Design a digital IIR low-pass filter, which satisfies the specifications:
$\left|H(e^{j\omega})\right|^2 \ge -1\,\text{dB}$ for $|\omega| \le 0.4\pi$ and $\left|H(e^{j\omega})\right|^2 \le -15\,\text{dB}$ for $0.5\pi < |\omega| < \pi$

SOLUTION: Our design will be based on the Butterworth analog filter, and the final digital filter will have the corresponding form in terms of its response to frequencies.

The frequency response of the Butterworth analog filter is given by:

$$|H(j\Omega)|^2 = \frac{1}{1+\left(\dfrac{\Omega}{\Omega_c}\right)^{2N}}$$

For example, for $\Omega_c = 100$ and N=4, the filter will be of the form:

```
W=0:0.1:200;
N=4;
Wc=100;
H2=1./(1+(W/Wc).^(2*N));
plot(W,H2)
```

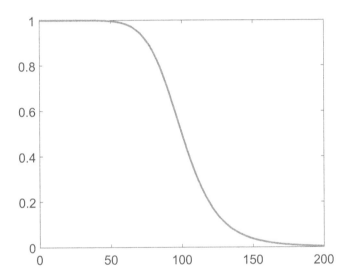

$|H(j\Omega)|^2$ gets to the half of its maximum value for $\Omega = \Omega_c$.

- First step: Define and understand the specifications.

    Our design will be based on the Butterworth analog filter. Because of the monotony of the filter, the relations concerning the specifications are transformed equivalently into:

$$|H(e^{j\omega})|^2 = -1\text{dB} \quad \text{for} \quad \omega = \omega_1 = 0.4\pi$$

$$|H(e^{j\omega})|^2 = -15\text{dB} \quad \text{for} \quad \omega = \omega_2 = 0.5\pi$$

- Second step: Transform the specifications into analog.

    We already mentioned that, as an intermediate design step, we will create an analog filter. For this reason, our specifications should be transformed into the corresponding analog ones. Using the method of bilinear transform, the conversion from digital to analog will be accomplished.

The relation of the bilinear transform which connects the analog Ω with the digital ones ω is: $\Omega = \dfrac{2}{T}\tan\left(\dfrac{\omega}{2}\right)$. Let the sampling period $T = 1$ sec.

Thus, the specifications are now transformed as follows:

$$\left|H(j\Omega)\right|^2 = -1\,\mathrm{dB} \ \ \text{for} \ \ \Omega = \Omega_1 \ \ \text{and} \ \ \Omega_1 = \frac{2}{T}\tan\left(\frac{\omega_1}{2}\right)$$

$$\left|H(j\Omega)\right|^2 = -15\,\mathrm{dB} \ \ \text{for} \ \ \Omega = \Omega_2 \ \ \text{and} \ \ \Omega_2 = \frac{2}{T}\tan\left(\frac{\omega_2}{2}\right)$$

We now write the following code in MATLAB:

```
T=1;
w1=0.4*pi;
w2=0.5*pi;
W1=2*tan(w1/2);
W2=2*tan(w2/2);
```

• Third step: Define N, Ωc.

The new system based on the new analog specifications is of the form:

$$10\log_{10}\frac{1}{1+\left(\dfrac{\Omega_1}{\Omega_c}\right)^{2N}} = -R_p$$

and

$$10\log_{10}\frac{1}{1+\left(\dfrac{\Omega_2}{\Omega_c}\right)^{2N}} = -A_s$$

where for the specific design $R_p = 1$, $A_s = 15$

By solving the system, we can compute N:

$$N = \frac{\log_{10}\left[(10^{R_p/10}-1)/(10^{A_s/10}-1)\right]}{2\log_{10}\dfrac{\Omega_1}{\Omega_2}}$$

Because the resulting N is usually decimal, we round it to the next integer value. To determine Ω_c, we have the following options:

$$\Omega_c = \frac{\Omega_1}{\sqrt[2N]{\left(10^{R_p/10}-1\right)}} \text{ or } \Omega_c = \frac{\Omega_2}{\sqrt[2N]{\left(10^{A_s/10}-1\right)}}.$$

We use the first of the above relations in our code:

```
Rp=1;
As=15;
numN=log10(((10^(Rp/10)-1)/(10^(As/10)-1));
denN=2*log10(W1/W2);
N=ceil(numN/denN);   % ceil rounds to the next integer
denWc=(10^(Rp/10)-1)^(1/(2*N));
Wc=W1/denWc;
```

- Fourth step: Define the poles of the analog filter.

    If we run the code, we find that N=8. Laplace transform will be of the form:

    $$H(s) = \frac{\Omega_c^N}{(s-s_1)\dots(s-s_8)}, \text{ where } s_i \text{ represents the poles of the filter. These poles}$$
    are the roots of the equation $1+\left(\dfrac{s}{j\Omega_c}\right)^{2N} = 0$ that have a negative real part.

    In our code, we will find the roots of this equation, and then we will keep the N of them that have a negative real part. This is because we want to make sure that our system is casual and stable. The MATLAB code is:

```
%Design the polynomial
denanal=[(1/(j*Wc))^(2*N) zeros(1,2*N-1) 1];
polesanal=roots(denanal);
polesanalneg=polesanal(real(polesanal)<0);
```

- Fifth step: Apply the bilinear transform.

    By setting $s = \dfrac{2}{T}\dfrac{1-z^{-1}}{1+z^{-1}}$, and after some calculations, we define the final digital filter. By using the following MATLAB *function*, we can make the process described above very simple. So, we add the next line to our code:

```
[Zdig,Pdig,Kdig]=bilinear([],polesanalneg,Wc^N,1/T);
```

The **bilinear** *function* accepts as input arguments: the analog filter zeros (not present in Butterworth therefore []), the analog filter poles (**polesanalneg**), the multiplication constant (*Wc^N*) and the sampling frequency (*1/T*). We run it (so the bilinear transform is applied), and the output will be: the digital filter zeros (**Zdig**), the digital filter poles (**Pdig**) and the multiplication constant (**Kdig**).

So the filter obtained will be of the form:

$$H(z) = Kdig \frac{(z - z_{z1})...(z - z_{z8})}{(z - z_{p1})...(z - z_{p8})}$$

where **Kdig** is real. Due to rounding errors MATLAB returns an extremely small fantastic part. For this reason, we add the following line in our code:

```
Kdig=real(Kdig);
```

- Sixth step: Define the digital filter coefficients.

At this point, we have already fully defined the digital filter. If we were just asked to determine it, we would have finished. But we are interested in seeing the filter operate and in practice to filter out some sinusoidal components. In MATLAB this process is done with the **filter** *function*, which however requires knowledge of the filter coefficients. So we have to bring $H(z)$ in the form of a quotient of two polynomials with respect to z^{-1}. The coefficients of this polynomial are the coefficients we have to define. The code is:

```
b=Kdig*poly(Zdig);   % We compute the nominator coefficients
a=poly(Pdig);        % We compute the denominator coefficients
b=real(b);
a=real(a);
```

Since **filter** *function* requires a(1) to equal to one, we add the following lines to the code:

```
b=b/a(1);
a=a/a(1); % Divide the numerator and the denominator with a(1)
```

In order to design the frequency response, we use the commands:

```
H=freqz(b,a,[0:0.01:pi]);
plot([0:0.01:pi],abs(H).^2)
```

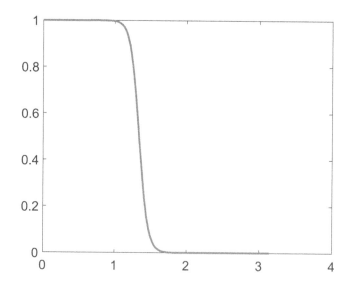

We can check whether the original digital specifications are satisfied using the commands:

```
Hw1w2=freqz(b,a,[w1 w2]);
10*log10(abs(Hw1w2).^2)
```

ans =

 -1.0000 -16.4299

We note that, with respect to the first frequency, the initial condition is satisfied, whereas for the second frequency, there is a difference. This is due both to the rounding we made when choosing N and to the equation we chose to calculate Ωc. The (unavoidable) difference in the second condition should not disappoint us. We were asking for a filter that would have an attenuation of 15 dB at ω_2. We finally designed a filter that has more attenuation, i.e., 16.4299 dB, at the same frequency. So we designed a better filter than the one the original specifications required.

We will filter now two sinusoidal functions, one of which is in the pass band and the other in stop band. The code is as follows:

```
n=0:200;
x1=sin(0.1*pi*n);
x2=sin(0.7*pi*n);
x=x1+x2;
y=filter(b,a,x);
subplot(211)
stem(n,x)
subplot(212)
stem(n,y)
```

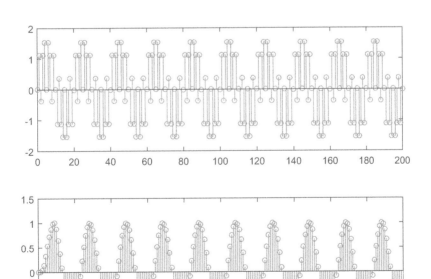

Using the following code, we can plot graphs with respect to frequencies:

```
X=fft(x);
Y=fft(y);
subplot(211)
plot(n*(2*pi)/length(x),abs(X))
subplot(212)
plot(n*(2*pi)/length(y),abs(Y))
```

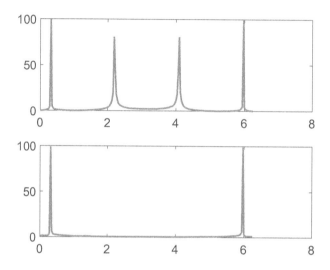

6.17.14 Develop the FIR design process of digital filters using MATLAB.

SOLUTION: The basic idea behind successfully designing FIR filters is to choose an ideal filter (which has always a non-causal and of infinite-length impulse response) and to cut off the impulse response to make the filter an FIR one. Consider an ideal linear phase filter with bandwidth $\omega_c < \pi$

$$H_d(e^{-j\alpha\omega}) = \begin{cases} 1 \cdot e^{-j\alpha\omega}, |\omega| \le \omega_c \\ 0, \omega_c < |\omega| \le \pi \end{cases}$$

where ω_c represents the filter cut-off frequency.

The infinite-length impulse response of the filter is:

$$h_d[n] = \frac{\sin\left[\omega_c(n-a)\right]}{\pi(n-a)}, n = 0, \pm1, \pm2, \cdots$$

We create the *function* **ideal_lp(wc,N)** for the computation of the impulse response of the filter we described.

```
function [hd,n]=ideal_lp(wc,N)
a=(N-1)/2;
m=[0:1:(N-1)];
n=m-a+eps;
hd=sin(wc*n)./(pi*n);
```

We then run it for *N*=21 and for a cut-off frequency of 0.25π to get the impulse response:

```
[hd,n]=ideal_lp(0.25*pi,21);
stem(n,hd);
```

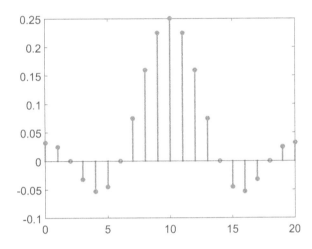

Now, in order to implement an FIR filter, we must:

a. Approximate *hd*[*n*] with a sequence *h'*[*n*] of finite length, *N*, using a window.

b. Shift *h'*[*n*] in time by $n_0 = (N-1)/2$ samples, $h[n] = h'[n - n_0]$, so as $h[n] = 0, n < 0$, for the filter to be casual. This is the equivalent of considering $a = n_0$.

The most common windows are described by the following equations:

$$RECTANGULAR \ \ w[n] = \begin{cases} 1, 0 \leq |n| \leq N-1 \\ 0, |n| \geq N \end{cases}$$

$$BARTLETT \ \ w[n] = \begin{cases} \dfrac{2n}{N}, 0 \leq n \leq \dfrac{N-1}{2} \\ 2 - \dfrac{2n}{N}, \dfrac{N-1}{2} \leq n \leq N-1 \end{cases}$$

$$HANNING \ \ \ w[n] = \frac{1}{2} \left| 1 - \cos\left(\frac{2\pi n}{N} \right) \right|, 0 \leq n \leq N-1$$

$$HAMMING \quad w[n] = 0.54 - 0.46\cos\left(\frac{2\pi n}{N}\right), 0 \leq n \leq N-1$$

In order to create these windows, we use the following MATLAB commands:

```
W=boxcar(N)
W=Bartlett(N)
W=hamming(N)
W=hanning(N)
```

To see the effect of a rectangular window of length, N, on a FIR filter with $\omega_c = 0.25\pi$, we write the code:

```
wc=0.25*pi;
N=11;
n=0:1:N-1;
hd=ideal_lp(wc,N);
w_tet=(boxcar(N))';
h=hd.*w_tet;
fr=0:0.01:pi;
H=freqz(h,1,fr);
figure
subplot(121)
semilogy(fr,abs(H))
subplot(122)
plot(fr,abs(H))
N=21;
n=0:1:N-1;
hd=ideal_lp(wc,N);
w_tet=(boxcar(N))';
h=hd.*w_tet;
fr=0:0.01:pi;
H=freqz(h,1,fr);
subplot(121)
hold on
semilogy(fr,abs(H),'k:')
subplot(122)
hold on
plot(fr,abs(H),'k:')
N=31;
n=0:N-1;
hd=ideal_lp(wc,N);
w_tet=(boxcar(N))';
h=hd.*w_tet;
fr=0:0.01:pi;
H=freqz(h,1,fr);
subplot(121)
semilogy(fr,abs(H),'r--')
subplot(122)
plot(fr,abs(H),'r--')
```

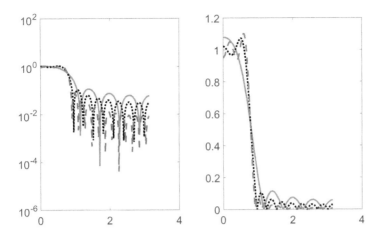

The phase of the filter is linear. We can design it for a frequency range, i.e., [-0.15π, 0.15π], using the following commands:

```
w=-0.15*pi:0.01:0.15*pi;
H=freqz(h,1,w);
plot(w,angle(H))
```

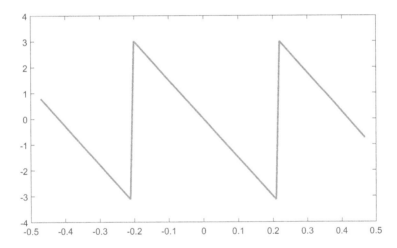

We notice that: The lobe frequency increases with N and the width of the first lobe remains almost constant. We also notice that Gibbs ringing has been introduced.

Using the following code we obtain the response of the low-pass filter for $N=21$, and for the rectangular and Hanning windows:

```
wc=0.25*pi;
N=21;
n=0:1:N-1;
hd=ideal_lp(wc,N);
w_tet=(boxcar(N))';
```

```
h=hd.*w_tet;
fr=0:0.01:pi;
[H,w]=freqz(h,1,fr);
figure
subplot(1,2,1);
semilogy(fr,abs(H))
subplot(1,2,2);
plot(fr,abs(H))
N=21;
n=0:1:N-1;
hd=ideal_lp(wc,N);
w_han=(hanning(N))';
h=hd.*w_han;
fr=0:0.01:pi;
[H,w]=freqz(h,1,fr);
subplot(1,2,1)
hold on
semilogy(fr,abs(H),'r')
subplot(1,2,2)
hold on
plot(fr,abs(H),'r')
```

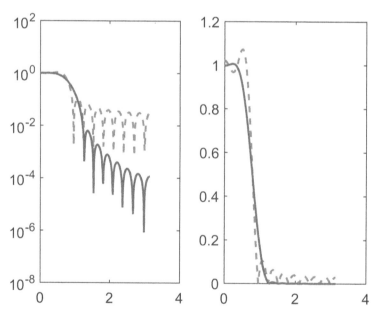

We notice that:

1. The lobes, including the main, have a much smaller width for the Hanning window.

2. Passing from the pass band to the stop band is much faster in the case of the rectangular window. This results from the fact that the main lobe of the Fourier transform of the Hanning window is of larger width.

Next, we study an application of real-case filtering. Let a signal consisting of a sum of two cosines with frequencies of 30Hz, 50Hz and a sine of 60Hz. We want to apply a filter that cuts the frequencies higher than 40Hz.

```
fs=300;
n=0:1/fs:0.2;
x=cos(2*pi*30*n)+cos(2*pi*50*n)+sin(2*pi*60*n);
wc=pi*40/(fs/2);
N=21;
n=0:1:N-1;
hd=ideal_lp(wc,N);
w_tet=(boxcar(N))';
h=hd.*w_tet;
y=filter(h,1,x);
fr=0:0.01:pi;
Xw=freqz(x,1,fr);
Hw=freqz(h,1,fr);
Yw=freqz(y,1,fr);
subplot(311)
plot(fr,abs(Xw))
subplot(312)
plot(fr,abs(Hw))
subplot(313)
plot(fr,abs(Yw))
```

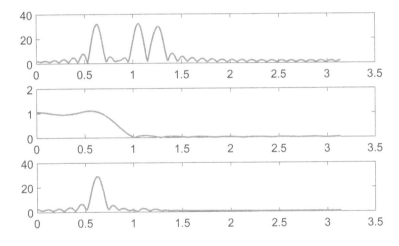

6.17.15 Create *functions* in MATLAB to calculate the frequency response amplitude of linear phase FIR filters of Type I and III.

SOLUTION:

a. Type I linear FIR filters

In this case: $H(e^{jw}) = \left[\displaystyle\sum_{n=0}^{(M-1)/2} a(n)\cos wn \right] e^{-jw(M-1)/2}$

where sequence $a[n]$ is derived from $h[n]$ as follows:

$$a(o) = h\left(\frac{M-1}{2}\right) \qquad : \quad \textit{the central sample}$$

$$a(n) = 2h\left(\frac{M-1}{2} - n\right) \quad , \quad 1 \le n \le \frac{M-3}{2}$$

Thus, it is $H_r(w) = \displaystyle\sum_{n=0}^{(M-1)/2} a(n)\cos wn$

The *function* below specifies the amplitude of the frequency response of the FIR, linear, Type I filter.

```
function [Hr,w,a,L]=Hr_Type1(h);
% [Hr,w,a,L]=Hr_Type1(h)
% Hr = Amplitude Response
% w = 500 frequencies between [0,pi] over which Hr is computed
% a = Type-1 LP filter coefficients
% L = Order of Hr
% h = Type-1 LP filter impulse response
M = length(h);
L= (M-1)/2;
a = [h(L+1) 2*h(L:-1:1)]; % 1x(L+1) row vector
n = [0:1:L]          % (L+1)x1 column vector
w = [0:1:500]'*pi/500;
Hr = cos(w*n)*a';
```

b. Type III linear FIR filters

In this case:

$$H(e^{jw}) = \left[\sum_{n=1}^{(M-1)/2} c(n)\sin wn\right] e^{-j\left[\frac{\pi}{2}-w(M-1)/2\right]}$$

where the sequence c[n] is derived from h[n] as follows:

$$c(n) = 2h\left(\frac{M-1}{2}-n\right) \quad , \quad n = 1,2,...,\frac{M-1}{2}$$

Thus, it is $H_r(w) = \displaystyle\sum_{n=1}^{(M-1)/2} c(n)\sin wn$

Note that, at $w = 0$ and $w = \pi$ it is $H_r(w) = 0$, regardless of $c(n)$ or $h(n)$. It is also $e^{j\pi/2} = j$, which means that $jH_r(w)$ has only an imaginary part. This type of filter is not suitable for the design of high- and low-pass filters. This behavior is suitable for differentiators and ideal Hilbert transformers.

Using the following *function*, the amplitude of the frequency response of Type III, linear, FIR filter is derived.

```
function [Hr,w,c,L]=Hr_Type3(h);
% [Hr,w,c,L]=Hr_Type3(h)
% Hr = Amplitude Response
% w = frequencies between [0,pi] over which Hr is computed
% b = Type-3 LP filter coefficients
% L = Order of Hr
```

```
% h = Type-3 LP filter impulse response
M = length(h);
L = (M-1)/2;
c = [2*h(L+1:-1:1)];
n = [0:1:L];
w = [0:1:500]'*pi/500;
Hr = sin(w*n)*c';
```

Type II and IV filters (FIR and Linear phase) are studied next.

Type II linear FIR filters

In this case, it is:

$$H(e^{jw}) = \left[\sum_{n=1}^{(M)/2} b(n)\cos\left\{ w\left(n-\frac{1}{2} \right) \right\} \right] e^{-jw(M-1)/2}$$

where the sequence $a[n]$ is derived from $h[n]$ as follows:

$$b(n) = 2h\left(\frac{M}{2} - n \right) \quad , \quad n = 1, 2, ..., \frac{M}{2}$$

Thus, $H_r(w) = \displaystyle\sum_{n=1}^{(M)/2} b(n)\cos\left\{ w\left(n-\frac{1}{2} \right) \right\}.$

For $w = \pi$, it is $H_r(w) = \displaystyle\sum_{n=1}^{(M)/2} b(n)\cos\left\{ \pi\left(n-\frac{1}{2} \right) \right\} = 0$ regardless of $b[n]$ or $h[n]$.

This Type is suitable for high- and low-pass filters.

Type IV linear FIR filters

In this case, it is:

$$H(e^{jw}) = \left[\sum_{n=1}^{(M)/2} d(n)\sin\left\{ w\left(n-\frac{1}{2} \right) \right\} \right] e^{j\left[\frac{\pi}{2}-w(M-1)/2 \right]}$$

where the sequence $d[n]$ is derived from $h[n]$ as follows:

$$d(n) = 2h\left(\frac{M}{2} - n \right) \quad , \quad n = 1, 2, ..., \frac{M}{2}$$

Thus, $H_r(w) = \displaystyle\sum_{n=1}^{(M)/2} d(n)\sin\left\{ w\left(n-\frac{1}{2} \right) \right\}$

Note that, at $w = 0$ and $w = \pi$, it is $H_r(0) = 0$ and $e^{j\pi/2} = j$.

This behavior is also suitable for differentiators and approximation of ideal Hilbert transformers.

6.17.16 Design the digital pass band filter with the following specifications:

$$lower\ stopband\ edge:\quad w_{1s} = 0.2\pi, \quad A_s = 60dB$$
$$lower\ passband\ edge:\quad w_{1p} = 0.35\pi \quad R_p = 1dB$$

$$upper\ passband\ edge:\quad w_{2p} = 0.65\pi, \quad R_p = 1dB$$
$$upper\ stopband\ edge:\quad w_{2s} = 0.8\pi \quad A_s = 60dB$$

The specifications above are illustrated in the following figure.

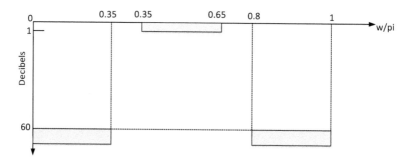

SOLUTION: The requested code in MATLAB for designing the digital pass band filter is:

```
ws1 = 0.2*pi;    wp1 = 0.35*pi;
wp2 = 0.65*pi;   ws2 = 0.8*pi;
As = 60;
tr_width = min((wp1-ws1),(ws2-wp2));
M = ceil(11*pi/tr_width) +1;
n = [0:1:M-1];
wc1 = (ws1+wp1)/2;   wc2 = (wp2+ws2)/2;
hd = ideal_lp(wc2,M) - ideal_lp(wc1,M);
w_bla = (blackman(M))';
h = hd.*w_bla;
[db,mag,pha,grd,w] = freqz_m(h,[1]);
delta_w =2*pi/1000;
Rp = -min(db(wp1/delta_w+1:1:wp2/delta_w))    % Actual Pass band Ripple
As = -round(max(db(ws2/delta_w+1:1:501)))    % Min Stop band Attenuation

%  Plots
subplot(2,2,1); stem(n,hd); title('Ideal Impulse Response')
axis([0 M-1 -0.4 0.5]); xlabel('n');   ylabel('hd(n)')

subplot(2,2,2); stem(n,w_bla); title('Blackman Window')
axis([0 M-1 0 1.1]); xlabel('n');   ylabel('w(n)')

subplot(2,2,3); stem(n,h); title('Actual Impulse Response')
axis([0 M-1 -0.4 0.5]); xlabel('n');   ylabel('h(n)')
```

```
subplot(2,2,4); plot(w/pi,db); axis([0 1 -150 10]);
title('Magnitude Response in db');grid;
xlabel('Frequency in pi units');   ylabel('Decibels')
```

In this code, the *function* **freqz_m** was used, which returns the amplitude of the frequency response in dB, the phase response and the group delay response.

```
function [db,mag,pha,grd,w] = freqz_m(b,a)
% [db,mag,pha,grd,w] = freqz_m(b,a)
% db = relative magnitude in dB computed over 0 to pi radians
% mag = absolute magnitude computed over 0 to pi radians
% pha = Phase response in radians over 0 to pi radians
% grd = Group delay over 0 to pi radians
%      w = 501 frequency samples between 0 to pi radians
%      b = numerator polynomial of H(z)   (for FIR: b=h)
%      a = denominator polynomial of H(z)   (for FIR: a=[1])
%
[H,w] = freqz(b,a,1000,'whole');
    H = (H(1:1:501))';       w = (w(1:1:501))';
  mag = abs(H);
   db = 20*log10((mag+eps)/max(mag));
  pha = angle(H);
  grd = grpdelay(b,a,w);
```

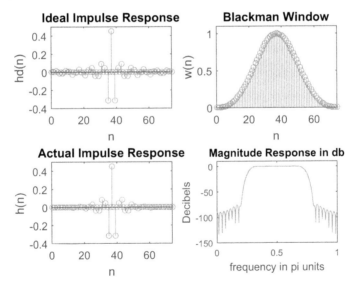

6.17.17 Let $H_d(e^{jw}) = \begin{cases} jw, & 0 \prec w \le \pi \\ -jw, & -\pi \prec w \prec 0 \end{cases}$ represent the frequency response of an ideal digital differentiator. Design a digital FIR differentiator using a Hamming window of length 21.

SOLUTION: Note that M is an even number, so $\alpha = (M-1)/2$ is integer and $h_i(n)$ will be zero for all n. Here, M must be an odd number, and that would be the case of an FIR, linear phase, Type III filter.

The code in MATLAB for designing the digital FIR filter is:

```
M = 21; alpha = (M-1)/2;
n = 0:1:M-1;
hd = (cos(pi*(n-alpha)))./(n-alpha);     hd(alpha+1) = 0;
w_ham = (hamming(M))';
h = hd.*w_ham;
[Hr,w,P,L]=Hr_Type3(h);


%  Plots
%     subplots(1,1,1);
subplot(2,2,1); stem(n,hd); title('Ideal Impulse Response')
axis([-1 M -1.2 1.2]); xlabel('n');    ylabel('hd(n)')
subplot(2,2,2); stem(n,w_ham); title('Hamming Window')
axis([-1 M 0 1.2]); xlabel('n');    ylabel('w(n)')
subplot(2,2,3); stem(n,h); title('Actual Impulse Response')
axis([-1 M -1.2 1.2]); xlabel('n');    ylabel('h(n)')
subplot(2,2,4); plot(w/pi,Hr/pi);   title('Amplitude Response');grid;
xlabel('Frequency in pi units');   ylabel('slope in pi units');
axis([0 1 0 1]);
```

In the MATLAB code, above, the *function* **ideal_lp** was used to calculate an ideal low-pass filter of impulse response $h_i(n)$.

```
function hd=ideal_lp(wc,M);
% [hd] = ideal_lp(wc,M);
% hd = ideal impulse response between 0 to M-1
% wc = cutoff frequency in radians
% M = length of the ideal filter
%
alpha = (M-1)/2;
n = [0:1:(M-1)];
m = n - alpha +eps;     % add smallest number to avoid division by zero
hd = sin(wc*m)./(pi*m);
```

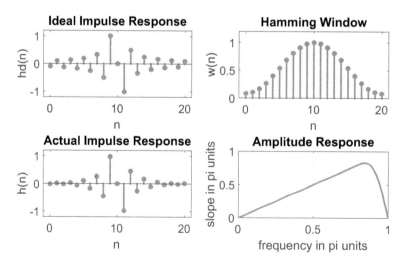

6.17.18 Create a MATLAB code for designing a low-pass filter using the equiripple method.

SOLUTION: The MATLAB *function* **firpm,** which is based on Parks-McClellan algorithm, can be used to design a FIR filter.

Syntax of **firpm** *function*: firpm(N,F,A,W) where N: the order of the filter, F: Frequency vector [F1 F2 F3 ...], A: desired amplitude for each frequency [A1 A2 A3...], W: weighting factor for each band in order to succeed equalization. Each W value indicates how much emphasis will be given to reduce the error of the transition or stop band with regard to other bands (i.e., pass or stop bands). The length of W is half of F. Example: If F=[F1 F2 F3 F4] : Two bands: F1-F2, F3-F4 with F2-F3 the transition band, W=[W1 W2].

If we want to design Hilbert transformers or differentiators, we can use the *functions*: **firpm**(N,F,A,W,'Hilbert') or **firpm** (N,F,A,W, 'differentiator').

The MATLAB code is:

```
%Define sampling frequency
fs=40000;
%Define filter order
N=40;
%Define cut-off frequency
f1=10000;
%Define transition frequency, 500Hz
ftr=500;
F1=f1/fs;
Ftr=ftr/fs;
%Compute h coefficients for the filter
h=firpm(N,[0 2*F1 2*(F1+Ftr) 0.5*2],[1 1 0 0],[1 1]);
%Compute frequency response
[H,W]=freqz(h,1,256);
%Design in Hz
df=(0:length(W)-1)*(fs/2)/length(W);
plot(df,20*log10(abs(H)));
```

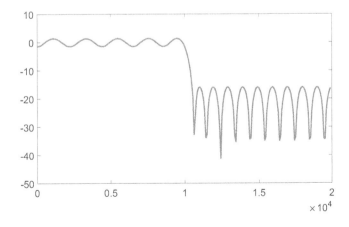

6.17.19 Describe the process of designing an FIR filter using a rectangular window.

SOLUTION: Suppose we want to calculate $h[n]$ coefficients for an ideal low-pass filter with the cut-off frequency $\omega_s = \dfrac{\pi}{5}$. Since the filter is low-pass, the coefficients $h[n]$ will be $h[n] = \dfrac{\sin(n\pi / 5)}{n\pi}$

MATLAB code:

```
n=-20:1:20; % N=41
h=sin(n*pi/5)./(n*pi);
stem(n,h);
```

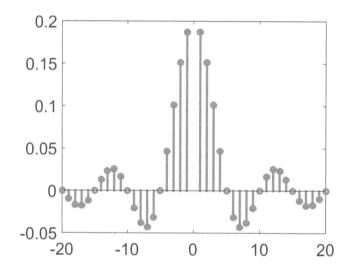

The ideal frequency response for $\omega_s = \dfrac{\pi}{5}$ and $N = 41$ is illustrated in the following figure:

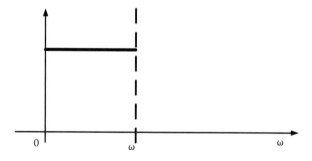

For the finite length filter (FIR), we need to keep a finite number of the $h[n]$ coefficients. This leads to truncation that distorts the original ideal low-pass function.

This truncation is best expressed by the use of a window. In the following figures, the window sequence $w[n]$ is illustrated, as well as its frequency response.

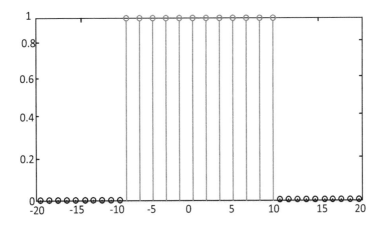

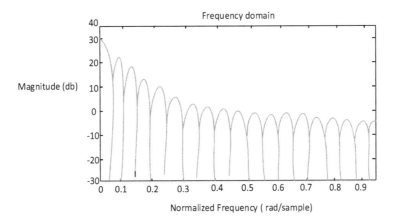

The transform from IIR to FIR is done by multiplying the (infinite) ideal sequence $h[n]$ with a rectangular window $w[n]$ of finite length N or else, in the frequency domain, by convolving $h(\omega)$ with a rectangular window, $w(\omega)$.

$$\text{Multiplication}: h[n] = h_{ideal}[n]w[n]$$

$$\text{Convolution}: h(\omega) = h_{ideal}(\omega) * w(\omega)$$

In the figures below, we can see the result of multiplying the ideal sequence, $h[n]$, with the window, $w[n]$, and the convolution of the ideal $h(\omega)$ with $w(\omega)$, respectively.

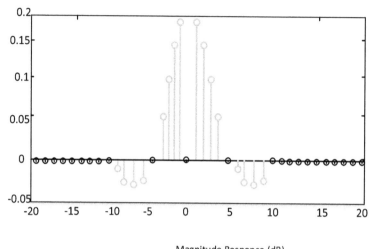

Magnitude Response (dB)

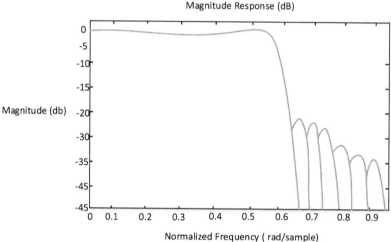

Normalized Frequency (rad/sample)

6.17.20 Design the frequency response of a 5-point low-pass moving average filter.

SOLUTION: First suggested way:

We shall compute the frequency response based on $Y(\omega) = H(\omega)X(\omega)$ by calculating the DTFT of $x[n-k]$ ($k = 0,1,2...N-1$). The MATLAB code is:

```
syms X w
N=5;
X1=X*exp(-j*w); %X1=DTFT{x(n-1)}.
X2=X*exp(-j*2*w); %X2=DTFT{x(n-2)}.
X3=X*exp(-j*3*w); %X3=DTFT{x(n-3)}.
X4=X*exp(-j*4*w); %X4=DTFT{x(n-4)}.
Y=(1/N)*(X+X1+X2+X3+X4); %Output.
H=Y/X; %Filter response.
H=simplify(H);
% Design of amplitude using ezplot command
ezplot(abs(H),[0 pi]);
```

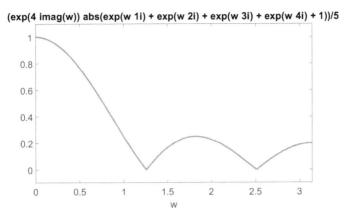

(exp(4 imag(w)) abs(exp(w 1i) + exp(w 2i) + exp(w 3i) + exp(w 4i) + 1))/5

Second suggested way: Computation of the frequency response using:

$$H(\omega) = (1/N)\frac{1-e^{-j\omega N}}{1-e^{-jw}}$$

```
N=5; %Filter order.
omega = 0:pi/400:pi; %.
Hw = (1/N)*(1-exp(-i*omega*N))./(1-exp(-i*omega));
plot(omega,abs(Hw)); %Design of magnitude .
xlabel('Frequency ω(rad/s) [0 - π] );
ylabel('Magnitude');
title(Frequency response m.a. filter') ;
```

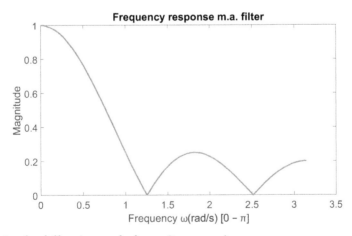

Frequency response m.a. filter

We write the following code for unit conversion:

```
fs=40000; %Define sampling frequency fs.
N=5; %Filter order.
omega = 0:pi/400:pi;
Hw = (1/N)*(1-exp(-i*omega*N))./(1-exp(-i*omega));
df=(0:length(omega)-1)*(fs/2)/length(omega); %unit conversion of y-axis
in Hz
plot(df,abs(Hw)); %Design magnitude.
xlabel('frequency f (KHz)');
ylabel('Magnitude');
title('Frequency response of m.a. filter);
```

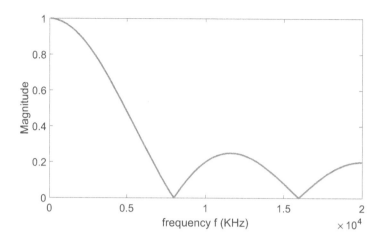

Third suggested way: Compute and design the frequency response using the FFT of *h*[*n*]

```
N = 5; %Filter order.
hn =(1/N)* ones(1,N); %Define impulse response
hF = fft(hn,1024); %Define number of samples of FFT of hn
plot(([0:511]/512)*pi, abs((hF(1:512)))); %Plot magnitude
xlabel('Normalized frequency ω ,[0 - π]');
ylabel('Magnitude');
title('Frequency response of m.a. filter) ;
```

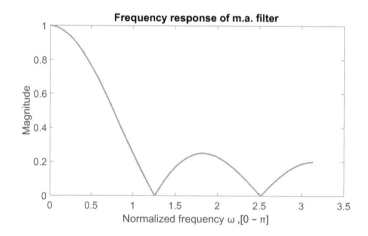

6.17.21 Design a moving average filter with $N = 5$ to smooth the noise of the input:

$x =$ [0 0.1 $-$0.2 0.3 0 $-$0.1 0 0.2 $-$0.1 0.1 $-$0.2 0 0 $-$0.1 0 0.1 0.5 1 1.2 0.9 1 1 1.1 1
0.8 1.1 0.8 1.1 0.9 1 0.8 0.5 0.1 $-$0.2 0.3 0 $-$0.1 0 0.2 $-$0.1 0.1 $-$0.2 0 0 $-$0.1 0 0.1]

SOLUTION:

```
N = 5; %Filter order.
hn = (1/N)*ones(1,N); % Define impulse response hn=1/N.
a = 1; %Define a: a(1)*y(n) = b(1)*x(n) + b(2)*x(n-1) + ....
x= [ 0 0.1 -0.2 0.3 0 -0.1 0 0.2 -0.1 0.1 -0.2 0 0 -0.1 0 0.1 0.5 1
    1.2 0.9 1 1 1.1 1 0.8 1.1 0.8 1.1 0.9 1 0.8 0.5 0.1 -0.2 0.3 0
    -0.1 0 0.2 -0.1 0.1 -0.2 0 0 -0.1 0 0.1]; %Define b: a(1)*y(n) =
    b(1)*x(n) + b(2)*x(n-1) + ....
y = filter(hn,a,x); %Compute output y.
t = 1:length(x); %.
plot(t,x,'-.',t,y,'-'); %Plot input and output.
legend('Real signal+ Noise', 'smoothing of noise',2);
```

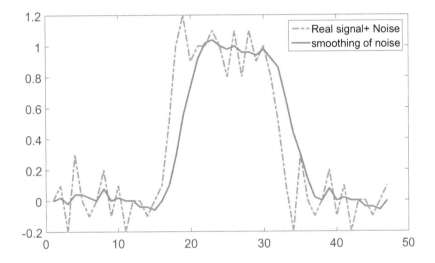

6.17.22 Design a FIR filter of length $N = 15$ with 0-10KHz frequency pass band and sampling frequency $f_s = 40KHz$.

SOLUTION: The distance between the samples is $\dfrac{40KHz}{15} = 2,66KHz$.

Then, we define the desired magnitude $H(k)$: $H(0$ to $3) = 1$ and $H(4$ to $7)=0$.

We use the following type (N=odd):

$$h(i) = \frac{H(0)}{N} + \frac{1}{N}\sum_{k=1}^{\frac{N-1}{2}}|H(k)|2(-1)^k \cos\left(\frac{(\pi k + 2\pi ki)}{N}\right)$$

The desired frequency response for the FIR filter of length $N = 15$ is illustrated in the following figure.

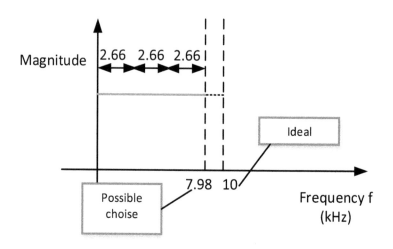

The code for design follows:

```
N=16; % Filter order
H0=1; % H(0)=1, since H(0to4) = 1
Hk(1:4)=1; % H(1to4)=1, since H(0to4) = 1
Hk(5:8)=0; % H(5to8)=0
for n=1:N
S=0;
for k=1:((N/2)-1)
S=S+2*Hk(k)*(-1)^k*cos((2*pi*k*(n-1)+pi*k)/N);
end
h(n)=H0/N+S/N;
end
fs=40000; % Define sampling frequency fs
[H,W]=freqz(h,1,256); % Compute frequency response
df=(0:255)*(fs/2)/256; % Unit conversion of y-axis in Hz
plot(df,abs(H)); % Design magnitude
xlabel('Frequency f, KHz');
ylabel('Magnitude');
title('F.I.R. Filter with the F.S. method N=16,Fc=10KHz,Fs=40KHz') ;
```

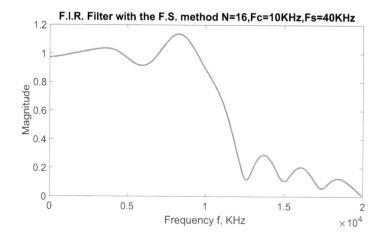

6.17.23 Design a FIR filter of length $N=20$ of $\omega_s = \pi/2$ (normalized frequency) and sampling frequency $f_s = 40KHz$.

SOLUTION: We calculate the distance between the samples $\dfrac{2\pi}{20} = \pi/10$.

We, then, define the desired magnitude $H(k)$: $H(0 \text{ to } 5) = 1$ and $H(6 \text{ to } 10) = 0$.

We use the following formula (N=even):

$$h(i) = \frac{H(0)}{N} + \frac{1}{N}\sum_{k=1}^{\frac{N}{2}-1}|H(k)|2(-1)^k\cos\left(\frac{(\pi k + 2\pi k i)}{N}\right)$$

The design code follows:

```
N=20; %Filter order.
H0=1; %H(0)=1.
Hk(1:5)=1; %H(1to5)=1.
Hk(6:10)=0; % H(6to10) = 1.
for n=1:N
S=0;
for k=1:(N/2)-1;
S=S+2*Hk(k)*(-1)^k*cos((2*pi*k*(n-1)+pi*k)/N);
end
h(n)=H0/N+S/N;
end
fs=40000; % Define sampling frequency fs
[H,W]=freqz(h,1,256); % Compute frequency response
plot(W,angle(H)); % Design phase
xlabel('Frequency f, KHz');ylabel('Phase');
title('F.I.R. Filter with F.S. method N=20,ωc=π/2,Fs=40KHz');
```

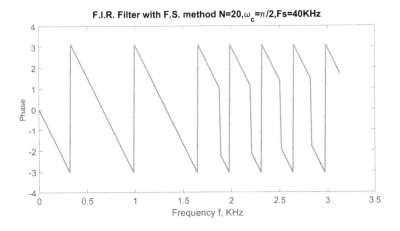

```
plot(df,abs(H)); % Design Magnitude
xlabel('Frequency f, KHz');ylabel('Magnitude');
title('F.I.R. Filter with F.S. method N=20,ωc=π/2,Fs=40KHz');
```

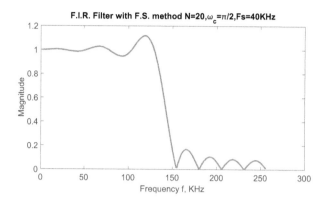

With the following commands, we can design the coefficients *h(i)*:

```
n=[1:N];
stem(n,h);
```

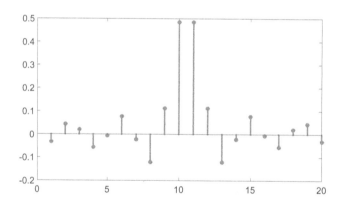

6.17.24 Design an FIR filter with the specifications of the following plot and with sampling frequency $f_s = 40KHz$, for *N*=16, *N*=24, *N*=32. Design the phases for each *N*.

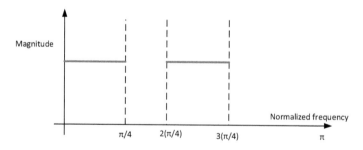

SOLUTION: We calculate the distance between the samples:

For *N*=16, the distance between the samples is: $\frac{2\pi}{16} = \pi/8$; for *N*=24, the distance between the samples is: $\frac{2\pi}{24} = \pi/12$; and for *N*=32, the distance between the samples is $\frac{2\pi}{32} = \pi/16$.

We define the desired magnitude, $H(k)$: For $N=16$, it is: $H(0$ to $2) = 1$, $H(3$ to $4)=0$, $H(5$ to $8)=1$; for $N=24$, it is: $H(0$ to $3) = 1$, $H(4$ to $6)=0$, $H(7$ to $12)=1$; and for $N=32$, it is $H(0$ to $4) = 1$, $H(5$ to $8)=0$, $H(9$ to $16)=1$.

We use the following formula (N=even):

$$h(i) = \frac{H(0)}{N} + \frac{1}{N} \sum_{k=1}^{\frac{N}{2}-1} |H(k)| 2(-1)^k \cos\left(\frac{(\pi k + 2\pi k i)}{N} \right)$$

The following figure depicts the desired frequency response of a FIR filter for $N=16$, $N=24$ and $N=32$.

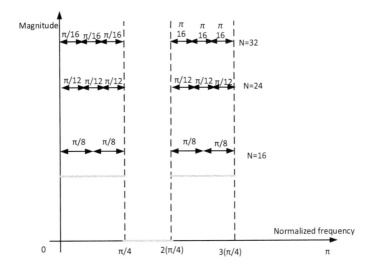

The design code is:

```
N=16; %Filter order.
H0=1; %H(0)=1.
Hk(1:2)=1; %H(1toç2) = 1.
Hk(3:4)=0; %H(3to4) = 0.
Hk(5:8)=1; %H(5to8) = 1.
for n=1:N
S=0;
for k=1:(N/2)-1;
S=S+2*Hk(k)*(-1)^k*cos((2*pi*k*(n-1)+pi*k)/N);
end
h(n)=H0/N+S/N;
end
fs=40000; %Define sampling frequency fs
[H,W]=freqz(h,1,256); %Compute frequency response
plot(W,abs(H),'-.'); %Design magnitude for N=16 order
hold on
N=24; %Filter order
```

```
H0=1; %H(0)=1.
Hk(1:3)=1; %H(1to3) = 1
Hk(4:6)=0; %H(4to6) = 0
Hk(7:9)=1; %H(7to9) = 1
Hk(10:12)=0; %H(10to12) = 0
for n=1:N
S=0;
for k=1:(N/2)-1;
S=S+2*Hk(k)*(-1)^k*cos((2*pi*k*(n-1)+pi*k)/N);
end
h(n)=H0/N+S/N;
end
fs=40000;
[H,W]=freqz(h,1,256); %Compute freq. response
plot(W,abs(H),'--'); % Design magnitude for N=24 order
hold on
N=32; % Filter order
H0=1; %H(0)=1
Hk(1:4)=1; %H(1to4) = 1
Hk(5:8)=0; %H(5to8) = 0
Hk(9:12)=1; %H(9to12) = 1
Hk(13:16)=0; %H(13to16) = 0
for n=1:N
S=0;
for k=1:(N/2)-1;
S=S+2*Hk(k)*(-1)^k*cos((2*pi*k*(n-1)+pi*k)/N);
end
h(n)=H0/N+S/N;
end
[H,W]=freqz(h,1,256); %Compute frequency response.
xlabel('Normalized frequency ω,[0 to π]');
ylabel('Amplitude');
title('F.I.R. filter with F.S. method N=16 ,N=24 and N=32,Fs=40KHz') ;
plot(W,abs(H),'-'); %Design magnitude for N=32 order.
legend('N=16','N=24','N=32',3);
```

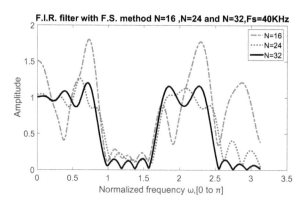

6.17.25 Design a low-frequency FIR filter with the window method with the cut-off
frequency $f_c = 10\text{KHz}$, the sampling frequency $f_s = 40\text{KHz}$, a transition band of
725Hz, and a maximum loss of the gain in the transition band equal to 20dB.

SOLUTION: Based on the filter specifications, we choose the rectangular window, $w(n)$, since the requested gain loss rate of the transition band is equal to 20dB. Thus, $K=0.9$.

We compute the order, N, of the filter using the formula $N \geq \dfrac{0.9}{\dfrac{10.725 KHz - 10 KHz}{40 KHz}}$, i.e., $N \geq 26,1$ or $N=27$ (integer).

We calculate the coefficients of the ideal impulse response of the filter $h[n]$ obtained by the formula $\dfrac{\sin(n2\pi F_1)}{n\pi}$. In this way, the frequencies are transformed from analog (f) to normalized digital (F).

$$f_1 \rightarrow F_1 : F_1 = \frac{f_1}{f_s} = \frac{10 KHz}{40 KHz} = 0.25 \text{ or } \omega_1 = 2\pi F_1 = 2\pi * 0.25$$

We shift the ideal $h[n]$ to the right by $(N-1)/2$ according to: $h_{shifted} = h_{ideal}\left(n - \dfrac{N-1}{2}\right), n = 0,1,2,\ldots,(N-1)$

We compute the FIR filter of the finite sequence $h(n)$ by multiplying the ideal $h[n]$ with the window $w[n]$: $h(n) = h_{shifted} \, w(n), 0 \leq n \leq (N-1)$.

The design code is:

```
fs=40000; %Define sampling frequency fs
N=27; %Filter order
w=ones(1,27); %Define rectangular window
F1=0.25; %F1=f1/fs, where f1 cut-off frequency

  w1=0.25*2*pi; %w1=F1*2π

M=(N-1)/2; %Shift by (N-1)/2
n=-M:M;
h=2*F1*sinc(2*F1*n); %Compute the ideal impulse response h
h=h.*w; %Multiply h coefficients with the window
[H,W]=freqz(h,1,256); %Compute frequency response
df=(0:length(W)-1)*(fs/2)/length(W); % Unit conversion of y-axis in Hz
plot(df,20*log10(abs(H))); %Design magnitude in dB
xlabel('Frequency f(KHz)');
ylabel('Magnitude db');
title('Frequency Response of FIR filter (rectangular window) ') ;
```

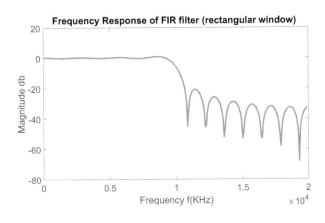

What follows is an alternative way of writing the code that designs the requested filter:

```
fs=40000; %Define sampling frequency fs
N=27; %Filter order
h=fir1(N,0.5,rectwin(N+1));%Compute h coefficients (rectangular window)
[H,W]=freqz(h,1,256); %Compute frequency response
df=(0:length(W)-1)*(fs/2)/length(W); % Unit conversion of y-axis (in Hz)
plot(df,20*log10(abs(H))); %Design magnitude
xlabel('Frequency f(KHz)');
ylabel('Magnitude dB');
title('Frequency response of FIR filter (rectangular window) 2nd way');
```

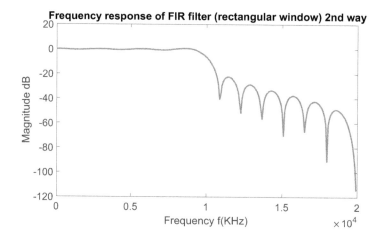

6.17.26 Design a band pass filter with frequency passing range from $f_1 = 4,3KHz$ to $f_2 = 5KHz$, transition band equal to 1.58KHz, in which the gain to be lost is at least 45dB and the sampling frequency is $f_s = 15KHz$.

SOLUTION: Based on the filter specifications we want to design, we choose Hamming window $w(n)$ and the desired rate of gain loss of the transition band equal to 45dB. Thus, K=3.3. The order of the filter N is computed by

$$N \geq \frac{3.3}{\dfrac{4,3KHz - (4,3 - 1.58)KHz}{15KHz}} = \frac{3.3}{1.58/15} = 31.32,\ \text{i.e.,}\ N \geq 31.32\ \text{or}\ N = 32\ \text{(we choose}$$

a larger number).

The coefficients of the ideal impulse response of the filter $h[n]$ are derived using the formula $\dfrac{\sin(n2\pi F_2)}{n\pi} - \dfrac{\sin(n2\pi F_1)}{n\pi}$. So, the frequencies are converted from analog (f) to normalized digital (F).

$$f_1 \rightarrow F_1 : F_1 = \frac{f_1}{f_s} = \frac{4,3KHz}{15KHz} \cong 0.287 \text{ or } \omega_1 = 2\pi F_1 = 2\pi * (4,3/15) \cong 1.8$$

$$f_2 \rightarrow F_2 : F_2 = \frac{f_2}{f_s} = \frac{5KHz}{15KHz} = 1/3 \text{ or } \omega_2 = 2\pi F_2 = 2\pi / 3 \cong 2,094$$

We shift the ideal $h[n]$ to the right by $(N\text{-}1)$ according to:

$$h_{shifted} = h_{ideal}\left(n - \frac{N-1}{2}\right), n = 0,1,2,\ldots,(N-1)$$

We compute the FIR filter of the finite sequence $h(n)$ by multiplying the ideal $h[n]$ with the window $w[n]$: $h(n) = h_{shifted} \, w(n), 0 \le n \le (N\text{-}1)$.

The design code is:

```
fs=15000; %Define sampling frequency fs
f1=4300; %Define fpass1
f2=5000; %Define fpass2
F1=f1/fs; %Normalized F1
F2=f2/fs; %normalized F2
N=32; %Filter order
w=hamming(N+1)'; %Define Hamming window
M=N/2; %Shift by N/2
n=-M:M;
A=2*F1*sinc(2*F1*n); %Compute the coefficients for low pass with Fc1=F1
B=2*F2*sinc(2*F2*n); % Compute the coefficients for low pass with Fc2=F2
h=B-A; %Subtract the Fc1 coefficients from Fc2
hh=h.*w; %Multiply band pass coefficients with the window
[H,W]=freqz(hh,1,256); %Compute frequency response
df=(0:length(W)-1)*(fs/2)/length(W); % Unit conversion of y-axis (in Hz)
plot(df,20*log10(abs(H))); %Design magnitude
xlabel('Frequency f(KHz)');
ylabel('Magnitude db');
title('Frequency response of FIR filter (hamming window) 1st way');
```

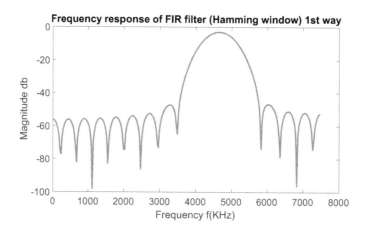

What follows is an alternative way of writing the code that designs the requested filter:

```
fs=15000; %Define sampling frequency fs
N=32; %Filter order
f1=4300; %Define fpass1
f2=5000; %Define fpass2
f=[f1 f2]/(fs/2); %Define frequency range from f1 to f2
h=fir1(N,f,hamming(N+1));%Compute coefficients using the hamming window
[H,W]=freqz(h,1,256); %Compute frequency response
df=(0:length(W)-1)*(fs/2)/length(W); % Unit conversion of y-axis (in Hz)
plot(df,20*log10(abs(H))); %Design magnitude
xlabel('Frequency f(KHz)');
ylabel('Magnitude dB');
title('Frequency response of FIR filter (hamming window) 2nd way ');
```

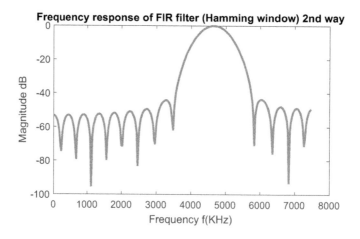

6.17.27 Design a band stop filter using a Kaiser window with frequencies from $f_1 = 150KHz$ to $f_2 = 250KHz$, transition band equal to 50Hz in which the gain that will be lost is at least 60dB and with the sampling frequency $f_s = 1KHz$

SOLUTION: Based on the filter specifications, we choose the Kaiser window $w(n)$ with γ equal to $\gamma = 0.1102(A–8.7)$. From the specifications $A = 60$ and $\gamma = 0.1102(A–8.7) \Rightarrow \gamma = 5.65$.

We compute the order of the filter N. Since:

$$A > 50 \Rightarrow N = \frac{A-7.95}{14.36\Delta F}+1 \Rightarrow N = \frac{60-7.95}{14.36\frac{50Hz}{1000Hz}}+1 = \frac{52.05}{0.718}+1 = 73.49 \text{ or } N = 74.$$

The coefficients of the ideal impulse response of the filter $h[n]$ are derived using the formula $\frac{\sin(n2\pi F_2)}{n\pi} - \frac{\sin(n2\pi F_1)}{n\pi}$. So, the frequencies are converted from analog (f) to normalized digital (F).

$$f_1 \to F_1 : F_1 = \frac{f_1}{f_s} = \frac{150Hz}{1000Hz} = 0.15 \text{ or } \omega_1 = 2\pi F_1 = 2\pi*(150/1000) \cong 0.9$$

$$f_2 \to F_2 : F_2 = \frac{f_2}{f_s} = \frac{250Hz}{1000Hz} = 0.25 \text{ or } \omega_2 = 2\pi F_2 = 2\pi.0.25 \cong 1.57$$

We shift the ideal $h[n]$ to the right by ($N/2$) according to:

$$h_{shifted} = h_{ideal}\left(n - \frac{N}{2}\right), n = 0,1,2,.....,(N-1)$$

We compute the FIR filter of the finite sequence $h(n)$ by multiplying the ideal $h[n]$ with the window $w[n]$: $h(n) = h_{shifted}w(n), 0 \le n \le (N-1)$.

The design code is:

```
fs=1000; % Define sampling frequency fs
f1=150; %Define f1
f2=250; %Define f2
F1=f1/fs; %Normalized frequency F1
F2=f2/fs; % Normalized frequency F2
N=74; %Filter order
w=kaiser(N+1,5.67)'; %Define window Kaiser (γ=5.65)
M=N/2; %Shift by N/2
n=-M:M;
A=2*F1*sinc(2*F1*n); %Low pass coefficients with Fc=Fc1
B=2*F2*sinc(2*F2*n); %Low pass coefficients with Fc=Fc2
C=2*0.5* sinc(2*0.5*n); %Coefficients of all pass filter
h=C-(B-A); %Compute coefficients of ideal stop band filter
hh=h.*w; %Multiply the coefficients with the window
[H,W]=freqz(hh,1,256); %Compute frequency response
df=(0:length(W)-1)*(fs/2)/length(W); % Unit conversion of y-axis (in Hz)
plot(df,20*log10(abs(H))); %Design magnitude
xlabel('Frequency f(Hz)');
ylabel('Magnitude dB');
title('Frequency response of FIR filter (kaiser window) 1st way');
```

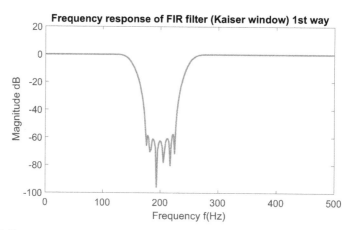

Frequency response of FIR filter (Kaiser window) 1st way

What follows is an alternative way of writing the code that designs the requested filter:

```
fs=1000; %Define sampling frequency fs
N=74; %Filter order
f1=150; %Define fpass1
f2=250; %Define fpass2
f=[f1 f2]/(fs/2); %Define frequency range from f1 to f2
h=fir1(N,f, 'stop', kaiser(N+1,5.67));%Compute coefficients using
the Kaiser window
[H,W]=freqz(h,1,256); %Compute frequency response
df=(0:length(W)-1)*(fs/2)/length(W); % Unit conversion of y-axis (in Hz)
plot(df,20*log10(abs(H))); %Design magnitude
xlabel('Frequency f(KHz)');
ylabel('Magnitude dB');
title('Frequency response of FIR filter (Kaiser(γ=5.67) window) 2nd
way ');
```

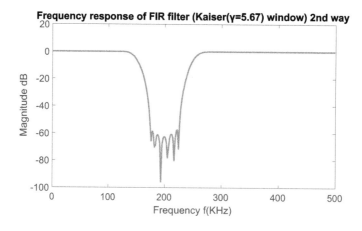

Frequency response of FIR filter (Kaiser(γ=5.67) window) 2nd way

6.17.28 Design a band pass FIR filter using the equiripple method to allow the signal to pass at the frequency range 10KHz to 13KHz, with transition band at 1KHz, with a gain loss of 50dB and ripple in the pass band at 0.25dB (sampling frequency $fs = 40$KHz)

SOLUTION: The design will be implemented with the following MATLAB code:

```
fs=40000; %Define sampling frequency fs
f1p=10000; %Define fpass1
ftr=1000; %Define transition width
f1s=f1p-ftr; %Define fstop1
f2p=13000; %Define fpass2.
f2s=f2p+ftr; %Define fstop2
F1p=f1p/fs; %Converse to normalized frequency.
F1s=f1s/fs;
F2p=f2p/fs;
F2s=f2s/fs;
Ftr=ftr/fs;
f=[0 2*F1s 2*F1p 2*F2p 2*F2s 0.5*2]; % Define vector [F1 F2 F3 …]
m=[0 0 1 1 0 0]; % Define desired width for each frequency
w1p=2*pi*F1p; % Converse to normalized frequency w=2π(f/fs).
w1s=2*pi*F1s;
w2p=2*pi*F2p;
w2s=2*pi*F2s;
wtr=2*pi*Ftr;
Rp=0.25; %Define ripple Rp
As=50; %Define As amplitude.
dp=(10^(Rp/20)-1)/(10^(Rp/20)+1); %Compute δp
ds=(1+dp)*(10^(-As/20)); %Compute δs
Weights=[1 ds/dp 1]; %Define weighting vector
DW=wtr/(2*pi);
%Compute N (rounding)
N=ceil( (-20*log10(sqrt(ds*dp))-13)/(14.6*DW)+1);
h=firpm(N,f,m ,Weights); %Compute filter coefficients
[H,W]=freqz(h,1,256); %Compute frequency response
df=(0:length(W)-1)*(fs/2)/length(W); % Unit conversion of y-axis (in Hz)
plot(df,20*log10(abs(H))); %Design magnitude
xlabel('Frequency f(KHz)');
ylabel('Magnitude dB');
title('Frequency response of FIR filter, method: equiripple ');
It is computed that N=85.
```

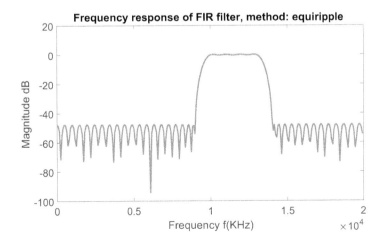

Section II

Statistical Signal Processing

7

Statistical Models

The estimation, detection and classification processes can be grouped altogether under the broader title of statistical evaluation, which is the process inferring properties about the distribution of a random variable, X, given a particular state, x, which is called a data sample, measurement or observation. One basic idea is that of the statistical model, which is simply a hypothetical probability distribution or a density function, $f(x)$, for the X variable. In general, statistics examine the possibility of matching a defined model in the set of x data. To simplify this task, $f(x)$ can be limited to a class of parametric models $\{f(x;\boldsymbol{\theta})\}_{\boldsymbol{\theta}\in\Theta}$, where $f(x;\boldsymbol{\theta})$ is a known function and $\boldsymbol{\theta}$ denotes an array of unknown parameters that take values from the parametric space, Θ. In this particular case, the process of statistical extraction shows the properties of the actual values of $\boldsymbol{\theta}$ by the parameterization of $f(x;\boldsymbol{\theta})$ that generated the sample x.

In this chapter, we will discuss several models related to (a) the well-known Gaussian distribution, (b) the general category of exponential distributions and (c) the fundamental concept of statistical sufficiency on obtaining the properties of $\boldsymbol{\theta}$.

7.1 The Gaussian Distribution and Related Properties

The Gaussian distribution plays an important role in parametric statistics due to the relatively simple Gaussian model and its broad spectrum of applications. Indeed, in engineering and science, the Gaussian distribution is probably the most jointly used distribution for random measurements. The Gaussian distribution is also called normal distribution. The probability density function (PDF) of a Gaussian random variable X (RV) is parameterized by two parameters θ_1 and θ_2, which is the position parameter μ ($\mu \in \mathbb{R}$) and the (squared) scale parameter $\sigma^2(\sigma^2 > 0)$, respectively. The PDF of the RV x is given by:

$$f(x;\mu;\sigma^2) = \frac{1}{\sqrt{2\pi}\sigma}e^{-\frac{(x-\mu)^2}{2\sigma^2}}, \quad -\infty < x < +\infty.$$

When $\mu = 0$ and $\sigma^2 = 1$, X is called standard Gaussian RV. A Gaussian RV with position parameter μ and scale parameter $\sigma^2 > 0$ can be expressed as:

$$X = \sigma Z + \mu \tag{7.1}$$

where Z is the standard Gaussian RV.

The cumulative distribution function (CDF) of a standard Gaussian RV is denoted as N(z) and is defined as:

$$\mathcal{N}(z) = \Pr[Z \le z]$$

$$= \int_{-\infty}^{z} \frac{1}{\sqrt{2\pi}} e^{-\frac{v^2}{2}} dv.$$

Using Relation 7.1, the CDF of a non-standard Gaussian RV X with parameters μ and σ^2 can be expressed in the form of the CDF of the standard Gaussian RV Z:

$$\Pr[X \le x] = \Pr\left[(X - \mu)/\sigma \le (x - \mu)/\sigma\right] = \mathcal{N}\left(\frac{x - \mu}{\sigma}\right)$$

The CDF of the standard Gaussian RV, N(z), is associated with the error function $\text{erf}(u) = \frac{2}{\sqrt{\pi}} \int_{0}^{u} e^{-t^2} dt, x \ge 0$, through the expression:

$$\mathcal{N}(x) = \begin{cases} \frac{1}{2}\left[1 + \text{erf}\left(|x|/\sqrt{2}\right)\right], & x \ge 0, \\ \frac{1}{2}\left[1 - \text{erf}\left(|x|/\sqrt{2}\right)\right], & x < 0. \end{cases}$$

When v is a positive integer, the statistical moments of a standard Gaussian RV Z are given as:

$$\mathbb{E}[Z^v] = \begin{cases} (v-1)(v-3)\cdots3\cdot1, & v \text{ is even} \\ 0, & v \text{ is odd} \end{cases}$$

where $\mathbb{E}[g(Z)] = \int_{-\infty}^{+\infty} g(z)f(z)dz$ denotes the statistically expected value of RV Z and $f(z)$ is the PDF of Z.

In particular, using Relation 7.1, the first and second statistical moments of a non-standard Gaussian RV X are $\mathbb{E}[X] = \mu$ and $\mathbb{E}[X^2] = \mu^2 + \sigma^2$, respectively. Therefore, with respect to a Gaussian RV X, the expected (mean) value $\mathbb{E}[X] = \mu$ and variance $\text{var}(X) = \mathbb{E}\left[\left(X - \mathbb{E}[X]\right)^2\right] = \mathbb{E}[X^2] - \mathbb{E}^2[X] = \sigma^2$ denote the position and scale parameters of the PDF $f(x; \mu, \sigma^2)$ of X, respectively. The following relation will then be used for denoting the non-central mean dispersion $\mathbb{E}\left[|X + \alpha|\right]$ for a Gaussian RV X:

$$\mathbb{E}\left[|X + \alpha|\right] = \sqrt{\frac{2}{\pi}} e^{-\alpha^2/2} + \alpha\left(1 - 2\mathcal{N}(-\alpha)\right).$$

Also, the following notations are useful:

- The RV X is Gaussian distributed with expected value μ and variance σ^2. Then, it is true that $X \sim N(\mu, \sigma^2)$.
- The RV X is equal to a Gaussian RV that changes and shifts as follows:

$$X = \alpha \underset{\mathcal{N}(0,1)}{Z} + b \Leftrightarrow X = \mathcal{N}(b, \alpha^2).$$

$$\sum_{i=1}^{n} \mathcal{N}(0,1) = \sum_{i=1}^{n} X_i.$$

- The CDF of $\mathcal{N}(0, 1)$ is equal to α when checked at point v:

$$\mathcal{N}(v) = \alpha.$$

- The inverse CDF of $N(0, 1)$ is equal to α when checked at point v:

$$\mathcal{N}^{-1}(v) = \alpha.$$

- Vector $\mathbf{x}$ is distributed as an n-dimensional vector with Gaussian RV, with expected values $\boldsymbol{\mu}$ and a covariance matrix $\mathbf{R}$:

$$\mathbf{x} \sim \mathcal{N}_n(\boldsymbol{\mu}, \mathbf{R}).$$

7.1.1 The Multivariate Gaussian Distribution

When a sequence of independent and identically distributed (IID) Gaussian RVs passes through a linear filter, the output remains Gaussian distributed but is no longer IID. The filter smooths the input and introduces a correlation between the values of the input vector. However, if the input sequence is a Gaussian RV, the output behaves the same. Consequently, the joint distribution of any p samples of the input sequence is a Gaussian multivariate. More specifically, a vector of random variables $\mathbf{x} = [x_1, ..., x_p]^T$ is multivariate Gaussian with a vector of expected values $\boldsymbol{\mu}$ and a covariance matrix $\boldsymbol{\Lambda}$, if and only if the joint PDF is in the form:

$$f(\mathbf{x}) = \frac{1}{(2\pi)^{p/2} |\boldsymbol{\Lambda}|^{1/2}} \exp\left(-\frac{1}{2}(\mathbf{x}-\boldsymbol{\mu})\boldsymbol{\Lambda}^{-1}(\mathbf{x}-\boldsymbol{\mu}) \right), \quad \mathbf{x} \in \mathbb{R}^p. \tag{7.2}$$

where $|\boldsymbol{\Lambda}|$ expresses the determinant of $\boldsymbol{\Lambda}$. The p-variate Gaussian distribution depends on $p(p + 3)/2$ parameters, which we can reconstruct in a parametric vector $\boldsymbol{\theta}$ containing the p elements of the vector of expected values:

$$\boldsymbol{\mu} = [\mu_1, \cdots, \mu_p]^T = \mathbb{E}[\mathbf{x}],$$

and on $p(p+1)/2$ discrete parameters of the symmetric positive-definite covariance matrix $p \times p$:

$$\mathbf{\Lambda} = \operatorname{cov}(\mathbf{x}) = \mathbb{E}\left[(\mathbf{x} - \mathbf{\mu})(\mathbf{x} - \mathbf{\mu})\right]^{T}.$$

- Unimodality and symmetry of a Gaussian PDF: The multivariate Gaussian distribution is unimodal (it has a single peak) and symmetric around its expected value.
- The uncorrelated Gaussian RVs are independent: When the covariance matrix $\mathbf{\Lambda}$ is diagonal, i.e., when it is true that $\operatorname{cov}(x_i, x_j) = 0$, $i \neq j$, then the multivariate Gaussian distribution is simplified as a product of the individual PDFs:

$$f(\mathbf{x}) = \prod_{i=1}^{p} f(x_i),$$

where

$$f(x_i) = \frac{1}{\sqrt{2\pi}\sigma_i} e^{-\frac{(x_i - \mu_i)^2}{2\sigma^2}},$$

the scalar Gaussian PDF with $\sigma_i^2 = \operatorname{var}(x_i)$
- The limit distribution of a Gaussian multivariate distribution remains Gaussian distributed: If $\mathbf{x} = [x_1, \cdots, x_m]^T$ is a multivariate Gaussian vector, then every subset of the elements of $\mathbf{x}$ is a multivariate Gaussian vector. For example, the RV x_1 is scalar Gaussian, while vector $[x_1, x_2]$ is bivariate Gaussian.
- The linear combination of Gaussian RV is a Gaussian RV: Let the Gaussian multivariate vector $\mathbf{x} = [x_1, \cdots, x_m]^T$ and a matrix $\mathbf{H}$ $p \times m$ with constant (non-random) values. Then, the vector $\mathbf{y} = \mathbf{H}\mathbf{x}$ consists of linear combinations of RV x_i. The distribution of $\mathbf{y}$ is p-variate Gaussian with expected values $\mathbf{\mu} = \mathbb{E}[\mathbf{y}] = \mathbf{H}\mathbf{\mu}$ and covariance matrix $p \times p$ $\mathbf{\Lambda}_\mathbf{y} = \operatorname{cov}(\mathbf{y}) = \mathbf{H}\operatorname{cov}(\mathbf{x})\mathbf{H}^T$.
- A vector IID with zero expected values remains unchanged in variations: Let the Gaussian multivariate vector $\mathbf{x} = [x_1, \cdots, x_m]^T$ with zero expected values ($\mathbf{\mu} = \mathbf{0}$) and covariance matrix $\operatorname{cov}(\mathbf{x}) = \sigma^2\mathbf{I}$. If $\mathbf{U}$ represents an orthogonal matrix $m \times m$, that is, $\mathbf{U}^T\mathbf{U} = \mathbf{I}$, then the vector $\mathbf{y} = \mathbf{U}^T\mathbf{x}$ follows the same distribution of $\mathbf{x}$.
- The conditional Gaussian distribution, given another Gaussian distribution, is also a Gaussian: Suppose the vector $\mathbf{z}^T = [\mathbf{x}^T, \mathbf{y}^T] = [x_1, \cdots, x_p, y_1, \cdots, y_q]^T$ is a Gaussian multivariate $((p + q)$-variate$)$ with $\mathbf{\mu}_\mathbf{z}^T = \left[\mathbf{\mu}_\mathbf{x}^T, \mathbf{\mu}_\mathbf{y}^T\right]$ expected values and covariance $\mathbf{\Lambda}_\mathbf{z}$. Then, the conditional PDF $f_{\mathbf{y}|\mathbf{x}}(\mathbf{y}|\mathbf{x})$ of $\mathbf{y}$, given $\mathbf{x}$, is a Gaussian multivariate (q-variate) of the form of Relation 7.2 with expected values of parameters and covariance matrix $\mathbf{\Lambda}$:

$$\mathbf{\mu}_{\mathbf{y}|\mathbf{x}}(\mathbf{x}) = \mathbb{E}[\mathbf{y}|\mathbf{x}] = \mathbf{\mu}_\mathbf{y} + \mathbf{\Lambda}_{\mathbf{x},\mathbf{y}}^T \mathbf{\Lambda}_\mathbf{x}^{-1}(\mathbf{x} - \mathbf{\mu}_\mathbf{x}), \tag{7.3}$$

and

$$\Lambda_{\mathbf{y}|\mathbf{x}} = \text{cov}(\mathbf{y}\,|\,\mathbf{x}) = \Lambda_{\mathbf{y}} - \Lambda_{\mathbf{x},\mathbf{y}}^{T}\Lambda_{\mathbf{x}}^{-1}\Lambda_{\mathbf{x},\mathbf{y}}.$$

- The conditional expected value of a Gaussian multivariate, given another Gaussian RV, is linear, and the conditional covariance is constant: For the former multivariate vector, $\mathbf{z}^{T}$, the partitioned covariance matrix is:

$$\Lambda_{z} = \begin{bmatrix} \Lambda_{x} & \Lambda_{x,y} \\ \Lambda_{x,y}^{T} & \Lambda_{y} \end{bmatrix},$$

where $\Lambda_{x} = \text{cov}(\mathbf{x}) = \mathbb{E}\left[(\mathbf{x}-\mathbf{\mu}_{\mathbf{x}})(\mathbf{x}-\mathbf{\mu}_{\mathbf{x}})^{T}\right]$ is $p \times p$, $\Lambda_{y} = \text{cov}(\mathbf{y}) = \mathbb{E}\left[(\mathbf{y}-\mathbf{\mu}_{\mathbf{y}})(\mathbf{y}-\mathbf{\mu}_{\mathbf{y}})^{T}\right]$ is $q \times q$ and $\Lambda_{x,y} = \text{cov}(\mathbf{x},\mathbf{y}) = \mathbb{E}\left[(\mathbf{x}-\mathbf{\mu}_{\mathbf{x}})(\mathbf{y}-\mathbf{\mu}_{\mathbf{y}})^{T}\right]$ is $p \times q$.
The vector of the expected values of the conditionssal PDF $f(\mathbf{y}|\mathbf{x})$ is linear to $\mathbf{x}$ and is given in Equation 7.3. In addition, the conditional covariance does not depend on $\mathbf{x}$ and is expressed in Equation 7.4.

7.1.2 The Central Limit Theorem

One of the most useful statistical tools is the central limit theorem (CLT). The CLT allows for the approximation of the sum of the IID of the finite RVs with the use of the Gaussian distribution. What follows is a general version of the CLT that applies to RV data estimated by vectors.

(Lindeberg-Levy) Central Limit Theorem: Let $\{\mathbf{x}_{i}\}_{i=1}^{n}$ IID random vectors in the $\mathbb{R}^{p}$ plane with a joint expected value $\mathbb{E}[\mathbf{x}_{i}] = \mathbf{\mu}$ and a finite positive covariance matrix $\Lambda = \text{cov}(\mathbf{x}_{i})$. In the case where $n \to +\infty$, the distribution of the RV vector $\mathbf{z}_{n} = n^{-1/2}\sum_{i=1}^{n}(\mathbf{x}_{i}-\mathbf{\mu})$ converges to the p-variate Gaussian distribution, with an expected value of zero and a covariance matrix Λ.

Also, the CLT can be expressed as a function of the expected sample value $\bar{\mathbf{x}} = \bar{\mathbf{x}}(n) = n^{-1}\sum_{i=1}^{n}\mathbf{x}_{i}$

$$\sqrt{n}\left(\bar{\mathbf{x}}(n)-\mathbf{\mu}\right) \to \mathbf{z}, \quad n \to +\infty,$$

where $\mathbf{z}$ is a vector with Gaussian RVs, expected values of zero and a covariance matrix Λ. Therefore, for a large but finite range of values, vector $\bar{\mathbf{x}}$ tightly approximates a Gaussian RV

$$\bar{\mathbf{x}} \approx \left(\mathbf{z}/\sqrt{n}+\mathbf{\mu}\right),$$

with an expected value $\mathbf{\mu}$ and covariance Λ/n. For example, in the case of a scalar value x_{i}, the CLT allows the useful approach for large n:

$$\Pr\left[n^{-1}\sum_{i=1}^{n}x_{i} \le y\right] \approx \int_{-\infty}^{y}\frac{1}{\sqrt{2\pi\sigma^{2}/n}}\exp\left(-\frac{(y-\mu)^{2}}{2\sigma^{2}/n}\right)dy.$$

7.1.3 The Chi-Squared RV Distribution

The central chi-squared PDF with k degrees of freedom is given by:

$$f_k(x) = \frac{1}{2^{k/2}\Gamma(k/2)} x^{k/2-1} e^{-x/2}, \quad x > 0, \tag{7.5}$$

where k is a positive integer and $\Gamma(u)$ denotes the Gamma-function:

$$\Gamma(u) = \int_0^{+\infty} x^{u-1} e^{-x}\, dx,$$

For integer values of n,

$$\Gamma(n+1) = n! = n(n-1)\cdots 1 \text{ and } \Gamma(n+1/2) = \frac{(2n-1)(2n-3)\cdots 5 \cdot 3 \cdot 1}{2^n}\sqrt{\pi}.$$

If $z_i \sim \mathcal{N}(0,1)$, $i = 1, \cdots, n$, is IID, then $x = \sum_{i=1}^{n} z_i^2$ is distributed as chi-squared with n degrees of freedom. The above case is denoted as:

$$\sum_{i=1}^{n} \left[\mathcal{N}(0,1) \right]^2 = \chi_n \tag{7.6}$$

This approach, i.e., expressing a RV in a chi-squared form, is also known as stochastic representation because it is expressed through operations reflected on other RVs.

Some useful properties of the chi-squared distribution are:

- $\mathbb{E}[\chi_n] = n, \quad \mathrm{var}(\chi_n) = 2n.$
- Asymptotic relation for large n:

$$\chi_n = \sqrt{2n}\,\mathcal{N}(0,1) + n.$$

- The notation χ_n expresses an exponential RV with an expected value 2 and PDF $f(x) = \frac{1}{2} e^{-x/2}$.

- $\sqrt{\chi_n}$ is a Rayleigh distributed RV.

7.1.4 Gamma Distribution

The PDF is defined as:

$$f(x) = \frac{\lambda^r}{\Gamma(r)} x^{r-1} e^{-\lambda x}, \quad x > 0,$$

with parameters $\lambda, r > 0$. Let $\{y_i\}_{i=1}^{n}$ be the IID RV that follows the exponential PDF with an expected value of $1/\lambda$:

$$f_\lambda(y) = \lambda\, e^{-\lambda y}, \quad y > 0.$$

It is true that the sum $x = \sum_{i=1}^{n} y_i$ follows the Gamma distribution. Other useful properties of the Gamma distribution of a RV x with parameters (λ, r) are:

- $\mathbb{E}[x] = r/\lambda$.
- $\mathrm{var}[x] = r/\lambda^2$.
- The chi-squared distribution of a RV with k degrees of freedom is a special case of the Gamma distribution when: $\lambda = 1/2$ and $r = k/2$.

7.1.5 The Non-Central Chi-Squared RV Distribution

The sum of squares of independent Gaussian RVs with unitary variance but non-zero expected values composes a non-central chi-squared RV. Specifically, if $z_i \sim \mathcal{N}(\mu_i, 1)$ are statistically independent RVs, then $x = \sum_{i=1}^{n} z_i^2$ is distributed as a non-central chi-squared RV with n degrees of freedom and a non-centrality parameter $\delta = \sum_{i=1}^{n} \mu_i^2$. We usually use:

$$\sum_{i=1}^{n} \left[\mathcal{N}(0,1) + \mu_i \right]^2 = \sum_{i=1}^{n} \left[\mathcal{N}(\mu_i, 1) \right]^2 = \chi_{n,\delta}. \tag{7.7}$$

The noncentral chi-squared PDF cannot be expressed in a simple closed-type form. Therefore, some of its asymptotic features are listed below:

- $\mathbb{E}[\chi_{n,\delta}] = n + \delta, \quad \mathrm{var}(\chi_{n,\delta}) = 2(n + 2\delta)$.
- $\sqrt{\chi_{2,\mu_1^2 + \mu_2^2}}$ is a Rician RV.

7.1.6 The Chi-Squared Mixed Distribution

The distribution of the sum of squares of independent RVs, with zero-mean (expected value) but with different variances, is not expressed in a closed form. However, due to the importance of this distribution in a variety of practical applications, the following notation will then be used to express it:

$$\sum_{i=1}^{n} \frac{c_i}{\sum_j c_j} z_i^2 = \bar{\chi}_{n,c}.$$

A useful asymptotic property is the following:

$$\mathbb{E}[\bar{\chi}_{n,c}] = 1, \quad \mathrm{var}(\bar{\chi}_{n,c}) = 2\sum_{i=1}^{n}\left(\frac{c_i}{\sum_j c_j}\right)^2.$$

Moreover, in the specific case where $\bar{\chi}_{n,1} = \dfrac{1}{n}\chi_n$, then the chi-squared mixed distribution is simplified in a scaled central chi-squared distribution.

7.1.7 The Student's t-Distribution

For $z \sim \mathcal{N}(0,1)$ and $y \sim \chi_n$ independent RVs, the ratio $x = z/\sqrt{y/n}$ is called Student-t $\mathcal{T}_n$, RV with n degrees of freedom.

The PDF of $\mathcal{T}_n$ is defined as:

$$f_n(x) = \frac{\Gamma\big([n+1]/2\big)}{\Gamma(n/2)\sqrt{2\pi}\,(1+x^2/n)^{(n+1)/2}}, \quad x \in \mathbb{R},$$

where n is a positive integer. Some important properties are presented next:

- $\mathbb{E}[\mathcal{T}_n] = 0\ (n>1), \quad \mathrm{var}(\mathcal{T}_n) = \dfrac{n}{n-2}\ (n>2).$
- $\mathcal{T}_n \approx \mathcal{N}(0,1), \quad n \to +\infty.$

For $n = 1$, the expected value of $\mathcal{T}_n$ does not exist, and for $n \le 2$, its variance is not finite.

7.1.8 The Fisher-Snedecor F-Distribution

For $u \sim \chi_m$ and $v \sim \chi_n$ independent RVs, the ratio $x = (u/m)\,/\,(v/n)$ is called Fisher/Snedecor-F (or Fisher-F) RV with m, n degrees of freedom:

$$\frac{\chi_m/m}{\chi_n/n} = \mathcal{F}_{m,n}.$$

The Fisher/Snedecor–F PDF is defined as:

$$f(x) = \frac{\Gamma\big([m+n]/2\big)}{\Gamma(m/2)\Gamma(n/2)}\left(\frac{m}{n}\right)^{m/2}\frac{x^{(m-2)/2}}{\left(1+\dfrac{m}{n}x\right)^{(m+n)/2}}, \quad x>0,$$

where n, m positive integers. It should be noted that the statistical moments $\mathbb{E}[x^k]$ of greater order than $k = n/2$ do not exist. A useful asymptotic relation is:

$$\mathcal{F}_{m,n} \approx \chi_m.$$

7.1.9 The Cauchy Distribution

The ratio of independent RVs N(0,1) u and v is called the standard Cauchy distribution and is defined as:

$$x = u/v \sim \mathcal{C}(0,1).$$

The PDF is expressed as follows:

$$f(x) = \frac{1}{\pi(1+x^2)}, \quad x \in \mathbb{R}.$$

If $\theta = [\mu, \sigma]$ the position and scale parameters ($\sigma > 0$), then $f_\theta(x) = f((x - \mu)/\sigma)$ denotes the shifted and scaled version of the standard PDF of the Cauchy distribution, $\mathcal{C}(\mu, \sigma^2)$.
Some notable properties are:

- The Cauchy distribution does not produce statistical moments for positive integer order.
- The Cauchy distribution is identical to the Student's t-distribution with 1 degree of freedom.

7.1.10 The Beta Distribution

For $u \sim \chi_m$ and $v \sim \chi_m$ independent chi-squared RVs with m and n degrees of freedom, respectively, the ratio $x = u/(u + v)$ is expressed by the Beta distribution:

$$\frac{\chi_m}{\chi_m + \chi_n} = \mathcal{B}(m/2, n/2),$$

where $\mathcal{B}(p, q)$ represents a RV with a Beta PDF and parameters $\theta = [p, q]$. It is of the form:

$$f(x) = \frac{1}{B(p,q)} x^{p-1}(1-x)^{q-1}, \quad x \in [0,1],$$

where $B(p,q) = \int_0^1 x^{p-1}(1-x)^{q-1}\, dx = \frac{\Gamma(p)\Gamma(q)}{\Gamma(p+q)}.$

Some notable properties are:

- $\mathbb{E}[\mathcal{B}(p,q)] = p/(p+q)$.
- $\text{var}(\mathcal{B}(p,q)) = pq/((p+q+1)(p+q)^2)$.

7.2 Reproducing Distributions

The RV x has a reproducing distribution if the sum of two independent x-variables, that is, x_1 and x_2, follows the same distribution as the individual RVs, probably though with different parameters. The Gaussian RVs are described by the reproducing distribution:

$$\mathcal{N}\left(\mu_1,\sigma_1^2\right)+\mathcal{N}\left(\mu_2,\sigma_2^2\right)=\mathcal{N}\left(\mu_1+\mu_2,\sigma_1^2+\sigma_2^2\right),$$

which results from the fact that the convolution of two independent Gaussian RVs produces a Gaussian RV. From Relations 7.5 and 7.7, it can be derived that the distributions of the central and the non-central chi-squared distribution are reproducing distributions:

- $\chi_n + \chi_m = \chi_{m+n}$, if χ_m and χ_n are independent RVs.
- $\chi_{m,\delta_1} + \chi_{n,\delta_2} = \chi_{m+n,\delta_1+\delta_2}$, if χ_{m,δ_1} and χ_{n,δ_2} are independent RVs.

The mixture distributions of chi-squared, Fisher-F and Student-t have no densities.

7.3 Fisher-Cochran Theorem

The Fisher-Cochran theorem is useful for deriving the distribution of the squares of Gaussian distributed RVs.

Let $\mathbf{x} = [x_1, \cdots, x_n]^T$ a vector of RVs distributed by $\mathcal{N}(0, 1)$ and an idempotent matrix $\mathbf{A}$ of p order (i.e., $\mathbf{A}\,\mathbf{A} = \mathbf{A}$). Then:

$$\mathbf{x}^T \mathbf{A} \mathbf{x} = \chi_p$$

Proof:
Let $\mathbf{A} = \mathbf{U}\,\Lambda\,\mathbf{U}^T$ be the eigendecomposition of $\mathbf{A}$. Then, all the eigenvalues of $\mathbf{A}$ matrix are either 0 or 1:

$$\mathbf{A}\,\mathbf{A} = \mathbf{U}\Lambda\underbrace{\mathbf{U}^T\mathbf{U}}_{=\mathbf{I}}\Lambda\mathbf{U}^T$$
$$= \mathbf{U}\Lambda^2\mathbf{U}^T = \mathbf{U}\Lambda\mathbf{U}^T$$

and, consequently:

$$\mathbf{x}^T \mathbf{A} \, \mathbf{x} = \mathbf{x}^T \mathbf{U} \, \boldsymbol{\Lambda} \, \underbrace{\mathbf{U}^T \mathbf{x}}_{\mathbf{z} = \mathcal{N}_n(0, \mathbf{I})}$$

$$= \sum_{i=1}^{n} \lambda_i z_i^2 = \sum_{i=1}^{p} \left[\mathcal{N}_n(0, 1) \right]^2.$$

7.4 Expected Value and Variance of Samples

Let x_i be an IID RV $\mathcal{N}(0,1)$. The expected value and the variance of samples approximate the position μ and the deviation σ of the set of samples, respectively.

- Expected value of samples: $\bar{x} = n^{-1} \sum_{i=1}^{n} x_i.$

- Variance: $s^2 = \dfrac{1}{n-1} \sum_{i=1}^{n} (x_i - \bar{x})^2.$

In the case of a Gaussian RV, the joint distribution of the expected value and variance of the samples is defined as:

1. $\bar{x} \sim \mathcal{N}(\mu, \sigma^2/n).$

2. $s^2 \sim \dfrac{\sigma^2}{n-1} \chi_{n-1}.$

3. $\bar{x}$ and s^2 are statistically independent RVs.

The latter results reflect the fact that the normalized ratio of the expected value and variance of samples is Student's t-distributed:

$$\frac{\bar{x} - \mu}{s/\sqrt{n}} = \mathcal{T}_{n-1}.$$

Proof of properties (2) and (3):
First of all, we will show that the expected value and variance of samples are independent RVs. Let the vector form of RVs $\mathbf{y} = [y_1, \cdots, y_n]^T$, which is defined as follows:

$$y_1 = \sqrt{n}\bar{x} = \mathbf{h}_1^T \mathbf{x},$$

where

$$\mathbf{h}_1 = \left[1/\sqrt{n}, \cdots, 1/\sqrt{n}\right]^T.$$

Note that the vector $\mathbf{h}_1$ has a unitary norm. The Gramm-Schmidt orthonormalization method (Gramm-Schmidt process) is then applied to the vector $\mathbf{h}_1$. According to this method, $n-1$ vectors $\mathbf{h}_2, \cdots, \mathbf{h}_n$ are created, which are orthonormal, orthogonal to each other and with respect to $\mathbf{h}_1$. The vector form of RVs $\mathbf{y}$ is defined as:

$$\mathbf{y} = \mathbf{H}^T \mathbf{x},$$

where $\mathbf{H} = [\mathbf{h}_1, \cdots, \mathbf{h}_n]$ depicts a $n \times n$ orthogonal matrix. Because $\mathbf{x} = \mathbf{H}\mathbf{y}$, the orthogonality of $\mathbf{H}$ introduces the following properties:

  a. $\mathbf{y} \sim N_n(\mathbf{0}, \mathbf{I})$
  b. $\mathbf{y}^T \mathbf{y} = \mathbf{x}^T \mathbf{x}$.

Since $\mathbf{y}_1 = \sqrt{n}\,\bar{x}$, considering (a), it is deduced that value $\bar{x}$ is independent of the values $y_2, \cdots, y_n$. Moreover, using the equivalence of terms:

$$\sum_{i=1}^{n}(x_i - \bar{x})^2 = \sum_{i=1}^{n} x_i^2 - n\,\bar{x}^2,$$

(b) and the RV y_1 definition yields:

$$\sum_{i=1}^{n}(x_i - \bar{x})^2 = \sum_{i=1}^{n} y_i^2 - y_1^2 = y_2^2 + \cdots + y_n^2. \tag{7.8}$$

Hence, the variance of the samples is a function of $y_2, \cdots, y_n$ and is therefore independent of the expected value y_1 of the samples. Also, since $y_2, \cdots, y_n$ are independent RVs $\mathcal{N}(0, 1)$, Relation 7.8 expresses the fact that the normalized variance of the samples follows a chi-squared distribution with $n-1$ degrees of freedom.

Property (3) is derived directly through the Fisher-Cochran theorem. The normalized variance of the samples on the left part of Relation 7.8 can also be expressed in a squared form:

$$[\mathbf{x} - \mathbf{1}\bar{x}]^T [\mathbf{x} - \mathbf{1}\bar{x}] = \mathbf{x}^T \underbrace{\left[\mathbf{I} - \mathbf{11}^T \frac{1}{n}\right]\left[\mathbf{I} - \mathbf{11}^T \frac{1}{n}\right]}_{\text{idempotent}} \mathbf{x}$$

$$= \mathbf{x}^T \underbrace{\left[\mathbf{I} - \mathbf{11}^T \frac{1}{n}\right]}_{\text{orthogonal projection}} \mathbf{x},$$

where $\mathbf{1} = [1, \cdots, 1]^T$.

NOTE: Since $\text{rank}\left[\mathbf{I} - \mathbf{11}^T\dfrac{1}{n}\right] = 1 - n$, then:

$$[\mathbf{x} - \mathbf{1}\bar{x}]^T[\mathbf{x} - \mathbf{1}\bar{x}] = (n-1)s^2 \sim \chi_{n-1}.$$

7.5 Statistical Sufficiency

Many detection/evaluation/classification problems have the following common structure.

A continuous time waveform $\{x(t): t \in \mathbb{R}\}$ is measured in n time-frames $t_1, \cdots, t_n$ creating the following vector:

$$\mathbf{x} = [x_1, \cdots, x_n]^T,$$

where $x_i = x(t_i)$. Vector $\mathbf{x}$ is modeled as a vector of RVs with a joint distribution, which is known, but is dependent on p unknown parameters $\boldsymbol{\theta} = [\theta_1, \cdots, \theta_p]^T$. The following notations are true:

- X is the sample space derived from the random vector $\mathbf{x}$.
- B is the event space of $\mathbf{x}$, that is, the Borel subset in the $\mathbb{R}^n$ field.
- $\boldsymbol{\theta} \in \Theta$ is the vector of unknown parameters of interest.
- Θ is the available parameter space of the current measurement.
- $P_{\boldsymbol{\theta}}$ is the probability of measurement in B for conditional $\boldsymbol{\theta}$.
- $\{P_{\boldsymbol{\theta}}\}_{\boldsymbol{\theta} \in \Theta}$ is defined as the statistical model of the measurement.

This probabilistic model produces the joint CDF associated with $\mathbf{x}$:

$$F_{\mathbf{x}}(\mathbf{x};\theta) = \Pr[x_1 < \mathsf{x}_1, \cdots, x_n < \mathsf{x}_n],$$

which is known for each $\boldsymbol{\theta} \in \Theta$. When $\mathbf{x}$ is a vector with continuous RVs, the joint CDF is determined by the corresponding joint PDF, which will then be listed next in different ways depending on the content, $f_{\boldsymbol{\theta}}(\mathbf{x})$, $f(\mathbf{x}; \boldsymbol{\theta})$, or $f_{\mathbf{x}}(\mathbf{x}; \boldsymbol{\theta})$.

Moreover, the statistical expected value of a RV z is defined, in relation to the PDF, as:

$$\mathbb{E}_{\boldsymbol{\theta}}[z] = \int z \, f_z(z;\boldsymbol{\theta}) \, dz.$$

The class of functions $\{f(\mathbf{x}; \boldsymbol{\theta})\}_{\mathbf{x}\in\chi, \boldsymbol{\theta}\in\Theta}$ determines the statistical model of the measurement.

The overall objective of the statistical inference can now be stated: Taking into account a vector $\mathbf{x}$, we can extract the properties of $\boldsymbol{\theta}$ knowing only the parametric form of the statistical model. Therefore, we wish to use a function that maps the values of $\mathbf{x}$ to subsets of the parametric space, e.g., a suitable estimator, classifier or detector of $\boldsymbol{\theta}$. As we will see next, there are several ways to form inference function, but the fundamental question is: Are there any general properties which such a function must have? One important property is that the statistical function must depend only on the **x** vector of length n, through a vector of smaller length, called sufficient statistical metric.

7.5.1 Statistical Sufficiency and Reduction Ratio

Let us consider as a statistical measurement any function of the given data $T = T(\mathbf{x})$. There is a remarkable interpretation of a statistical feature regarding memory storage requirements. Consider a special computing system that records the time samples of the vector $\mathbf{x} = [x_1, \cdots, x_n]^T, x_k = x(t_k)$, e.g., in a "byte" of the storage memory as well as the time tag t_k in another "byte" of the memory. Any non-invertible function T, which maps the data domain from $\mathbb{R}^n$ to $\mathbb{R}^m$, represents a dimension reduction of the data samples. We can evaluate the rate of reduction of the T-function by introducing the concept of the reduction ratio (RR):

$$RR = \frac{\#\,\text{bytes of storage required for } T(\mathbf{x})}{\#\,\text{bytes of storage required for } \mathbf{x}}.$$

This ratio is a measure of the data compression caused by a particular transformation through T. The number of bytes required to store vector **x** with the corresponding time tags is:

$$\#\,\text{bytes}\{\mathbf{x}\} = \#\,\text{bytes}\{x_1, \cdots, x_n\}^T = \#\,\text{bytes}\{\text{timestamps}\} + \#\,\text{bytes}\{\text{values}\}$$
$$= 2n.$$

Next, some examples are presented.

Let $x_{(i)}$ be the ith largest (sorted) value of **x** so that $x_{(1)} \geq x_{(2)} \geq \cdots \geq x_{(n)}$. These values are called sorted statistics values and do not include time tag information. The Table 7.1 below illustrates the reduction ratio (RR) for some important cases.

A discussion could arise about the maximum RR value applied so that no information about the parametric vector $\boldsymbol{\theta}$ is lost. The answer is given by a specific metric known as the minimal sufficient statistic.

TABLE 7.1

Reduction Ratio (RR) for Some Cases

Statistical Data	RR
$T(\mathbf{x}) = [x_1, \ldots, x_n]^T$	1
$T(\mathbf{x}) = [x_{(1)}, \ldots, x_{(n)}]^T$	½
$T(\mathbf{x}) = \bar{x}$	$1/(2n)$
$T(\mathbf{x}) = [\bar{x}, s^2]^T$	$1/n$

7.5.2 Definition of Sufficient Condition

The metric $T = T(\mathbf{x})$ is sufficient for the parameter $\boldsymbol{\theta}$ if it contains all the necessary information regarding the data samples. Therefore, when the metric T has been calculated, it can be stored for further processing without the need for storing the actual data samples.

The metric $T = T(\mathbf{x})$ is sufficient for the parameter $\boldsymbol{\theta}$ if it contains all the necessary information regarding the data samples. Therefore, when the metric T has been calculated, it can be stored for further processing without the need for storing the actual data samples.

More specifically, let $\mathbf{x}$ have a CDF $F_\mathbf{x}(\mathbf{x}, \boldsymbol{\theta})$, $\boldsymbol{\theta}$-dependent. The T statistic is sufficient for vector $\boldsymbol{\theta}$ if the conditional CDF of $\mathbf{x}$, which is $T = t$-dependent, is not a function of $\boldsymbol{\theta}$:

$$F_{\mathbf{x}|T}(\mathbf{x}\,|\,T = t, \boldsymbol{\theta}) = G(\mathbf{x}, t), \tag{7.9}$$

where G is a $\boldsymbol{\theta}$-independent function.

With respect to the discrete values of the vector $\mathbf{x}$ with a probability mass function $p_\theta(\mathbf{x}) = P_\theta[\mathbf{x} = \mathbf{x}]$, $T = T(\mathbf{x})$ is sufficient for the $\boldsymbol{\theta}$ if:

$$P_\boldsymbol{\theta}[\mathbf{x} = \mathbf{x}\,|\,T = t] = G(x, t). \tag{7.10}$$

For continuous values of $\mathbf{x}$ with PDF $f(\mathbf{x}; \boldsymbol{\theta})$, Equation 7.9 is transformed in

$$f_{\mathbf{x}|T}(\mathbf{x}\,|\,t, \boldsymbol{\theta}) = G(\mathbf{x}, t). \tag{7.11}$$

In some cases, the only sufficient statistical quantity is a vector, e.g. $T(\mathbf{x}) = [T_1(\mathbf{x}), \cdots, T_k(\mathbf{x})]^T$. Then, it is true that T_k are jointly statistic sufficient for $\boldsymbol{\theta}$.

Equation 7.9 is often difficult to implement because it involves extracting the conditional distribution of $\mathbf{x}$ dependent on T. When the RV $\mathbf{x}$ is continuous or discrete, a simple way to investigate the sufficient condition is through the Fisher factorization property.

Fisher factorization: $T = T(\mathbf{x})$ is a sufficient statistical quantity if the PDF $f(\mathbf{x}; \boldsymbol{\theta})$ of $\mathbf{x}$ can be expressed as:

$$f_\mathbf{x}(\mathbf{x}; \boldsymbol{\theta}) = g(T, \boldsymbol{\theta})\, h(\mathbf{x}), \tag{7.12}$$

for some non-negative functions g and h. Fisher factorization can be used as the functional definition of a sufficient statistic T element. A significant effect of Fisher's factorization follows: when the PDF of the vector sample $\mathbf{x}$ satisfies Equation 7.12, then the PDF $f_T(t; \boldsymbol{\theta})$ of the sufficient statistic T is equal to $g(t, \boldsymbol{\theta})$ with a range of values of up to a $\boldsymbol{\theta}$-independent constant $q(t)$:

$$f_T(t; \boldsymbol{\theta}) = g(t, \boldsymbol{\theta})\, q(t).$$

Examples of sufficiency follow:

Example 1: Use the Entire Symbolic Sequence of Data Samples

Vector $\mathbf{x} = [x_1, \cdots, x_n]^T$ is clearly sufficient, but it does not constitute an efficient method.

Example 2: Use of Sample Classification

Vector $[x_{(1)}, \cdots, x_{(n)}]$ is sufficient when x_i are IID.
Proof:
If x_i are IID, then the joint PDF is expressed as follows:

$$f_{\boldsymbol{\theta}}(x_1,\cdots,x_n) = \prod_{i=1}^{n} f_{\theta_i}(x_i) = \prod_{i=1}^{n} f_{\theta_{(i)}}(x_{(i)}).$$

The sufficiency is obtained by Fisher factorization.

Example 3: Binary Likelihood Ratios

Let $\boldsymbol{\theta}$ take only two values, $\boldsymbol{\theta}_0$ and $\boldsymbol{\theta}_1$, for example, a bit with a value range "0" or "1" at an information transmission channel. Then, the PDF $f(\mathbf{x}; \boldsymbol{\theta})$ can be expressed as $f(\mathbf{x}; \theta_0)$ or $f(\mathbf{x}; \theta_1)$, with a range of parameters $\theta \in \Theta = \{0, 1\}$. Therefore, the problem of a binary decision arises: "Decide between $\theta = 0$ and $\theta = 1$." Thus, if there is a finite number for all values of $\mathbf{x}$, the likelihood ratio $\Lambda(\mathbf{x}) = f_1(\mathbf{x})/f_0(\mathbf{x})$ is sufficient for θ, where $f_1(\mathbf{x}) \triangleq f(\mathbf{x};1)$ and $f_0(\mathbf{x}) \triangleq f(\mathbf{x};0)$.

Proof:
We express $f_{\theta}(\mathbf{x})$ as a function of θ, f_0, f_1 and then determine the likelihood ratio Λ and apply the Fisher factorization method:

$$f_{\theta}(\mathbf{x}) = \theta f_1(\mathbf{x}) + (1-\theta)f_0(\mathbf{x})$$
$$= \underbrace{\left(\theta\Lambda(\mathbf{x}) + (1-\theta)\right)}_{g(T,\theta)} \underbrace{f_0(\mathbf{x})}_{h(\mathbf{x})}.$$

Therefore, in order to distinguish between the two states $\boldsymbol{\theta}_0$ and $\boldsymbol{\theta}_1$ of the parametric vector $\boldsymbol{\theta}$, we can exclude the individual data and take into account only the scalar sufficient statistic $T = \Lambda(\mathbf{x})$.

Example 4: Discrete Likelihood Ratios

Let $\Theta = [\boldsymbol{\theta}_1,\cdots,\boldsymbol{\theta}_P]$ and the following vector of $p-1$ probability ratios

$$T(\mathbf{x}) = \left[\frac{f_{\theta_1}(\mathbf{x})}{f_{\theta_p}(\mathbf{x})}, \cdots, \frac{f_{\theta_{p-1}}(\mathbf{x})}{f_{\theta_p}(\mathbf{x})}\right]^T$$
$$= \left[\Lambda_1(\mathbf{x}),\cdots,\Lambda_{p-1}(\mathbf{x})\right]^T,$$

be finite for all $\mathbf{x}$ values. The specific vector is sufficient for θ parameter. An alternative way to determine this vector is by using the sequence $\{\Lambda_{\theta}(\mathbf{x})\}_{\theta\in\Theta} = \Lambda_1(\mathbf{x}),\cdots,\Lambda_{p-1}(\mathbf{x})$, called likelihood trajectory, around θ.

Proof:
We define vector $p-1$ of the data $\mathbf{u}_{\theta} = \mathbf{e}_k$ when $\theta = \theta_k$, $k = 1,\cdots, p-1$ where $\mathbf{e}_k = [0,\cdots, 0, 1, 0,\cdots, 0]^T$ is the column vector of the $(p-1) \times (p-1)$ identity matrix. Consequently, for all $\theta \in \Theta$, the joint PDF is given by:

$$f_{\theta}(\mathbf{x}) = \underbrace{\mathbf{u}_{\theta}^T T}_{g(T,\theta)} \underbrace{f_{\theta_p}(\mathbf{x})}_{h(\mathbf{x})},$$

which satisfies the sufficiency according to Fisher factorization.

Example 5: Likelihood Ratio Trajectory

In the case when Θ is the sum of the single-variant parameters θ, the likelihood ratio trajectory

$$\Lambda(\mathbf{x}) = \left\{ \frac{f_\theta(\mathbf{x})}{f_{\theta_0}(\mathbf{x})} \right\} \theta \in \Theta, \qquad (7.13)$$

is sufficient for θ. In the above relation, θ_0 denotes any reference point in Θ for which the trajectory is finite for the vector $\mathbf{x}$. When θ is not single-variant, Relation 7.13 is transformed to a likelihood ratio surface, which is also statistical sufficient.

7.5.3 Minimal Sufficiency

What is the maximum reduction that can be achieved in a set of data samples without losing information with regard to the dependence of the model on $\boldsymbol{\theta}$? The answer to this question lies in the concept of minimal sufficiency. Such data cannot be reduced without loss of information. Alternatively, all other sufficient quantity data may be reduced to the value determined by the minimal sufficiency without loss of essential information.

Definition: $T_{\min}$ expresses minimal sufficiency if it can be calculated from any sufficient statistic quantity T by applying a suitable transformation to T. In other words, if T is a sufficient statistic quantity, there is a function q so that $T_{\min} = q(T)$ is true.

Minimal sufficient statistics are not unique: if $T_{\min}$ expresses minimal sufficiency, then $h(T_{\min})$ also expresses minimal sufficiency in case h is an invertible function. The minimal sufficient statistics can be derived by many ways. One way is by finding a complete, sufficient statistic quantity. A sufficient statistic T is complete if

$$\mathbb{E}_{\boldsymbol{\theta}}[g(T)] = 0, \quad \forall \boldsymbol{\theta} \in \Theta,$$

which implies that g function is uniformly zero ($g(t) = 0, \forall t$).

To prove that completeness involves minimal sufficiency, we quote the following discussion. Suppose M represents minimal sufficiency and C complete sufficiency. Since M is minimal, then it is a function of C. Therefore, $g(C) \triangleq C - \mathbb{E}_{\boldsymbol{\theta}}[C \mid M]$ is a function of C, because the conditional expected value $\mathbb{E}_{\boldsymbol{\theta}}[C \mid M]$ is a function of M. Since $\mathbb{E}_{\boldsymbol{\theta}}[g(C)] = 0, \quad \forall \theta, C$ is statistically complete, then $C = \mathbb{E}_{\boldsymbol{\theta}}[C \mid M], \quad \forall \theta$. Therefore, C is minimal since it is a function of M, which is also expressed as a function of any sufficient statistic. Therefore, C 'inherits' the property of being minimal from M.

Another way to yield minimal sufficiency is by the reduction of data on the likelihood ratio surface. As in Example 5, $\theta_0 \in \Theta$ a reference point so that the following ratio reduction function is finite for each $\mathbf{x} \in \chi$ and $\theta \in \Theta$:

$$\Lambda_{\boldsymbol{\theta}}(\mathbf{x}) = \frac{f_{\boldsymbol{\theta}}(\mathbf{x})}{f_{\theta_0}(\mathbf{x})}.$$

For a conditional $\mathbf{x}$, consider a set of reduction ratios:

$$\Lambda(\mathbf{x}) = \left\{ \Lambda_{\boldsymbol{\theta}}(\mathbf{x}) \right\}_{\boldsymbol{\theta} \in \Theta}.$$

Definition 1: A (θ-independent) function of $\mathbf{x}$, $\tau = \tau(\mathbf{x})$, adjusts the likelihood ratios Λ when the following conditions apply concurrently:

1. $\Lambda(\mathbf{x}) = \Lambda(\tau)$, that is, Λ depends only on $\mathbf{x}$ through $\tau = \tau(\mathbf{x})$.
2. $\Lambda(\mathbf{x}) = \Lambda(\tau')$, i.e., $\tau = \tau'$, so the mapping $\tau \rightarrow \Lambda(\tau)$ is invertible.

Condition 1 specifies another way of expressing the fact that $\tau(\mathbf{x})$ is a sufficient statistic for θ.

Theorem: If $\tau = \tau(\mathbf{x})$ readjusts the likelihood ratios $\Lambda(\mathbf{x})$, then $T_{min} = \tau(\mathbf{x})$ expresses minimal sufficiency for θ.

Proof: The proof concerns the case of a continuous $\mathbf{x}$. The same result, however, also applies to discrete values of $\mathbf{x}$. First of all, Condition 1, in Definition 1, implies that $\tau = \tau(\mathbf{x})$ is a sufficient quantity. This is because, according to Fisher factorization, $f_\theta(\mathbf{x}) = \Lambda_\theta(\tau) f_{\theta_0}(\mathbf{x}) = g(\tau; \theta)h(\mathbf{x})$. Next, let T be a sufficient statistic. Then, through Fisher factorization, $f_\theta(\mathbf{x}) = g(T; \theta)h(\mathbf{x})$, hence:

$$\Lambda(\tau) = \left\{ \frac{f_\theta(\mathbf{x})}{f_{\theta_0}(\mathbf{x})} \right\}_{\theta \in \Theta} = \left\{ \frac{g(T, \boldsymbol{\theta})}{g(T, \boldsymbol{\theta}_0)} \right\}_{\theta \in \Theta}.$$

Therefore, $\Lambda(\tau)$ is a function of T. But, from Condition 2, the mapping $\tau \rightarrow \Lambda(\tau)$ is invertible, and τ is a function of T.

An equally important issue in practical applications is the size of the finite dimension of a sufficient statistic.

Definition: A sufficient statistic $T(\mathbf{x})$ is of finite dimension if its dimension is not a function of the number of data samples n. Often, the minimum sufficient statistics are finite.

Example 6: Minimal Sufficiency of the Expected Value of the Gaussian Distribution

Let $x \sim \mathcal{N}(0, \sigma^2)$ with known σ^2. Find a minimal sufficient statistic for $\theta = \mu$ for the IID samples $\mathbf{x} = [x_1, \cdots, x_n]^T$.

SOLUTION:

The joint PDF is given by:

$$f_\theta(\mathbf{x}) = \left(\frac{1}{\sqrt{2\pi\sigma^2}} \right)^n e^{-\frac{1}{2\sigma^2} \sum_{i=1}^n (x_i - \mu)^2}$$

$$= \left(\frac{1}{\sqrt{2\pi\sigma^2}} \right)^n e^{-\frac{1}{2\sigma^2} \left(\sum_{i=1}^n x_i^2 - 2\mu \sum_{i=1}^n x_i + n\mu^2 \right)}$$

$$= e^{-\frac{n\mu^2}{2\sigma^2}} \underbrace{e^{\frac{\mu}{\sigma} \overbrace{\sum_{i=1}^n x_i}^{T(\mathbf{x})}}}_{g(T, \theta)} \underbrace{\left(\frac{1}{\sqrt{2\pi\sigma^2}} \right)^n e^{-\frac{1}{2\sigma^2} \sum_{i=1}^n x_i^2}}_{h(\mathbf{x})},$$

Consequently, using Fisher factorization:

$$T = \sum_{i=1}^{n} x_i,$$

which represents a sufficient statistic for μ. Moreover, since $q(T) = n^{-1}T$ is a 1-1 function of T, then $\bar{x}$ presents, equally, statistical sufficiency.

It is proved, next, that the expected sample value is minimal sufficient, and depicts that it indexes the likelihood ratio trajectory $\Lambda(x) = \{\Lambda_\theta(x)\}_{\theta \in \Theta}$ with $\theta = \mu$, $\Theta = \mathcal{R}$. The reference point $\theta_0 = \mu_0 = 0$ is chosen such that:

$$\Lambda_\mu(x) = \frac{f_\mu(x)}{f_0(x)} = \exp\left(\frac{\mu}{\sigma^2} \sum_{i=1}^{n} x_i - \frac{n\mu^2}{2\sigma^2} \right).$$

By stating $\tau = \sum_{i=1}^{n} x_i$, condition 1 of definition 1 is satisfied since $\Lambda_\mu(x) = \Lambda_\mu \sum_i x_i$. Also, according to Condition 2, the function $\Lambda_\mu \sum_i x_i$ is invertible to $\sum_i x_i$ for any non-zero value of μ. Consequently, the expected value of the samples indexes the trajectories, and expresses minimal sufficiency.

Example 7: Minimal Sufficiency of the Expected Value and Variance of the Gaussian Distribution

Let $x \sim \mathcal{N}(0, \sigma^2)$ with a known μ and σ^2. Find a minimal sufficient statistic for $\theta = [\mu, \sigma^2]^T$ and for the IID samples $x = [x_1, \cdots, x_n]^T$.

SOLUTION:

$$f_\theta(x) = \left(\frac{1}{\sqrt{2\pi\sigma^2}} \right)^n e^{-\frac{1}{2\sigma^2}\sum_{i=1}^{n}(x_i-\mu)^2}$$

$$= \left(\frac{1}{\sqrt{2\pi\sigma^2}} \right)^n e^{-\frac{1}{2\sigma^2}\left(\sum_{i=1}^{n}x_i^2 - 2\mu\sum_{i=1}^{n}x_i + n\mu^2 \right)}$$

$$= \underbrace{\left(\frac{1}{\sqrt{2\pi\sigma^2}} \right)^n e^{-\frac{n\mu^2}{2\sigma^2}} e^{\left[\frac{\mu}{\sigma^2}, -\frac{1}{2\sigma^2} \right] \overbrace{\left[\sum_{i=1}^{n}x_i, \sum_{i=1}^{n}x_i^2 \right]^T}^{T(x)}}}_{g(T,\theta)} \underbrace{1}_{h(x)}.$$

It is true that

$$T(x) = \left[\underbrace{\sum_{i=1}^{n} x_i}_{T_1}, \underbrace{\sum_{i=1}^{n} x_i^2}_{T_2} \right]^T,$$

is a (jointly) sufficient statistic for μ, σ^2. Also, since $q(T) = \left[n^{-1}T_1, (n-1)^{-1}(T_2 - T_1^2) \right]$ is a 1-1 function of T, then $s = [\bar{x}, \sigma^2]$ expresses sufficiency.

As in Example 6, the minimal sufficiency of this statistical quantity results from the trajectory of the likelihood ratio $\{\Lambda_\theta(x)\}_{\theta \in \Theta}$ with $\theta = [\mu, \sigma^2]$, $\Theta = \mathbb{R} \times \mathbb{R}^+$. Choosing arbitrarily a reference point $\boldsymbol{\theta}_0 = \left[\mu_0, \sigma_0^2\right] = [0, 1]$, we obtain:

$$\Lambda_\theta(\mathbf{x}) = \frac{f_\theta(\mathbf{x})}{f_{\theta_0}(\mathbf{x})} = \left(\frac{\sigma_0}{\sigma}\right)^n e^{\left(-\frac{n\mu^2}{2\sigma^2}\right)} e^{\left[\frac{\mu}{\sigma^2}, -\frac{\delta}{2}\right]\left[\sum_{i=1}^n x_i, \sum_{i=1}^n x_i^2\right]^T},$$

where $\delta = \dfrac{\sigma_0^2 - \sigma^2}{\sigma^2 \sigma_0^2}$. Setting $\boldsymbol{\tau} = \left[\sum_{i=1}^n x_i, \sum_{i=1}^n x_i^2\right]$, Condition 1 is satisfied again. Condition 2 needs further analysis. Although $\Lambda_\theta(\boldsymbol{\tau})$ is no longer an invertible function to τ for each value of $\boldsymbol{\theta} = [\mu, \sigma^2]$, we can obtain two values $\boldsymbol{\theta} \in \{\theta_1, \theta_2\}$ in Θ, for which the function of vectors $\left[\Lambda_{\theta_1}(\boldsymbol{\tau}), \Lambda_{\theta_2}(\boldsymbol{\tau})\right]$ is invertible to $\boldsymbol{\tau}$. Since this vector is defined by $\Lambda(\boldsymbol{\tau})$, vector τ 'follows' the likelihood ratios.

In order to formulate the invertibility relation, let $\boldsymbol{\lambda} = [\lambda_1, \lambda_2]^T$ represent a pair of observed samples, resulting from the vector $\left[\Lambda_{\theta_1}(\boldsymbol{\tau}), \Lambda_{\theta_2}(\boldsymbol{\tau})\right]^T$ at $\Lambda(\boldsymbol{\tau})$ surface. Now the problem is to define $\boldsymbol{\tau}$ from equation $\boldsymbol{\lambda} = \left[\Lambda_{\theta_1}(\boldsymbol{\tau}), \Lambda_{\theta_2}(\boldsymbol{\tau})\right]^T$. By talking the logarithm of each side of the equation, we find that it can equally be expressed as a 2×2 linear system of equations of the form $\boldsymbol{\lambda}' = \mathbf{A}\boldsymbol{\tau}$, where $\mathbf{A}$ is a matrix of the elements $\theta_0, \theta_1, \theta_2$ and $\boldsymbol{\lambda}'$ is a linear function of $\ln \boldsymbol{\lambda}$. As a typical example, we verify the above if we choose $\theta_0 = [0, 1]$, $\theta_1 = [1, 1]$, $\theta_2 = [0, 1/2]$. Then we have, $\delta = 0$ or 1 for $\boldsymbol{\theta} = \theta_1$ or θ_2, respectively, and $\mathbf{A} = \text{diag}(1, -1/2)$ is an invertible matrix. Therefore, we conclude that the vector [expected value, variance] of samples indexes the trajectories, thus expresses minimal sufficiency.

Example 8: Minimal Sufficiency of the Position Parameters of The Cauchy Distribution

Let $x_i \sim f(x;\theta) = \dfrac{1}{\pi\left[1 + (x - \theta)^2\right]}$ and $\mathbf{x} = [x_1, \cdots, x_n]^T$ be the IID samples.

It is true that:

$$f(\mathbf{x};\theta) = \prod_{i=1}^n \frac{1}{\pi\left[1 + (x_i - \theta)^2\right]} = \frac{1}{\pi^n} \frac{1}{\prod_{i=1}^n 1 + (x_i - \theta)^2}.$$

We can see that the denominator consists of a polynomial of degree of 2 at θ, whose coefficients cannot be determined without determining the set of all possible cross products $x_{i_1}, \cdots, x_{i_p}, p = 1, 2, \cdots, n$, of x_i. Since this requires the whole set of sample values to be determined, there is no finite dimensional sufficient statistic. However, each of the individual cross product is independent of the ordering sequence, and thus, the ordered statistic $[x_{(1)}, \cdots, x_{(n)}]^T$ expresses minimal sufficiency.

7.5.4 Exponential Distributions Category

Let vector $\boldsymbol{\theta} = [\theta_1, \cdots, \theta_p]^T$ receive values in the parametric space Θ. The PDF f_θ of the RV x is a member of the category of the exponential p-parametric distributions if for all $\boldsymbol{\theta} \in \Theta$ it is

$$f_\theta(x) = a(\boldsymbol{\theta})b(x)e^{c^T(\boldsymbol{\theta})t(x)}, \quad -\infty < x < +\infty, \tag{7.14}$$

for some scalar functions a, b and p-parameter vector functions $\mathbf{c}$ and $\mathbf{t}$. A similar definition applies to vector RVs $\mathbf{x}$.

It is noteworthy that for each f_θ of the exponential distribution categories, the supported set of values $\{x : f_\theta(x) > 0\}$ does not depend on the $\boldsymbol{\theta}$. Also, the same length of the vector $\mathbf{c}(\boldsymbol{\theta}) = \mathbf{t}(x) = p$ must be respected. This ensures that the size of sufficiency is of the same dimension with the parametric vector θ.

Parametrization of the exponential distribution category is not unique, i.e., these distributions are not affected by any changes in parametrization. For example, if f_θ, $\theta > 0$ belongs to the category of exponential distributions, then if $a = 1/\theta$ and $g_a = f_{1/\theta}$, it is true that g_a, $a > 0$ also belongs to the category of exponential distributions, possibly with different functions $a(\cdot)$, $b(\cdot)$, $c(\cdot)$ and $t(\cdot)$.

As a typical example, let the redefinition of the parameters use the mapping $\mathbf{c}: \boldsymbol{\theta} \to \boldsymbol{\eta}$ and let $\mathbf{c}(\boldsymbol{\theta}) = \boldsymbol{\eta}$ be a reversible function. Then, using Relation 7.14, f_θ can be expressed as:

$$f_{\boldsymbol{\eta}}(x) = \tilde{a}(\boldsymbol{\eta})b(x)e^{\boldsymbol{\eta}^T \mathbf{t}(x)}, \quad -\infty < x < +\infty, \tag{7.15}$$

with $a(\boldsymbol{\eta}) = a(\mathbf{c}^{-1}(\boldsymbol{\eta}))$. Therefore, $f_{\boldsymbol{\eta}}$ remains in the category of exponential distributions. When it is depicted as in Relation 7.15, it is expressed in normal form with normal parameterization $\boldsymbol{\eta}$. In this case, the expected value and covariance of the sufficient statistic $\mathbf{T} = \mathbf{t}(x)$ results as:

$$\mathbb{E}_\theta[\mathbf{T}] = \nabla \ln \tilde{a}(\boldsymbol{\eta}),$$

and

$$\mathrm{cov}_\theta[\mathbf{T}] = \nabla^2 \ln \tilde{a}(\boldsymbol{\eta}).$$

Another useful parameterization of the exponential distribution class is the parameterization of the expected value. In this case, the functions $a(\cdot)$, $b(\cdot)$, $c(\cdot)$ and $t(\cdot)$ of Equation 7.14 are manipulated in an appropriate way so that:

$$\mathbb{E}_\theta[\mathbf{T}] = \boldsymbol{\theta}. \tag{7.16}$$

As we will note in detail in the next chapter, when a category of exponential distributions is expressed through the parameterization of the expected value, the sufficiency T, is a minimal variance estimator of $\boldsymbol{\theta}$. Thus, the parameterizations of the expected value are absolutely advantageous.

Examples of distributions belonging to the exponential category include: the Gaussian distribution with unknown predicted value or variance, the Poisson distribution with unknown expected value, the exponential distribution with unknown expected value, the Gamma distribution, the Bernoulli distribution with unknown probability of success, the binomial distribution with unknown probability of success, and the polynomial distribution with unknown cell probabilities.

Distributions not belonging to the exponential category are: the Cauchy distribution with unknown expected value, the uniform distribution with unknown support, the Fisher-F distribution with an unknown degree of freedom.

When a statistical model belongs to this category, the sufficient statistics, for the model parameters, have a simple yet specific form:

$$f_{\boldsymbol{\theta}}(x) = \prod_{i=1}^{n} a(\boldsymbol{\theta})b(x_i)e^{c^T(\boldsymbol{\theta})t(x_i)}$$

$$= \underbrace{a^n(\boldsymbol{\theta})e^{c^T(\boldsymbol{\theta})\overbrace{\sum_{i=1}^{n}t(x_i)}^{\mathbf{T}}}}_{g(\mathbf{T},\boldsymbol{\theta})} \underbrace{\prod_{i=1}^{n} b(x_i)}_{h(x)}.$$

For this reason, the following vector, of p-dimension, is sufficient for $\boldsymbol{\theta}$:

$$\sum_{i=1}^{n} \mathbf{t}(x_i) = \left[\sum_{i=1}^{n} t_1(x_i), \cdots, \sum_{i=1}^{n} t_p(x_i) \right]^T.$$

In particular, the above vector is of finite dimensional sufficient statistic, which is complete and minimal.

7.5.5 Checking Whether a PDF Belongs to the Exponential Distribution Category

Due to the many attractive properties of the exponential distribution category, in many cases, the first question to be answered becomes: Is the PDF of the data x a member of this particular case? This question can arise, for example, if the input to a known filter or another system has a known PDF, and one can calculate a mathematical representation of the PDF of the filter output. To check if the output PDF is exponential, it must be transformed in an exponential form. If this is difficult, the next step is to try and show that density does not belong to that specific category. Certain properties can be checked immediately, e.g., that the parametric space, Θ, does not depend on the x range (as in the case of the uniform PDF with an unknown support region of the distribution boundaries). Another way is by computing $\partial^2/\partial\theta\partial x \ln f_\theta(x)$ and verifying that it cannot be expressed in the "discrete" form $c'(\theta)t'(x)$ for some functions c and t.

8

Fundamental Principles of Parametric Estimation

In Chapter 7, we investigated the basic principles of statistical evaluation: the formulation of a statistical model and the statistical sufficiency for the model parameters. In this chapter, we continue with the analysis and derivation of methods for estimating the parameters from random samples of the model, paying particular attention to how accurate these estimates remain in different sampling realizations.

We shall begin with the basic mathematical formulation of estimation, and then, specifying for the case of scalar parameters, we will study two different cases: the random parameters and the non-random parameters. For random parameters, one can estimate the mean accuracy of the estimator easier and define procedures for deriving optimal estimators, called Bayes estimators, providing the highest possible accuracy. More specifically, three different optimization criteria are defined: the mean square error (MSE), mean absolute error (MAE) and mean uniform error, also called high error probability (P_e). Afterwards, the deterministic scalar parameters are studied focusing on how unbiased the estimator is and on variance, which measures the estimation accuracy. This leads to the concept of Fisher information and Cramer-Rao lower bound to the variance of unbiased estimators. Finally, we generalize this analysis to the case of multiple (vector) parameters.

8.1 Estimation: Basic Components

In this chapter, the following notations apply:

- $x \in \chi$ is a vector of observed RVs.
- X is the sample space of realization of the x vector.
- $\theta \in \Theta$ is a vector of unknown parameters of interest.
- $\Theta \subset \mathbb{R}^p$ is the parametric space.
- $f(x;\theta)$ is the PDF of x for conditional θ (known function).

The goal of parameter estimation is to design an estimator function $\hat{\theta} = \hat{\theta}(x)$, which correlates χ at $\mathbb{R}^p \supset \Theta$. This is illustrated in Figure 8.1.

It is important to distinguish between an estimator, which is a function of sample x, and an estimate, which is an evaluation of the function in a particular realization of x, i.e.:

- The function $\hat{\theta}$ is an estimator.
- The point $\hat{\theta}(x)$ is an estimate.

A reasonable question arises: What is the appropriate design criterion for constructing an estimator? There are many possible approaches that give an answer to this. In this

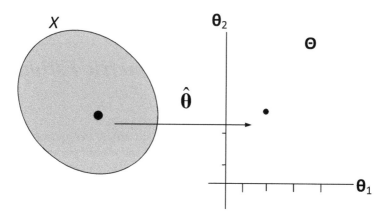

FIGURE 8.1
An estimator of vector $\boldsymbol{\theta}$ with conditional sample vector of RV **x** is matched as χ at $\mathbb{R}^p$.

chapter, we will describe two basic approaches. The first one assumes that θ is random, while the second assumes that it is predefined (known). Common in both approaches is the description of a loss function, also called a risk function, which is associated with an estimator that measures the estimation error as a function of both the sample value and the parameters.

Let $c\!\left(\hat{\boldsymbol{\theta}}(x);\boldsymbol{\theta}\right)$ be a risk function associated with $\hat{\boldsymbol{\theta}}$ for conditional $\boldsymbol{\theta}$ and **x**. The optimal estimator, if any, can be derived by minimizing the mean loss, $\mathbb{E}[C]$, where the capital letter C denotes the RV $c\!\left(\hat{\boldsymbol{\theta}}(\mathbf{x});\boldsymbol{\theta}\right)$.

8.2 Estimation of Scalar Random Parameters

When $\boldsymbol{\theta}$ is scalar, i.e., θ, we have direct access to the following:

- $f(\theta)$: a priori PDF of θ.
- $f(\mathbf{x}|\theta)$: conditional PDF of x at θ.
- $f(\theta|\mathbf{x})$: a posteriori PDF of θ determined by the Bayes rule:

$$f(\theta\,|\,\mathbf{x}) = \frac{f(\mathbf{x}\,|\,\theta)f(\theta)}{f(\mathbf{x})}.$$

- $f(\mathbf{x})$: marginal PDF, which results by marginalization over θ:

$$f(\mathbf{x}) = \int_{\Theta} f(\mathbf{x}\,|\,\theta)f(\theta)\mathrm{d}\theta.$$

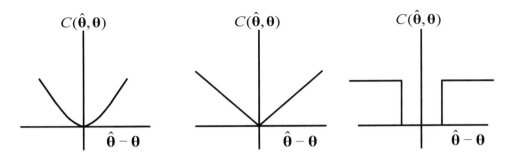

FIGURE 8.2
Loss functions for estimating scalar parameters: (a) square error, (b) absolute error, (c) uniform error.

Based on the above, the mean loss or Bayes risk function can be calculated as:

$$\mathbb{E}[C] = \int_{\Theta} \int_{\mathcal{X}} c\left(\hat{\boldsymbol{\theta}}(\mathbf{x}); \boldsymbol{\theta}\right) f(\mathbf{x} \mid \boldsymbol{\theta}) f(\boldsymbol{\theta}) d\mathbf{x} d\boldsymbol{\theta}.$$

At this point, we are able to define the optimal estimator. A scalar estimator, $\hat{\theta}$, which minimizes the average loss, is called the Bayes estimator. Some loss functions for this estimation problem are listed as follows:

- $c(\hat{\theta}; \theta) = |\hat{\theta} - \theta|^2$: Square error.
- $c(\hat{\theta}; \theta) = |\hat{\theta} - \theta|$: Absolute error.
- $c(\hat{\theta}; \theta) = I\left(|\hat{\theta} - \theta| > \varepsilon\right)$: Uniform error.

Figure 8.2 depicts these three cases.
For each of these three loss functions, the mean loss and Bayes risk functions can be calculated:

- MSE (mean square error) estimator: $\text{MSE}(\hat{\theta}) = \mathbb{E}\left[|\hat{\theta} - \theta|^2\right]$.
- MAE (mean absolute error) estimator: $\text{MAE}(\hat{\theta}) = \mathbb{E}\left[|\hat{\theta} - \theta|\right]$.
- Error probability: $P_e(\hat{\theta}) = \Pr\left[|\hat{\theta} - \theta| > \varepsilon\right]$.

The optimal estimators $\hat{\theta}$ that minimize the above criteria are still to be found.

8.2.1 Estimation of Mean Square Error (MSE)

The MSE is the estimation criterion most widely applied and undoubtedly the one with the longest history. The minimum mean square error estimator is the conditional mean estimator (CME) defined as:

$$\hat{\theta}(\mathbf{x}) = \mathbb{E}[\theta \mid \mathbf{x}] = \text{mean}_{\theta \in \Theta}\left\{f(\theta \mid \mathbf{x})\right\},$$

where

$$\text{mean}_{\theta \in \Theta}\{f(\theta \,|\, \mathbf{x})\} = \int_{-\infty}^{+\infty} \theta f(\theta \,|\, \mathbf{x}) \mathrm{d}\theta.$$

The CME has an intuitive mechanical interpretation as the center of mass (first moment of inertia) of mass density $f(\theta|\mathbf{x})$ (Figure 8.3). The CME represents the a posteriori mean value of the parameter when the corresponding data sample has been obtained.

The CME ensures the orthogonality: the error of the Bayes estimator is orthogonal to any (linear or non-linear) data function. This condition is mathematically expressed as:

$$\mathbb{E}\left[\left(\theta - \hat{\theta}(\mathbf{x})\right) g(\mathbf{x})^* \right] = 0.$$

for every function g of x, where g^* implies a complex conjugate of g.

Proof:
The MSE is expressed as:

$$\mathbb{E}[\,|\hat{\theta} - \theta|^2] = \mathbb{E}\left[\left|\left(\hat{\theta} - \mathbb{E}[\theta\,|\,\mathbf{x}]\right) - \left(\theta - \mathbb{E}[\theta\,|\,\mathbf{x}]\right)\right|^2 \right]$$
$$= \mathbb{E}\left[\left|\hat{\theta} - \mathbb{E}[\theta\,|\,\mathbf{x}]\right|^2 \right] + \mathbb{E}\left[\left|\theta - \mathbb{E}[\theta\,|\,\mathbf{x}]\right|^2 \right]$$
$$- \mathbb{E}\left[g(\mathbf{x})^*\left(\theta - \mathbb{E}[\theta\,|\,\mathbf{x}]\right)\right] - \mathbb{E}\left[g(\mathbf{x})\left(\theta - \mathbb{E}[\theta\,|\,\mathbf{x}]\right)^* \right],$$

where $g(\mathbf{x}) = \hat{\theta} - \mathbb{E}[\,\theta|\mathbf{x}\,]$ is an exclusive function of $\mathbf{x}$.

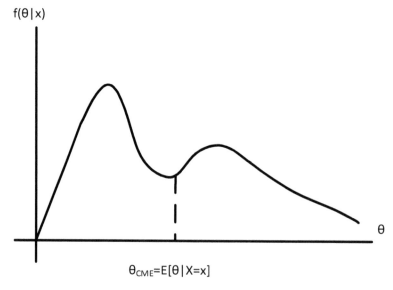

FIGURE 8.3
The CME minimizes the MSE.

Step 1: Express the orthogonality condition.

$$\mathbb{E}\Big[g(x)\big(\theta - \mathbb{E}[\theta|x]\big)\Big] = \mathbb{E}\Big[\mathbb{E}\big[g(x)\big(\theta - \mathbb{E}[\theta|x]\big)^* | x\big]\Big]$$
$$= \mathbb{E}\Big[g(x)\underbrace{\mathbb{E}\big[\theta - \mathbb{E}[\theta|x]|x\big]}_{=0}\Big] = 0.$$

Step 2: $\mathbb{E}\big[\theta|x\big]$ minimizes the MSE.

$$\mathbb{E}[|\hat{\theta}-\theta|^2] = \mathbb{E}\Big[|\hat{\theta}-\mathbb{E}[\theta|x]|^2\Big] + \mathbb{E}\Big[|\theta-\mathbb{E}[\theta|x]|^2\Big]$$
$$\geq \mathbb{E}\Big[|\theta-\mathbb{E}[\theta|x]|^2\Big],$$

where the equality in the last relation occurs if and only if $\hat{\theta} = \mathbb{E}[\theta|x]$.

8.2.2 Estimation of Minimum Mean Absolute Error

For simplicity reasons, consider a scalar real value of θ as well as $F(\theta|x) = \int^{\theta} f(\theta'|x)d\theta'$, a continuous function of θ. The minimum mean absolute error estimator is the conditional median estimator (CmE) defined as:

$$\hat{\theta}(x) = \mathbb{E}[\theta|x] = \text{median}_{\theta\in\Theta}\{f(\theta|x)\}, \tag{8.1}$$

where

$$\text{median}_{\theta\in\Theta}\{f(\theta|x)\} = \min\Big\{u: \int_{-\infty}^{u} f(\theta|x)d\theta = 1/2\Big\}$$
$$= \min\Big\{u: \int_{-\infty}^{u} f(x|\theta)f(\theta)d\theta = \int_{u}^{+\infty} f(x|\theta)f(\theta)d\theta\Big\}. \tag{8.2}$$

The median of a PDF separates the function into two parts of equal mass density (Figure 8.4).
When $F(\theta|x)$ is strictly increasing in Θ, the term "min" in the definition of the median is not obligatory, but is needed in the case of areas of Θ in which $f(\theta|x)$ is equal to zero. If $f(\theta|x)$ is continuous at θ, the CmE satisfies the orthogonality:

$$\mathbb{E}\Big[\text{sgn}\big(\theta - \hat{\theta}(x)\big)g(x)\Big] = 0,$$

and therefore, for the estimation of the minimum MAE, the sign of the optimal error estimate is that which remains orthogonal to each function of the data sample.

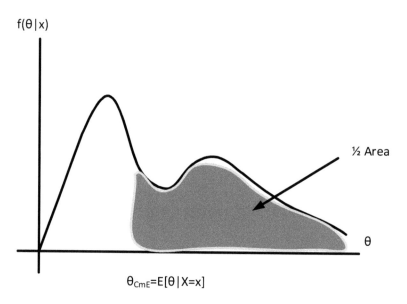

FIGURE 8.4
The conditional median estimator minimizes the MAE.

Proof:
Let $\hat{\theta}_m$ be the median of $f(\theta|\mathbf{x})$. Therefore, from the definition of the median, it is:

$$\mathbb{E}\left[\operatorname{sgn}(\theta-\hat{\theta}_m)|\mathbf{x}\right]=\int_{\Theta}\operatorname{sgn}\left(\theta-\hat{\theta}_m(\mathbf{x})\right)f(\theta|\mathbf{x})d\theta$$

$$=\int_{\theta>\hat{\theta}_m(\mathbf{x})}f(\theta|\mathbf{x})d\theta-\int_{\theta\leq\hat{\theta}_m(\mathbf{x})}f(\theta|\mathbf{x})d\theta$$

$$=0.$$

Step 1: Express the orthogonality:

$$\mathbb{E}\left[\operatorname{sgn}(\theta-\hat{\theta}_m)g(\mathbf{x})\right]=\mathbb{E}\left[\underbrace{\mathbb{E}\left[\operatorname{sgn}(\theta-\hat{\theta}_m)|\mathbf{x}\right]}_{=0}g(\mathbf{x})\right]$$

Step 2: For arbitrary $\hat{\theta}$, it is:

$$\operatorname{MAE}(\hat{\theta})=\mathbb{E}\left[\left|\underbrace{\theta-\hat{\theta}_m}_{a}+\underbrace{\hat{\theta}_m-\theta}_{\Delta}\right|\right]$$

$$=\mathbb{E}\left[\left|\theta-\hat{\theta}_m\right|\right]+\underbrace{\mathbb{E}\left[\operatorname{sgn}(\theta-\hat{\theta}_m)\Delta\right]}_{=0}$$

$$+\underbrace{\mathbb{E}\left[\operatorname{sgn}(a+\Delta)-\operatorname{sgn}(a)\right](a+\Delta)}_{\geq\left[\operatorname{sgn}(a+\Delta)-1\right](a+\Delta)\geq0}$$

$$\geq\mathbb{E}[|\theta-\hat{\theta}_m|].$$

Useful property:

$$|a + \Delta| = |a| + \mathrm{sgn}(a)\Delta + \left[\mathrm{sgn}|a + \Delta| - \mathrm{sgn}(a)\right](a + \Delta)$$

8.2.3 Estimation of Mean Uniform Error (MUE)

Unlike the MSE or the MAE, MUE considers only the errors that exceed a certain threshold $\varepsilon > 0$, where this approach is uniform. For small values of ε, the optimal estimator is the maximum a posteriori estimator (MAP) (Figure 8.5):

$$\hat{\theta}(\mathbf{x}) = \mathrm{argmax}_{\theta \in \Theta}\left\{f(\theta \mid \mathbf{x})\right\}$$
$$= \mathrm{argmax}_{\theta \in \Theta}\left\{\frac{f(\mathbf{x} \mid \theta)f(\theta)}{f(\mathbf{x})}\right\} \qquad (8.3)$$
$$= \mathrm{argmax}_{\theta \in \Theta}\left\{f(\mathbf{x} \mid \theta)f(\theta)\right\}.$$

It is noteworthy that the last equality of Equation 8.3 is most preferred because it does not require the marginal PDF $f(\mathbf{x})$, which may be difficult to calculate.

Proof:
Let ε be a small positive value. The probability of the estimation error being greater than ε is expressed as:

$$P_e(\hat{\theta}) = 1 - P\left(|\theta - \hat{\theta}| \le \varepsilon\right)$$
$$= 1 - \int_{\mathcal{X}} d\mathbf{x} f(\mathbf{x}) \int_{\{\theta: |\theta - \hat{\theta}| \le \varepsilon\}} f(\theta \mid \mathbf{x}) d\theta.$$

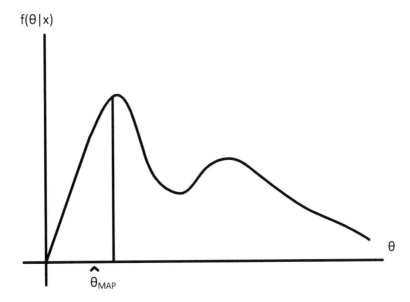

FIGURE 8.5
The MAP estimator minimizes P_e.

Now consider the inner integral (at θ) of the above relation. This is an integral at θ inside a window, called 2ε-length window, centered at $\hat{\theta}$. With regard to Figure 8.6, if the value ε is sufficiently small, the integer is maximized by centering the 2ε-length window to the value of θ that maximizes the embedded function $f(\theta|x)$. This value is, by default, the MAP estimator for the parameter $\hat{\theta}$.

After the detailed description of the aforementioned estimation criteria (and their respective optimal estimators), some general observations follow:

1. The CME may not exist for a discrete Θ because the median may not be properly defined.

2. Only the CME requires the (often difficult) computation of the normalized function $f(x)$ for the derivation of $f(\theta|x) = f(x|\theta)/f(x)$.

3. All three estimators depend on x only via the function $f(\theta|x)$.

4. When the MAP estimator is continuous, unimodal and symmetric then each of the above estimators is identical. This is illustrated in Figure 8.7.

5. If $T = T(x)$ is a sufficient statistic, the a posteriori estimation depends on x only through the median of T. So, if $f(x|\theta) = g(T;\theta)h(x)$, according to Bayes, it is true that:

$$f(\theta|x) = \frac{f(x|\theta)f(\theta)}{\int_\Theta f(x|\theta)f(\theta)d\theta} = \frac{g(T;\theta)f(\theta)}{\int_\Theta g(T;\theta)f(\theta)d\theta},$$

which is a function of x through T. Therefore, with regards to the operation of the optimal estimation, absolutely no information is lost by compressing x to a sufficient statistic.

6. The CME has the following linearity property: For each RV θ_1 and θ_2: $\mathbb{E}[\theta_1 + \theta_2|x] = \mathbb{E}[\theta_1|x] + \mathbb{E}[\theta_2|x]$. This property is not true for CmE and MAP estimators.

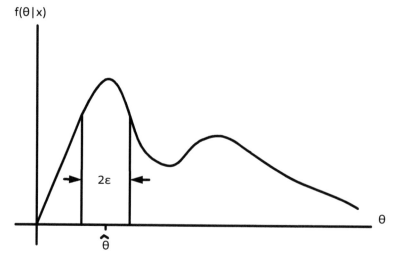

FIGURE 8.6
The a posteriori PDF, which is integrated in the 2ε-length window.

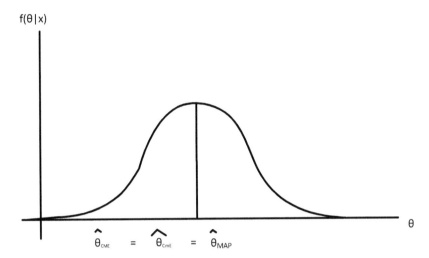

FIGURE 8.7
Symmetric and continuous a posteriori PDF.

8.2.4 Examples of Bayesian Estimation

In this section, four typical examples of statistical models will be studied and optimal estimators will be suggested based on various criteria.

- Estimation of the width of the uniform PDF.
- Estimation of a Gaussian signal.
- Estimation of the size of a Gaussian signal.
- Estimation of a binary signal with Gaussian noise.

Example 9: Estimation of the Uniform PDF Amplitude

Consider the following problem: A network terminal requires random time to connect to another terminal after sending a login request at the time point $t = 0$. You, the user, wish to schedule a transaction with a potential customer as soon as possible after sending the request. However, if your machine is not connected within the scheduled time, the customer will address someone else. Assuming that the connection delay is a random variable x that is uniformly distributed in the time interval $[0, \theta]$, you can assure your client that the delay will not exceed θ. The problem is that you do not know θ, so it must be estimated from previous experience, e.g., the order of delayed connections previously observed $x_1, \cdots, x_n$. Assuming an a priori distribution at θ, the optimal estimate can be obtained using the theory developed above.

Let $x_1, \cdots, x_n$ be the IID samples, where each sample has a conditional PDF:

$$f(x_1 \mid \theta) = \frac{1}{\theta} I_{[0,\theta]}(x_1).$$

Based on past experience, we consider that:

$$f(\theta) = \theta\, e^{-\theta}, \quad \theta > 0.$$

Figure 8.8 illustrates these two PDFs.
Next, we will express the CME, CmE and MAP estimators for θ.

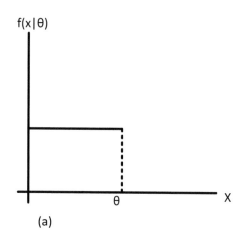

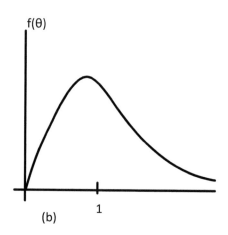

FIGURE 8.8
(a) A uniform PDF of the unknown amplitude θ, (b) a priori PDF at θ.

Step 1: Find the a posteriori PDF $f(\theta|\mathbf{x}) = f(\mathbf{x}|\theta)f(\theta)/f(\mathbf{x})$:

$$f(\mathbf{x}|\theta)f(\theta) = \left(\prod_{i=1}^{n}\frac{1}{\theta}I_{[x_i,+\infty]}(\theta)\right)\left(\theta\,e^{-\theta}\right)$$

$$= \frac{e^{-\theta}}{\theta^{n-1}}\underbrace{\prod_{i=1}^{n}I_{[x_i,+\infty]}(\theta)}_{I_{[x_{(1)},+\infty]}(\theta)}$$

$$= \frac{e^{-\theta}}{\theta^{n-1}}I_{[x_{(1)},+\infty]}(\theta),$$

where $x_{(1)} = \max\{x_i\}$. It can be noted that the function is monotonically decreasing for $\theta > 0$ (because the log derivative is negative).
Also:

$$f(\mathbf{x}) = \int_{0}^{+\infty} f(\mathbf{x}|\theta)f(\theta)\,d\theta$$

$$= q_{-n+1}(x_{(1)}),$$

where q_n is a monotonically decreasing function:

$$q_n \triangleq \int_{x}^{+\infty}\theta^n e^{-\theta}\,d\theta.$$

Repeatable formula: $q_{-n+1}(x) = \dfrac{1}{n}\left(\dfrac{1}{x^n}e^{-x} - q_{-n}(x)\right), n = 0, -1, -2, \ldots$

Step 2: Derive the functions of optimal estimator.

$$\hat{\theta}_{MAP} = x_{(1)}$$

$$\hat{\theta}_{CME} = q_{-n+2}(x_{(1)})/q_{-n+1}(x_{(1)})$$

$$\hat{\theta}_{CmE} = q_{-n+1}^{-1}\left(\frac{1}{2}q_{-n+1}\left(x_{(1)}\right)\right).$$

Note that only the MAP estimator is a simple function of $\mathbf{x}$, while the other two require more difficult calculations of the integers q_n and/or of the inverse function q_n^{-1}. The a priori PDF $f(\theta|\mathbf{x})$, using these estimators, is depicted in Figure 8.9.

Example 10: Estimation of a Gaussian Signal

A common assumption that arises in many signal extracting problems is that of a Gaussian distributed signal observed with additional Gaussian noise. For example, a radar system can transmit a pulse to a detector for locating potential targets in an area. If there is a strong reflective target, then it reflects some energy from the pulse back to the radar, resulting in a high-energy signal, called signal-return.

The range of this signal may contain useful information about the identity of the target. Radar signal-return evaluation is complicated by the presence of noise produced in the radar receiver (thermal noise) or by interference from other sources in the coverage area. Based on radar field tests, the a priori average and variance of the received signal and noise may be available.

Let two jointly distributed Gaussian RVs: s, x with known expected values, variance and covariance:

$$\mathbb{E}[s] = \mu_s, \quad \mathbb{E}[x] = \mu_x,$$

$$\mathrm{var}(s) = \sigma_s^2, \quad \mathrm{var}(x) = \sigma_x^2,$$

$$\mathrm{cov}(s, x) = \rho\sigma_s\sigma_x.$$

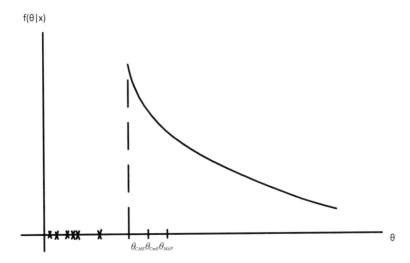

FIGURE 8.9
The CME, CmE, MAP estimations for the amplitude parameter θ based on the a priori PDF, as derived from Figure 8.8 (b).

s represents the signal and x the measurement. Of course, the specific form of the covariance function will depend on the structure of the receiver—for example, it is reduced to a simple function of σ_s and σ_x for the additive noise model.

We will find the optimal estimator of s with a conditional x. As with the previous example, the process of finding the CME, CmE and MAP estimators is presented in two parts.

Step 1: Find the a posteriori PDF.

A fundamental fact for jointly distributed Gaussian RVs is that if one of the variables is conditional, the other variable is also a Gaussian, but with a different expected value and variance, equal to their respective conditional values (Figure 8.10).

More specifically, the conditional PDF of s with a conditional x is a Gaussian with expected value,

$$\mu_{s|x}(x) = \mathbb{E}[s \mid x] = \mu_s + \rho \frac{\sigma_s}{\sigma_x}(x - \mu_x),$$

and variance,

$$\sigma_{s|x}^2 = \mathbb{E}\left[\left(s - \mathbb{E}[s \mid x]\right)^2 \mid x\right] = (1 - \rho^2)\sigma_s^2,$$

so that the conditional PDF is of the form:

$$f_{s|x}(s \mid x) = \frac{f_{x|s}(x \mid s) f_s(s)}{f_x(x)}$$

$$= \frac{1}{\sqrt{2\pi\sigma_{s|x}^2}} \exp\left\{-\frac{\left(s - \mu_{s|x}(x)\right)^2}{2\sigma_{s|x}^2}\right\}.$$

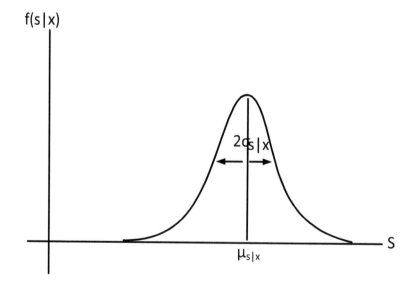

FIGURE 8.10
The a posteriori $f(s|x)$ is a Gaussian PDF when x, s are jointly Gaussian.

Step 2: Derive the optimal estimator.

As the a posteriori PDF is continuous, symmetric and unimodal, the CME, CmE and MAP estimators are of the same form. By deriving the dependence of the estimator $\hat{s}$ from the observed realization x, it is:

$$\hat{s}(x) = \mu_{s|x}(x) = \text{linear at } x.$$

An interesting special case, related to the aforementioned example of the radar, is the model of independent additive noise, where $x = s + v$. In this case, $\sigma_x^2 = \sigma_s^2 + \sigma_v^2$, $\rho^2 = \sigma_s^2 / (\sigma_s^2 + \sigma_v^2)$, and therefore:

$$\sigma_x^2 = \mu_s + \frac{\sigma_s^2}{\sigma_s^2 + \sigma_v^2}(x - \mu_x).$$

Consequently, it is true that:

- Minimum MSE: $\mathbb{E}\left[(s - \hat{s})^2\right] = (1 - \rho^2)\sigma_s^2$.
- Minimum MAE: $\mathbb{E}\left[(s - \hat{s})^2\right] = \sqrt{(1 - \rho^2)\sigma_s^2}\sqrt{2/\pi}$.
- Minimum P_e: $P\left[|s - \hat{s}| > \varepsilon\right] = 1 - \text{erf}\left(\varepsilon / \sqrt{2(1 - \rho^2)\sigma_s^2}\right)$.

Example 11: Estimating the Size of a Gaussian Signal

We are now slightly changing Example 10. What if the radar operator was only interested in the energy of the received signal and not in its sign (phase)? Then, we would aim to estimate the amplitude or magnitude, $|s|$, instead of the magnitude and the phase of s. Of course, a quick estimation process would simply be to use the previous estimator $\hat{s}$ and its magnitude $|\hat{s}|$ to estimate the real $|s|$. But is this approach the ideal one?

We define again the two jointly Gaussian RVs: s, x with expected values, variance and covariance:

$$\mathbb{E}[s] = \mu_s, \quad \mathbb{E}[x] = \mu_x,$$
$$\text{var}(s) = \sigma_s^2, \quad \text{var}(x) = \sigma_x^2,$$
$$\text{cov}(s, x) = \rho\sigma_s\sigma_x.$$

Now, the question is to estimate the conditional RV $y = |s|$ in x.

NOTE: y, x RVs no longer represent a jointly Gaussian model. The first step is to calculate the a posteriori PDF $f_{y|x}$. Since we know $f_{s|x}$ from the previous example, the problem is essentially presented as a simple transformation of the variables into a known function. Using the difference method (Figure 8.11) we express the following relation, which is valid for small Δ:

$$f_{y|x}(y \mid x)\Delta = f_{s|x}(y \mid x)\Delta + f_{s|x}(-y \mid x)\Delta, \quad y \geq 0,$$

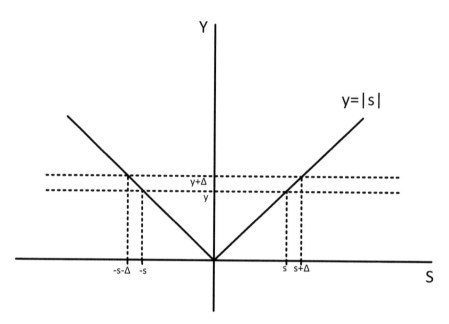

FIGURE 8.11
Schematic representation of the difference method for deriving the conditional PDF of the RV $y = |s|$ with conditional x by the probability: $P(y < y \leq y + \Delta | x) \approx f_{y|x}(y|x)\, \Delta, 0 < \Delta \ll 1$.

or:

$$f_{y|x}(y|x) = \frac{1}{\sqrt{2\pi\sigma_{s|x}^2}}\left(\exp\left\{-\frac{\left(y - \mu_{s|x}(x)\right)^2}{2\sigma_{s|x}^2}\right\} + \exp\left\{-\frac{\left(y + \mu_{s|x}(x)\right)^2}{2\sigma_{s|x}^2}\right\}\right)I_{[0,+\infty]}(y). \quad (8.4)$$

In contrast to Example 10, the specific a posteriori PDF, is no longer symmetric to y. Therefore, we expect that the CME, CmE and MAP estimators will be different.

The CME is derived in a closed mathematical expression, integrating the function $yf_{y|x}(y|x)$ in $y\in[0,+\infty)$, according to the Relation 8.4, above, as follows:

$$\hat{y}_{CME}(x) = \mathbb{E}[y|x] = \left|\mu_{s|x}(x)\right| \mathrm{erf}\left(\frac{\left|\mu_{s|x}(x)\right|}{\sigma_{s|x}\sqrt{2}}\right) + \sqrt{\frac{2}{\pi}}\sigma_{s|x}\, e^{-\mu_s^2/\left(2\sigma_{s|x}^2\right)}.$$

On the other hand, from the equation of the minimum absolute mean error $\int_{\hat{y}}^{+\infty} f_{y|x}(y|x)dy = \int_0^{\hat{y}} f_{y|x}(y|x)dy$, the CmE, $\hat{y} = \hat{y}_{CmE}$, is derived by solving the following equation:

$$\mathrm{erf}\left(\frac{\hat{y} - \mu_{s|x}(x)}{\sigma_{s|x}\sqrt{2}}\right) + \mathrm{erf}\left(\frac{\hat{y} + \mu_{s|x}(x)}{\sigma_{s|x}\sqrt{2}}\right) = \frac{1}{2}.$$

Finally, since the function $f_{y|x}(y|x)$ is concave and smooth in y, the MAP estimator, $\hat{y} = \hat{y}_{MAP}$, occurs at a stationary point at y:

$$0 = \frac{\partial f(y|x)}{\partial y}.$$

Using Relation 8.4, we obtain:

$$\hat{y}(x) = \mu_{s|x}(x) \frac{\exp\left\{-\dfrac{\left(\hat{y}-\mu_{s|x}(x)\right)^2}{2\sigma_{s|x}^2}\right\} - \exp\left\{-\dfrac{\left(\hat{y}+\mu_{s|x}(x)\right)^2}{2\sigma_{s|x}^2}\right\}}{\exp\left\{-\dfrac{\left(\hat{y}-\mu_{s|x}(x)\right)^2}{2\sigma_{s|x}^2}\right\} + \exp\left\{-\dfrac{\left(\hat{y}+\mu_{s|x}(x)\right)^2}{2\sigma_{s|x}^2}\right\}}.$$

It can be confirmed that when $\mu_{s|x}/\sigma_{s|x} \to +\infty$, the three estimators converge to a common point:

$$\hat{y}(x) \to \left| \mu_{s|x}(x) \right|.$$

This happens because the a posteriori PDF is transformed to the Dirac function, with a point of concentration $y = \mu_{s|x}(x)$ as $\mu_{s|x}/\sigma_{s|x} \to +\infty$. It is noteworthy that none of the aforementioned estimators of $|s|$ are expressed as a function of $|\hat{s}|$, where $\hat{s}$ the estimation of s by CME, CmE, MAP estimators, as derived in Example 10. This fact indicates that the estimation of random parameters is not invariant with respect to the functional operators.

Example 12: Sign Estimation of a Gaussian Signal

In the previous examples, optimal estimators were studied for deriving the magnitude of a Gaussian RV, based on Gaussian observations. This example examines the case where only the phase of a signal interests, e.g., when a radar evaluates the sign of a signal.

Consider the following observation model:

$$x = \theta + w,$$

where w is a Gaussian variable with zero expected value and variance σ^2, which represents the additive noise. Also, the RV θ is a binary value with an equal probability of the two discrete states: $P(\theta = 1) = P(\theta = -1) = 1/2$, $\Theta = \{-1,1\}$. This model is consistent with the radar problem with an a priori expected value μ_s and additive noise.

Here, the a posteriori PDF is the mass probability function, since the signal θ has discrete values:

$$p(\theta|x) = \frac{f(x|\theta)p(\theta)}{f(x)},$$

where $p(\theta) = 1/2$. For convenience, we have eliminated the indices regarding the follow-ing probability functions. Also, it is true that:

$$f(x\,|\,\theta) = \begin{cases} \dfrac{1}{\sqrt{2\pi\sigma^2}}\exp\!\left(\dfrac{(x-1)^2}{2\sigma^2}\right), & \theta = 1 \\[4mm] \dfrac{1}{\sqrt{2\pi\sigma^2}}\exp\!\left(\dfrac{(x+1)^2}{2\sigma^2}\right), & \theta = 0. \end{cases}$$

Hence:

$$f(x) = f(x\,|\,\theta = 1)\frac{1}{2} + f(x\,|\,\theta = -1)\frac{1}{2}.$$

By following the steps below, it is evident that the MAP estimator constitutes a deci-sion rule with regard to the minimum distance so that the value $\hat{\theta}$ would be chosen as the actual θ, which is the closest to the measured value of x:

$$\begin{aligned} \hat{\theta}_{MAP} &= \arg\max_{\theta=1,-1} f(x\,|\,\theta) \\ &= \arg\min_{\theta=1,-1}\left\{(x-\theta)^2\right\} \\ &= \begin{cases} 1, & x \geq 0 \\ -1, & x < 0. \end{cases} \end{aligned}$$

On the other hand, CME is expressed as:

$$\begin{aligned} \hat{\theta}_{CME} &= (1)P(\theta = 1\,|\,x) + (-1)P(\theta = -1\,|\,x) \\ &= \frac{\exp\!\left(-\dfrac{(x-1)^2}{2\sigma^2}\right) - \exp\!\left(-\dfrac{(x+1)^2}{2\sigma^2}\right)}{\exp\!\left(-\dfrac{(x-1)^2}{2\sigma^2}\right) + \exp\!\left(-\dfrac{(x+1)^2}{2\sigma^2}\right)}. \end{aligned}$$

Based on the above examples, the following useful conclusions are drawn:

1. Different error criteria reflect different optimal estimators.
2. The estimators of optimal random parameters are not invariant regarding the functional operators. That is, if $\hat{g}(\theta)$ is an optimal estimator of $g(\theta)$, and $\hat{\theta}$ is an optimal estimator of θ, it is true that:

$$\hat{g}(\theta) \neq g(\hat{\theta}).$$

3. When CmE and MAP exist, they receive values within the parametric space Θ. The corresponding values of the CME may also be outside this area, e.g., if the values are discrete or if they do not belong to a convex set.
4. The stationary point condition (MAP equation) $\partial f(\theta\,|\,x)/\partial\theta = 0$ at $\theta = \hat{\theta}_{MAP}$ is useful only for continuous PDFs, which are differentiable and concave at con-tinuous values of θ parameters.

8.3 Estimation of Random Vector Parameters

We define a parametric vector $\boldsymbol{\theta} \in \Theta \subset \mathbb{R}^p$, $\boldsymbol{\theta} = \left[\theta_1, \cdots, \theta_p\right]^T$ and a norm-s in the domain Θ as:

$$\|\boldsymbol{\theta}\|_s = \left(\sum_{i=1}^{p} |\theta_i|^s\right)^{1/s}.$$

Note that when $s = +\infty$, the norm is equal to the maximum value of $|\theta_i|$.

The estimation criterion $\mathbb{E}\left[c(\hat{\theta}, \theta)\right]$ described above should be generalized to contain vector parameters. Some characteristic generalizations of the above three criteria are:

- Estimator MSE:

$$\text{MSE}(\hat{\boldsymbol{\theta}}) = \mathbb{E}\left[\|\hat{\boldsymbol{\theta}} - \boldsymbol{\theta}\|_2^2\right] = \sum_{i=1}^{p} \mathbb{E}\left[(\hat{\theta}_i - \theta_i)^2\right].$$

- Estimator MAE:

$$\text{MAE}(\hat{\boldsymbol{\theta}}) = \mathbb{E}[\|\hat{\boldsymbol{\theta}} - \boldsymbol{\theta}\|_1] = \sum_{i=1}^{p} \mathbb{E}[|\hat{\theta}_i - \theta_i|].$$

- Estimator of error probability $-(0 < p < +\infty)$:

$$P_e(\hat{\boldsymbol{\theta}}) = 1 - P\left(\|\hat{\boldsymbol{\theta}} - \boldsymbol{\theta}\|_p \le \varepsilon\right).$$

When $p = +\infty$, P_e is the probability where the magnitude of at least one of the elements of the vector $\hat{\boldsymbol{\theta}} - \boldsymbol{\theta}$ exceeds the value ε.

The MAE criterion, also known as the final norm deviation, does not often offer unique optimal estimates for vector values. Below, we will focus on MSE and P_e.

8.3.1 Squared Vector Error

As the function $\text{MSE}(\hat{\boldsymbol{\theta}}) = \sum_{i=1}^{p} \text{MSE}(\hat{\theta}_i)$ is additive, the minimum MSE vector estimator achieves the minimum of each individual value $\text{MSE}(\hat{\theta}_i)$, $i = 1, \cdots, p$. Therefore, it is deduced that this minimization is achieved by a vector that receives the individual scalar values of the CMEs for each element:

$$\hat{\boldsymbol{\theta}}_{\text{CME}} = \mathbb{E}\left[\boldsymbol{\theta} \mid x\right] = \begin{bmatrix} \mathbb{E}\left[\theta_1 \mid x\right] \\ \vdots \\ \mathbb{E}\left[\theta_p \mid x\right] \end{bmatrix}.$$

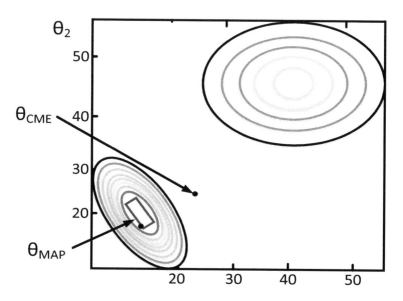

FIGURE 8.12
The MAP estimate is located at the total peak of the function while the CME at the center of the probability mass.

As in the case of the scalar estimation, the MSE estimator is the mass probability center of the multivariate a posteriori PDF (Figure 8.12).

8.3.2 Uniform Vector Error

For small values of ε, similar to the scalar case, it is true that:

$$\hat{\boldsymbol{\theta}}_{MAP} = \arg\max_{\boldsymbol{\theta} \in \Theta} f(\boldsymbol{\theta} \,|\, x).$$

8.4 Estimation of Non-Random (Constant) Parameters

In order to calculate random parameters, we consider an a priori distribution, and we can define an error criterion for the total estimate, the Bayes risk criterion, which depends on the a priori information but not on a specific current parameter value. In the estimation of non-random parameters process, there is no a priori distribution. We can, of course, consider the problem of estimating non-random parameters as an estimate of random parameters that depend on the value of the parameter, which we can call a real value. However, the derivation of the optimal non-random parameter estimation requires a completely different approach. This is because if we do not have a priori distribution of the parameter, almost any reasonable estimation error criterion will be local, that is, it will depend on the actual parameter value. Therefore, we have to define weaker optimality properties than Bayes' minimum risk, such as the unbiased criterion, according to which an efficient estimator of non-random parameters should work.

As in previous sections, we shall first study the case of the scalar parameters θ. At this point, it is obvious that we shall not use the conditional PDF symbol $f(x|\theta)$ and, therefore, the formula $f_\theta(\mathbf{x}) = f(\mathbf{x};\theta)$ will be employed.

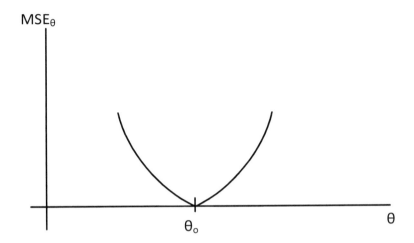

FIGURE 8.13
The MSE curve as a function of θ for the simple estimator $\hat{\theta} = \theta_0$ of a non-random parameter.

So, what could be the design criteria for scalar parameter estimators θ? An approach could be the minimization of the MSE:

$$\mathrm{MSE}_\theta = \mathbb{E}_\theta\left[(\hat{\theta} - \theta)^2\right].$$

Here, however, the following difficulty is encountered: If the actual value of θ is θ_0, the constant estimator $\hat{\theta} = c$ approaches zero (Figure 8.13).

8.4.1 Scalar Estimation Criteria for Non-Random Parameters

Some possible criteria for designing efficient estimators are the minimax criteria.

1. Minimization of the worst MSE case. Choose $\hat{\theta}$ so that:

$$\max_\theta \mathrm{MSE}_\theta(\hat{\theta}) = \max_\theta \mathbb{E}_\theta\left[(\hat{\theta} - \theta)^2\right].$$

2. Minimization of the worst case of probability estimation error:

$$\max_\theta P_e = \max_\theta P_\theta[\,|\hat{\theta} - \theta| > \varepsilon].$$

If we are satisfied by minimizing an upper limit of the maximum P_e, then we could quote Chebychev's inequality:

$$P_\theta[\,|\hat{\theta} - \theta| > \varepsilon] \le \frac{\mathbb{E}_\theta\left[(\hat{\theta} - \theta)^2\right]}{\varepsilon^2}, \tag{8.5}$$

and work on minimizing the worst case of MSE.

Then, we give some weaker conditions that an efficient estimator must satisfy: being consistent and unbiased.

Definition: $\hat{\theta}_n = \hat{\theta}(x_1, \cdots, x_n)$ is (weakly) consistent if, for all θ and all $\varepsilon > 0$, it is true that:

$$\lim_{n \to +\infty} P_\theta(|\hat{\theta}_n - \theta| > \varepsilon) = 0.$$

This means that $\hat{\theta}_n$ converges with regards to probability in the real value θ. Also, it implies that the PDF of the estimator is located around θ (Figure 8.14). In addition, from the Chebychev inequality of Equation 8.5, if the MSE tends to zero with $n \to +\infty$, then $\hat{\theta}_n$ is consistent.

Usually, MSE can be calculated more easily than P_e, while indicating that MSE converges to zero is a classic way for checking the consistency of an estimator.

For an estimator, $\hat{\theta}$, we define the bias of the estimator at a point, θ:

$$b_\theta(\hat{\theta}) = \mathbb{E}_\theta[\hat{\theta}] - \theta.$$

Similarly, the variance of the estimator is:

$$\mathrm{var}_\theta(\hat{\theta}) = \mathbb{E}_\theta\left[\left(\hat{\theta} - \mathbb{E}_\theta[\hat{\theta}]\right)^2\right].$$

It makes sense to require an efficient estimator to be unbiased, that is, $b_\theta(\hat{\theta}) = 0 \ \forall \theta \in \Theta$. This suggests the following design approach: restricting the category of estimators so that they are unbiased and trying to find a suitable estimator that minimizes variance compared to the other estimators of the category. In some cases, such an approach leads to a very efficient, in fact optimal, unbiased estimator called uniform minimum variance unbiased (UMVU), as depicted in Figure 8.15. However, there are cases where being unbiased is not

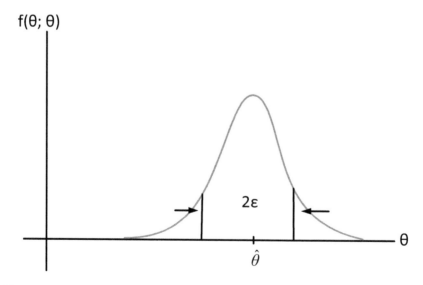

FIGURE 8.14

The function $f(\hat{\theta}; \theta)$ calculates concentration of $\hat{\theta}$ around the real value θ.

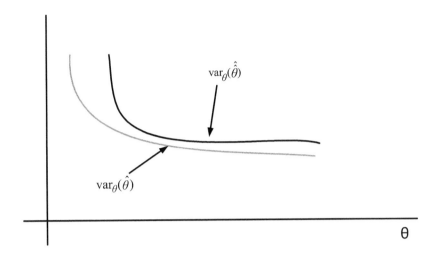

FIGURE 8.15

The UMVU estimator is an unbiased estimator $\hat{\theta}$ that has less variance than any other unbiased estimator, $\hat{\hat{\theta}}$.

a desirable property for an estimator. For example, there are models for which there is no unbiased estimator of the parameter and others for which the biased estimator arises too high MSE. Fortunately, such models do not often occur in signal processing applications.

Definition: $\hat{\theta}$ is a UMVU estimator in the case when for all $\theta \in \Theta$ has the smallest variance compared to any other unbiased estimator $\hat{\hat{\theta}}$. Therefore, a UMVU estimator satisfies the relation:

$$\mathrm{var}_\theta(\hat{\theta}) \leq \mathrm{var}_\theta(\hat{\hat{\theta}}), \, \theta \in \Theta.$$

Unfortunately, the UMVU estimators rarely exist and only for a finite range of n samples $x_1, \cdots, x_n$. Therefore, in most practical applications it is preferable to sacrifice the unbiased property in favor of designing efficient estimation procedures. For such estimators, there is an important relation between MSE, variance and bias:

$$\mathrm{MSE}_\theta(\hat{\theta}) = \mathbb{E}_\theta\left[(\hat{\theta}-\theta)^2\right] = \mathbb{E}_\theta\left[\left((\hat{\theta}-\mathbb{E}_\theta[\hat{\theta}]) + (\mathbb{E}_\theta[\hat{\theta}]-\theta)\right)^2\right]$$

$$= \underbrace{\mathbb{E}_\theta\left[(\hat{\theta}-\mathbb{E}_\theta[\hat{\theta}])^2\right]}_{\mathrm{var}_\theta(\hat{\theta})} + \underbrace{(\mathbb{E}_\theta[\hat{\theta}]-\theta)^2}_{b_\theta(\hat{\theta})} + 2\underbrace{\mathbb{E}_\theta\left[\hat{\theta}-\mathbb{E}_\theta[\hat{\theta}]\right]}_{=0}b_\theta(\hat{\theta})$$

$$= \mathrm{var}_\theta(\hat{\theta}) + b_\theta^2(\hat{\theta}).$$

The above relation implies that for a given MSE there will always be a tradeoff between bias and variance.

Two important categories with regard to the design of efficient non-random parameter estimators are:

- The method of statistical moments.
- The Maximum-Likelihood Estimation method.

8.4.2 The Method of Statistical Moments for Scalar Estimators

This method consists of deriving the parameter that achieves the best match between empirically calculated moments and real moments. More specifically, for a positive integer, k, let $m_k = m_k(\theta)$ the k-order statistical moment:

$$m_k = \mathbb{E}_\theta[x^k] = \int x^k f(x;\theta)dx.$$

It is desirable to create a set of K moments so that a certain function, **h**, would exist such that the following condition is satisfied:

$$\theta = \mathbf{h}\big(m_1(\theta),\cdots,m_K(\theta)\big).$$

For example, consider that a closed-type expression $g(\theta)$ for the k-order moment $\mathbb{E}_\theta[x^k]$ can be computed and the function $g(\theta)$ is invertible. Then, if only the value m_k is known, without determining the value θ for which it was calculated, we could recover θ by applying the inverse function:

$$\theta = g^{-1}(m_k).$$

As function g^{-1} recovers θ from the moment of x, if we have access to IID samples $x_1,\cdots,x_n$ of $f(x;\theta)$, then we can estimate θ by using g^{-1} at an estimated moment, such as the empirical mean value:

$$\hat{m}_k = \frac{1}{n}\sum_{i=1}^n x_i^k,$$

which implies the following estimator:

$$\hat{\theta} = g^{-1}(\hat{m}_k).$$

However, it is often difficult to extract a single moment that derives a corresponding invertible function of θ. Indeed, using only the k-order moment we can locate a restricting equation that $g(\theta) = \hat{m}_k$ offers multiple solutions, $\hat{\theta}$. In these cases, other statistical moments are calculated to create more restricting equations aiming to a single solution. We will discuss more about this phenomenon in the examples that follow.

Next, some asymptotic optimization properties are described for moment estimators. When the moments m_k are smooth functions with respect to parameter θ and the inverse function g^{-1} exists, then:

- Moment estimators are asymptotically unbiased as $n \to +\infty$.
- Moment estimators are consistent.

At this point, it is worth noting that moment estimators are not always unbiased for finite samples. There are, however, some difficulties that we sometimes encounter using this method, which are summarized below.

1. Moment estimators are not unique. They are depending on the order of the moment used.
2. Moment estimators are not applicable in cases where statistical moments do not exist (e.g., Cauchy PDF), or when they are not stable.

An alternative approach that solves the above problems is to match empirical moments with real moments m_k when k is a positive real value less than one. In this case, fractional moments may exist (when integer moments do not exist) and are very useful.

Example 13: IID RV Bernoulli Distributed

Bernoulli type measurements occur when (binary) quantized versions of continuous RV are used, e.g., radar threshold signals (above or below a certain value), failure indication data or digital media, e.g., measurements from the Internet. In these cases, the parameter of interest is usually the probability of success, i.e., the probability of the variable measured to be "logical 1."
The model is expressed by $\mathbf{x} = [x_1, \cdots, x_n]$, to be IID and:

$$x_i \sim f(x;\theta) = \theta^x (1-\theta)^{1-x}, \, x = 0,1.$$

It is true that $\theta \in [0,1]$ or, more specifically, $\theta = P(x_i = 1)$, $1 - \theta = P(x_i = 0)$.
Aim: Derivation of moment estimator for θ.
For each $k > 0 \, \mathbb{E}\left[x_i^k\right] = P(x_i = 1) = \theta$ so that all moments are the same and function g that matches the moments of θ represents the identity map. Therefore, a simple moment estimator of θ is the expected (mean) value of the samples:

$$\hat{\theta} = \bar{x}.$$

Obviously, $\hat{\theta}$ is an unbiased statistic, since $\mathbb{E}_\theta[\bar{x}] = m_1 = \theta$. In addition, the variance gets a maximum value $\theta = 1/2$ (Figure 8.16):

$$\text{var}_\theta(\bar{x}) = \left(m_2 - m_1^2\right)/n = \theta(1-\theta)/n.$$

Repeating, for this Bernoulli example, the order of the moment used in the moment matching process leads to identical 'method-of-moments' estimators. This behavior is very unusual.

Example 14: IID RV Poisson Distributed

Poisson measurements are found in a wide range of applications where sample values are available. For example, in positron emission tomography (PET) the decay of an isotope at a particular position within a patient's body produces a gamma ray that is recorded as a single "count" in a detector. The time recording of the periods in which these measurements occur and are recorded to the detector is a Poisson process.

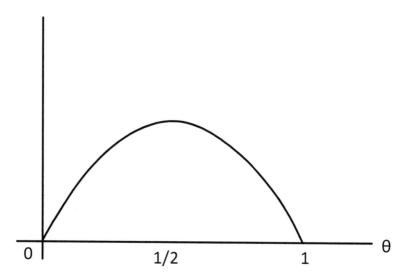

FIGURE 8.16
Variance of the moment estimator for Bernoulli RV.

The total number of records over a finite time period is a Poisson RV with an input rate parameter determined by the mean concentration of the isotope. The aim of a PET system is the reconstruction, i.e., to estimate the distribution of rates over the imaging volume. Poisson distribution is also often used as a model for the number of components or degrees of freedom that produce the measured values, e.g., the number of molecules in a mass spectroscopy measurement, the number of atoms in a molecule or the number of targets in a radar detection coverage area.

Considering IID measurement samples, the model is presented as:

$$x_i \sim p(x;\theta) = \frac{\theta^x}{x!}e^{-\theta}, \quad x = 0,1,2,\ldots,$$

where $\theta > 0$ is the unknown rate. Also, it is true that $m_1 = \theta$. Therefore, as in the previous example, the moment estimator of θ is the mean value of the samples:

$$\hat{\theta}_1 = \bar{x}.$$

Alternatively, as relation $m_2 = \theta + \theta^2$ is satisfied by the second moment, another moment estimator is the (positive) value of $\hat{\theta}_2$ that satisfies the relation $\hat{\theta}_2 + \hat{\theta}_2^2 = \frac{1}{n}\sum_{i=1}^{n} x_i^2 \triangleq \bar{x^2}$, that is:

$$\hat{\theta}_2 = \frac{-1 \pm \sqrt{1 + 4\bar{x^2}}}{2}.$$

As another example, m_2 can be expressed as $m_2 = \theta + m_1^2$ or $\theta = m_2 - m_1^2 = \mathrm{var}_\theta(x_i)$. Therefore, the moment estimator is:

$$\hat{\theta}_3 = \overline{x^2} - \overline{x}^2 = n^{-1} \sum_{i=1}^{n} (x_i - \overline{x})^2.$$

Among the aforementioned estimators, only the estimator of the mean sample value is unbiased for finite n:

$$\mathbb{E}_\theta[\hat{\theta}_1] = \theta, \qquad \mathrm{var}_\theta(\hat{\theta}_1) = \theta/n,$$

$$\mathbb{E}_\theta[\hat{\theta}_3] = \frac{n-1}{n}\theta, \quad \mathrm{var}_\theta(\hat{\theta}_3) \approx (2\theta^2 + \theta)/n.$$

Closed-form expressions for bias and variance of $\hat{\theta}_2$ do not exist. The following conclusions are summarized:

1. $\hat{\theta}_1$ is unbiased for all n.
2. $\hat{\theta}_2, \hat{\theta}_3$ are asymptotically unbiased for $n \to +\infty$.
3. The consistency of $\hat{\theta}_1$ and $\hat{\theta}_3$ is directly derived from the above expressions for the mean value and variance, as well as by using Chebychev inequality.

8.4.3 Scalar Estimators for Maximum Likelihood

Maximum likelihood (ML) is undoubtedly the most widespread principle of parametric estimation in signal processing. Unlike other methods, ML typically results in unique estimators that can be applied to almost all problems.

For a given measurement, x, the likelihood function for θ is defined as:

$$L(\theta) = f(\mathbf{x};\theta),$$

and the logarithmic likelihood function is defined as:

$$l(\theta) = \ln f(\mathbf{x};\theta).$$

These expressions are functions of θ for a constant value of $\mathbf{x}$ (Figure 8.17).

The ML estimator, $\hat{\theta}$, is defined as the value of θ, which makes the data $\mathbf{x}$ "more probable," i.e., $\hat{\theta}$ determines the greatest probability for the $\mathbf{x}$ vector to be derived from the function $f(\mathbf{x};\theta)$. Mathematically the following relations are derived:

$$\hat{\theta} = \arg\max_{\theta \in \Theta} f(\mathbf{x};\theta)$$
$$= \arg\max_{\theta \in \Theta} L(\theta)$$
$$= \arg\max_{\theta \in \Theta} l(\theta).$$

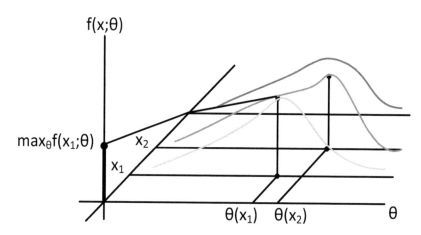

FIGURE 8.17
The likelihood function for θ.

In practice, the ML estimate can be obtained by maximizing each monotonically increasing function $L(\theta)$.

Some important properties of ML estimators are:

1. ML evaluators are asymptotically unbiased.

2. ML estimators are consistent.

3. Unlike the MAP and UMVU estimators, ML estimators are invariant at each transformation of the parameters:

$$\varphi = g(\theta) \Rightarrow \hat{\varphi} = g(\hat{\theta}).$$

This is depicted in Figure 8.18 for monotonic transformations, but in fact it is generally valid for each transformation.

4. ML estimators are UMVU asymptotically in the sense that:

$$\lim_{n \to +\infty} n \, \text{var}_\theta(\hat{\theta}) = \frac{1}{F_1(\theta)},$$

where F_1 expresses a mathematical concept known as Fisher information (described in detail below), and $1/F_1$ identifies the fastest asymptotic rate of variance reduction of any invariant estimator.

5. ML estimators are asymptotic Gaussian:

$$\sqrt{n}(\hat{\theta}_n - \theta) \to z,$$

where $z \sim N(0, 1/F_1(\theta))$. In the above relation, the symbol $(\to)$ implies convergence of the distribution. This means that the CDF of RV $\sqrt{n}(\hat{\theta}_n - \theta)$ converges to the (standard Gaussian) CDF of RV z.

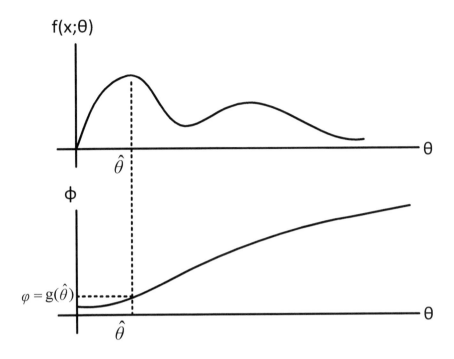

FIGURE 8.18
Invariance of ML estimator under a functional transformation *g*.

6. The ML estimator is equivalent to MAP for a uniform a priori condition $f(\theta) = c$.
7. If the ML estimator is unique, then it is expressed as a function of the data only through a sufficient statistic.

Example 15: IID RV Bernoulli Distributed

We can work in two ways to obtain the ML estimator: (1) by considering the whole observation sample **x** and (2) by considering only the sufficient statistic, $T(\mathbf{x})$.

First way:
The likelihood function is the following product:

$$L(\theta) = f(\mathbf{x};\theta) = \prod_{i=1}^{n} \theta^{x_i} (1-\theta)^{1-x_i}.$$

It is desirable to rewrite it in the form:

$$L(\theta) = \theta^{\sum_{i=1}^{n} x_i} (1-\theta)^{n-\sum_{i=1}^{n} x_i} \tag{8.6}$$
$$= \theta^{n\bar{x}_i} (1-\theta)^{n-n\bar{x}_i}.$$

Since this function is smooth and concave at θ, factorization at θ yields the stationary point condition for the ML estimator, $\hat{\theta}$:

$$0 = \frac{\partial}{\partial \hat{\theta}} f(\mathbf{x};\hat{\theta}) = n\left[\frac{(1-\hat{\theta})\overline{x_i} - \hat{\theta}(1-\overline{x_i})}{\hat{\theta}(1-\hat{\theta})}\right] f(\mathbf{x};\hat{\theta}).$$

Solving the equation $(1-\hat{\theta})\overline{x_i} - \hat{\theta}(1-\overline{x_i}) = 0$ results in the ML estimator:

$$\hat{\theta} = \overline{x}, \tag{8.7}$$

which is identical to the moment estimator analyzed previously.

Second way:

Using the Fisher factorization of Equation 7.12 to the PDF of Equation 8.6, it is true that $T(\mathbf{x}) = \sum_{i=1}^{n} x_i$ is a sufficient statistic for θ. The distribution of T is binomial with parameter θ:

$$f_T(t;\theta) = \binom{n}{t}\theta^t(1-\theta)^{n-t}, \quad t = 0,\cdots,n,$$

where f_T expresses the PDF of T. Defining $t = n\overline{x}$ reveals the fact that it is exactly of the same form (except of a fixed multiplication factor), as in Equation 8.6. Consequently, the ML equation remains the same as before, so that the ML estimator is identical to the one expressed by the Equation 8.7.

Example 16: IID RVs Distributed by Poisson

To find the ML estimator of rate of parameter θ, we express the PDF of the samples as:

$$f(\mathbf{x};\theta) = \prod_{i=1}^{n} \frac{\theta^{x_i}}{x_i!}e^{-\theta}.$$

The likelihood function $L(\theta) = f(\mathbf{x};\theta)$ must be maximized at θ for deriving the ML estimator. It is convenient to analyze the logarithmic relation:

$$\hat{\theta}_{ML} = \arg\max_{\theta>0} \ln L(\theta),$$

where it is:

$$l(\theta) = \ln f(\mathbf{x};\theta)$$

$$= \ln \prod_{k=1}^{n} \frac{\theta^{x_k}}{x_k!} e^{-\theta}$$

$$= \sum_{k=1}^{n} x_k \ln\theta - n\theta - \underbrace{\sum_{k=1}^{n} \ln x_k!}_{constant\,in\,\theta}$$

$$= \overline{x_i} n \ln\theta - n\theta + c,$$

and c is an independent coefficient.

It can be easily verified (for example, through the second derivative) that the above logarithmic function is a smooth and strictly concave function of θ. Therefore, the ML estimate finds the unique solution $\theta = \hat{\theta}$ from the equation:

$$0 = \partial \ln f / \partial\theta = \frac{n\,\overline{x_i}}{\theta} - n.$$

The above relation is the same as the solution derived for the first moment estimator:

$$\hat{\theta} = \overline{x},$$

for which we already know that it is an unbiased estimate with variance equal to θ.

Next, we study the asymptotic Gaussian property:

$$\sqrt{n}(\overline{x} - \theta) = \sqrt{n}\left(\frac{1}{n}\sum_{i=1}^{n}(x_i - \theta)\right)$$

$$= \frac{1}{\sqrt{n}}\sum_{i=1}^{n}(x_i - \theta).$$

Based on the central limit theorem, the above RV converges to a distribution with a Gaussian RV:

$$\mathbb{E}_\theta\left[\sqrt{n}(\overline{x} - \theta)\right] = 0,$$

$$\mathrm{var}_\theta\left(\sqrt{n}(\overline{x} - \theta)\right) = \theta.$$

8.4.4 Cramer-Rao Bound (CRB) in the Estimation Variance

Cramer-Rao (Cramer-Rao bound - CRB) can be defined for random and non-random parameters. However, CRB is more useful for non-random parameters, as it can be used to determine the optimal or almost optimal performance of a prospective unbiased estimator. Unlike the non-random case, for the random parameters the optimum estimator, and its MSE are functions of the known joint PDF of θ and $\mathbf{x}$. Thus, there are more accurate alternative solutions than the CRB for approximating the MSE estimator, most of which apply the approximation of an integral that represents the relation of the minimum mean square error. Consequently, the case of the non-random parameters will be discussed next.

Let $\theta \in \Theta$ describe a non-random scalar value and:

1. Θ describe an open subset in the $\mathbb{R}$ domain.
2. $f(\mathbf{x};\theta)$ describe a smooth and differentiable in θ function.

What follows is the definition of CRB for a scalar, θ.
For each unbiased estimator $\hat{\theta}$ of θ:

$$\text{var}_\theta(\hat{\theta}) \geq 1/F(\theta), \tag{8.8}$$

where the equivalence occurs if and only if for some non-random value k_θ:

$$\frac{\partial}{\partial\theta}\ln f(\mathbf{x};\theta) = k_\theta(\hat{\theta}-\theta). \tag{8.9}$$

Here, k_θ expresses a constant that depends on θ and not on $\mathbf{x}$. CRB is considered a very tight bound, and the Relation 8.9 reflects the so-called CRB tightness condition.

The parameter $F(\theta)$, in the CRB, is the Fisher information, which is expressed in two equivalent relations:

$$F(\theta) = \mathbb{E}_\theta\left[\left(\frac{\partial}{\partial\theta}\ln f(\mathbf{x};\theta)\right)^2\right]$$
$$= -\mathbb{E}_\theta\left[\frac{\partial^2}{\partial\theta^2}\ln f(\mathbf{x};\theta)\right].$$

The latter can be used to indicate that the value k_θ in the tightness assurance condition (8.9) is actually equal to $F(\theta)$.

Before proceeding with some examples, consider a simpler version of CRB. There are three steps to express the CRB. The first step is to notice that the expected value of the derivative of the logarithmic function is equal to zero:

$$\mathbb{E}_\theta\left[\partial\ln f_\theta(\mathbf{x})/\partial\theta\right] = \mathbb{E}_\theta\left[\frac{\partial f_\theta(\mathbf{x})/\partial\theta}{f_\theta(\mathbf{x})}\right]$$
$$= \int\frac{\partial}{\partial\theta}f_\theta(\mathbf{x})d\mathbf{x}$$
$$= \frac{\partial}{\partial\theta}\underbrace{\int f_\theta(\mathbf{x})d\mathbf{x}}_{=1}$$
$$= 0.$$

The next step is to show that the correlation between the derivative of the logarithmic like-lihood function and the estimator is a constant:

$$\mathbb{E}_\theta\left[\left(\hat{\theta}(\mathbf{x})-\mathbb{E}_\theta[\hat{\theta}]\right)\left(\partial\log f_\theta[\mathbf{x}]/\partial\theta\right)\right]=\int\left(\hat{\theta}(\mathbf{x})-\mathbb{E}_\theta[\hat{\theta}]\right)\frac{\partial}{\partial\theta}f_\theta(\mathbf{x})\,d\mathbf{x}$$

$$=\frac{\partial}{\partial\theta}\underbrace{\int\hat{\theta}(\mathbf{x})f_\theta(\mathbf{x})\,d\mathbf{x}}_{=\mathbb{E}_\theta[\hat{\theta}]=\theta}$$

$$=1,$$

where the result of the second equation of the previous step was used. Finally, we apply the Cauchy-Schwarz $\left(\mathbb{E}^{\,2}[uv]\le\mathbb{E}\,[u^2]\mathbb{E}\,[v^2]\right)$ inequality so that:

$$1=\mathbb{E}_\theta^2\left[\left(\hat{\theta}(\mathbf{x})-\mathbb{E}_\theta[\hat{\theta}]\right)\left(\partial\log f_\theta[\mathbf{x}]/\partial\theta\right)\right]$$

$$\le\mathbb{E}_\theta\left[\left(\hat{\theta}(\mathbf{x})-\mathbb{E}_\theta[\hat{\theta}]\right)^2\right]\mathbb{E}_\theta\left[\left(\partial\log f_\theta[\mathbf{x}]/\partial\theta\right)^2\right]$$

$$=\mathrm{var}_\theta(\hat{\theta})F(\theta).$$

Equality is true if and only if $u = kv$ for some non-random constant, k. This result is consistent with the Relation 8.8 and gives the CRB.

Example 17: The CRB for the Poisson Rate

Let $\mathbf{x} = [x_1,\cdots,x_n]$, a vector with IID Poisson RV:

$$x_i \sim f(x;\theta)=\frac{\theta^x}{x!}e^{-\theta},\quad x=0,1,2,\dots$$

In order to find the CRB, we must first calculate the Fisher information. So it is:

$$\ln f(\mathbf{x};\theta)=\sum_{k=1}^n x_k\ln\theta-n\theta-\underbrace{\sum_{k=1}^n\ln x_k!}_{\text{constant in }\theta}$$

and factorizing twice:

$$\partial\ln f(\mathbf{x};\theta)/\partial\theta=\frac{1}{\theta}\sum_{k=1}^n x_k-n,$$

$$\partial^2\ln f(\mathbf{x};\theta)/\partial\theta^2=-\frac{1}{\theta^2}\sum_{k=1}^n x_k.$$

(8.10)

As $\mathbb{E}\left[\sum_{k=1}^{n} x_k\right] = n\theta$, the Fisher information, with conditional n samples, is expressed as:

$$F_n(\theta) = \frac{n}{\theta}.$$

The CRB for each unbiased estimator of Poisson rate θ yields:

$$\mathrm{var}_\theta(\hat{\theta}) \geq \frac{\theta}{n}$$

We note the following:

1. From Example 14 we know that the mean value of samples $\bar{x}$ is an unbiased statistical quantity with variance $\mathrm{var}\,\theta(\bar{x}) = \theta/n$. This coincides with the CRB.
2. We can understand if the impartial estimator achieves the CRB, alternatively. From the first equivalence of Relation 8.10, we note the tightness assurance condition of Relation 8.9:

$$\partial \ln f(\mathbf{x};\theta)/\partial\theta = \frac{1}{\theta}\sum_{k=1}^{n} x_k - n = \underbrace{\frac{n}{\theta}}_{k_\theta}(\bar{x} - \theta). \qquad (8.11)$$

Then, the variance of $\bar{x}$ can be expressed by calculating the CRB. This alternative way may be simpler than the direct estimation of the estimator variance.

3. The expected value on the right part of Relation 8.11 is zero because $\hat{\theta}$ is unbiased. This implies that:

$$\mathbb{E}_\theta\left[\partial \ln f(\mathbf{x};\theta)/\partial\theta\right] = 0.$$

This is because the derivative of the logarithmic function at θ is an unbiased estimator at the zero when θ is the real parameter. This observation is general and applies to any smooth and differentiable PDF.

GENERAL PROPERTIES OF THE SCALAR CRB

First Property:
The Fisher information is a metric with respect to the mean (negative) curvature of the logarithmic likelihood function $\ln f(\mathbf{x};\theta)$ around the real value of the θ parameter (Figure 8.19).

Second Property:
Let $F_n(\theta)$ the Fisher information for an IID RV vector, $\mathbf{x} = [x_1,\cdots,x_n]$. It is true that:

$$F_n(\theta) = nF_1(\theta).$$

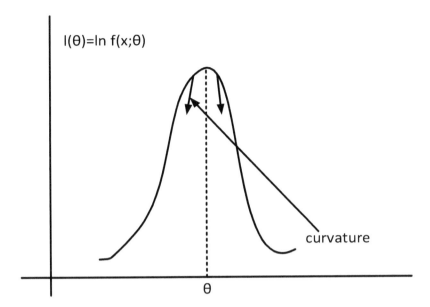

FIGURE 8.19
The curvature of the logarithmic likelihood function $\ln f(x; \theta)$ around the real value of the parameter θ.

Therefore, for smooth likelihood functions and unbiased estimators, the variance $\mathrm{var}_\theta(\hat{\theta})$ cannot be attenuated faster than $1/n$ rate.

Proof of the second property:
Since $\mathbf{x} = [x_1, \cdots, x_n]$ is IID, it is true that

$$f(\mathbf{x};\theta) = \prod_{i=1}^{n} f\left(x_i;\theta\right),$$

so as:

$$
\begin{aligned}
F_n(\theta) &= -\mathbb{E}\left[\frac{\partial^2}{\partial\theta^2}\ln f(\mathbf{x};\theta)\right] \\
&= -\mathbb{E}\left[\sum_{i=1}^{n}\frac{\partial^2}{\partial\theta^2}\ln f(x_i;\theta)\right] \\
&= \sum_{i=1}^{n}\underbrace{-\mathbb{E}\left[\frac{\partial^2}{\partial\theta^2}\ln f(x_i;\theta)\right]}_{F_1(\theta)}.
\end{aligned}
$$

For unbiased estimators, the CRB determines an unreachable range of variance as a function of n (Figure 8.20). Efficient unbiased estimators $\hat{\theta} = \hat{\theta}(x_1, \cdots x_n)$ present a variance ranging up to $\mathrm{var}_\theta(\hat{\theta}) = \mathcal{O}(1/n)$.

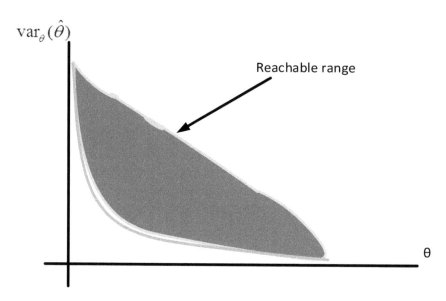

FIGURE 8.20
The CRB determines a unreachable range of variance, which is located under the CRB curve, appearing as the non-shaded area. Efficient unbiased estimators, with constant parameters, exhibit a rate of variance attenuation $1/n$.

Third Property:

If $\hat{\theta}$ is unbiased and $\mathrm{var}_{\theta}(\hat{\theta})$ approaches the CRB for all θ, $\hat{\theta}$ is called an efficient estimator. Efficient estimators are always UMVU (note: the reverse is not true). In addition, if an estimator is asymptotically unbiased and its variance is attenuated at the optimum rate:

$$\lim_{n\to+\infty} b_{\theta}(\hat{\theta}) = 0, \quad \lim_{n\to+\infty} n\,\mathrm{var}_{\theta}(\hat{\theta}) = 1/F_1(\theta),$$

where F_1 is the Fisher information given the sample x_i, $\hat{\theta}$ is considered an asymptotically efficient estimator.

The exponential distribution family (or category) plays a major role in this efficiency. More specifically, if x is a sample of a PDF belonging to the exponential distribution family with a scalar parameter θ, then:

$$\theta = \mathbb{E}_{\theta}[t(x)], \tag{8.12}$$

$$F(\theta) = 1/\mathrm{var}_{\theta}\big(t(x)\big), \tag{8.13}$$

where $F(\theta)$ is the the Fisher information with conditional sample x. Therefore, if we have a vector of samples of IID RVs from such a PDF, then $\hat{\theta} = n^{-1}\sum_{i=1}^{n} t(x_i)$ is an unbiased and efficient estimator of θ.

Fourth Property:
Efficient estimators for θ exist only when the probabilistic models describing them belong to the exponential distribution family:

$$f(x;\theta) = a(\theta)b(x)e^{-c(\theta)t(x)},$$

as well as when $\mathbb{E}_\theta[t(x)] = \theta$, i.e., the PDF is subject to a parameterization of its expected value.

Proof of fourth property:
For simplicity (and without loss of generalization of property) we study the case of a single sample $n = 1$ and $\Theta = (-\infty, +\infty)$. Let us recall, at this point, that in order for the equality to be true at the CRB, the PDF must be expressed in the form of:

$$\frac{\partial}{\partial\theta}\ln f(x;\theta) = k_\theta(\hat{\theta} - \theta). \tag{8.14}$$

For constant θ_0, we integrate the left part of Equation 8.14 in $\theta \in [\theta_0, \theta']$:

$$\int_{\theta_0}^{\theta'} \frac{\partial}{\partial\theta}\ln f(x;\theta)\,dx = \ln f(x;\theta') - \ln f(x;\theta_0).$$

On the other hand, integrating the right-hand part of Equation 8.14 yields:

$$\int_{\theta_0}^{\theta'} k_\theta(\hat{\theta} - \theta)\,d\theta = \hat{\theta}\underbrace{\int_{\theta_0}^{\theta'} k_\theta\,d\theta}_{c(\theta')} - \underbrace{\int_{\theta_0}^{\theta'} k_\theta\theta\,d\theta}_{d(\theta')}.$$

By combining the above equations, it is obtained that:

$$f(x;\theta) = \underbrace{e^{-d(\theta)}}_{a(\theta)} \underbrace{f(x;\theta_0)}_{b(x)} e^{-c(\theta)\overset{t(x)}{\overset{\frown}{\hat{\theta}}}}.$$

The above properties are studied in more detail through the following examples.

Example 18: Parameter Estimation for Exponential PDF

A non-negative RV x has an exponential distribution with an expected value θ if the PDF of x is of the form $f(x;\theta) = \theta^{-1}\exp(-x/\theta)$, where $\theta > 0$. This exponential RV is often used in practice to model service time or waiting time in communications networks and other queuing systems. We can easily show that this PDF belongs to the exponential family of distributions substituting $a(\theta) = \theta^{-1}$, $b(x) = I_{[0,+\infty)}(x)$, $c(\theta) = -\theta^{-1}$ and $t(x) = x$. Since $\mathbb{E}_\theta[x] = \theta$, the PDF is subject to expected value parameterization, and therefore it can be concluded that the mean sample value $\bar{x}$ is an unbiased estimator of θ. Moreover, it is an efficient estimator and therefore is UMVU when n IID samples $\mathbf{x} = [x_1, \cdots, x_n]$ are available.

NOTE: From the above we cannot conclude that the statistical quantity $1/\bar{x}$ is an efficient estimator of $1/\theta$.

Example 19: Let the Vector IID x with $x_i \sim \mathcal{N}(\theta, \sigma^2)$

The "bell" curve of the Gaussian PDF is met in a variety of applications, making it a benchmark. The use of this model is usually justified by invoking the central limit theorem, describing measurements or measurement noise as the sum of many small components, e.g., random atomic collisions, light diffraction, and aggregation of repeated measurements.

Our first objective will be to find the ML and CRB estimator for estimating the mean value θ of the scalar Gaussian RV with a known variance, σ^2. As the Gaussian PDF with an unknown mean value belongs to the exponential family, we can follow the same approach as above to find effective estimators. But let's follow an alternative process to verify this fact.

$$f(\mathbf{x};\theta) = \left(\frac{1}{\sqrt{2\pi\sigma^2}}\right)^n \exp\left(-\frac{1}{2\sigma^2}\sum_{i=1}^{n}(x_k - \theta)^2\right),$$

or

$$\ln f(\mathbf{x};\theta) = -\frac{n}{2}\ln(\sigma^2) - \frac{1}{2\sigma^2}\sum_{i=1}^{n}(x_k - \theta)^2 + c,$$

where c is a constant. Calculating the first derivative yields

$$\partial \ln f / \partial\theta = \frac{1}{\sigma^2}\sum_{i=1}^{n}(x_k - \theta)$$

$$= \underbrace{\frac{n}{\sigma^2}}_{k_\theta}\left(\bar{x}_i - \theta\right). \tag{8.15}$$

Therefore, the CRB tightness condition is valid. It is also concluded that $\bar{x}_i$ is the optimal estimator for the mean value of a Gaussian sample.

By taking the first derivative of the logarithmic likelihood function with respect to θ and reversing it, we confirm what we already know about the variance of the mean value of the sample:

$$\text{var}_\theta(\bar{x}) = 1/F_n(\theta) = \sigma^2/n.$$

The first equality occurs because we already know that $\bar{x}$ is an efficient estimator.

8.5 Estimation of Multiple Non-Random (Constant) Parameters

In this section, we are studying the general problem of multiple unknown parameters. This problem is different from the previous case of multiple random parameters since there is no joint a priori PDF for further processing. First, we model all unknown parameters in a vector:

$$\boldsymbol{\theta} = [\theta_1, \cdots, \theta_p]^T,$$

and we describe the problem as the derivation of a vector estimator $\hat{\boldsymbol{\theta}}$ for $\boldsymbol{\theta}$.

The joint PDF for the measured vector $\mathbf{x}$ is defined as:

$$f(\mathbf{x}; \theta_1, \cdots, \theta_p) = f(\mathbf{x}; \boldsymbol{\theta}).$$

Criteria for estimator performance evaluation:
We define the biased vector estimator:

$$b_{\boldsymbol{\theta}}(\hat{\boldsymbol{\theta}}) = \mathbb{E}_{\boldsymbol{\theta}}[\hat{\boldsymbol{\theta}}] - \boldsymbol{\theta},$$

and the symmetric matrix of estimator covariance:

$$\text{cov}_{\boldsymbol{\theta}}(\hat{\boldsymbol{\theta}}) = \mathbb{E}_{\boldsymbol{\theta}}\left[\left(\hat{\boldsymbol{\theta}} - \mathbb{E}[\hat{\boldsymbol{\theta}}]\right)\left(\hat{\boldsymbol{\theta}} - \mathbb{E}[\hat{\boldsymbol{\theta}}]\right)^T \right]$$

$$= \begin{bmatrix} \text{var}_{\boldsymbol{\theta}}(\hat{\boldsymbol{\theta}}_1) & \text{cov}_{\boldsymbol{\theta}}(\hat{\boldsymbol{\theta}}_1, \hat{\boldsymbol{\theta}}_2) & \cdots & \text{cov}_{\boldsymbol{\theta}}(\hat{\boldsymbol{\theta}}_1, \hat{\boldsymbol{\theta}}_p) \\ \text{cov}_{\boldsymbol{\theta}}(\hat{\boldsymbol{\theta}}_2, \hat{\boldsymbol{\theta}}_1) & \text{var}_{\boldsymbol{\theta}}(\hat{\boldsymbol{\theta}}_2) & \ddots & \vdots \\ \vdots & \ddots & \ddots & \vdots \\ \text{cov}_{\boldsymbol{\theta}}(\hat{\boldsymbol{\theta}}_p, \hat{\boldsymbol{\theta}}_1) & \cdots & \cdots & \text{var}_{\boldsymbol{\theta}}(\hat{\boldsymbol{\theta}}_p) \end{bmatrix}.$$

This matrix is often called a variance-covariance matrix.

In many cases, only the diagonal entries of the covariance matrix of the estimator are a point of interest. However, as we will see below, the entire covariance matrix of the estimator is very useful for generalizing the scalar CRB case.

Also, the estimator concentration is defined as:

$$P_{\boldsymbol{\theta}}\left(\|\hat{\boldsymbol{\theta}} - \boldsymbol{\theta}\| > \varepsilon\right) = \int\limits_{\|\hat{\boldsymbol{\theta}} - \boldsymbol{\theta}\| > \varepsilon} f(\hat{\boldsymbol{\theta}}; \boldsymbol{\theta}) d\hat{\boldsymbol{\theta}}$$

$$= \int\limits_{\{\mathbf{x}: \|\hat{\boldsymbol{\theta}} - \boldsymbol{\theta}\| > \varepsilon\}} f(\mathbf{x}; \boldsymbol{\theta}) d\mathbf{x}.$$

The first step is the generalization of the CRB from the scalar case (of one parameter) to the vector case, called a CRB matrix.

8.5.1 Cramer-Rao (CR) Matrix Bound in the Covariance Matrix

Let $\boldsymbol{\theta} \in \Theta$ be a $p \times 1$ vector and:

1. Θ be an open subset in $\mathbb{R}^p$.
2. $f(\mathbf{x};\boldsymbol{\theta})$ be smooth and differentiable at $\boldsymbol{\theta}$.
3. The coefficients $\mathrm{cov}_{\boldsymbol{\theta}}(\hat{\boldsymbol{\theta}})$ and $\mathbf{F}(\boldsymbol{\theta})$ (to be subsequently specified) are non-singular matrices.

The CRB matrix for vector parameters is defined below. For each unbiased estimator $\hat{\boldsymbol{\theta}}$ of $\boldsymbol{\theta}$:

$$\mathrm{cov}_{\boldsymbol{\theta}}(\hat{\boldsymbol{\theta}}) \geq \mathbf{F}^{-1}(\boldsymbol{\theta}), \tag{8.16}$$

where equality is true if and only if the following relation is satisfied for a non-random matrix $\mathbf{K}_{\boldsymbol{\theta}}$:

$$\mathbf{K}_{\boldsymbol{\theta}} \nabla_{\boldsymbol{\theta}} \ln f(\mathbf{x};\boldsymbol{\theta}) = \hat{\boldsymbol{\theta}} - \boldsymbol{\theta}. \tag{8.17}$$

If the tightness condition of Relation 8.17 is satisfied, the estimator $\hat{\boldsymbol{\theta}}$ is called an efficient vector estimator. In the CRB matrix of Relation 8.16, $\mathbf{F}(\boldsymbol{\theta})$ is the Fisher information matrix, which takes one of two following formats:

$$\mathbf{F}(\boldsymbol{\theta}) = \mathbb{E}\left[\left(\nabla_{\boldsymbol{\theta}} \ln f(\mathbf{x};\boldsymbol{\theta})\right)\left(\nabla_{\boldsymbol{\theta}} \ln f(\mathbf{x};\boldsymbol{\theta})\right)^T\right]$$
$$= -\mathbb{E}\left[\nabla_{\boldsymbol{\theta}}^2 \ln f(\mathbf{x};\boldsymbol{\theta})\right].$$

where the factorization operator is defined:

$$\nabla_{\boldsymbol{\theta}} = \left[\frac{\partial}{\partial \theta_1}, \cdots, \frac{\partial}{\partial \theta_p}\right]^T,$$

and the Hessian symmetric matrix operator:

$$\nabla_{\boldsymbol{\theta}}^2 = \begin{bmatrix} \dfrac{\partial^2}{\partial \theta_1^2} & \dfrac{\partial^2}{\partial \theta_1 \theta_2} & \cdots & \dfrac{\partial^2}{\partial \theta_1 \theta_p} \\ \dfrac{\partial^2}{\partial \theta_2 \theta_1} & \dfrac{\partial^2}{\partial \theta_2^2} & \ddots & \vdots \\ \vdots & \ddots & \ddots & \vdots \\ \dfrac{\partial^2}{\partial \theta_p \theta_1} & \cdots & \cdots & \dfrac{\partial^2}{\partial \theta_p^2} \end{bmatrix}.$$

The CRB matrix of Relation 8.16 has some other properties in relation to the scalar CRB.

First Property:
The inequality in the matrix bound must be interpreted in the sense of positive definiteness. In particular, if $\mathbf{A}$, $\mathbf{B}$ are $p \times p$ matrices:

$$\mathbf{A} \geq \mathbf{B} \Leftrightarrow \mathbf{A} - \mathbf{B} \geq 0,$$

where $\mathbf{A} - \mathbf{B} \geq 0$ reflects that the matrix $\mathbf{A} - \mathbf{B}$ is non-negative defined. This implies that:

$$\mathbf{z}^T (\mathbf{A} - \mathbf{B})\mathbf{z} \geq 0,$$

for each vector $\mathbf{z} \in \mathbb{R}^p$ and all $\mathbf{A} - \mathbf{B}$ eigenvalues are non-negative. For example, choosing $\mathbf{z} = [1,0,\cdots,0]^T$ and $\mathbf{z} = [1,1,\cdots,1]^T$, respectively, $\mathbf{A} \geq \mathbf{B}$, $\mathbf{A} \geq \mathbf{B}$ implies that:

$$a_{ii} \geq b_{ii} \text{ and } \sum_{i,j} a_{i,j} \geq \sum_{i,j} b_{i,j}.$$

However, $\mathbf{A} \geq \mathbf{B}$ does not mean that $a_{ij} \geq b_{ij}$ is generally true. A simple numerical example is the following. Suppose that $0 < \rho < 1$ and:

$$\underbrace{\begin{bmatrix} 2 & 0 \\ 0 & 2 \end{bmatrix}}_{A} - \underbrace{\begin{bmatrix} 1 & \rho \\ \rho & 1 \end{bmatrix}}_{B} = \begin{bmatrix} 1 & -\rho \\ -\rho & 1 \end{bmatrix},$$

where two eigenvalues $1 - \rho > 0$ and $1 + \rho > 0$ are present. Therefore, it is derived that $\mathbf{A} - \mathbf{B} > 0$, as $a_{1,2} = 0 \neq \rho$.

Second Property:
The inequality matrix of Relation 8.16 is also true for the scalar CRB that expresses the variance of the i-th element of the unbiased vector estimator $\hat{\boldsymbol{\theta}}$:

$$\text{var}_\theta(\hat{\theta}_i) \geq \left[\mathbf{F}^{-1}(\boldsymbol{\theta}) \right]_{ii},$$

where the right part of the above inequality denotes the i-th element of the diagonal of the inverse Fisher information matrix.

Third Property:
The Fisher information matrix is a metric that expresses the mean curvature of the logarithmic likelihood function around $\boldsymbol{\theta}$.

Fourth Property:
Let $\mathbf{F}_n(\boldsymbol{\theta})$ be the Fisher information for a vector sample with IID measurements $x_1, \cdots, x_n$. Then, it is true that:

$$\mathbf{F}_n(\boldsymbol{\theta}) = n\,\mathbf{F}_1(\boldsymbol{\theta}).$$

Thus:

$$\mathrm{var}_\theta(\hat{\boldsymbol{\theta}}) = \mathcal{O}(1/n).$$

Fifth Property:
Efficient vector estimators exist only for multivariate parameters belonging to the exponential distribution family with expected value parameterization:

$$f(\mathbf{x};\boldsymbol{\theta}) = a(\boldsymbol{\theta})b(x)e^{-\left[c(\theta)^T\right]\left[t(x)\right]},$$

and

$$\mathbb{E}_\theta\left[\mathbf{t}(x)\right] = \boldsymbol{\theta}.$$

In addition, in this case, $\mathbb{E}\left[n^{-1}\sum_{i=1}^{n}\mathbf{t}(x_i)\right] = \theta$, and $\hat{\boldsymbol{\theta}} = n^{-1}\sum_{i=1}^{n}\mathbf{t}(x_i)$ is an unbiased and efficient estimator of $\boldsymbol{\theta}$.

Sixth Property:
$\hat{\boldsymbol{\theta}}$ satisfies the relation:

$$\nabla_\theta \ln f = \mathbf{K}_\theta(\hat{\boldsymbol{\theta}} - \boldsymbol{\theta}),$$

for a non-random matrix, $\mathbf{K}_\theta$, we conclude that:

1. $\hat{\boldsymbol{\theta}}$ is an unbiased statistical quantity as it is true that:

$$\mathbb{E}_\theta\left[\nabla_\theta \ln f(\mathbf{x};\boldsymbol{\theta})\right] = 0.$$

2. $\hat{\boldsymbol{\theta}}$ is an efficient estimator, and therefore its elements are UMVU.
3. The covariance $\hat{\boldsymbol{\theta}}$ is given by the inverse Fisher information matrix.
4. The matrix $\mathbf{K}_\theta$ is the Fisher information $\mathbf{F}(\boldsymbol{\theta})$ since it is true that:

$$\mathbb{E}_\theta\left[\nabla_\theta^2 \ln f(x;\boldsymbol{\theta})\right] = \mathbb{E}_\theta\left[\nabla_\theta^T\nabla_\theta \ln f(x;\boldsymbol{\theta})\right] = \mathbb{E}_\theta\left[\nabla_\theta\{\mathbf{K}_\theta(\hat{\boldsymbol{\theta}} - \boldsymbol{\theta})\}\right],$$

and via the chain rule and the unbiased property of $\hat{\boldsymbol{\theta}}$:

$$\mathbb{E}_\theta\left[\nabla_\theta\{\mathbf{K}_\theta(\hat{\boldsymbol{\theta}} - \boldsymbol{\theta})\}\right] = \nabla_\theta\{\mathbf{K}_\theta\}\mathbb{E}_\theta\left[\hat{\boldsymbol{\theta}} - \boldsymbol{\theta}\right] + \mathbf{K}_\theta\mathbb{E}_\theta\left[\nabla_\theta(\hat{\boldsymbol{\theta}} - \boldsymbol{\theta})\right] = -\mathbf{K}_\theta.$$

5. The covariance of the estimator is expressed as:

$$\text{cov}_{\boldsymbol{\theta}}(\hat{\boldsymbol{\theta}}) = \mathbf{K}_{\boldsymbol{\theta}}^{-1}.$$

Proof of the CRB matrix:
There are three steps to the proof, which, with one exception, is the direct generalization of proof of the scalar CRB: (1) deducing that the derivative of the logarithmic likelihood function is zero, (2) deducing that the correlation of the derivative of the logarithmic likelihood function and estimator is a constant, and (3) deducing that the derivative covariance matrix and estimator error implies a specific relation between the Fisher information and the estimator covariance.

Step 1: Verification of $\mathbb{E}_{\boldsymbol{\theta}}\left[\nabla_{\boldsymbol{\theta}} \ln f(\mathbf{x};\boldsymbol{\theta})\right] = 0.$

$$\Rightarrow \mathbb{E}_{\boldsymbol{\theta}}\left[\frac{1}{f(\mathbf{x};\boldsymbol{\theta})} \nabla_{\boldsymbol{\theta}} f(\mathbf{x};\boldsymbol{\theta})\right] = \int_{\mathcal{X}} \nabla_{\boldsymbol{\theta}} f(\mathbf{x};\boldsymbol{\theta}) d\mathbf{x}$$

$$= \nabla_{\boldsymbol{\theta}} \underbrace{\int_{\mathcal{X}} f(\mathbf{x};\boldsymbol{\theta}) d\mathbf{x}}_{1} = 0.$$

Step 2: Verification of $\mathbb{E}_{\boldsymbol{\theta}}\left[\nabla_{\boldsymbol{\theta}} \ln f(\mathbf{x};\boldsymbol{\theta})(\hat{\boldsymbol{\theta}} - \boldsymbol{\theta})^T\right] = \mathbf{I}.$
We note that:

$$\mathbb{E}_{\boldsymbol{\theta}}\left[\nabla_{\boldsymbol{\theta}} \ln f(\mathbf{x};\boldsymbol{\theta})(\hat{\boldsymbol{\theta}} - \boldsymbol{\theta})^T\right] = \mathbb{E}_{\boldsymbol{\theta}}\left[\frac{1}{f(\mathbf{x};\boldsymbol{\theta})} \nabla_{\boldsymbol{\theta}} f(\mathbf{x};\boldsymbol{\theta}) \hat{\boldsymbol{\theta}}^T\right]$$

$$= \int_{\mathcal{X}} \nabla_{\boldsymbol{\theta}} f(\mathbf{x};\boldsymbol{\theta}) \hat{\boldsymbol{\theta}}^T \mathbf{x} \, d\mathbf{x}$$

$$= \nabla_{\boldsymbol{\theta}} \underbrace{\int_{\mathcal{X}} f(\mathbf{x};\boldsymbol{\theta}) \hat{\boldsymbol{\theta}}^T \mathbf{x} \, d\mathbf{x}}_{\mathbb{E}_{\boldsymbol{\theta}}[\hat{\boldsymbol{\theta}}^T] = \boldsymbol{\theta}^T} = \mathbf{I}.$$

Therefore, combining the above result with that of the previous step, it is derived that:

$$\mathbb{E}_{\boldsymbol{\theta}}\left[\nabla_{\boldsymbol{\theta}} \ln f(\mathbf{x};\boldsymbol{\theta})(\hat{\boldsymbol{\theta}} - \boldsymbol{\theta})^T\right]$$

$$= \underbrace{\mathbb{E}_{\boldsymbol{\theta}}\left[\nabla_{\boldsymbol{\theta}} \ln f(\mathbf{x};\boldsymbol{\theta}) \hat{\boldsymbol{\theta}}^T\right]}_{=\mathbf{I}} - \underbrace{\mathbb{E}_{\boldsymbol{\theta}}\left[\nabla_{\boldsymbol{\theta}} \ln f(\mathbf{x};\boldsymbol{\theta})\right] \boldsymbol{\theta}^T}_{=0}.$$

Step 3: Let a $2p \times 1$ RV vector:

$$\mathbf{u} = \begin{bmatrix} \hat{\boldsymbol{\theta}} - \boldsymbol{\theta} \\ \nabla_{\boldsymbol{\theta}} \ln f(\mathbf{x};\boldsymbol{\theta}) \end{bmatrix}. \tag{8.18}$$

Since any matrix expressed as an outer product of two vectors is positively defined (i.e., $\mathbb{E}_{\theta}\left[\mathbf{u}\mathbf{u}^{T}\right] \geq 0$), and using the results of the two previous steps, it is derived that:

$$\mathbb{E}_{\theta}[\mathbf{u}\mathbf{u}^{T}] = \begin{bmatrix} \operatorname{cov}_{\theta}(\hat{\theta}) & \mathbf{I} \\ \mathbf{I} & \mathbf{F}(\theta) \end{bmatrix} \geq 0.$$

We still have to apply the result of Appendix I to the above partitioned matrix so that:

$$\operatorname{cov}_{\theta}(\hat{\theta}) - \mathbf{F}^{-1}(\theta) \geq 0.$$

We will then prove that the tightness criterion ensures the equality in the CRB. First of all, it is true that if $\operatorname{cov}_{\theta}(\hat{\theta}) = \mathbf{F}^{-1}$, then the class of matrix $\mathbb{E}[\mathbf{u}\mathbf{u}^{T}]$ is p. This can only happen if the RV vector of Relation 8.18 has p independent linear coefficients. As $\operatorname{cov}_{\theta}(\hat{\theta})$ and $\mathbf{F}(\theta)$ are considered non-singular parameters, vector $\hat{\theta} - \theta$ as well as $\nabla_{\theta} \ln f$ has no linear correlations.

Therefore, it is:

$$\mathbf{K}_{\theta} \nabla_{\theta} \ln f = \hat{\theta} - \theta,$$

for a non-random matrix. It follows that the derivative of the logarithmic likelihood function is located in the error range of the estimator.

8.5.2 Methods of Vector Estimation through Statistical Moments

Let $m_{k} = m_{k}(\theta)$ a moment of k-th order of $f(x;\theta)$. The vector that expresses the process of estimating statistical moments involves K moments so that the vector function of $\theta \in \mathbb{R}^{p}$:

$$\mathbf{g}(\theta) = \left[m_{1}(\theta), \cdots, m_{p}(\theta) \right],$$

can be reversed, i.e., a unique θ exists that satisfies:

$$\theta = \mathbf{g}^{-1}(m_{1}, \cdots, m_{K}).$$

As in the scalar case, the moment estimator is constructed by substituting m_{k} with its corresponding empirical estimation:

$$\hat{\theta} = \mathbf{g}^{-1}\left(\hat{m}_{1}, \cdots, \hat{m}_{K} \right).$$

where

$$\hat{m}_{k} = \frac{1}{n} \sum_{i=1}^{n} x_{i}^{k}.$$

8.5.3 Maximum Likelihood Vector Estimation

The ML vector is an obvious generalization of the corresponding scalar case:

$$\hat{\boldsymbol{\theta}} = \arg\max_{\boldsymbol{\theta} \in \Theta} f(\mathbf{x}; \boldsymbol{\theta}).$$

For smooth likelihood functions, the ML vector has several important properties:

1. The ML vector estimator is asymptotically unbiased.
2. The ML vector estimator is consistent.
3. The ML vector estimator is invariant to any transformation:

$$\boldsymbol{\varphi} = \mathbf{g}(\boldsymbol{\theta}) \Rightarrow \hat{\boldsymbol{\varphi}} = \mathbf{g}(\hat{\boldsymbol{\theta}}).$$

4. The ML vector estimator is asymptotically efficient and, therefore, its individual elements (estimates) are UMVU asymptotic.
5. The ML vector estimators is asymptotic Gaussian:

$$\sqrt{n}(\hat{\boldsymbol{\theta}} - \boldsymbol{\theta}) \to \mathbf{z}, \quad \mathbf{z} \sim \mathcal{N}_p\left(\mathbf{0}, \mathbf{F}_1^{-1}(\boldsymbol{\theta})\right),$$

where $\mathbf{F}_1(\boldsymbol{\theta})$ is the Fisher information matrix of the sample:

$$\mathbf{F}_1(\boldsymbol{\theta}) = -\mathbb{E}_{\boldsymbol{\theta}}\left[\nabla_{\boldsymbol{\theta}}^2 \log f(x_1; \boldsymbol{\theta})\right].$$

Example 20: A Joint Estimation of the Expected Value and of the Variance of a Gaussian Sample

Let a IID sample vector $\mathbf{x} = [x_1, \cdots, x_n]$ that consists of a Gaussian RV $x_i \sim \mathcal{N}(\mu, \sigma^2)$. The unknown parameters are $\boldsymbol{\theta} = [\mu, \sigma^2]$.
The likelihood function is defined as:

$$l(\boldsymbol{\theta}) = \ln f(\mathbf{x}; \boldsymbol{\theta}) = -\frac{n}{2}\ln(\sigma^2) - \frac{1}{2\sigma^2}\sum_{k=1}^{n}(x_k - \mu)^2 + c. \tag{8.19}$$

Step 1: Approach via the statistical moments method.
It is known that $m_1 = \mu$, $m_2 = \sigma^2 + \mu^2$, so:

$$\mu = m_1, \quad \sigma^2 = m_2 - m_1^2.$$

Therefore, the moment estimator for $\boldsymbol{\theta}$ is expressed as:

$$\begin{aligned}
\hat{\boldsymbol{\theta}} &= \left[\hat{\mu}, \hat{\sigma}^2\right] \\
&= \left[\hat{m}_1, \hat{m}_2 - \hat{m}_1^2\right] \\
&= \left[\bar{x}, \overline{x^2} - \bar{x}^2\right] \\
&= \left[\bar{x}, \overline{(x - \bar{x})^2}\right].
\end{aligned}$$

As usual, we define:

$$\bar{x} = n^{-1} \sum_{k=1}^{n} x_k$$

$$\overline{(x - \bar{x})^2} = n^{-1} \sum_{k=1}^{n} (x_k - \bar{x})^2 = \frac{n-1}{n} s^2,$$

where

$$s^2 = (n-1)^{-1} \sum_{k=1}^{n} (x_k - \bar{x})^2.$$

is the sample variance.

Step 2: Approach via ML method.
Since Relation 8.19 is a concave function (because $-\nabla_\theta^2 \ln f$ is positively defined), the likelihood equation (stationary point condition) can be used to find $\theta = \hat{\theta}$:

$$0 = \nabla_\theta \ln f(\mathbf{x};\theta) = \begin{bmatrix} \dfrac{1}{\theta_2} \sum_{k=1}^{n} (x_k - \theta_1) \\[4mm] \dfrac{n/2}{\theta_2} - \dfrac{1}{2\theta_2^2} \sum_{k=1}^{n} (x_k - \theta_1)^2 \end{bmatrix}.$$

Hence:

$$\hat{\theta}_1 = \hat{\mu} = \bar{x}, \quad \hat{\theta}_2 = \hat{\sigma}^2 = \frac{n-1}{n} s^2,$$

and therefore, the statistical moment estimators and ML estimators are the same.
Bias and covariance are easy to calculate:

$$\underbrace{\mathbb{E}_\theta[\hat{\mu}] = \mu}_{\text{unbiased}},$$

$$\underbrace{\mathbb{E}_\theta[\widehat{\sigma^2}] = \left(\frac{n-1}{n}\right)\sigma^2}_{\text{biased}},$$

$$\text{var}_\theta(\bar{x}) = \sigma^2 / n,$$

$$\text{var}_\theta(\widehat{\sigma^2}) = \left(\frac{n-1}{n}\right)^2 \text{var}_\theta(s^2) = \left(\frac{2\sigma^4}{n}\right)\left(\frac{n-1}{n}\right).$$

Since the expected value and sample variance are uncorrelated to each other (Appendix I):

$$\text{cov}_\theta(\hat{\boldsymbol{\theta}}) = \begin{bmatrix} \sigma^2/n & 0 \\ 0 & \left(\dfrac{2\sigma^4}{n}\right)\left(\dfrac{n-1}{n}\right) \end{bmatrix}. \tag{8.20}$$

Then, we calculate the Fisher information using the expected value of the Hessian matrix $-\nabla_\theta^2 \ln f(\mathbf{x};\boldsymbol{\theta})$

$$\mathbf{F}(\boldsymbol{\theta}) = \begin{bmatrix} \sigma^2/n & 0 \\ 0 & n/(2\sigma^4) \end{bmatrix}, \tag{8.21}$$

by deriving the CRB:

$$\text{cov}_\theta(\hat{\boldsymbol{\theta}}) \geq \begin{bmatrix} \sigma^2/n & 0 \\ 0 & 2\sigma^4/n \end{bmatrix}. \tag{8.22}$$

Some useful comments are to follow:

1. Moments and ML estimators have covariance matrices that violate the CRB (this can be derived by comparing the data between the matrices of Relations 8.20 and 8.22). This is expected because the ML estimator is biased.
2. Let the (biased) corrected estimator for $[\mu,\sigma^2]^T$:

$$\hat{\theta} = \left[\bar{x}, s^2\right]^T.$$

This estimator is unbiased. Since it is true that: $s^2 = \left(\dfrac{n-1}{n}\right)\widehat{\sigma^2}$

$$\text{var}_\theta(s^2) = \left(\dfrac{n-1}{n}\right)^2 \text{var}_\theta(\widehat{\sigma^2}),$$

$$\text{cov}_\theta(\hat{\boldsymbol{\theta}}) \geq \begin{bmatrix} \sigma^2/n & 0 \\ 0 & \left(\dfrac{2\sigma^4}{n}\right)\left(\dfrac{n-1}{n}\right) \end{bmatrix} \geq \mathbf{F}^{-1}(\boldsymbol{\theta}).$$

We conclude that the covariance matrix of the corrected estimator no longer violates the CRB. Indeed, it is an effective estimator of value μ since:

$$\text{var}_\theta(\hat{\mu}) = [\mathbf{F}^{-1}]_{11} = \sigma^2/n.$$

However, s^2 is not an efficient estimator of σ^2 because:

$$\operatorname{var}_\theta\left(s^2\right) > \left[\mathbf{F}^{-1}\right]_{22}.$$

3. The estimator ML is asymptotic efficient when $n \to +\infty$:

$$n\operatorname{cov}_\theta(\hat{\boldsymbol{\theta}}) \geq \begin{bmatrix} \sigma^2 & 0 \\ 0 & 2\sigma^4\left(\dfrac{n-1}{n}\right) \end{bmatrix} \to \begin{bmatrix} \sigma^2 & 0 \\ 0 & 2\sigma^4 \end{bmatrix} = \mathbf{F}_1^{-1}(\boldsymbol{\theta}).$$

4. We can verify that the vector $[\hat{\mu}, \widehat{\sigma^2}]$ is asymptotic Gaussian simply by noting that:

- $\hat{\mu}, \widehat{\sigma^2}$ are independent RVs.
- $\sqrt{n}(\hat{\mu}-\mu) \sim \mathcal{N}(0,\sigma^2)$.
- $\sqrt{n}(s^2-\sigma^2) \sim \sigma^2\sqrt{n}\left(\chi_{n-1}^2/(n-1)-1\right)$.
- $\chi_v^2 \sim \mathcal{N}(v,2v), \quad v \to +\infty$.

5. Based on the CRB equality condition, it is possible to recover an efficient vector estimator (but not for $\boldsymbol{\theta}$):

$$\nabla_\theta \ln f(\mathbf{x};\boldsymbol{\theta}) = \mathbf{K}_\theta \begin{bmatrix} \bar{x}-\mu \\ \overline{x^2}-(\sigma^2+\mu^2) \end{bmatrix},$$

where

$$\mathbf{K}_\theta \triangleq \begin{bmatrix} n/\sigma^2 & 0 \\ 0 & n/(2\sigma^4) \end{bmatrix}\begin{bmatrix} 1 & 0 \\ 2\mu & 1 \end{bmatrix}^{-1}.$$

Since the sampling moments are unbiased estimates, we conclude that $\bar{x}, \overline{x^2}$ are efficient estimates of the first order moment $\mathbb{E}[x]=\mu$ and of the second order $\mathbb{E}[x^2]=\sigma^2+\mu^2$, respectively.

Example 21: $\mathbf{n} = [N_1,\cdots,N_p]^T$ is a Polynomial Random Vector

The polynomial model is a generalization of the binomial, with more than two output states "0" and "1." Let the output vector $\mathbf{z}$ one of the elementary vectors belonging to $\mathbb{R}^p$, $\mathbf{e}_1 = [1,0,\cdots,0]^T,\cdots,\mathbf{e}_p = [0,0,\cdots,1]^T$, with probabilities $\theta_1,\cdots,\theta_p$, respectively. The vector $\mathbf{e}_k$ may be a tag related to the fact that, for example, a random roll of a dice resulted in the number 6 ($p = 6$). The polynomial model describes the distribution of the sum

$$\mathbf{n} = \left[N_1,\cdots,N_p\right]^T = \sum_{i=1}^{n} z_i,$$

of the vectors obtained after n IID trials.

The probability of a particular polynomial result **n** has a mass probability function:

$$p(\mathbf{n};\boldsymbol{\theta}) = \frac{n!}{N_1! \cdots N_p!} \theta_1^{N_1} \cdots \theta_1^{N_p},$$

where $N_i \geq 0$ are integers that satisfy the equations $\sum_{i=1}^{p} N_i = n$, $\sum_{i=1}^{p} \theta_i = 1$, $\theta_i \in [0,1]$.

The moment estimator of $\boldsymbol{\theta}$ is obtained by matching the first empirical moment **n** with the first statistical moment, $\mathbb{E}_{\boldsymbol{\theta}}[\mathbf{n}] = \boldsymbol{\theta} n$. This has as a result the estimator $\hat{\boldsymbol{\theta}} = \mathbf{n}/n$ or:

$$\hat{\boldsymbol{\theta}} = \left[\frac{N_1}{n}, \cdots, \frac{N_p}{n} \right].$$

The ML estimator should then be modeled. The p parameters exist in a $p-1$ subspace in $\mathbb{R}^p$ domain due to the limitation of the total probability. In order to find the ML estimator, a re-parameterization of the problem can be used or, alternatively, the Lagrange multipliers technique. We adopt the second technique as follows.

We form the equation by including the restrictive terms as follows:

$$J(\boldsymbol{\theta}) = \ln f(\mathbf{n};\boldsymbol{\theta}) - \lambda \left(\sum_{i=1}^{p} \theta_i - 1 \right),$$

where λ is the Lagrange multiplier, to be defined below.

Since the function J is smooth and concave, we set its derivative to zero to determine the ML estimator:

$$0 = \nabla_{\boldsymbol{\theta}} J(\boldsymbol{\theta}) = \nabla_{\boldsymbol{\theta}} \left[\sum_{i=1}^{p} N_i \ln \theta_i - \lambda \theta_i \right]$$

$$= \left[\frac{N_1}{\theta_1} - \lambda, \cdots, \frac{N_p}{\theta_p} - \lambda \right].$$

Thus,

$$\hat{\theta}_i = N_i/\lambda, \quad i = 1, \cdots, p.$$

Finally, we determine λ making the estimator $\hat{\boldsymbol{\theta}}$ to satisfy the condition:

$$\sum_{i=1}^{p} N_i/\lambda = 1 \Rightarrow \lambda = \sum_{i=1}^{p} N_i = n.$$

From the solution of the equation, the ML estimator is obtained, which is the same as the moment estimator.

8.6 Handling of Nuisance Parameters

In many cases only one parameter, θ_1, is of interest, while the other unknown parameters $\theta_2,\cdots,\theta_p$ are nuisance parameters that are not a point of interest. For example, in Example 20, with unknown values of both mean and variance, the variance may not be of direct interest. In this example, we found that the covariance of the estimator is a diagonal matrix, which means that there is no correlation between the estimation errors of the mean value and the variance parameter estimation errors. As we will see below, this means that variance is a nuisance parameter, as knowledge or lack of knowledge of it does not affect the ML estimator variance for the mean value. We separate the analysis of the nuisance parameters in cases of random and non-random parameters.

Handling of random nuisance parameters:
For random values, the average loss affects the estimation errors only for $\hat{\theta}_1$:

$$\mathbb{E}\left[c(\hat{\theta}_1,\theta_1)\right] = \int_{\Theta_1} d\theta_1 \int_{\mathcal{X}} dx c(\hat{\theta}_1,\theta_1) f(x\,|\,\theta_1) f(\theta_1).$$

The a priori PDF of θ_1 is calculated from the a priori PDF at $\boldsymbol{\theta}$:

$$f(\theta_1) = \int d\theta_2 \cdots \int d\theta_p f(\theta_1,\theta_2,\cdots,\theta_p).$$

The conditional PDF of x at $\boldsymbol{\theta}$ is derived:

$$f(x\,|\,\theta_1) = \int d\theta_2 \cdots \int d\theta_p f(x\,|\,\theta_1,\theta_2,\cdots,\theta_p) f(\theta_2,\cdots,\theta_p\,|\,\theta_1),$$

resulting in the a posteriori PDF of θ_1:

$$f(\theta_1\,|\,x) = \int d\theta_2 \cdots \int d\theta_p f(\theta_1,\theta_2,\cdots,\theta_p\,|\,x).$$

Note that the estimates of $\theta_2,\cdots,\theta_p$ are not necessary for the calculation of the a posteriori PDF of θ_1. However, integration (marginalization) of the conditional PDF in the range of the parameters $\theta_2,\cdots,\theta_p$ is required but may lead to significant difficulties in realization.

Handling of non-random nuisance parameters:
This case is quite different. The average loss concerns $\hat{\theta}_1$, again, but also depends on all the other parameters:

$$\mathbb{E}_\theta[C] = \int_{\mathcal{X}} c(\hat{\theta}_1,\theta_1) f(x\,|\,\theta) dx.$$

The ML estimator of θ_1 is expressed as:

$$\hat{\theta}_1 = \arg\max_{\theta_1}\left\{\max_{\theta_2,\cdots,\theta_p} \log f(x\,|\,\theta_1,\theta_2,\cdots,\theta_p)\right\}.$$

It is worth noting that, in this case, the estimates of nuisance parameters are necessary because the aforementioned maximization process is realized by comparing all other parameters $\theta_2, \cdots, \theta_p$.

CRB predictions for non-random nuisance parameters:
As before, we consider only the case of the unbiased estimation of the first parameter θ_1 in the vector of the unknown parameters $\boldsymbol{\theta}$. The extraction of the CRB matrix in Relation 8.16 made it possible for us to assume that there were unbiased estimators of all parameters. It can be proved that this limitation is unnecessary when we are interested only in θ_1.

Let $\boldsymbol{\theta} = [\theta_1, \cdots, \theta_p]^T$ be the parameter vector of unknown values. The variance of an independent estimator $\hat{\theta}_1$ of θ_1 obeys the following rule:

$$\text{var}_\theta(\hat{\theta}) \geq \left[\mathbf{F}^{-1}(\boldsymbol{\theta}) \right]_{11}, \tag{8.23}$$

where the equality is true if and only if there is a non-random vector $\mathbf{h}_\theta$ such that:

$$\mathbf{h}_\theta^T \nabla_\theta \ln f(\mathbf{x};\boldsymbol{\theta}) = (\hat{\theta}_1 - \theta_1).$$

In Equation 8.23, it is true that:

$$\mathbf{F}(\boldsymbol{\theta}) = -\mathbb{E}\begin{bmatrix} \dfrac{\partial^2 l(\boldsymbol{\theta})}{\partial \theta_1^2} & \dfrac{\partial^2 l(\boldsymbol{\theta})}{\partial \theta_1 \partial \theta_2} & \cdots & \dfrac{\partial^2 l(\boldsymbol{\theta})}{\partial \theta_1 \partial \theta_p} \\[2ex] \dfrac{\partial^2 l(\boldsymbol{\theta})}{\partial \theta_2 \partial \theta_1} & \dfrac{\partial^2 l(\boldsymbol{\theta})}{\partial \theta_2^2} & \ddots & \vdots \\[2ex] \vdots & \ddots & \ddots & \vdots \\[2ex] \dfrac{\partial^2 l(\boldsymbol{\theta})}{\partial \theta_p \partial \theta_1} & \cdots & \cdots & \dfrac{\partial^2 l(\boldsymbol{\theta})}{\partial \theta_p^2} \end{bmatrix},$$

and $l(\boldsymbol{\theta}) = \ln f(\mathbf{x}; \boldsymbol{\theta})$.

Let the partitioned Fisher information matrix:

$$\mathbf{F}(\boldsymbol{\theta}) = \begin{bmatrix} a & \mathbf{b}^T \\ \mathbf{b} & \mathbf{C} \end{bmatrix},$$

where:
$a = -\mathbb{E}_\theta \left[\partial^2 \ln f(\mathbf{x};\boldsymbol{\theta}) / \partial \theta_1^2 \right] =$ The Fisher information for θ_1, without nuisance parameters.
$\mathbf{b} = -\mathbb{E}_\theta \left[\partial \nabla_{\theta_2, \cdots, \theta_p} \ln f(\mathbf{x};\boldsymbol{\theta}) / \partial \theta_1 \right] =$ The Fisher coupling of θ_1, with nuisance parameters.
$\mathbf{C} = -\mathbb{E}_\theta \left[\nabla^2_{\theta_2, \cdots, \theta_p} \ln f(\mathbf{x};\boldsymbol{\theta}) \right] =$ The Fisher information for the nuisance parameters.

Using the inverse property of the partitioned matrix (Appendix I), the right part of Relation 8.23 is expressed as:

$$\left[\mathbf{F}^{-1}(\boldsymbol{\theta}) \right]_{11} = \frac{1}{a - \mathbf{b}^T \mathbf{C}^{-1} \mathbf{b}}.$$

We can draw some useful conclusions from this result:

- $\left[\mathbf{F}^{-1}(\boldsymbol{\theta})\right]_{11} \geq 1/a = 1/\left[\mathbf{F}(\boldsymbol{\theta})\right]_{11}$. Consequently, the presence of nuisance parameters may reduce the performance of the estimator.
- The magnitude of the performance reduction of the estimator is proportional to the amount of the information coupling between θ_1 and $\theta_2, \cdots, \theta_p$.
- When the Fisher information matrix is diagonal, there is no reduction in the performance of the estimator.

9

Linear Estimation

In Chapter 8, several approaches to the parameter estimation process were discussed, given the PDF $f(x; \theta)$ of the measurements. All these approaches require accurate knowledge of the model, which may not always be available and/or feasible. In addition, even when the model is very reliable, these approaches usually yield estimators, which are non-linear functions of the measurements and whose implementation may be difficult; e.g., they may include the analytical maximization of a complex PDF or the analysis of its statistical moments as a function of θ. In this chapter, we present an alternative linear estimation approach that only requires knowledge of the first two moments or their empirical estimates. Although linear methods do not have the desirable properties of the Bayes optimal estimators, such as MAP and CME, they are very attractive due to their simplicity and robustness to unknown variations in higher order moments.

The theory of linear estimation begins with the assumption of random parameters by adopting a square error loss function and then seeks to minimize the mean square error in all the processes of the estimator, which are defined as linear or *affine* functions of the considered measurements. It can be proven that the linear minimum mean square error (LMMSE) problem can be redefined as the minimization of a norm in a linear vector space. This leads to an elegant geometric interpretation of the optimal LMMSE estimator via the projection theorem and the orthogonality principle of the norm minimization problems in the linear vector space. The resulting LMMSE estimator is dependent on the expected value and the variance of the measurement, the average of the parameter and the covariance of the measurements and the parameters. It is no surprise that when the measurements and parameters are jointly Gaussian distributed, the affine estimator is equivalent to the conditional optimal mean estimator. When the predicted value and variance are not a priori known, then a non-statistical linear least squares theory (LLS) can be developed a priori, leading to the problem of linear regression.

As usual, the main components for the linear estimation will be the measurement vector $\mathbf{x} = [x_1, \cdots, x_p]^T$ and the parametric vector $\boldsymbol{\theta} = [\theta_1, \cdots, \theta_p]^T$. The cases in which the latter vectors are RV realizations with known (i.e., given) or unknown their first two statistical moments are studied in the following two sections.

9.1 Constant MSE Minimization, Linear and Affine Estimation

Our objective is to derive the solution of the following minimization problem:

$$\min_{\hat{\boldsymbol{\theta}}} \mathrm{MSE}(\hat{\boldsymbol{\theta}}) = \min_{\hat{\boldsymbol{\theta}}} \mathbb{E}\left[\left\| \boldsymbol{\theta} - \hat{\boldsymbol{\theta}}(\mathbf{x}) \right\|^2 \right],$$

where the expected value is over both $\boldsymbol{\theta}$ and $\mathbf{x}$, as the minimization is limited to constant, linear, or affine functions $\hat{\boldsymbol{\theta}}$ of $\mathbf{x}$. The norm in the above equation is the Euclidean norm $\|\mathbf{u}\| = \sqrt{\mathbf{u}^T \mathbf{u}}$. Then, the simple case of scalar parameters is discussed first.

9.1.1 Optimal Constant Estimator of a Scalar RV

This is the simplest estimator because the constant estimator $\hat{\theta} = c$ does not depend on the measurements. It can be derived that the optimal constant estimator depends only on the mean value of the parameter as no additional information about the parameter distribution is necessary.

The problem is modeled as the derivation of the constant $\hat{\theta} = c$ that minimizes the MSE:

$$\text{MSE}(c) = \mathbb{E}\left[(\theta - c)^2\right].$$

SOLUTION:

The optimal constant estimator is defined as $\hat{\theta} = \mathbb{E}[\theta]$.

Since the MSE is a squared function of c, the above relation can easily be proved by setting the derivative $\dfrac{d}{dc}\text{MSE}(c)$ equal to zero. Alternatively, we can show that the following is true:

$$\begin{aligned}
\text{MSE}(\hat{\theta}) &= \mathbb{E}\left[\left((\theta - \mathbb{E}[\theta]) - (c - \mathbb{E}[\theta])\right)^2\right] \\
&= \mathbb{E}\left[(\theta - \mathbb{E}[\theta])^2\right] + (\mathbb{E}[\theta] - c)^2 - 2(\mathbb{E}[\theta] - c)\underbrace{\mathbb{E}[\theta - \mathbb{E}[\theta]]}_{=0} \\
&= \mathbb{E}\left[(\theta - \mathbb{E}[\theta])^2\right] + (\mathbb{E}[\theta] - c)^2.
\end{aligned}$$

Since only the second term of the last equation depends on c, which is a non-negative value, it is obvious that $\hat{\theta} = \mathbb{E}[\theta]$ is the optimal constant estimator. Thus,

$$\min_c = \text{MSE}(c) = \mathbb{E}\left[(\theta - \mathbb{E}[\theta])^2\right].$$

which represents the a priori variance of θ, var(θ). Since the constant estimator does not use information about the metrics, it is expected that any estimator, depending on the x-vector, will have a smaller MSE than variance var(θ).

9.2 Optimal Linear Estimator of a Scalar Random Variable

The next estimator (which is a more complicated case) depends linearly on vector **x**:

$$\hat{\theta} = \mathbf{h}^T \mathbf{x},$$

where $\mathbf{h} = [h1, \cdots, h_n]^T$ is a set of linear coefficients to be defined. For the implementation of this estimator, the second moment matrix is necessary:

$$\mathbf{M}_x = \mathbb{E}[\mathbf{x}\mathbf{x}^T],$$

as well as the cross-moment vector:

$$\mathbf{m}_{x,\theta} = \mathbb{E}[\mathbf{x}\theta].$$

Let $\mathbf{M}_x$ be the invertible matrix.
The problem is to derive the coefficient vector $\mathbf{h}$ that minimizes the MSE:

$$\text{MSE}(\mathbf{h}) = \mathbb{E}\left[(\theta - \mathbf{h}^T\mathbf{x})^2\right].$$

SOLUTION:
The LMMSE estimator is expressed as $\hat{\theta} = \mathbf{m}_{x,\theta}^T\mathbf{M}_x^{-1}\mathbf{x}$.
To derive the above result, we model the square form of the MSE with respect to $\mathbf{h}$:

$$\text{MSE}(\hat{\theta}) = \mathbb{E}\left[(\theta - \hat{\theta})^2\right] = \mathbb{E}\left[(\theta - \mathbf{h}^T\mathbf{x})^2\right]$$
$$= \mathbf{h}^T\mathbb{E}\left[\mathbf{x}\mathbf{x}^T\right]\mathbf{h} + \mathbb{E}[\theta^2] - \mathbf{h}^T\mathbb{E}\left[\mathbf{x}\theta\right] - \mathbb{E}\left[\theta\mathbf{x}^T\right]\mathbf{h}.$$

The vector $\mathbf{h}$, which minimizes this square function, can be solved by factorization:

$$\mathbf{0}^T = \nabla_{\mathbf{h}}\text{MSE}(\mathbf{h}) = \left[\frac{\partial}{\partial h_1}, \cdots, \frac{\partial}{\partial h_n}\right]\text{MSE}(\hat{\theta})$$
$$= 2\left(\mathbf{h}^T\mathbb{E}[\mathbf{x}\mathbf{x}^T] - \mathbb{E}[\theta\mathbf{x}^T]\right).$$

Therefore, the optimum vector $\mathbf{h}$ satisfies the equation:

$$\mathbb{E}[\mathbf{x}\mathbf{x}^T]\mathbf{h} = \mathbb{E}[\mathbf{x}\theta].$$

Considering the matrix $\mathbf{M}_x = \mathbb{E}[\mathbf{x}\mathbf{x}^T]$ is non-singular, this is equivalent to:

$$\mathbf{h} = \mathbf{M}_x^{-1}\mathbf{m}_{x\theta},$$

and the optimal linear estimator is derived:

$$\hat{\theta} = \mathbf{m}_{x\theta}^T\mathbf{M}_x^{-1}\mathbf{x}.$$

By substituting the above result in the expression of the MSE($\mathbf{h}$), the minimum MSE for linear estimators is defined as:

$$\text{MSE}_{min} = \mathbb{E}[\theta^2] - \mathbf{m}_{x\theta}^T\mathbf{M}_x^{-1}\mathbf{m}_{x\theta}.$$

It is noteworthy that, as the matrix $\mathbf{M}_x^{-1}$ is positively defined, the MSE cannot exceed the a priori second moment $\mathbb{E}[\theta^2]$ of θ. If the parameter has a zero mean, then $\mathbb{E}[\theta^2] = \mathbb{E}[(\theta - \mathbb{E}[\theta])^2] = \text{var}(\theta)$, i.e., the second moment is equal to the a priori variance, and the

LMMSE estimator performance is superior to that of the constant estimator $\hat{\theta} = \mathbb{E}[\theta] = 0$. However, if $\mathbb{E}[\theta] \neq 0$ then $\mathbb{E}[\theta^2] > \mathbb{E}[(\theta - \mathbb{E}[\theta])^2]$ and the LMMSE estimator may perform worse than the constant estimator. Basically, this problem occurs because the LMMSE estimator is a biased estimator of θ, in the sense that its biased mean is $\mathbb{E}[\hat{\theta}] - \mathbb{E}[\theta] \neq 0$, except if $\mathbb{E}[\theta] = 0$. The way to deal with this bias is by generalizing the category of linear estimators to that of the affine estimators.

9.3 Optimal Affine Estimator of a Scalar Random Variable θ

The affine estimator also depends linearly on the x-vector, but also includes a constant term to control bias:

$$\hat{\theta} = \mathbf{h}^T \mathbf{x} + b = \mathbf{h}^T (\mathbf{x} - \mathbb{E}[\mathbf{x}]) + c,$$

where b and $c = b + \mathbf{h}^T \mathbb{E}[\mathbf{x}]$ constitute a parameterization of the constant term and the remaining term, respectively. It is easier to use constant c in the sequel. Our objective is to derive the optimal coefficients $\{\mathbf{h} = [h_1, \cdots, h_n]^T, c\}$. To implement an affine MSE minimization estimator, the knowledge of the expected values $\mathbb{E}[\mathbf{x}]$, $\mathbb{E}[\theta]$ of the (invertible) covariance matrix is required:

$$\mathbf{R}_x = \mathrm{cov}(\mathbf{x}) = \mathbb{E}\left[\left(\mathbf{x} - \mathbb{E}[\mathbf{x}] \right) \left(\mathbf{x} - \mathbb{E}[\mathbf{x}] \right)^T \right],$$

and the cross-correlation vector:

$$\mathbf{r}_{x,\theta} = \mathrm{cov}(\mathbf{x}, \theta) = \mathbb{E}\left[\left(\mathbf{x} - \mathbb{E}[\mathbf{x}] \right) \left(\theta - \mathbb{E}[\theta] \right) \right].$$

The problem is the derivation of the vector $\mathbf{h}$ and of the constant c that minimize the MSE:

$$\mathrm{MSE}(\mathbf{h}, c) = \mathbb{E}\left[\left(\theta - \mathbf{h}^T \left(\mathbf{x} - \mathbb{E}[\mathbf{x}] \right) - c \right)^2 \right].$$

SOLUTION:

The relation $\hat{\theta} = \mathbb{E}[\theta] + \mathbf{r}_{x,\theta}^T \mathbf{R}_x^{-1} (\mathbf{x} - \mathbb{E}[\mathbf{x}])$ represents the optimal affine estimator.

To obtain the above desired result we again use the square form of the MSE with unknown $\mathbf{h}$ and c:

$$\mathrm{MSE}(\hat{\theta}) = \mathbb{E}\left[\left(\theta - \hat{\theta} \right)^2 \right] = \mathbb{E}\left[\left((\theta - c) - \mathbf{h}^T (\mathbf{x} - \mathbb{E}[\mathbf{x}]) \right)^2 \right]$$

$$= \mathbf{h}^T \underbrace{\mathbb{E}\left[\left(\mathbf{x} - \mathbb{E}[\mathbf{x}] \right) \left(\mathbf{x} - \mathbb{E}[\mathbf{x}] \right)^T \right]}_{= \mathbf{R}_x} \mathbf{h} + \mathbb{E}[(\theta - c)^2]$$

$$- \mathbf{h}^T \underbrace{\mathbb{E}\left[(\mathbf{x} - \mathbb{E}[\mathbf{x}])(\theta - c)^T \right]}_{= \mathbf{r}_{x,\theta}} - \underbrace{\mathbb{E}\left[(\theta - c)(\mathbf{x} - \mathbb{E}[\mathbf{x}])^T \right]}_{= \mathbf{r}_{\theta,x}} \mathbf{h}$$

$$= \mathbf{h}^T \mathbf{R}_x \mathbf{h} + \mathbb{E}\left[(\theta - c)^2 \right] - 2\mathbf{h}^T \mathbf{r}_{x,\theta}.$$

Note that the only dependence on constant c is via a term that is minimized by choosing $c = \mathbb{E}[\theta]$.

With respect to vector $\mathbf{h}$, minimization occurs through factorization:

$$0 = \nabla_{\mathbf{h}}\text{MSE} = \mathbf{h}^T \mathbf{R}_x - \mathbf{r}_{\theta,x},$$

leading to the minimization equation for $\mathbf{h}$:

$$\mathbf{R}_x \mathbf{h} = \mathbf{r}_{x,\theta}.$$

When the matrix is non-singular, this is equivalent to:

$$\mathbf{h} = \mathbf{R}_x^{-1} \mathbf{r}_{x,\theta},$$

and the optimal affine estimator is expressed as:

$$\hat{\theta} = \mathbb{E}[\theta] + \mathbf{r}_{x,\theta}^T \mathbf{R}_x^{-1}(\mathbf{x} - \mathbb{E}[\mathbf{x}]).$$

Unlike the linear estimator, the affine estimator is unbiased (for the mean value), in the sense that $\mathbb{E}[\hat{\theta}] = \mathbb{E}[\theta]$.

The minimum MSE obtained by this estimator is calculated as:

$$\text{MSE}_{\text{min}} = \text{var}(\theta) - \mathbf{r}_{x,\theta}^T \mathbf{R}_x^{-1} \mathbf{r}_{x,\theta}.$$

Thus, based on the bias handling, the optimal affine estimator has an MSE that never exceeds the value $\text{var}(\theta)$, that is, the MSE of the constant estimator.

9.3.1 Superposition Property of Linear/Affine Estimators

Let ψ and φ represent two RVs. The optimal linear (affine) estimator of the sum $\theta = \psi + \varphi$, with conditional $\mathbf{x}$, is defined as:

$$\hat{\theta} = \hat{\psi} + \hat{\varphi}, \tag{9.1}$$

where $\hat{\psi}$ and $\hat{\varphi}$ are the optimal linear (affine) estimators of ψ and φ, respectively, with a conditional $\mathbf{x}$.

9.4 Geometric Interpretation: Orthogonality Condition and Projection Theorem

There is a deeper geometric interpretation of the structure for the affine or linear minimum mean square error estimators. In order to study this geometric interpretation, we re-define the problem of the affine estimation into a linear approach problem in the vector space.

9.4.1 Reconsideration of the Minimum MSE Linear Estimation

The key to embedding this problem into vector space is to locate the right space for the approximation problem. There are two kinds of space to consider: the space $\mathcal{H}$ containing quantities to be estimated (e.g., θ) and the space $\mathcal{S}$, called the solution space, in which the proper approach is structured (e.g., linear combinations of x_i). The problem then involves deriving a linear vector combination in space $\mathcal{S}$ closest to the quantity we want to approximate in $\mathcal{H}$. For the efficiency of this approach, it is absolutely necessary for $\mathcal{S} \subset \mathcal{H}$ to be true. Once we identify these spaces, what remains is the modeling of an inner product, which causes the appropriate norm that expresses an approximation error, such as the one of the MSE approach.

In a similar basis as the minimum MSE problem, we try to approach the RV θ through a linear combination of measured RVs $x_1, \cdots, x_n$, and we define H as the space that all scalar RVs with zero mean belong to and $\mathcal{S}$ as the linearly produced space span $\{x_1, \cdots, x_n\}$ of the measurements. For technical reasons, we will require all RVs in H to have a finite second moment (otherwise we can end up with vectors with infinite norms).

The MSE between vectors $\eta, v \in \mathcal{H}$ can be expressed as the squared norm:

$$\left\| \eta - v \right\|^2 = \mathbb{E}\left[(\eta - v)^2 \right],$$

which results from the inner product:

$$\langle \eta, v \rangle = \mathbb{E}[\eta\, v].$$

Since $\hat{\theta} = \mathbf{h}^T \mathbf{x} = \sum_{i=1}^{n} h_i x_i$ is located in S space, the estimation of the linear minimum MSE for θ is the vector $\hat{\theta} \in \mathcal{S}$, which minimizes the squared norm $\left\| \theta - \hat{\theta} \right\|^2$.

Linear Projection Theorem:

The optimal linear estimator of θ, based on $x_1, \cdots, x_n$ is the projection of θ in $\mathcal{S} = \mathrm{span}\{x_1, \cdots, x_n\}$.

Using the orthogonality condition in the projection theorem, the optimal linear estimator $\hat{\theta}$ must ensure (Figure 9.1):

$$\langle \theta - \hat{\theta}, u \rangle = 0, \quad \forall u \in \mathcal{S}.$$

Equivalently, if $u_1, \cdots, u_{n'}$ constitutes a basis of values for S:

$$\langle \theta - \hat{\theta}, u_i \rangle = 0, \quad i = 1, \cdots, n'. \tag{9.2}$$

When $\{x_i\}_{i=1}^{n}$ are linearly independent, the dimension of S is equal to n and each basis of values must have $n' = n$ linearly independent elements.

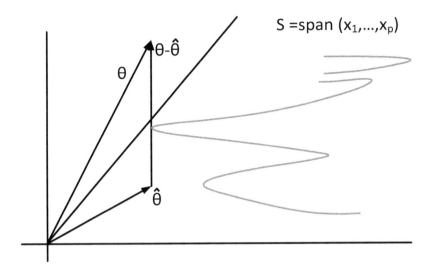

FIGURE 9.1
The ortogonality principle for the optimal linear estimator $\hat{\theta}$ of the RV θ, with conditional $x_1, \cdots, x_n$.

Now we just need to adopt a specific basis to model the optimal linear estimator. Perhaps the most reasonable basis of values is the set of measurements themselves, $u_i = x_i$, $i = 1, \cdots, n$ (assuming they are linearly independent), so that:

$$\mathbb{E}\left[(\theta - \hat{\theta}) \mathbf{x}^T \right] = \mathbf{0}.$$

Equivalently:

$$\mathbb{E}\left[(\theta - \mathbf{h}^T \mathbf{x}) \mathbf{x}^T \right] = \mathbf{M}_{x,\theta}^T - \mathbf{h}^T \mathbf{M}_x = \mathbf{0}.$$

Since the linear independence of $\{x_i\}$ implies that the matrix $\mathbf{M}_x = \mathbb{E}[\mathbf{x}\mathbf{x}^T]$ is invertible, it leads to the optimal solution developed earlier: $\mathbf{h} = \mathbf{M}_x^{-1} \mathbf{M}_{x,\theta}$. It is derived that:

$$
\begin{aligned}
\left\| \theta - \hat{\theta} \right\|^2 &= \left\langle \theta - \hat{\theta}, \theta - \hat{\theta} \right\rangle \\
&= \left\langle \theta - \hat{\theta}, \theta \right\rangle - \left\langle \theta - \hat{\theta}, \underset{\in \mathcal{S}}{\hat{\theta}} \right\rangle \\
&= \left\langle \theta - \hat{\theta}, \theta \right\rangle \\
&= \mathbb{E}[\theta^2] - \mathbf{h}^T \mathbf{M}_{x,\theta} \\
&= \mathbb{E}[\theta^2] - \mathbf{M}_{x,\theta}^T \mathbf{M}_x^{-1} \mathbf{M}_{x,\theta},
\end{aligned}
$$

where, from the second equality and on, the fact that the optimal error is orthogonal to each vector of the space S was used.

9.4.2 Minimum Affine MSE Estimation

The problem of minimum affine MSE estimation can also be transformed into a norm minimization problem of a vector space. One way to do so is to subtract the average from the parameter and the average from the measurements, and proceed as in the linear estimation (adding the mean parameter back to the solution at the end). A more direct approach is to include the degenerate constant RV "1" in the measurement vector (we can always add a virtual sensor to the measuring system that measures a constant).

To see how this could work, we first re-define the equation of the affine estimator as:

$$\hat{\theta} = \mathbf{h}^T \mathbf{x} + b$$
$$= [\mathbf{h}^T, b] \begin{bmatrix} \mathbf{x} \\ 1 \end{bmatrix}.$$

Next, the subspace of the solution is defined:

$$\mathcal{S} \triangleq \text{span}\{x_1, \cdots x_n, 1\},$$

with the help of which the following theorem of affine projection is presented.

The Affine Projection Theorem:
The optimal affine estimator of θ, based on RV $x_1, \cdots x_n$, is the projection of θ to the produced space span $\{x_1, \cdots x_n, 1\}$.

9.4.3 Optimization of the Affine Estimator for the Linear Gaussian Model

The additional assumption that $\mathbf{x}, \theta$ are jointly Gaussian distributed is introduced. It is true that the minimum MSE estimator is in fact affine:

$$\mathbb{E}[\theta \mid \mathbf{x}] = \mathbb{E}[\theta] + \mathbf{r}_{x,\theta}^T \mathbf{R}_x^{-1}(\mathbf{x} - \mathbb{E}[\mathbf{x}]).$$

One way to show the validity of the above relation is to calculate the CME in such a way that then it can be expressed in the form of this relation. Alternatively, we can develop the following approach. Without loss of generality, let us consider the case of $\mathbf{x}$ and θ with zero expected values. Let the LMMSE estimator, which is the same as the affine estimator in this case. From the linear projection theorem it is known that the error of the optimal estimator is orthogonal to the measurement samples:

$$\mathbb{E}\left[(\theta - \hat{\theta}_l)\mathbf{x}\right] = \mathbf{0}.$$

However, since $\theta - \hat{\theta}_l$ constitutes a combination of linear Gaussian RVs, it implies that it is Gaussian distributed. Moreover, as the Gaussian RVs that are orthogonal to each other are independent RVs: $\mathbb{E}\left[(\theta - \hat{\theta}_l) \mid \mathbf{x}\right] = \mathbb{E}[\theta - \hat{\theta}_l] = 0$.
Therefore, since $\hat{\theta}_l$ is a function of $\mathbf{x}$, it is:

$$0 = \mathbb{E}\left[(\theta - \hat{\theta}_l) \mid \mathbf{x}\right] = \mathbb{E}[\theta \mid \mathbf{x}] - \hat{\theta}_l,$$

or

$$\mathbb{E}[\theta \mid \mathbf{x}] = \hat{\theta}_l,$$

that describes the desired result.

9.5 Optimal Affine Vector Estimator

When the parameter $\boldsymbol{\theta} = [\theta_1, \cdots, \theta_p]^T$ is a vector, the aforementioned analysis for the scalar parametric values can easily be extended if the sum of individual MSEs is designated as the error criterion. We define the a priori vector of the mean value $\mathbb{E}[\boldsymbol{\theta}]$ and the cross-correlation matrix:

$$\mathbf{R}_{x,\theta} = \operatorname{cov}(\mathbf{x},\boldsymbol{\theta}) = \mathbb{E}\left[(\mathbf{x} - \mathbb{E}[\mathbf{x}])(\boldsymbol{\theta} - \mathbb{E}[\boldsymbol{\theta}])^T\right].$$

The error criterion of the MSE sum is defined as:

$$\operatorname{MSE}(\hat{\boldsymbol{\theta}}) = \sum_{i=1}^{p} \operatorname{MSE}(\hat{\theta}_i)$$

$$= \sum_{i=1}^{p} \mathbb{E}\left[\left|\theta_i - \hat{\theta}_i\right|^2\right] = \operatorname{trace}\left(\mathbb{E}\left[(\boldsymbol{\theta} - \hat{\boldsymbol{\theta}})(\boldsymbol{\theta} - \hat{\boldsymbol{\theta}})^T\right]\right).$$

Let $\hat{\theta}_i$ be the affine estimator of the i-th element of $\boldsymbol{\theta}$:

$$\hat{\theta}_i = \mathbf{h}_i^T \mathbf{x} + b_i, \quad i = 1, \cdots, p.$$

The corresponding vector estimator is defined as:

$$\hat{\boldsymbol{\theta}} = \left[\hat{\theta}_1, \cdots, \hat{\theta}_p\right]^T = \mathbf{H}^T \mathbf{x} + \mathbf{b},$$

$$\mathbf{H} = [\mathbf{h}_1, \cdots, \mathbf{h}_p].$$

The problem of the minimum affine vector MSE estimation is reflected in the derivation of the appropriate $\mathbf{H}, \mathbf{b}$ as the sum of MSE $(\mathbf{H},\mathbf{b})$ is minimized.

The solution to this problem is the optimal affine vector estimator:

$$\hat{\boldsymbol{\theta}} = \mathbb{E}[\boldsymbol{\theta}] + \mathbf{R}_{\theta,x}\mathbf{R}_x^{-1}(\mathbf{x} - \mathbb{E}[\mathbf{x}]). \tag{9.3}$$

The above relation is based on the fact that each pair of values $\mathbf{h}_i$ and b_i appears separately in each summand $\mathrm{MSE}(\hat{\boldsymbol{\theta}})$. Therefore, minimizing the MSE is equivalent to minimizing each $\mathrm{MSE}(\hat{\theta}_i)$:

$$\min_{\boldsymbol{\theta}} \mathrm{MSE}(\mathbf{H}, \mathbf{b}) = \sum_{i=1}^{p} \min_{\mathbf{h}_i, b_i} \mathrm{MSE}(\mathbf{h}_i, b_i).$$

Therefore, the solution for the minimum MSE is the sequence of multiple optimal scalar affine estimators for the individual θ_i:

$$\begin{bmatrix} \hat{\theta}_1 \\ \vdots \\ \hat{\theta}_p \end{bmatrix} = \begin{bmatrix} \mathbb{E}[\theta_1] \\ \vdots \\ \mathbb{E}[\theta_p] \end{bmatrix} + \begin{bmatrix} \mathbf{r}_{\theta_1, x} \mathbf{R}_x^{-1}(\mathbf{x} - \mathbb{E}[\mathbf{x}]) \\ \vdots \\ \mathbf{r}_{\theta_p, x} \mathbf{R}_x^{-1}(\mathbf{x} - \mathbb{E}[\mathbf{x}]) \end{bmatrix},$$

which is identical to Equation 9.3.

The final minimum sum MSE can be defined as:

$$\mathrm{MSE}_{\min} = \mathrm{trace}\left(\mathbf{R}_{\boldsymbol{\theta}} - \mathbf{R}_{\boldsymbol{\theta}, x} \mathbf{R}_x^{-1} \mathbf{R}_{x, \boldsymbol{\theta}} \right).$$

9.5.1 Examples of Linear Estimation

The aforementioned results of the minimum MSE are applied in the following examples.

Example 22: Linear Prediction of the Minimum MSE

In the linear prediction, we consider that a part $\{x_{k-p}, \cdots, x_{k-1}\}$ of the time sequence of samples $\{x_i\}_{i=-\infty}^{+\infty}$ is being measured, aiming to the creation of a linear p-class prediction, which is in the form of:

$$\hat{x}_k = \sum_{i=1}^{p} a_i x_{k-i}.$$

We also consider that $\{x_i\}$ is a zero-mean wide sense stationary random sequence, with an autocorrelation function:

$$r(k) \triangleq \mathbb{E}\left[x_i x_{i-k} \right].$$

The problem lies (Figure 9.2) in deriving the prediction components $\mathbf{a} = [a_1, \cdots, a_p]^T$ that minimize the mean square prediction error: $\mathrm{MSE}(\mathbf{a}) = \mathbb{E}\left[\left(x_k - \hat{x}_k \right)^2 \right]$.

To find the optimum solution, we set $\theta = x_k$ as the random scalar parameter, $\mathbf{x}$, as the time interval that the measurement occurs and $\mathbf{h} = \mathbf{a}$, the vector of coefficients to be determined.

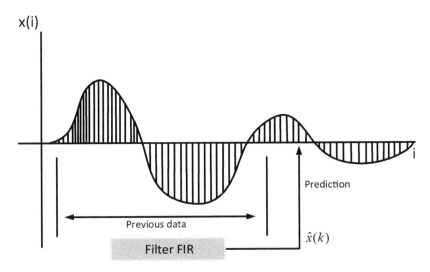

FIGURE 9.2
The linear prediction mechanism as an FIR filter.

Step 1: Define of the prediction equation in vector form.

$$\hat{x}_k = \mathbf{a}^T \mathbf{x},$$

where

$$\mathbf{x} = \left[x_{k-1}, \cdots, x_{k-p} \right]^T, \quad \mathbf{a} = \left[a_1, \cdots, a_p \right]^T.$$

Step 2: Determine the orthogonality principle:

$$\mathbb{E}\left[(x_k - \mathbf{a}^T \mathbf{x}) x_{k-i} \right] = 0, \quad i = 1, \cdots, p,$$

or, in a vector form:

$$\mathbf{0}^T = \begin{bmatrix} \mathbb{E}\left[(x_k - \mathbf{a}^T \mathbf{x}) x_{k-1} \right] \\ \vdots \\ \mathbb{E}\left[(x_k - \mathbf{a}^T \mathbf{x}) x_{k-p} \right] \end{bmatrix} = \mathbb{E}\left[(x_k - \mathbf{a}^T \mathbf{x}) \mathbf{x}^T \right].$$

Therefore, the optimal prediction coefficients $\mathbf{a} = \hat{\mathbf{a}}$ are depicted as:

$$\hat{\mathbf{a}} = \mathbf{R}^{-1} \mathbf{r},$$

where the correlation vector is introduced:

$$\mathbf{r}^T = [r_1, \cdots, r_p] = \mathbb{E}[\mathbf{x}\, x_k],$$

and the covariance matrix (Toeplitz):

$$\mathbf{R} = (r_{i,j})_{i,j=1,p} = \mathbb{E}[\mathbf{x}\mathbf{x}^T].$$

Finally, the prediction has a minimum MSE:

$$\begin{aligned}
\mathrm{MSE}_{\min} &= \left\langle x_k - \hat{\mathbf{a}}^T \mathbf{x}, x_k \right\rangle \\
&= r_0 - \hat{\mathbf{a}}^T \mathbf{r} \\
&= r_0 - \mathbf{r}^T \mathbf{R}^{-1} \mathbf{r}.
\end{aligned}$$

Example 23: Inversion Problem

Consider the following measurement model:

$$\mathbf{x} = \mathbf{A}\boldsymbol{\theta} + \mathbf{n}$$

where

- $\boldsymbol{\theta} = [\theta_1, \cdots, \theta_p]^T$ are the unknown random measurements.
- $\mathbf{n} = [n_1, \cdots, n_m]^T$ is the the random noise of zero expected value and covariance matrix $\mathbf{R}_n$.
- $\mathbf{A}$ is an $m \times p$ matrix with known elements.
- $\boldsymbol{\theta}, \mathbf{n}$ are uncorrelated.
- $\mathbf{x} = [x_1, \cdots, x_m]^T$ are the random measurements.

The problem is to derive an affine estimator min MSE $\hat{\boldsymbol{\theta}}$ of $\boldsymbol{\theta}$ (Figure 9.3).

SOLUTION:

From the previous results for the minimum MSE vector values, it is obtained that:

$$\hat{\boldsymbol{\theta}} = \mathbb{E}[\boldsymbol{\theta}] + \mathbf{R}_{\boldsymbol{\theta},x} \mathbf{R}_x^{-1} (\mathbf{x} - \mathbb{E}[\mathbf{x}]).$$

We only have to determine the form of the optimal affine estimator as a function of $\mathbf{A}$ and $\mathbf{R}_n$:

$$\begin{aligned}
E[\mathbf{x}] &= \mathbb{E}[\mathbf{A}\boldsymbol{\theta} + \mathbf{n}] = \mathbf{A}\mathbb{E}[\boldsymbol{\theta}], \\
\mathbf{R}_x &= \mathrm{cov}\left(\underbrace{\mathbf{A}\boldsymbol{\theta} + \mathbf{n}}_{uncorrelated}\right) = \mathbf{A}\mathbf{R}_\theta \mathbf{A}^T + \mathbf{R}_n, \\
\mathbf{R}_{x,\theta} &= \mathrm{cov}((\mathbf{A}\boldsymbol{\theta} + \mathbf{n}), \boldsymbol{\theta}) = \mathbf{A}\mathbf{R}_\theta.
\end{aligned}$$

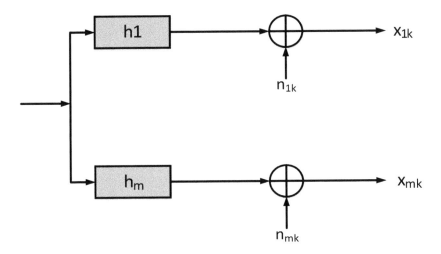

FIGURE 9.3
Block diagram of the inversion problem in an FIR filter assembly with p taps.

Consequently, the final result can be expressed as:

$$\hat{\boldsymbol{\theta}} = \mathbb{E}[\boldsymbol{\theta}] + \mathbf{R}_{\boldsymbol{\theta}} \mathbf{A}^T \left(\mathbf{A} \mathbf{R}_{\boldsymbol{\theta}} \mathbf{A}^T + \mathbf{R}_n \right)^{-1} (\mathbf{x} - \mathbf{A}\, \mathbb{E}[\boldsymbol{\theta}]),$$

and the minimum sum MSE as:

$$\mathrm{MSE}_{\min} = \mathrm{trace}\!\left(\mathbf{R}_{\boldsymbol{\theta}} - \mathbf{R}_{\boldsymbol{\theta}} \mathbf{A}^T \left(\mathbf{A} \mathbf{R}_{\boldsymbol{\theta}} \mathbf{A}^T + \mathbf{R}_n \right)^{-1} \mathbf{A} \mathbf{R}_{\boldsymbol{\theta}} \right).$$

Note that:
1. When $\mathbf{R}_n$ dominates $\mathbf{A}\mathbf{R}_{\boldsymbol{\theta}}\mathbf{A}^T$, then $\mathrm{MSE}_{\min} \approx \mathrm{trace}(\mathbf{R}_{\boldsymbol{\theta}})$.
2. When $\mathbf{A}\mathbf{R}_{\boldsymbol{\theta}}\mathbf{A}^T$ dominates $\mathbf{R}_n$ and Matrix $\mathbf{A}$ is full rank, then $\mathrm{MSE}_{\min} \approx 0$.

9.6 Non-Statistical Least Squares Technique (Linear Regression)

In some cases, the model being measured is not sufficient for a clear calculation $\mathbf{R}$ and $\mathbf{R}_{x,\,\theta}$, which are the necessary statistics for the minimum MSE linear estimators described above.

In these cases, we must turn to the training data for the estimation of these parameters. However, the following question arises: To what extent is it preferable to replace the empirical averages in the above formulas? The answer depends, of course, on the definition of optimization. The non-statistical least squares technique is a re-definition of this problem, for which the optimal solutions are in the same basis as the previous ones, but with empirical estimates, by appropriately substituting $\mathbf{R}$ and $\mathbf{R}_{x,\theta}$.

Let the available pair of measurements ($n \geq p$):

$$y_i, \quad \mathbf{x}_i = [x_{i1}, \cdots, x_{ip}]^T, \quad i = 1, \cdots, n.$$

where x_{ip} may be equal to x_{i-p}, but this is not necessary.

Let the following relationship between input-output in the model (Figure 9.4):

$$y_i = \mathbf{x}_i^T \mathbf{a} + v_i, \quad i = 1, \cdots, n,$$

where
- y_i is the response or output or conditional variable,
- $\mathbf{x}_i$ is the input or independent variable,
- $\mathbf{a}$ is a $p \times 1$ vector of unknown coefficients to be estimated:

$$\mathbf{a} = [a_1, \cdots, a_p]^T.$$

Objective

Derive the least squares linear estimator $\hat{\mathbf{a}}$ of $\mathbf{a}$ that minimizes the sum of squared errors:

$$SSE(\mathbf{a}) = \sum_{i=1}^{n} \left(y_i - \mathbf{x}_i^T \mathbf{a} \right)^2.$$

This model is expressed in $n \times 1$ vector formulation as:

$$\begin{bmatrix} y_1 \\ \vdots \\ y_n \end{bmatrix} = \begin{bmatrix} x_1^T \\ \vdots \\ x_n^T \end{bmatrix} \mathbf{a} + \begin{bmatrix} v_1 \\ \vdots \\ v_n \end{bmatrix},$$

$$\mathbf{y} = \mathbf{X}\mathbf{a} + \mathbf{v},$$

where $\mathbf{X}$ is a non-random $n \times p$ input matrix.

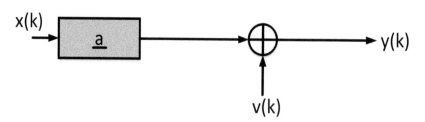

FIGURE 9.4
Block diagram of the linear regression.

The estimation criterion is expressed as:

$$\text{SSE}(\mathbf{a}) = (\mathbf{y} - \mathbf{X}\mathbf{a})^T (\mathbf{y} - \mathbf{X}\mathbf{a}).$$

SOLUTION:
Step 1: Determine the vector space containing vector **y**: $\mathcal{H} = \mathbb{R}^n$.
Inner Product: $\langle \mathbf{y}, \mathbf{z} \rangle = \mathbf{y}^T \mathbf{z}$.

Step 2: Determine the subspace that defines the solution of the problem, which contains **Xa**.

$$\mathcal{S} = \text{span}\{\text{columns of } \mathbf{X}\},$$

which contains vectors of the form:

$$\mathbf{X}\mathbf{a} = \sum_{i=1}^{p} a_k \left[x_{1k}, \cdots, x_{nk} \right]^T.$$

Step 3: Apply the projection theorem.
Orthogonality principle: The optimal linear estimator $\hat{\mathbf{a}}$ satisfies the following relation:

$$\langle \mathbf{y} - \mathbf{X}\hat{\mathbf{a}}, \mathbf{u}_i \rangle = 0, \quad i = 1, \cdots, n,$$

where $\mathbf{u}_i$ are the columns of matrix **X**, or equivalently:

$$\begin{aligned} \mathbf{0}^T &= (\mathbf{y} - \mathbf{X}\hat{\mathbf{a}})^T \mathbf{X} \\ &= \mathbf{y}^T \mathbf{X} - \hat{\mathbf{a}}^T \mathbf{X}^T \mathbf{X}, \end{aligned}$$

or, if matrix **X** is of full p-rank, then $\mathbf{X}^T\mathbf{X}$ is invertible, and:

$$\begin{aligned} \hat{\mathbf{a}} &= \left[\mathbf{X}^T \mathbf{X} \right]^{-1} \mathbf{X}^T \mathbf{y} \\ &= \left[n^{-1} \mathbf{X}^T \mathbf{X} \right]^{-1} \left[n^{-1} \mathbf{X}^T \right] \mathbf{y} \\ &= \hat{\mathbf{R}}_x^{-1} \hat{\mathbf{r}}_{x,y}. \end{aligned}$$

It is true that:

$$\hat{\mathbf{R}}_x \triangleq \frac{1}{n} \sum_{i=1}^{n} \mathbf{x}_i \mathbf{x}_i^T, \quad \hat{\mathbf{r}}_{x,y} \triangleq \frac{1}{n} \sum_{i=1}^{n} \mathbf{x}_i \mathbf{y}_i.$$

The form of the projection operator is then determined to predict the output response as:

$$\hat{\mathbf{y}} = \mathbf{X}\,\hat{\mathbf{a}},$$

where, as before, it can be expressed as an orthogonal projection of $\mathbf{y}$ to S:

$$\hat{\mathbf{y}} = \mathbf{X}\,\hat{\mathbf{a}}$$
$$= \underbrace{\mathbf{X}[\mathbf{X}^T\mathbf{X}]^{-1}\mathbf{X}^T}_{\text{orthogonal projection}}\ \mathbf{y}.$$

Properties of the orthogonal projection operation:

$$\Pi_X = \mathbf{X}[\mathbf{X}^T\mathbf{X}]^{-1}\mathbf{X}^T.$$

First Property: The operator Π_X projects vectors to the column space of $\mathbf{X}$.
We define the decomposition of the vector $\mathbf{y}$ in the coefficients $\mathbf{y}_X$ in the column space of $\mathbf{X}$ and $\mathbf{y}_X^{\perp}$, which is orthogonal to the column space of $\mathbf{X}$(Figure 9.5):

$$\mathbf{y} = \mathbf{y}_X + \mathbf{y}_X^{\perp}.$$

For a vector $\mathbf{a} = [a_1,\cdots,a_p]^T$, it is true that:

$$\mathbf{y}_X = \mathbf{X}\,\mathbf{a}, \quad \mathbf{X}^T\mathbf{y}_X^{\perp} = 0.$$

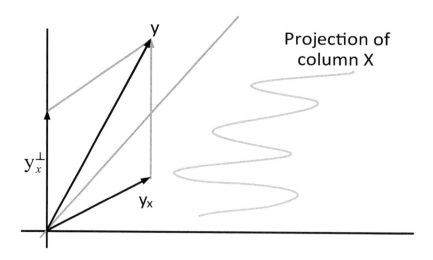

FIGURE 9.5
The decomposition of vector **y** in the column space.

Thus:

$$\Pi_{xy} = \Pi_x\left(y_x + y^{\perp}\right)$$
$$= X\underbrace{[X^TX]^{-1}X^TX}_{=I}\,a + X[X^TX]^{-1}\,\underbrace{X^Ty_X^{\perp}}_{=0}$$
$$= X\,a$$
$$= y_X,$$

so that the operator Π_x derives the column space for the coefficients of the **y**-vector. Thus, $y_x = \Pi_x y$ can be determined such that the following result is obtained:

$$y = \Pi_x y + \underbrace{\left(I - \Pi_x\right)y}_{y_X^{\perp}}.$$

Second Property: From the previous property it is true that the operator $I - \Pi_x$ is projected into a space orthogonal to span{columns of **X**}.

Third Property: The operator Π_x is symmetric and idempotent: $\Pi_x^T\Pi_x = \Pi_x$.

Fourth Property: $(I - \Pi_x)\,\Pi_x = 0$.

Example 24: Least Squares Optimality of Sample Mean

Let the measurement: $x = \left[x_1, \cdots, x_n\right]^T$.

    Objective: Derive the appropriate constant c, which minimizes the sum of the squares:

$$\sum_{i=1}^{n}(x_i - c)^2 = (x - c1)^T(x - c1),$$

where $1 = [1, \cdots, 1]^T$.

Step 1: Define the solution space.
Space S is the diagonal line: $\left\{y : y = a1, \quad a \in \mathbb{R}\right\}$, as illustrated in Figure 9.6.

Step 2: Apply the orthogonality principle.

$$(x - c1)^T 1 = 0 \Leftrightarrow c = \frac{x^T 1}{1^T 1} = n^{-1}\sum_{k=1}^{n}x_i.$$

Example 25: Let the Measurement Sequence $\{z_i\}$

The sequence of training data consists of $n + p$ samples of z_i:

$$\{z_i\}_{i=1}^{n+p}, \quad i = 1, \cdots, n.$$

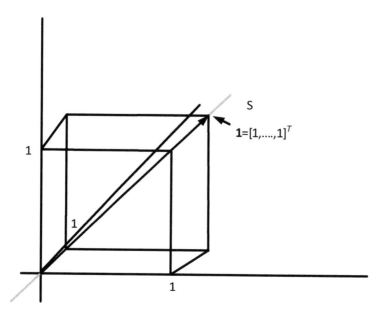

FIGURE 9.6
The diagonal line is the subspace produced for the scalar least squares case.

Using an auto-regressive model for the training data process, it should be true that:

$$z_k = \sum_{i=1}^{p} a_i z_{k-i} + v_k, \quad k = p+1, \cdots n,$$

so, as the SSE gets minimized (Figure 9.7):

$$\text{SSE}(n) = \sum_{i=1}^{n} \left(z_{k+p} - \sum_{i=1}^{p} a_i z_{k+p-i} \right)^2.$$

SOLUTION:

Step 1: Determine response variables $y_k = z_k$ and input vectors $z_k = [z_{k-1}, \cdots, z_{k-p}]^T$.

$$\begin{bmatrix} z_{n+p} \\ \vdots \\ y_{p+1} \end{bmatrix} = \begin{bmatrix} z_{n+p}^T \\ \vdots \\ z_{p+1}^T \end{bmatrix} \mathbf{a} + \begin{bmatrix} v_{n+p} \\ \vdots \\ v_{p+1} \end{bmatrix},$$

$$\mathbf{y} = \mathbf{X}\mathbf{a} + \mathbf{v},$$

Step 2: Apply the orthogonality principle. The prediction of p-th order LLS is of the form:

$$\hat{z}_k = \sum_{i=1}^{p} \hat{a}_i z_{k-i},$$

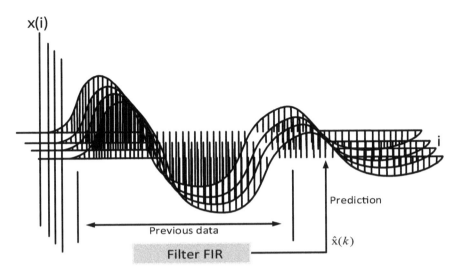

FIGURE 9.7
Implementation of the LLS prediction from a sequence of training data.

where vector $\mathbf{a} = [\hat{a}_1, \cdots, \hat{a}_p]^T$ is derived by the equation:

$$\hat{\mathbf{a}} = [\mathbf{X}^T\mathbf{X}]^{-1}\mathbf{X}^T\mathbf{y} = \hat{\mathbf{R}}^{-1}\hat{\mathbf{r}},$$

We also introduce the following quantities:

$$\hat{\mathbf{r}} = \left[\hat{r}_1, \cdots, \hat{r}_p\right]^T,$$

$$\hat{\mathbf{R}} = (\hat{r}(i-j))_{i,j=1,p},$$

$$\hat{r}_j = n^{-1} \sum_{i=1}^{n} z_{i+p} z_{i+p-j}, \quad j = 0, \cdots, p.$$

9.7 Linear Estimation of Weighted LLS

As before, we consider a linear model that expresses the input-output relation:

$$\begin{bmatrix} y_1 \\ \vdots \\ y_n \end{bmatrix} = \begin{bmatrix} x_1^T \\ \vdots \\ x_n^T \end{bmatrix} \mathbf{a} + \begin{bmatrix} v_1 \\ \vdots \\ v_n \end{bmatrix},$$

$$\mathbf{y} = \mathbf{X}\mathbf{a} + \mathbf{v},$$

The linear minimum weighted least squares-LMWLS estimator $\hat{\mathbf{a}}$ of $\mathbf{a}$ minimizes the following relation:

$$\text{SSE}(\mathbf{a}) = (\mathbf{y} - \mathbf{X}\mathbf{a})^T \mathbf{W}(\mathbf{y} - \mathbf{X}\mathbf{a}),$$

where W is a symmetric positive defined matrix, $n \times n$.

Solution to LMWLS problem:
Step 1: Define the vector space that contains the vector **y**: $\mathcal{H} = \mathbb{R}^n$.
Inner product: $\langle \mathbf{y}, \mathbf{z} \rangle = \mathbf{y}^T \mathbf{W}\mathbf{z}$.

Step 2: Define the subspace S that constitutes the solution of the problem:

$$\mathbf{X}\mathbf{a} = \text{span}\{\text{columns of } \mathbf{X}\}.$$

Step 3: Apply the projection theorem.

Orthogonality principle: The optimal linear estimator $\hat{\mathbf{a}}$ satisfies the following relation:

$$\mathbf{0}^T = \left(\mathbf{y} - \mathbf{X}\hat{\mathbf{a}}\right)^T \mathbf{W}\mathbf{X}$$
$$= \mathbf{y}^T \mathbf{W}\mathbf{X} - \hat{\mathbf{a}}^T \mathbf{X}^T \mathbf{W}\mathbf{X},$$

or, if the matrix **X** is full p-rank, then $\mathbf{X}^T\mathbf{W}\mathbf{X}$ is invertible and:

$$\hat{\mathbf{a}} = \left[\mathbf{X}^T \mathbf{W}\mathbf{X}\right]^{-1} \mathbf{X}^T \mathbf{W}\mathbf{y}.$$

Alternative interpretation: The vector $\hat{\mathbf{y}}$ of the least square predictions $\hat{y}_i = \mathbf{x}_i^T \hat{\mathbf{a}}$ of the real output **y** is:

$$\hat{\mathbf{y}} = \mathbf{X}\mathbf{a},$$

which represents a non-orthogonal (oblique) projection of **y** in space H (Figure 9.8):

$$\hat{\mathbf{y}} = \underbrace{\mathbf{X}\left[\mathbf{X}^T\mathbf{W}\mathbf{X}\right]^{-1} \mathbf{X}^T \mathbf{W}}_{\text{oblique projection } \Pi_{X,W}} \mathbf{y}.$$

Then, the weighted sum of square error (WSSE) is derived:

$$\text{WSSE}_{\min} = \mathbf{y}^T \left[\mathbf{I} - \mathbf{X}[\mathbf{X}^T\mathbf{W}\mathbf{X}]^{-1}\mathbf{X}^T\mathbf{W}\right]\left[\mathbf{I} - \mathbf{X}[\mathbf{X}^T\mathbf{W}\mathbf{X}]^{-1}\mathbf{X}^T\mathbf{W}\right]^T \mathbf{y}$$
$$= \mathbf{y}^T \left[\mathbf{I} - \Pi_{X,W}\right]^T \left[\mathbf{I} - \Pi_{X,W}\right]\mathbf{y}.$$

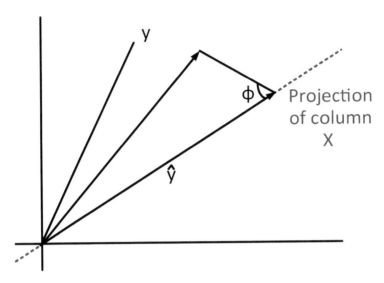

FIGURE 9.8
Oblique projection of WLS estimator.

The LMWLS prediction technique can be depicted as a linear least squares (not weighted) prediction on the status of the pre- and post-processing data (Figure 9.9).

As the matrix $\mathbf{W}$ is symmetric positive definite, it is subjected to a quadratic root factorization that is of the form:

$$\mathbf{W} = \mathbf{W}^{1/2}\mathbf{W}^{1/2},$$

and

$$\hat{\mathbf{y}} = \mathbf{W}^{-1/2}\underbrace{\mathbf{W}^{1/2}\mathbf{X}\left[\mathbf{X}^T\mathbf{W}^{1/2}\mathbf{W}^{1/2}\mathbf{X}\right]^{-1}\mathbf{X}^T\mathbf{W}^{1/2}}_{\text{orthogonal projection }\Pi_{\mathbf{W}^{1/2}\mathbf{X}}}\left[\mathbf{W}^{1/2}\mathbf{y}\right]$$

$$= \mathbf{W}^{-1/2}\Pi_{\mathbf{W}^{1/2}\mathbf{X}}\mathbf{W}^{1/2}\mathbf{y}.$$

Example 26: *Adaptive Linear Prediction.*

Let us model the auto-regressive p-order model:

$$z_k = \sum_{i=1}^{p} a_i z_{k-i} + v_k, \quad k = 1, 2, \ldots$$

so that, at time n, the weighted least squares criterion is minimized:

$$\text{WSSE}(n) = \sum_{k=1}^{n} \rho^{n-k}\left(z_{k+p} - \sum_{i=1}^{p} a_i z_{k+p-i}\right)^2,$$

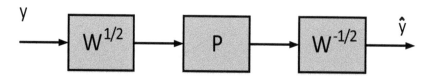

FIGURE 9.9
Implementation of the LMWLS estimator using pre-processing and post-processing units for orthogonal projection.

and the coefficient $\rho \in [0,1]$, as illustrated in Figure 9.10.

SOLUTION:
As before, we define the response variables $y_k = z_k$ and the input vectors, $z_k = [z_{k-1}, \cdots, z_{k-p}]^T$. We also define the weighted matrix:

$$\mathbf{W} = \begin{bmatrix} \rho^0 & 0 & 0 \\ 0 & \ddots & 0 \\ 0 & 0 & \rho^{n-1} \end{bmatrix}.$$

In this way, the prediction coefficients for LMWLS are derived as:

$$\hat{\mathbf{a}} = [\mathbf{X}^T \mathbf{W} \mathbf{X}]^{-1} \mathbf{X}^T \mathbf{W} \mathbf{y}$$
$$= \widehat{\tilde{\mathbf{R}}}^{-1} \widehat{\tilde{\mathbf{r}}},$$

where the *smoothed* correlation coefficients of the samples are given:

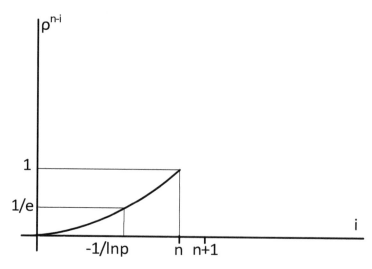

FIGURE 9.10
The coefficient $\rho \in [0,1]$.

$$\hat{\mathbf{r}} = \left[\hat{\tilde{r}}_1, \cdots, \hat{\tilde{r}}_p\right]^T,$$

$$\hat{\mathbf{R}} = \left(\hat{\tilde{r}}(i-j)\right)_{i,j=1,p},$$

$$\hat{\tilde{r}}_j \triangleq \sum_{i=1}^{n} \rho^{n-i} z_{i+p} z_{i+p-j}, \quad j = 0, \cdots, p.$$

The minimum weighted sum of the square error (WSSE) is expressed as:

$$\text{WSSE}_{\min} = \hat{\tilde{r}}_0 - \hat{\mathbf{r}}^T \hat{\mathbf{R}}^{-1} \hat{\mathbf{r}}.$$

9.8 Optimization of LMWLS in Gaussian Models

We know that the linear MMSE (LMMSE) estimator is the best among all others (linear and non-linear) for Gaussian models with respect to the measurement samples and the parameters. In this section, a similar case of the linear WSSE estimator is studied.

Let the following Gaussian model:

$$\mathbf{y} = \mathbf{X}\mathbf{a} + \mathbf{v},$$

where we consider that

- $\mathbf{v} \sim \mathcal{N}_n(\mathbf{0}, \mathbf{R})$.
- The correlation matrix $\mathbf{R}$ is known.
- The matrix $\mathbf{X}$ expresses the known and non-random measurements samples.

Based on the aforementioned model, for each conditional matrix $\mathbf{X}$ or vector $\mathbf{a}$, the PDF of vector $\mathbf{y}$ is a Gaussian multivariable given by:

$$f(\mathbf{y};\mathbf{a}) = \frac{1}{\sqrt{(2\pi)^n |\mathbf{R}|}} \exp\left(-\frac{1}{2}(\mathbf{y} - \mathbf{X}\mathbf{a})^T \mathbf{R}^{-1}(\mathbf{y} - \mathbf{X}\mathbf{a})\right).$$

This implies that the maximum likelihood (ML) estimator of $\mathbf{a}$ is the same as the LMWLS. This is more clearly stated by the following relation:

$$\hat{a}_{\text{ML}} = \arg\max_{\mathbf{a}} \ln f(\mathbf{y};\mathbf{a})$$
$$= \arg\min_{\mathbf{a}} (\mathbf{y} - \mathbf{X}\mathbf{a})^T \mathbf{R}^{-1}(\mathbf{y} - \mathbf{X}\mathbf{a}).$$

Consequently,

$$\hat{\mathbf{y}} = \mathbf{X}\hat{\mathbf{a}}_{\text{ML}} = \mathbf{X}[\mathbf{X}^T \mathbf{R}^{-1} \mathbf{X}]^{-1} \mathbf{X}^T \mathbf{R}^{-1} \mathbf{y} = \Pi_{\mathbf{X},\mathbf{W}} \mathbf{y}.$$

With reference to this model, we can evaluate its performance by examining whether it satisfies the equality condition of CRB:

$$(\nabla_{\mathbf{a}} \ln f)^T = (\mathbf{y} - \mathbf{X}\mathbf{a})^T \mathbf{R}^{-1}\mathbf{X} = \left(\underbrace{\mathbf{y}^T \mathbf{R}^{-1}\mathbf{X}\left[\mathbf{X}^T\mathbf{R}^{-1}\mathbf{X}\right]^{-1}}_{\hat{\mathbf{a}}^T} - \mathbf{a}^T \right) \underbrace{\mathbf{X}^T\mathbf{R}^{-1}\mathbf{X}}_{K_{\mathbf{a}}}$$

We conclude that when the matrix $\mathbf{X}$ is non-random and the (known) noise covariance matrix $\mathbf{R}$ is equal to the weighted LS matrix $\mathbf{W}^{-1}$, then the following conditions are true:

- The LMWLS estimator $\hat{\mathbf{a}}$ is unbiased.
- The LMWLS estimator is efficient and, consequently, UMVU.
- Based on the property 5 of the CRB (Chapter 8), since $K_{\mathbf{a}}$ is not a function of $\mathbf{a}$, the covariance of the estimator is:

$$\mathrm{cov}_{\mathbf{a}}(\hat{\mathbf{a}}) = K_{\mathbf{a}}^{-1} = \left[\mathbf{X}^T\mathbf{R}^{-1}\mathbf{X}\right]^{-1} = \frac{1}{n}\widehat{\mathbf{R}}^{-1}.$$

10

Fundamentals of Signal Detection

In this chapter, the problem of signal detection is analyzed. This is equivalent to the problem of the estimation studied in the previous chapters if we consider that there is a limited range of possible values of the unknown parameter θ. For this reason, the theoretical concepts of detection are generally simpler than the concepts that construct the problem of estimation. As discussed below, these theoretical concepts are related to each other in the case of unknown nuisance parameters.

In this chapter, the following concepts are discussed:

- Optimal detection theory
- Detection using the Bayes approach
- Frequentist Approach for Detection
- Receiver Operating Characteristic (ROC)
- Multiple hypotheses testing

Example 27: Typical Radar Application

Let a continuous time $x(t)$ be over the time interval $[0, T]$. We want to decide if $x(t)$ contains only noise:

$$x(t) = w(t), \quad 0 \le t \le T,$$

or if there is a signal along with the additive noise:

$$x(t) = \theta\, s(t - \tau) + w(t), \quad 0 \le t \le T.$$

The following apply:

- $s(t)$ is a known signal, which may be present or not.
- $w(t)$ indicates white Gaussian noise with zero expected value and known level of power spectral density $N_0/2$.
- τ is a known time delay, $0 \le \tau \ll T$
- $\displaystyle\int_0^T |s(t-\tau)|^2\, dt = \int_0^T |s(t)|^2\, dt$ is the signal energy.
- $\theta \in [0,1]$ is an unknown nuisance parameter.

The objective of the detection process is (a) to decide on the presence of the signal or not and (b) to make the decision with the minimum (average) number of errors.

We will use the following symbols for the detection case: "Signal Presence -H_1" and "Signal Absence -H_0":

$$H_0 : x(t) = w(t) \qquad\qquad H_0 : \theta = 0$$

$$\Leftrightarrow$$

$$H_1 : x(t) = s(t-\tau) + w(t) \qquad\qquad H_1 : \theta = 1.$$

Two typical methods for detecting signals are illustrated in Figure 10.1 and are the following:

- Energy threshold detection:

$$y = \int_0^T |x(t)|^2 \, dt \overset{\overset{H_1}{>}}{\underset{\underset{H_0}{<}}{}} \eta.$$

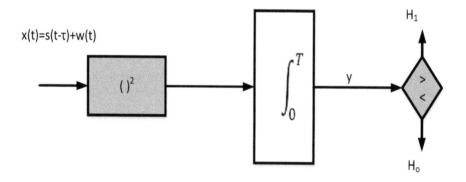

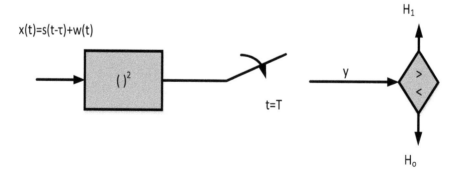

FIGURE 10.1
(a) Energy threshold detector, (b) threshold-filter detector.

- Threshold-Filter Detection:

$$y = \int_0^T h(T-t)x(t)\,dt \underset{H_0}{\overset{H_1}{\gtrless}} \eta.$$

Error analysis:
According to Figure 10.2, there are two categories of the decision error:

- False alarm: $y > \eta$ when there is no signal presence.
- Miss detection: $y < \eta$ when there is a signal presence.

We can easily calculate the conditional probabilities of these error cases, respectively, as:

$$P_F = P(\text{signal detection} \mid \text{no actual signal}) = \int_{y>\eta} f(y \mid \text{no actual signal})\,dy$$

$$P_M = P(\text{non signal detection} \mid \text{signal}) = \int_{y\le\eta} f(y \mid \text{signal})\,dy$$

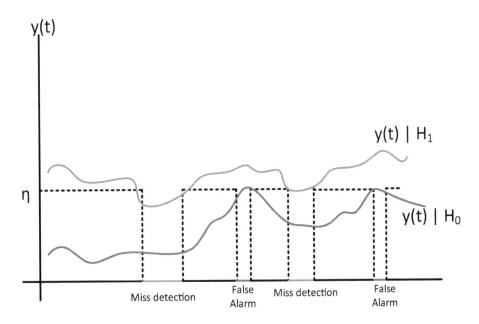

FIGURE 10.2
Repeated radar tests in received sequences y_i. One sequence contains a signal (H_1), while the second one contains its absence (H_0).

The following two fundamental questions arise:

Q1: Is there an optimal way of matching P_M and P_F?

Q2: Can the filter $h(\cdot)$ in the threshold-filter detector be optimized to provide the optimal tradeoff?

The answer to Q1 will be given later. At this point, the filter of the threshold-filter detector can be designed in an appropriate way so as to optimize a given design criterion. As shown in Figure 10.3, there is a relatively large overlap between the two PDFs. This fact should be taken into account in the design of the optimum filter in order to minimize this overlapping. One measure of the degree of this overlap is the deflection, which is defined as:

$$d^2 = \frac{\left| \mathbb{E}\left[y \mid \text{signal presence} \right] - \mathbb{E}\left[y \mid \text{signal absence} \right] \right|^2}{\text{var}(y \mid \text{signal absence})}.$$

Large values of d^2 are translated into well separated PDFs $f(y|H_0)$ and $f(y|H_1)$, with little overlapping. Thus, by increasing d^2, the overlap is reduced.

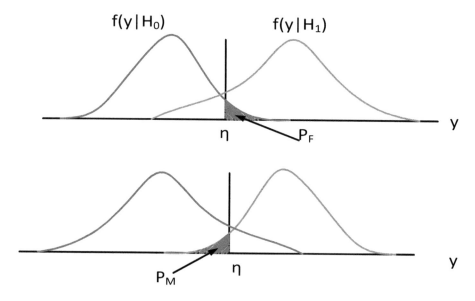

FIGURE 10.3
Likelihoods of miss detection (P_M) and false alarm (P_F) for the radar example. We notice that the reduction in the error of one of the two PDFs (by changing the decision threshold η) causes an increase in the error of the other and vice versa

Deflection can be easily calculated for the radar example. We notice that the presence of a signal produces some shift in the expected value, but not in the variance:

$$\mathbb{E}[y \mid \text{signal absence}] = 0,$$

$$\mathbb{E}[y \mid \text{signal presence}] = \int_0^T h(T-t)\,s(t-\tau)\,dt,$$

$$\text{var}[y \mid \text{signal absence}] = \frac{N_0}{2}\int_0^T |h(t)|^2\,dt.$$

Thus, be applying the Cauchy-Schwarz inequality, it is derived that:

$$d^2 = \frac{2}{N_0}\frac{\left|\int_0^T h(T-t)\,s(t-\tau)\,dt\right|^2}{\int_0^T |h(T-t)|^2\,dt} \leq \frac{2}{N_0}\underbrace{\int_0^T |s(t-\tau)|^2\,dt}_{\int_0^T |s(t)|^2\,dt=\|s\|^2},$$

where '=' is true if and only if $h(T-t) = a\,s(t-\tau)$ for a constant a.

Consequently, the optimal deflection filter is the matched filter:

$$h(t) = s(T+\tau-t).$$

If $s(\tau)$ is a short duration pulse:

- $\int_0^T |s(t-\tau)|^2\,dt$ does not depend on τ.
- The optimal detector is expressed as:

$$y = \int_0^T s(t-\tau)\,x(t)\,dt$$

$$= \int_{-\infty}^{+\infty} s(t-\tau)\,x(t)\,dt$$

$$= s(-t) * x(t)\big|_{t=\tau}.$$

In Figures 10.4 and 10.5, the matched filter detector is illustrated in different signal cases.

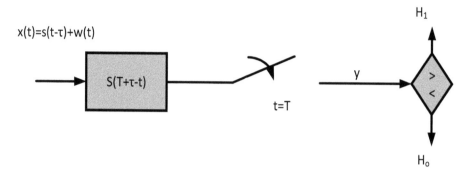

FIGURE 10.4
Application of the matched filter to the optimal receiver when receiving a delayed signal and noise.

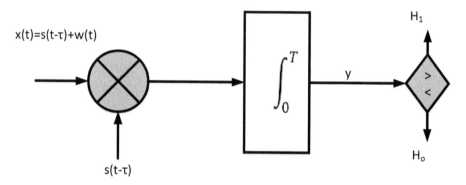

FIGURE 10.5
Application of the correlated receiver to receive delayed signal and noise.

10.1 The General Detection Problem

First of all, the following terms are defined:

- $x \in \mathcal{X}$ is a RV, the sample to be measured.
- $\theta \in \Theta$ is an unknown parameter.
- $f(x; \theta)$ is the PDF of the RV x, which is considered known.

We have two discrete hypotheses regarding parameter θ:

$$\theta \in \Theta_0 \quad \text{or} \quad \theta \in \Theta_1.$$

Θ_0, Θ_1 form a partition of Θ in two non-overlapping regions (Figure 10.6):

$$\Theta_0 \cup \Theta_1 = \Theta, \quad \Theta_0 \cap \Theta_1 = \{\varnothing\}.$$

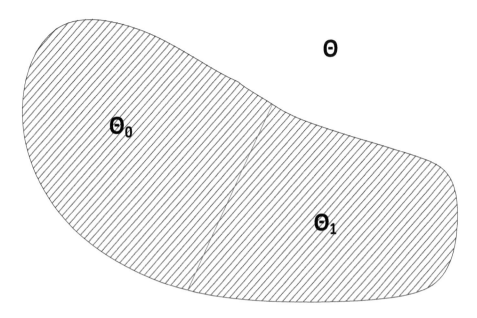

FIGURE 10.6
The detector decides on the range of values of Θ which contains the possible values for the parameter θ.

It is true that:

$$H_0 : \theta \in \Theta_0 \qquad\qquad H_0 : x \sim f(x;\theta), \quad \theta \in \Theta_0$$

$$\Leftrightarrow$$

$$H_1 : \theta \in \Theta_1 \qquad\qquad H_1 : x \sim f(x;\theta), \quad \theta \in \Theta_1.$$

where:

- H_0 is the null hypothesis and the existence of only noise.
- H_1 is the alternative hypothesis with the existence of both noise and signal.

10.1.1 Simple and Composite Hypotheses

When the θ parameter can only get two values and Θ_0, Θ_1 are monotone sets, the hypotheses are called simple:

$$\Theta = \{\theta_0, \theta_1\}, \quad \Theta_0 = \{\theta_0\}, \quad \Theta_1 = \{\theta_1\}.$$

In this case, the PDF $f(x; \theta)$ is fully known with either H_0 or H_1 being conditional.

If the hypotheses are not simple, at least one of Θ_0, Θ_1 is not a monotone set and, in this case, it is called composite.

10.1.2 The Decision Function

The purpose of detection is to efficiently design the decision rule (test function):

$$\varphi(x) = \begin{cases} 1, \text{ decide } H_1 \\ 0, \text{ decide } H_0 \end{cases}.$$

The test function $\varphi(x)$ assigns the field $\mathcal{X}$ to the decision area $\{0,1\}$ for the choice of H_0 and H_1, respectively. More specifically, $\varphi(x)$ divides $\mathcal{X}$ in decision areas (Figure 10.7):

$$\mathcal{X}_0 = \{x : \varphi(x) = 0\}, \quad \mathcal{X}_1 = \{x : \varphi(x) = 1\}.$$

Errors of miss detection and false alarm:
The probabilities of false alarm and miss detection associated with the test function φ are expressed as:

$$P_F = \mathbb{E}_\theta[\varphi] = \int_{\mathcal{X}} \varphi(x) f(x;\theta) dx, \quad \theta \in \Theta_0,$$

$$P_M = \mathbb{E}_\theta[1-\varphi] = \int_{\mathcal{X}} (1-\varphi(x)) f(x;\theta) dx, \quad \theta \in \Theta_1.$$

Equivalently, it is true that:

$$P_F = \int_{\mathcal{X}_1} f(x|\theta) dx, \quad \theta \in \Theta_0,$$

$$P_M = 1 - \int_{\mathcal{X}_1} f(x|\theta) dx, \quad \theta \in \Theta_1.$$

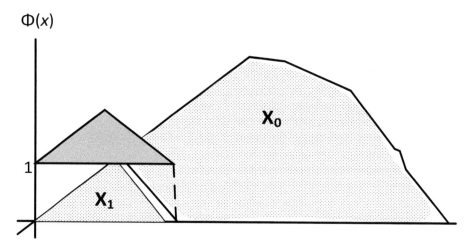

FIGURE 10.7
The test function splits the measurement space in two decision areas.

The probability of a right decision for H_1 is called detection probability:

$$P_D(\theta) = 1 - P_M(\theta) = \mathbb{E}_\theta[\varphi], \quad \theta \in \Theta_1,$$

and is illustrated in Figure 10.8.
    Figure 10.9 illustrates the above metrics.
    Next, the cases of random or non-random θ are examined separately.

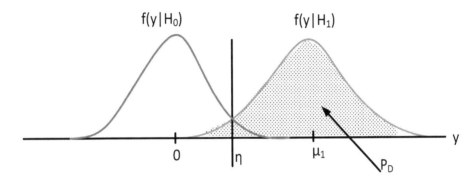

FIGURE 10.8
The probability of detection $P_D = 1 - P_M$

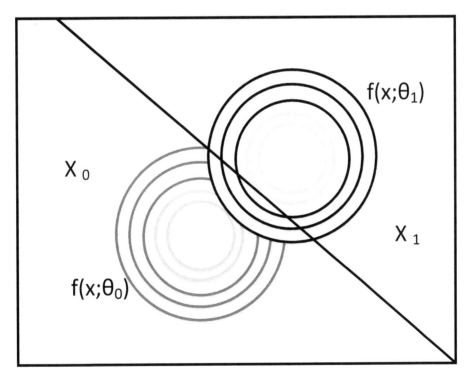

FIGURE 10.9
Graphic representation of decision areas $\mathcal{X}_0, \mathcal{X}_1$ with respect to decisions H_0, H_1 for an observation of RV x. The probability of a false alarm, P_F, is given by the integral of $f(x;\theta_0)$ in $\mathcal{X}_1$, the probability of miss detection, P_M, is given by the integral of $f(x;\theta_1)$ in $\mathcal{X}_0$, and the probability of detection, P_D, is given by the integral of $f(x;\theta_1)$ in $\mathcal{X}_1$.

10.2 Bayes Approach to the Detection Problem

In order to implement the Bayes approach, the following must be considered:

1. Assign an a priori PDF $f(\theta)$ for θ.
2. Assign cost or risk to miss decisions: c_{ij} = cost when H_i is decided, but H_j is really true.
3. Find and implement a decision rule that exhibits the minimum average risk.

10.2.1 Assign a Priori Probabilities

The a priori probabilities for H_0, H_1 are defined as:

$$P(H_0) = P(\theta \in \Theta_0) = \int_{\Theta_0} f(\theta)d\theta,$$

$$P(H_1) = P(\theta \in \Theta_1) = \int_{\Theta_1} f(\theta)d\theta.$$

with $P(H_0) + P(H_1) = 1$.

The calculation of conditional PDFs, of H_0 and H_1, is derived by integration at θ:

$$f(x|H_0) = \frac{\int_{\Theta_0} f(x|\theta)f(\theta)d\theta}{P(H_0)},$$

$$f(x|H_0) = \frac{\int_{\Theta_1} f(x|\theta)f(\theta)d\theta}{P(H_1)}.$$

10.2.2 Minimization of the Average Risk

We define the cost or risk matrix as follows:

$$\mathbf{C} \triangleq \begin{bmatrix} c_{11} & c_{10} \\ c_{01} & c_{00} \end{bmatrix}.$$

We consider $c_{ii} \leq c_{ij}$, i.e., the cost of a right decision is less than the cost of a miss decision. The actual cost for a given RV x, denoted as C, is a function of (a) the output $\varphi(x)$ of the test and (b) the real case H_0 or H_1. Consequently, the cost $C \in \{c_{11}, c_{10}, c_{01}, c_{00}\}$ is a RV and thus, we are looking for decision rules that minimize its average value, also called the "average risk" associated with the decision function.

We accept the following Bayes criterion:
Choice of the appropriate φ (equivalently $\mathcal{X}_0, \mathcal{X}_1$), that minimizes the average risk, which is equal to the statistical expected value $\mathbb{E}[C]$ of the cost, C:

$$\mathbb{E}[C] = c_{11}P(H_1 \mid H_1)P(H_1) + c_{00}P(H_0 \mid H_0)P(H_0)$$
$$+ c_{10}P(H_1 \mid H_0)P(H_0) + c_{01}P(H_0 \mid H_1)P(H_1). \qquad (10.1)$$

Definition of Bayes' false alarm and miss detection:

$$P_F = \int_{\mathcal{X}_1} f(x \mid H_0)\,dx = P(H_1 \mid H_0),$$
$$P_M = 1 - \int_{\mathcal{X}_1} f(x \mid H_1)\,dx = P(H_0 \mid H_1). \qquad (10.2)$$

The above expressions differ from the probabilities $P_F(\theta)$ and $P_M(\theta)$, as mentioned earlier, as they represent error probabilities involving integration of θ at Θ_0 and Θ_1. Therefore, Relation 10.1 can be expressed as:

$$\mathbb{E}[C] = c_{11}P(H_1) + c_{00}P(H_0)$$
$$+ (c_{10} - c_{00})P(H_0)P_F + (c_{01} - c_{11})P(H_1)P_M.$$

NOTE: $\mathbb{E}[C]$ is linear with respect to $P_M, P_F, P(H_1), P(H_0)$ for any decision rule φ.

10.2.3 The Optimal Bayes Test Minimizes $\mathbb{E}[C]$

According to Relation 10.2, $\mathbb{E}[C]$ can be expressed as a function of the decision area $\mathcal{X}_1$:

$$\mathbb{E}[C] = c_{01}P(H_1) + c_{00}P(H_0)$$
$$+ \int_{\mathcal{X}_1} \left[(c_{10} - c_{00})P(H_0)f(x \mid H_0) - (c_{01} - c_{11})P(H_1)f(x \mid H_1) \right] dx.$$

The solution is now obvious: if we had to choose to assign a point x to x to $\mathcal{X}_1$ or $\mathcal{X}_0$, then we would only choose $\mathcal{X}_1$ if it reduced the average risk (i.e., it would make the integral negative). Thus, we assign x to $\mathcal{X}_1$ if

$$(c_{10} - c_{00})P(H_0)f(x \mid H_0) < (c_{01} - c_{11})P(H_1)f(x \mid H_1),$$

and x to $\mathcal{X}_0$, otherwise.

When $c_{10} > c_{00}$ and $c_{01} > c_{11}$, the optimal test is expressed by the Bayes likelihood ratio test (BLRT):

$$\Lambda_B(x) \triangleq \frac{f(x|H_1)}{f(x|H_0)} \underset{H_0}{\overset{H_1}{\underset{<}{>}}} \eta,$$

where η is the optimal Bayes threshold:

$$\eta = \frac{(c_{10} - c_{00})P(H_0)}{(c_{01} - c_{11})P(H_1)}.$$

We notice that the costs and the a priori probability $p = P(H_0) = 1 - P(H_1)$ affect the BLRT only via the threshold η, since the BLRT does not depend on p.

10.2.4 Minimum Probability of the Error Test

Let the special case, where $c_{00} = c_{11} = 0$ and $c_{01} = c_{10} = 1$. This changes the average risk of the probability error criterion:

$$\mathbb{E}[C] = P_M P(H_1) + P_F P(H_0) = P_e,$$

which is minimized by the likelihood ratio test:

$$\frac{f(x|H_1)}{f(x|H_0)} \underset{H_0}{\overset{H_1}{\underset{<}{>}}} \frac{P(H_0)}{P(H_1)}.$$

Using Bayes' rule, we can easily see that it is equivalent to the maximum a posteriori (MAP) test:

$$\frac{P(H_1|x)}{P(H_0|x)} \underset{H_0}{\overset{H_1}{\underset{<}{>}}} 1.$$

10.2.5 Evaluation of the Performance of Bayes Likelihood Ratio Test

For the sake of simplicity of symbolism, let $\bar{C} \triangleq \mathbb{E}[C]$. Consider $\bar{C}^*$ to be the minimum average risk derived by the BLRT:

$$\bar{C}^* = c_{11}P(H_1) + c_{00}P(H_0)$$

$$+ (c_{10} - c_{00})P(H_0)P_F^*(\eta) + (c_{01} - c_{11})P(H_1)P_M^*(\eta).$$

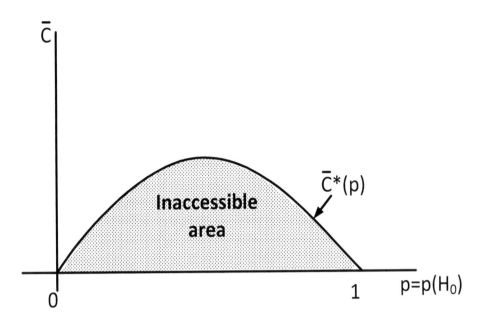

FIGURE 10.10
The minimum (average) risk curve, corresponding to the optimal BLRT, describes a realizable lower threshold at the average risk of any test

where

$$P_F^*(\eta) = P(\Lambda_B > \eta \mid H_0), \quad P_M^*(\eta) = P(\Lambda_B \leq \eta \mid H_1).$$

Interpreting $\bar{C}^* = \bar{C}^*(p)$ as a function of p, the minimum average risk is represented as the performance curve illustrated in Figure 10.10, which is a function of $p = P(H_0)$. Note that the average risk of any test is linear at p. The minimum risk curve describes a lower threshold at the average risk that is encountered to any test and any value of p.

10.2.6 The Minimax Bayes Detector

In many cases, the real value of p is not known. Therefore, the BLRT optimal threshold cannot be applied. Since all tests present a linear average risk, this average risk may take very large (unacceptable) values as p approaches either 0 or 1. This is represented by the straight line in Figure 10.11. It is true that:

$$\bar{C}_{\text{minimax}} = \max_{p \in [0,1]} \bar{C}(p).$$

It is evident from Figure 10.11 that the minimax test is the optimal Bayes test (i.e., a test with its average risk line tangent to the minimum risk curve), which is implemented using the

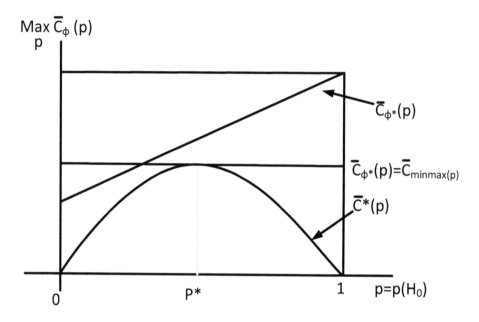

FIGURE 10.11
The curve of each test φ is a straight line. The optimal minimax test φ^* is depicted by a horizontal line that is tangential to the minimum risk, named $\bar{C}^*(p)$, at its maximum value.

threshold η^* that causes $\bar{C}$ to be a horizontal line (the slope of $\bar{C}$ must be zero). Therefore, we have the following optimization minimax condition:

$$\bar{C} = \underbrace{\left[c_{00}\left(1 - P_F^*(\eta)\right) + c_{10}P_F^*(\eta) - c_{11}\left(1 - P_M^*(\eta)\right) - c_{01}\left(1 - P_M^*(\eta)\right) \right]}_{=0} p$$
$$+ c_{11}\left(1 - P_M^*(\eta)\right) + c_{01}P_M^*(\eta),$$

where $P_F^*(\eta)$ and $P_M^*(\eta)$ are the Bayes probabilities for the false alarm and the miss detection respectively, of the BLRT with threshold η.

In the specific case that $\bar{C} = P_e : c_{00} = c_{11} = 0, c_{10} = c_{01} = 1$, the minimax condition for the MAP test is derived, which is expressed as:

$$\bar{C} = \underbrace{\left[P_F^*(\eta) - P_M^*(\eta) \right]}_{=0} p + P_M^*(\eta).$$

This implies that the threshold η should be chosen in a way such that the "equalization" condition is satisfied:

$$P_F^*(\eta) = P(\Lambda_B > \eta \mid H_0) = P(\Lambda_B \leq \eta \mid H_1) = P_M^*(\eta).$$

By substituting the minimax value of η with η^*, and by selecting the appropriate threshold by the choice of a value of p, the minimax threshold is then related to the choice of p^* through the condition $\eta^* = p^*/(1 - p^*)$.

10.2.7 Typical Example

Example 28: A Radar Application

Let a known output of the matched filter, y. The optimal Bayes estimator is to be derived. Let $P(H_0) = P(H_1) = 1/2$. It is true that:

$$y = \int_0^T s(t)x(t)\,dt,$$

which constitutes the realization of a Gaussian RV with an expected value and variance:

$$\mathbb{E}[y \mid H_0] = 0, \qquad \mathrm{var}(y \mid H_0) = \frac{N_0}{2}\int_0^T |s(t)|^2\,dt = \sigma_0^2,$$

$$\mathbb{E}[y \mid H_1] = \int_0^T |s(t)|^2\,dt = \mu_1, \qquad \mathrm{var}(y \mid H_1) = \frac{N_0}{2}\int_0^T |s(t)|^2\,dt = \sigma_0^2.$$

The likelihood ratio test is given by:

$$\Lambda_B(y) = \frac{\dfrac{1}{\sqrt{2\pi\sigma_0^2}}e^{-\frac{1}{2\sigma_0^2}(y-\mu_1)^2}}{\dfrac{1}{\sqrt{2\pi\sigma_0^2}}e^{-\frac{1}{2\sigma_0^2}y^2}}$$

$$= e^{y\mu_1/\sigma_0^2 - \frac{1}{2}\mu_1^2/\sigma_0^2} \underset{H_0}{\overset{H_1}{\gtrless}} \eta = 1.$$

Also, the statistic of the likelihood ratio test $\Lambda_B(y)$ is a monotonic function of y, since $\mu_1 > 0$. An equivalent test is via the threshold detector-filter:

$$y \underset{H_0}{\overset{H_1}{\underset{<}{\overset{>}{}}}} \gamma = \frac{1}{2}\mu_1.$$

For the example discussed here, the Bayes test performance is described as:

$$
\begin{aligned}
P_F &= P(y > \gamma | H_0) \\
&= P(\underbrace{y/\sigma_0}_{\mathcal{N}(0,1)} > \gamma/\sigma_0 | H_0) \\
&= \int_{\gamma/\sigma_0}^{+\infty} \frac{1}{\sqrt{2\pi}} e^{-u^2/2}\, du \\
&= 1 - \mathcal{N}(\gamma, \sigma_0) \triangleq Q(\gamma/\sigma_0),
\end{aligned}
$$

and

$$
\begin{aligned}
P_M &= P(y < \gamma | H_1) \\
&= P\left(\underbrace{(y-\mu_1)/\sigma_0}_{\mathcal{N}(0,1)} > (\gamma - \mu_1)/\sigma_0 \,|\, H_1 \right) \\
&= \mathcal{N}\left((\gamma - \mu_1)/\sigma_0 \right).
\end{aligned}
$$

NOTE: Since the standard Gaussian PDF is symmetric $\mathcal{N}(-u) = 1 - \mathcal{N}(u)$ and $\gamma = \mu_1/2$, thus:

$$P_M = P_F = 1 - \mathcal{N}\left(\mu_1/(2\sigma_0) \right).$$

We conclude that the Bayes threshold is minimax. Therefore, the error probability is reduced as follows:

$$
\begin{aligned}
P_e &= P_M^{1/2} + P_F^{1/2} \\
&= P_F.
\end{aligned}
$$

Figures 10.12 through 10.14 illustrate the above results.

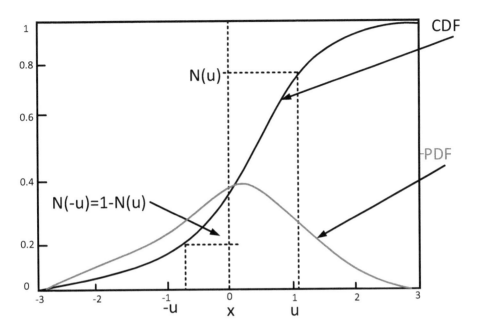

FIGURE 10.12
The CDF N(*u*) of the Gaussian distribution.

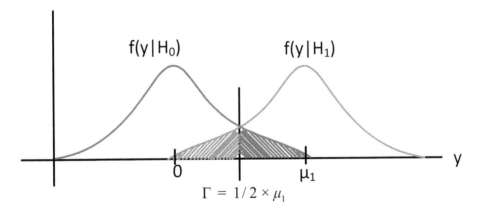

FIGURE 10.13
Hypotheses of equal likelihoods, which have a minimax threshold $\gamma = \frac{1}{2}\mu_1$ for the shift detection problem of the expected value of a Gaussian RV.

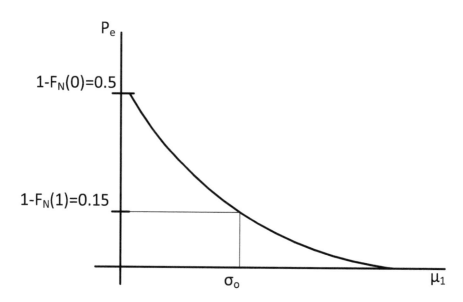

FIGURE 10.14
The probability error curve of the Bayes likelihood ratio test as a function of $\mu_1 = \|s\|^2$.

10.3 Multiple Hypotheses Tests

The objective is to measure the RV x, which has a conditional PDF $f(x|\theta)$, with $\theta \in \Theta$. Also, in this section, we consider that the total parametric space is divided into subspaces $\Theta_1, \cdots, \Theta_M$.
   For the test of M hypotheses for θ:

$$H_1 : \theta \in \Theta_1$$
$$\vdots$$
$$H_M : \theta \in \Theta_M$$

the decision function is defined as:

$$\boldsymbol{\varphi}(x) \triangleq \left[\varphi_1(x), \cdots, \varphi_M(x)\right]^T,$$

where

$$\varphi_i(x) \in \{0,1\}, \quad \sum_{i=1}^{M} \varphi_i(x) = 1.$$

NOTE: The decision function determines the partitioning of the measurement area $\mathcal{X}$ in M decision areas (Figures 10.15 and 10.16):

$$\mathcal{X} = \{x : \varphi_i = 1\}, \quad i = 1, \cdots, M.$$

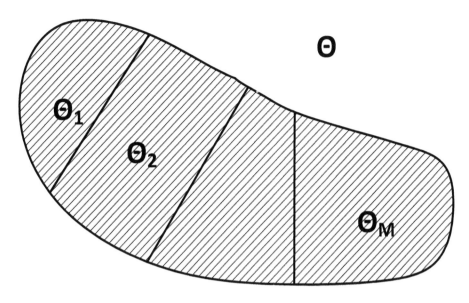

FIGURE 10.15
Space partitioning of Θ into M different regions.

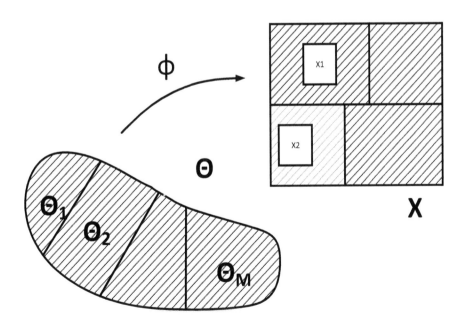

FIGURE 10.16
The partitioning of the parametric space in M hypotheses is equivalent (via function $\varphi(x)$) with the partitioning of X in M areas.

The Bayes approach is based on three factors:

1. Assign of an a priori PDF $f(\theta)$ for θ.
2. Assign cost to miss decisions: c_{ij} = cost when H_i is decided, but H_j is true.
3. Derive and implement a decision rule that presents the minimum average risk.

10.3.1 A priori Probabilities

The a priori probabilities of H_i $(i = 1, \cdots, M)$ are defined as:

$$P(H_i) = P(\theta \in \Theta_i) = \int_{\Theta_i} f(\theta) d\theta,$$

with $\sum_{i=1}^{M} P(H_i) = 1$.

The corresponding conditional PDF are derived:

$$f(x|H_i) = \frac{\int_{\Theta_i} f(x|\theta)f(\theta)d\theta}{P(H_i)},$$

thus, the above composite hypotheses are transformed in the following simple hypotheses:

$$H_1 : x \sim f(x|H_1)$$
$$\vdots$$
$$H_M : x \sim f(x|H_M)$$

where H_i has an a priori probability $P(H_i)$.

10.3.2 Minimization of the Average Risk

The dimensions of the cost or risk matrix is $M \times M$, and the matrix is given by:

$$\mathbf{C} = \begin{bmatrix} c_{11} & \cdots & c_{1M} \\ \vdots & \ddots & \vdots \\ c_{M1} & \cdots & c_{MM} \end{bmatrix}.$$

Design criterion:
Choose the appropriate φ (alternatively, $\{\mathcal{X}_i\}_{i=1}^{M}$) so as to minimize the average risk $\mathbb{E}[C] = \bar{C}$:

$$\bar{C} = \sum_{i,j=1}^{M} c_{ij} P(H_i|H_j) P(H_j).$$

Next, the specific case is studied, where:

- $c_{ii} = 0$,
- $c_{ij} = 1, i \neq j$

In this case, it is true that $\bar{C} = P_e$:

$$\bar{C} = \sum_{i,j:i\neq j} P(H_i|H_j)P(H_j)$$

$$= 1 - \sum_{i,j:i=j} P(H_i|H_j)P(H_j)$$

$$= 1 - \sum_{i,j:i=j} P(x \in \mathcal{X}_i|H_j)P(H_i)$$

$$= 1 - \sum_{i=1}^{M} \int_{\mathcal{X}_i} f(x|H_i)P(H_i)dx.$$

In order for $\bar{C}$ to get the lowest possible value, it should be true that:

$$x \in \mathcal{X}_i \Leftrightarrow f(x|H_i)P(H_i) \geq f(x|H_j)P(H_j), \quad i \neq j,$$

or, with respect to the decision function:

$$\varphi_i(x) = \begin{cases} 1, & f(x|H_i)P(H_i) \geq f(x|H_j)P(H_j) \\ 0, & f(x|H_i)P(H_i) < f(x|H_j)P(H_j) \end{cases}.$$

Introducing the following symbolism for simplicity:

$$\hat{H}_i = \hat{H}_i(x) = \text{argmax}_{H_j}\left\{f(x|H_j)P(H_j)\right\},$$

we can see that it is equivalent to the MAP rule:

$$\hat{H}_i = \text{argmax}_{H_j}\left\{P(H_j|x)\right\}.$$

NOTES:

- The MAP decision rule minimizes the average error probability P_e.
- The average error probability P_e is equal to:

$$P_e = 1 - \sum_{i=1}^{M} \mathbb{E}\left[\varphi_i(x)|H_i\right]P(H_i).$$

- The MAP decision rule for the RV x depends only on the likelihood ratio (which is statistically sufficient).
- For equiprobable H_i, $P(H_i) = 1/M$ and MAP are of the form:

$$\hat{H}_i = \mathrm{argmax}_{H_j}\left\{f(x\,|\,H_j)\right\},$$

which can be also expressed as: "the estimation $\hat{H}_i = H_1$ if $f(x|H_1) > f(x|H_0)$". This expresses the estimation of maximum likelihood of the true hypothesis H_j.

Example 29: Classifier of Gaussian Average Values

Let $\mathbf{x} = [x_1,\cdots,x_n]^T$ be an IUD RV $\mathcal{N}(\mu,\sigma^2)$ with a known variance σ^2.
Objective: Classification of μ to three possible values:

$$H_1 : \mu = \mu_1,$$
$$H_2 : \mu = \mu_2,$$
$$H_3 : \mu = \mu_3,$$

considering equiprobable hypotheses.
It is known that the MAP classifier depends on x, only via the statistic sufficiency for μ:

$$\bar{x} = n^{-1}\sum_{i=1}^{n} x_i,$$

which is a Gaussian RV with expected value, μ, and variance, σ^2/n.
Thus, the MAP test is of the following form:
Decision of the hypothesis H_k if and only if:

$$f(\bar{x}\,|\,H_k) \geq f(\bar{x}\,|\,H_j),$$

where

$$f(\bar{x}\,|\,H_k) = \frac{1}{\sqrt{2\pi\sigma^2/n}}\exp\left(-\frac{1}{2\sigma^2/n}(\bar{x}-\mu_k)^2\right),$$

or, alternatively, after some cancellations and taking the log of the above relation:

$$\bar{x}\mu_k - \frac{1}{2}\mu_k^2 \geq \bar{x}\mu_j - \frac{1}{2}\mu_j^2.$$

We conclude that these are linear decision areas.
Numerical example: Let $\mu_1 = -1, \mu_2 = +1, \mu_3 = 2$.

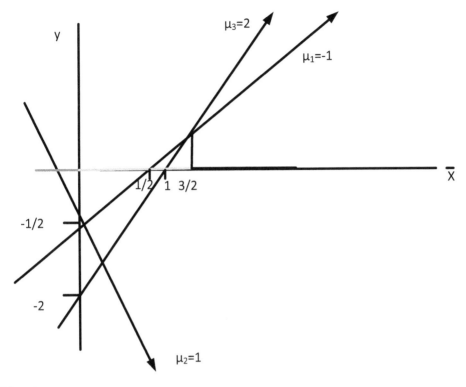

FIGURE 10.17
For three hypotheses of the Gaussian average, the corresponding decision areas are determined by the intersections of three lines $y = \bar{x}\mu_k - \frac{1}{2}\mu_k^2$, $k = 1, 2, 3$, which are depicted as a function of $\bar{x}$.

We design three lines as a function of $\bar{x}$ in order to locate the decision areas:

$$\mathcal{X}_1 = \{\mathbf{x} : \bar{x} \leq 0\},$$
$$\mathcal{X}_2 = \{\mathbf{x} : 0 \leq \bar{x} \leq 3/2\},$$
$$\mathcal{X}_3 = \{\mathbf{x} : \bar{x} \geq 3/2\}.$$

These areas are separated by specific levels in $\mathcal{X} = \mathbb{R}^n$ (Figure 10.17).

10.3.3 Disadvantages of Bayes Approach

The disadvantages of this approach are:

- A priori knowledge of $\theta, H_0, H_1, \ldots$ is required.
- It ensures only the optimal average performance with respect to the a priori chosen parameter.
- It does not guarantee protection of the metric of the false alarm and the miss detection.

10.4 Frequentist Approach for Detection

The Frequentist approach is not based on a priori knowledge of H_0, H_1. Consequently, the average probability of error or risk cannot be defined and minimized thereafter. Thus, an alternative criterion is adopted: limitation of the false alarm (up to a given value) and then minimization of the probability of miss detection. It is derived that, in order to find the optimal test satisfying this condition, we need to extend the aforementioned definition of the test function φ, so as to allow random decisions:

$$\varphi(x) = \begin{cases} 1, & \text{estimate } H_1 \\ q, & \text{probability of } H_1 = q. \\ 0, & \text{estimate } H_0 \end{cases}$$

We note that the above relation is interpreted as:

$$\varphi(x) = P(H_1 | \text{ given the observation of } x).$$

The probabilities of false alarm and miss detection are functions of θ:

$$\mathbb{E}_\theta[\varphi] = \int_\mathcal{X} \varphi(x) f(x;\theta) dx = \begin{cases} P_F(\theta), & \theta \in \Theta_0, \\ P_D(\theta), & \theta \in \Theta_1. \end{cases}$$

Definition: A test φ is at the level $a \in [0,1]$ regarding the false alarm when:

$$\max_{\theta \in \Theta_0} P_F(\theta) \le a.$$

Definition: The power function of a test φ is given by:

$$\beta(\theta) \triangleq P_D(\theta) = 1 - P_M(\theta), \quad \theta \in \Theta_1.$$

10.4.1 Case of Simple Hypotheses: $\theta \in \{\theta_0, \theta_1\}$

Let:

$$H_0 : x \sim f(x;\theta_0),$$
$$H_1 : x \sim f(x;\theta_1).$$

We follow the *Neyman-Pearson* strategy, which detects the most powerful (MP) test φ^* of level a:

$$\mathbb{E}_{\theta_1}[\varphi^*] \ge \mathbb{E}_{\theta_1}[\varphi],$$

for all trials that satisfy the condition $\mathbb{E}_{\theta_0}[\varphi] \le a$.

Lemma Neyman-Pearson: The MP test of level $a \in [0,1]$ is a random likelihood ratio test of the form:

$$\varphi^*(x) = \begin{cases} 1, & f(x;\theta_1) > \eta\, f(x;\theta_0), \\ q, & f(x;\theta_1) = \eta\, f(x;\theta_0), \\ 0, & f(x;\theta_1) < \eta\, f(x;\theta_0). \end{cases} \qquad (10.3)$$

where the η and q are appropriately chosen to satisfy:

$$\mathbb{E}_{\theta_0}[\varphi^*] = a.$$

Proof:
First way:
We use the Karush-Kuhn-Tucker method to maximize under constraints.

The MP test mentioned above maximizes the term $\mathbb{E}_{\theta_1}[\varphi(x)]$ under the constraint $\mathbb{E}_{\theta_0}[\varphi(x)] \le a$. This estimation problem under a constraint is equivalent to maximizing the following objective function without constraints:

$$\mathcal{L}(\varphi) \triangleq \mathbb{E}_{\theta_1}[\varphi(x)] + \lambda(a - \mathbb{E}_{\theta_0}[\varphi(x)]),$$

where $\lambda > 0$ is the Lagrange multiplier, which is chosen so that the resulting solution φ^* satisfies the equality relation to the aforementioned constraint (i.e., $\mathbb{E}_{\theta_0}[\varphi(x)] = a$).

Consequently, the resulting power can be expressed through the transformation of the expected value of the likelihood ratio, also known as "Girsanov representation":

$$\mathbb{E}_{\theta_1}[\varphi(x)] = \mathbb{E}_{\theta_0}\left[\varphi(x) \frac{f(x;\theta_1)}{f(x;\theta_0)}\right],$$

and thus:

$$\mathcal{L}(\varphi) \triangleq \mathbb{E}_{\theta_0}\left[\varphi(x)\left(\frac{f(x;\theta_1)}{f(x;\theta_0)} - \lambda\right)\right] + \lambda\, a.$$

According to the above relations, for a random realization of x, we only set $\varphi(x) = 1$ if the likelihood ratio exceeds the value λ. If the likelihood ratio is less than λ, we set $\varphi(x) = 0$. What remains is the case where the likelihood ratio is equal to λ. In this case, we make a random decision (set $\varphi(x) = q, 0 < q < 1$), so that the desired level of false alarm is satisfied.

Second way:
For any value of φ, φ^* satisfies:

$$\mathbb{E}_{\theta_1}[\varphi^*] \ge \mathbb{E}_{\theta_1}[\varphi], \text{ when } \mathbb{E}_{\theta_0}[\varphi^*] = a,\ \mathbb{E}_{\theta_0}[\varphi] \le a.$$

Step 1:
By listing all the possible cases ">", "<" and "=" among the terms of Equation 10.3, we have:

$$\varphi^*(x)\big[f(x;\theta_1)-\eta\,f(x;\theta_0)\big]\ge\varphi(x)\big[f(x;\theta_1)-\eta\,f(x;\theta_0)\big]. \tag{10.4}$$

Step 2:
We integrate Relation 10.4 to all possible values of x:

$$\int_{\mathcal{X}}\varphi^*(x)\big[f(x;\theta_1)-\eta\,f(x;\theta_0)\big]dx\ge\int_{\mathcal{X}}\varphi(x)\big[f(x;\theta_1)-\eta\,f(x;\theta_0)\big]dx$$

$$=\underbrace{\int_{\mathcal{X}}\varphi^*(x)f(x;\theta_1)dx}_{\mathbb{E}_{\theta_1}[\varphi^*]}-\eta\underbrace{\int_{\mathcal{X}}\varphi^*(x)f(x;\theta_0)dx}_{\mathbb{E}_{\theta_0}[\varphi^*]}$$

$$\ge\underbrace{\int_{\mathcal{X}}\varphi(x)f(x;\theta_1)dx}_{\mathbb{E}_{\theta_1}[\varphi]}-\eta\underbrace{\int_{\mathcal{X}}\varphi(x)f(x;\theta_0)dx}_{\mathbb{E}_{\theta_0}[\varphi]}.$$

Hence,

$$\mathbb{E}_{\theta_1}[\varphi^*]-\mathbb{E}_{\theta_1}[\varphi]\ge\eta\bigg(\underbrace{\mathbb{E}_{\theta_0}[\varphi^*]}_{=a}-\underbrace{\mathbb{E}_{\theta_0}[\varphi]}_{\le a}\bigg)\ge0,$$

which proves the result of Neyman-Pearson Lemma.

General remarks concerning the MP tests:

a. The definition of the likelihood ratio test:

$$\Lambda(x)=\frac{f(x;\theta_1)}{f(x;\theta_0)}\overset{H_1}{\underset{H_0}{\gtrless}}\eta.$$

b. The probability of false alarm of the MP test is $\big(\Lambda\triangleq\Lambda(x)\big)$:

$$P_F=\mathbb{E}_{\theta_0}\big[\varphi^*(x)\big]=\underbrace{P_{\theta_0}(\Lambda>\eta)}_{1-F_\Lambda(\eta|H_0)}+q\,P_{\theta_0}(\Lambda=\eta). \tag{10.5}$$

"Randomization" should only occur when it is impossible to find the appropriate value n, so that $P_{\theta_0}(\Lambda>\eta)=a$. This is achieved only in the case where the CDF $F_\Lambda(t|H_0)$ has jump discontinuities, i.e., when there are points $t>0$, where $P_{\theta_0}(\Lambda=t)>0$ and $\Lambda=\Lambda(x)$ is not a continuous RV. Otherwise, $q=0$ and therefore, such randomization is not necessary.

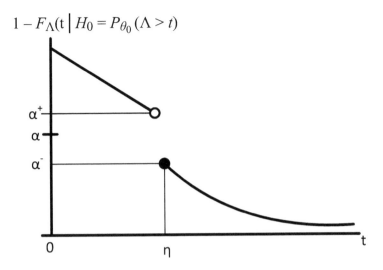

FIGURE 10.18
Randomization is necessary to achieve level *a*, when 1- *a* is not within the CDF domain of Λ.

When an appropriate value of *n* cannot be derived, so as to make $P_{\theta_0}(\Lambda > \eta) = a$ true, the implementation process is described as follows (Figure 10.18):

- We derive the smallest value of *t* for which the probability $P_{\theta_0}(\Lambda > \eta)$ is less than *a* when a jump discontinuity is detected in the CDF. We set this value as a^- and the threshold η equal to this value of *t*.
- We define $a^+ = P_{\theta_0}(\Lambda = \eta) + a^-$, where a^- and η are already defined above. From Relation 10.5 it is true that for any *q* value, the test has the following false alarm probability:

$$P_F = a^- + q(a^+ - a^-).$$

By setting $P_F = a$, the above equation can be solved for *q*:

$$q = \frac{a - a^-}{a^+ - a^-}. \tag{10.6}$$

c. The likelihood ratio is the same as the Bayes likelihood ratio in the case of simple hypotheses.

d. Unlike the BLRT, the threshold η is only determined by a value *a*.

e. If $T = T(x)$ is statistically sufficient for θ, the likelihood ratio test (LRT) depends on *x*, only via $T(x)$. Indeed, if $f(x;\theta) = g(T,\theta)h(x)$, then:

$$\Lambda(x) = \frac{g(T,\theta_1)}{g(T,\theta_0)} = \Lambda(T).$$

Conclusion: We can formulate LRT based on the PDF of *T*, instead of using the PDF of the entire *x* data sample.

10.5 ROC Curves for Threshold Testing

All threshold tests have P_F and P_D, which are parameterized by a value η. The receiver operating characteristic (ROC) curve is simply the depiction of the parametric curve $\{P_F(\eta,q),P_D(\eta,q)\}_{\eta,q}$. Equivalently, ROC curve is the depiction of $\beta = P_D$ with respect to $a = P_F$ (Figure 10.19).

Properties of the ROC curve:

1. The ROC curve for two equiprobable states, e.g., the coin flipping ($\varphi(x) = q$, independent sample data) is a diagonal straight line with a slope $= 1$.

$$a = P_F = \mathbb{E}_{\theta_0}[\varphi] = q,$$
$$\beta = P_D = \mathbb{E}_{\theta_1}[\varphi] = q.$$

2. The ROC curve of an MP test is always located over the diagonal: The MP test is objective (a test is objective if the probability of detecting β is at least equal to the probability of a false alarm $a{:}\beta \geq a$).
3. The ROC curve of each MP test is always concave.

   Figures 10.20 through 10.22 illustrate these properties.

   To determine the concave behavior of the ROC curve, let (a_1,β_1) be the level and the power of a test φ_1, and (a_2,β_2) be the level and the power of a test φ_2. We define the test:

$$\varphi_{12} = p\,\varphi_1 + (1-p)\varphi_2.$$

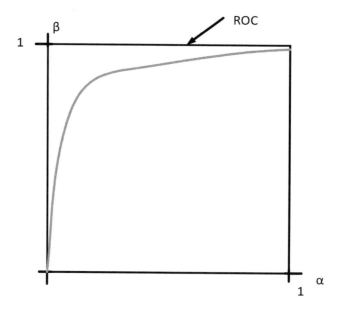

FIGURE 10.19
The ROC curve.

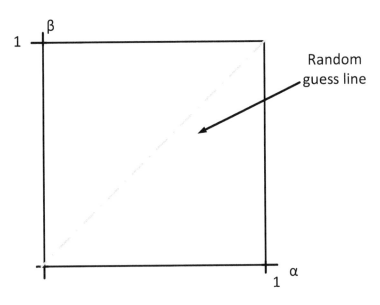

FIGURE 10.20
Illustration of the ROC curve for equiprobable detection.

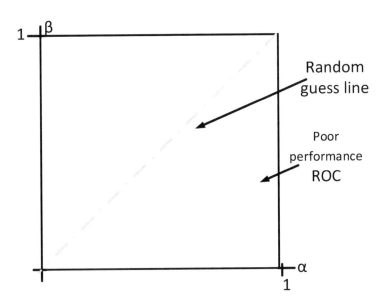

FIGURE 10.21
Plot of the ROC curve, which is always located above the diagonal for MP tests.

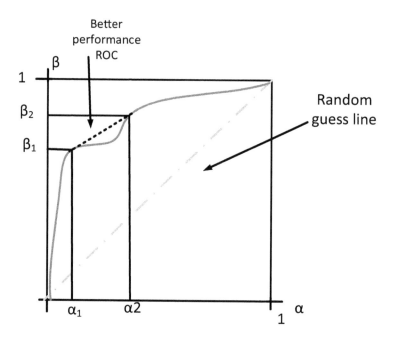

FIGURE 10.22
The ROC curve for each MP test is always concave. A non-smooth curvature test (bold line) can be improved by randomization, through which two reference points (a_1, β_1) and (a_2, β_2) are linked by a straight line.

This test can be applied by random choice of φ_1 and φ_2 with probability p and $1-p$, respectively. The level of this test is expressed as:

$$a_{12} = \mathbb{E}_0[\varphi_{12}] = p\,\mathbb{E}_0[\varphi_1] + (1-p)\mathbb{E}_0[\varphi_2] = p\,a_1 + (1-p)a_2,$$

and its power reads as:

$$\beta_{12} = \mathbb{E}_1[\varphi_{12}] = p\,\beta_1 + (1-p)\beta_2.$$

Therefore, as p varies between 0 and 1, φ_{12} has a performance (a_{12}, β_{12}) that varies on a straight line connecting the points a_1, β_1 and a_2, β_2.

4. If the ROC curve is differentiable, the threshold MP-LRT is necessary to approximate the minimax P_e, which can be calculated graphically via the ROC gradient at a:

$$\eta = \frac{d}{da} P_D(a).$$

5. When the hypotheses H_0 and H_1 are simple, the MP-LRT threshold necessary to approximate the minimax P_e can be calculated graphically by the intersection of the line $P_M = 1 - P_D = P_F$ with the ROC curve.

The above properties are illustrated in Figures 10.23 and 10.24.

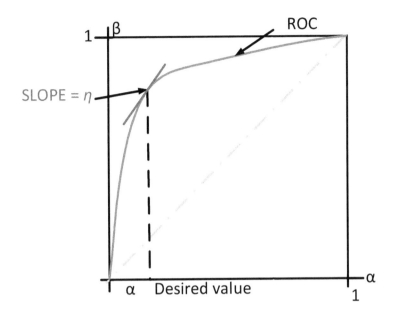

FIGURE 10.23
The MP-LRT threshold can be calculated by differentiating the ROC curve.

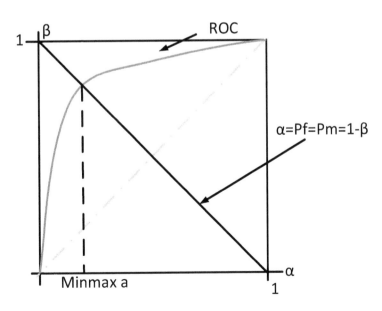

FIGURE 10.24
The threshold of the minimax Bayes test, which can be calculated by the appropriate line intersection.

Example 30: Test with Respect to an Identical PDF

Let two hypotheses for the scalar RV x:

$$H_0 : f(x) = f_0(x),$$
$$H_1 : f(x) = f_1(x),$$

where f_0 and f_1 represent two PDFs, as illustrated in Figure 10.25.
Objective: Derive the MP-LRT.

SOLUTION:

The LRT is expressed as:

$$\Lambda(x) = \frac{f_1(x)}{f_0(x)} \underset{\substack{< \\ H_0}}{\overset{\substack{H_1 \\ >}}{}} \eta,$$

or, alternatively:

$$f_1(x) \underset{\substack{< \\ H_0}}{\overset{\substack{H_1 \\ >}}{}} \eta\, f_0(x).$$

In Figure 10.26, it is obvious that for a conditional n, the decision area H_1 is:

$$\mathcal{X}_1 = \begin{cases} \{\eta/4 < x < 1-\eta/4\}, & 0 \le \eta \le 2 \\ \varnothing, & \text{otherwise} \end{cases}.$$

In order to choose the appropriate threshold η, we must consider the constraint $P_F = a$.

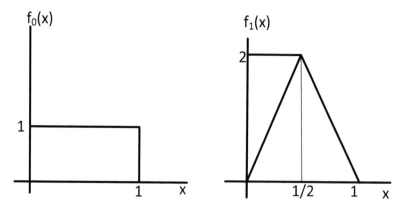

FIGURE 10.25
Two PDFs under test.

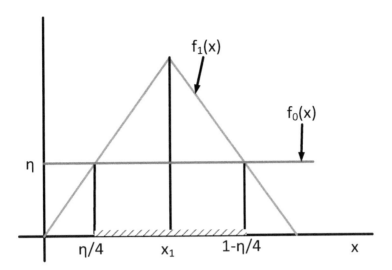

FIGURE 10.26

The region $\mathcal{X}_1$ for which MP-LRT decides on the hypothesis H_1 and is a set of x values for which the triangle exceeds the horizontal line of height η.

At first, we try to set a value to η without randomization ($q = 0$). Therefore, we consider $\eta \in [0,2]$:

$$a = P(x \in \mathcal{X}_1 \mid H_0) = \int_{\eta/4}^{1-\eta/4} f_0(x)\,dx$$

$$= 1 - \eta/2.$$

Thus, by solving with respect to η, it yields:

$$\eta = 2(1-a),$$

and, therefore, no randomization is required.

The MP-LRT power (probability of detection) is expressed as:

$$P_D = P(x \in \mathcal{X}_1 \mid H_1) = \int_{\eta/4}^{1-\eta/4} f_1(x)\,dx$$

$$= 2\int_{\eta/4}^{1/2} f_1(x)\,dx = 2\int_{\eta/4}^{1/2} 4x\,dx$$

$$= 1 - \eta^2/4.$$

By substituting $\eta = 2(1-a)$ in the above relation, the ROC curve is derived:

$$\beta = 1 - (1-a)^2.$$

Example 31: Detection of Increase in Poisson Rate

Let x be the readout of the number of photons collected by a charge array in a certain period of time. The average number of photons θ_0 that are falling upon the array is constant and known because of the environmental conditions. When there is a known photon source; the photon rate is increased to a known value θ_1, where $\theta_1 > \theta_0$. The objective of the photodetector is to detect the presence of activity based on the measurement of x. Typically, the RV x is Poisson-distributed:

$$x \sim f(x;\theta) = \frac{\theta^x}{x!}e^{-\theta}, \quad x = 0,1,\dots$$

and the problem lies in detecting the change from θ_0 in θ_1, with respect to the θ Poisson rate parameter, i.e., in modeling of the test using the following simple hypotheses:

$$H_0 : \theta = \theta_0,$$
$$H_1 : \theta = \theta_1,$$

where $\theta_1 > \theta_0 > 0$. At this point, we want to design the MP test at a predetermined level $a \in [0,1]$.

SOLUTION:

We already know that the MP test is modeled as the LRT:

$$\Lambda(x) = \left(\frac{\theta_1}{\theta_0}\right)^x e^{\theta_0 - \theta_1} \underset{H_0}{\overset{H_1}{\underset{<}{>}}} \eta.$$

Since the function of the logarithm is a monotonically increasing function and $\theta_1 > \theta_0$, the MP-LRT is equivalent to the linear test:

Since the Poisson CDF is not continuous, only a discrete set of values is reachable by the normalized LRT test:

$$x \underset{H_0}{\overset{H_1}{\underset{<}{>}}} \gamma.$$

where:

$$\gamma = \frac{\ln \eta + \theta_1 - \theta_0}{\ln(\theta_1/\theta_0)}.$$

First, we shall try to set the threshold γ without randomization:

$$a = P_{\theta_0}(x > \gamma) = 1 - Po_{\theta_0}(\gamma),$$

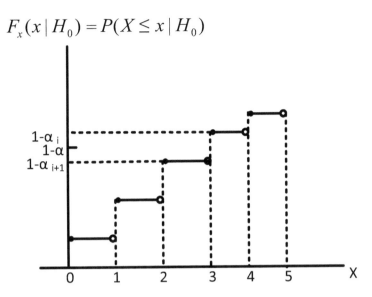

$$F_x(x \mid H_0) = P(X \le x \mid H_0)$$

FIGURE 10.27
The CDF of the LR test for the increase of the Poisson rate, which is presented as a staircase function.

where $Po_{\theta_0}(\bullet)$ is a CDF of a Poisson RV with θ rate. In this case, we encounter a difficulty, which is graphically described in Figure 10.27.

As the Poisson CDF is not continuous, only a discrete set of values is realizable by the normalized LRT test:

$$a \in \{a_i\}_{i=1}^{+\infty}, \quad a_i = 1 - Po_{\theta_0}(i).$$

Let $a \in (a_i, a_{i+1})$. In this case, we must randomize the LRT test with the choice of γ, q so that the following relation is satisfied:

$$a = P_{\theta_0}(x > \gamma) + q\, P_{\theta_0}(x = \gamma).$$

Following the aforementioned procedure and using Relation 10.6, we choose:

$$\gamma = \gamma^* \triangleq Po_{\theta_0}^{-1}(1 - a_i),$$

which yields: $P_{\theta_0}(x > \gamma^*) = a_i$. We set the random parameter according to:

$$q = q^* \triangleq \frac{a - a_i}{a_{i+1} - a_i}.$$

Based on the above relations, the probability of detection is obtained:

$$P_D = P_{\theta_1}(x > \gamma^*) + q^* P_{\theta_1}(x = \gamma^*),$$

which is illustrated in Figure 10.28.

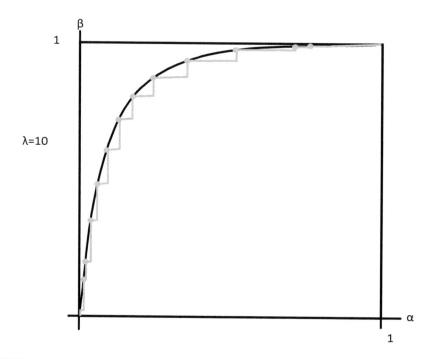

FIGURE 10.28
The smooth curve expresses the normalized MP test, while the staircase curve is the non-normalized LRT test.

Example 32: On-off Keying Technique (OOK) to Gaussian Noise

On-off keying (OOK) is a type of binary modulation used in many digital and optical communication systems. In the time interval for the transmission of one digit, the built-in output x of the receiver can be modulated as either only noise w (if the transmitted information is zero) or a constant, assumed to be 1, plus the additive noise. The detector must decide between the following hypotheses:

$$H_0 : x = w,$$
$$H_1 : x = 1 + w,$$

where it is true that $w \sim \mathcal{N}_1(0,1)$, i.e., the received SNR (Signal-to-Noise Ratio) is 0dB. The LR test is expressed as:

$$\Lambda(x) = \frac{\dfrac{1}{\sqrt{2\pi}} e - \dfrac{1}{2}(x-1)^2}{\dfrac{1}{\sqrt{2\pi}} e - \dfrac{1}{2}x^2} = e^{x - \frac{1}{2}}.$$

The ROC curve is derived, as usual, by the equation:

$$P_D = P(x > \lambda \mid H_1) = P(x - 1 > \lambda - 1 \mid H_1) = 1 - \mathcal{N}(\lambda - 1)$$
$$\Leftrightarrow \beta = P_D = 1 - \mathcal{N}(\lambda - 1) = 1 - \mathcal{N}\left(\mathcal{N}^{-1}(1-a) - 1\right).$$

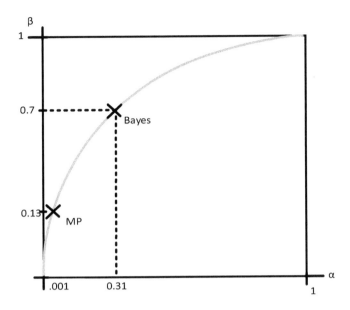

FIGURE 10.29
The ROC curve for the OOK Gaussian example.

This curve is illustrated in Figure 10.29 for three different modeling approaches of the LRT threshold: LRT Bayes, minimax LRT and MP-LRT.

1. Bayes test for the minimization of P_e, considering equiprobable hypotheses $H_0, H_1 (\eta = 1)$:

$$x \overset{H_1}{\underset{H_0}{\overset{>}{<}}} \ln \eta + \frac{1}{2} = \frac{1}{2}.$$

2. Minimax test:

$$x \overset{H_1}{\underset{H_0}{\overset{>}{<}}} \ln \eta + \frac{1}{2} \triangleq \lambda,$$

where the parameter λ is chosen to satisfy the condition:

$$P_F = 1 - \mathcal{N}(\lambda) = \mathcal{N}(\lambda - 1) = P_M.$$

The solution to this equation is, again, $\lambda = 1/2$, since $\mathcal{N}(-x) = 1 - \mathcal{N}(x)$.

3. MP test at level a:

$$x \overset{H_1}{\underset{H_0}{\overset{>}{<}}} \lambda,$$

TABLE 10.1

A Typical Performance Comparison for Three Different Threshold Settings in OOk Gaussian Example

	P_F	P_D	P_e
Bayes	0.31	0.69	0.31
Minimax	0.31	0.69	0.31
MP	0.001	0.092	0.5

where $a = P(x > \lambda \mid H_0) = 1 - \mathcal{N}(\lambda)$, or, alternatively:

$$\lambda = \mathcal{N}^{-1}(1 - a).$$

A typical performance comparison is given in Table 10.1, where it is determined that the false alarm level is $a = 0.001$ for MP-LRT (the corresponding MP-LRT threshold is $\lambda = 2.329$).

We can see from the above table that the Bayes and minimax tests perform the same since they use the same threshold. On the other hand, the MP test has a much lower error rate but also presents a significantly smaller probability of detection, P_D, and a greater error probability, P_e.

Appendix I: Introduction to Matrix Algebra and Application to Signals and System

Many computational engineering applications require the derivation of a signal or a parameter of interest from degraded measurements. To achieve this, it is often useful to deploy fine-grained statistical models; diverse sensors which acquire extra spatial, temporal, or polarization information; or, multidimensional signal representations, e.g., in time scale and time-frequency scale. When applied in combination, these approaches can be used to develop very sensitive algorithms for signal estimation, detection and monitoring that can utilize the small but lasting differences between signals, interference and noise. Additionally, these approaches can be used to develop algorithms to identify a channel or system that generates a signal under the influence of additive noise and interference, even when the channel input is unknown but has known statistical properties.

Generally, statistical signal processing involves the reliable estimation, detection and classification of signals subject to random variations. Statistical signal processing is based on probability theory, mathematical statistical analysis and, recently, system theory and statistical communication theory. The practice of statistical processing of signals includes:

1. Description of a mathematical and statistical model for measurement data, including sensor, signal and noise models.
2. Careful statistical analysis of the fundamental limitations of the data including deriving benchmarks on performance, e.g., the Cramer-Rao, Ziv-Zakai, Barankin, Rate Distortion, Chernov, or other lower bounds on average estimator/detector error.
3. Development of mathematics of optimal or sub-optimal estimation/detection algorithms.
4. Asymptotic analysis of error performance, which derives that the proposed algorithm is approaching a benchmark resulting from (2).
5. Simulations or experiments that compare the performance of the algorithm with the lower limit and with other competing algorithms.

Depending on the application, the algorithm may also be adaptive to variations in signal and noise conditions. This requires the application of flexible statistical models, of low complexity real-time estimation and filtering algorithms, as well as online performance monitoring.

A.I.1 Basic Principles of Vectors/Matrices

A.I.1.1 Row and Column Vectors

A vector is a list of n sorted values:

$$\mathbf{x} = \begin{bmatrix} x_1 \\ \vdots \\ x_n \end{bmatrix}$$

that belongs to the $\mathbb{R}^n$ domain. Its transpose matrix yields the row vector

$$\mathbf{x}^T = [x_1 \cdots x_n]$$

In the case where the data $x_i = u + jv$ are complex numbers, the Hermitian transpose (conjugate-transpose) is defined as follows:

$$\mathbf{x}^H = [x_1^* \cdots x_n^*],$$

where $x_i^* = u - jv$ is the conjugate of x_i.

Some vectors that are often encountered below are the vectors containing only ones and the elementary j-vector, which constitutes the j-th column of the unitary matrix (identity matrix):

$$\mathbf{1} = [1, \cdots, 1]^T, \qquad \mathbf{e}_j = \Big[0, \cdots, 0, \underbrace{1}_{j\text{-th element}}, 0, \cdots, 0\Big]^T$$

A.I.1.2 Vector Multiplication

The inner product of two vectors $\mathbf{x}$ and $\mathbf{y}$ with an equal number of elements n is calculated by:

$$\mathbf{x}^T\mathbf{y} = \sum_{i=1}^{n} x_i y_i$$

The norm-2 $\|\mathbf{x}\|_2$ of the vector $\mathbf{x}$ equals:

$$\|\mathbf{x}\| = \sqrt{\mathbf{x}^T\mathbf{x}} = \sqrt{\sum_{i=1}^{n} x_i^2}$$

The outer product of two vectors $\mathbf{x}$ and $\mathbf{y}$ with an equal or unequal number of elements, n, m, yields a matrix of $n \times m$ dimensions:

$$\mathbf{x}\mathbf{y}^T = (x_i y_i)_{i,j=1}^{n,m}$$
$$= \left[\mathbf{x}\,y_1, \cdots, \mathbf{x}\,y_m \right]$$
$$= \left[\begin{pmatrix} x_1 y_1 & \cdots & x_1 y_m \\ \vdots & \ddots & \vdots \\ x_n y_1 & \cdots & x_n y_m \end{pmatrix} \right].$$

A.I.2 Orthogonal Vectors

If $\mathbf{x}^T\mathbf{y} = 0$ is true, then the vectors $\mathbf{x}$ and $\mathbf{y}$ are said to be orthogonal. Moreover, if they are both of unit lengths, $\|\mathbf{x}\| = 1$ and $\|\mathbf{y}\| = 1$, then they are called orthonormal vectors.

A.I.2.1 Vector Matrix Multiplication

Let $\mathbf{A}$ be a matrix of $m \times n$ dimensions with columns $a_{*1}, \cdots, a_{*n}$ and $\mathbf{x}$ be a vector with n elements. The product $\mathbf{A}\mathbf{x}$ forms a column vector, which is derived from linear combinations of the columns of $\mathbf{A}$

$$Ax = \sum_{j=1}^{n} x_j a_{*j}.$$

Let $\mathbf{y}$ be a vector with m elements. The product $\mathbf{y}^T\mathbf{A}$ is a row vector resulting from linear combinations of the rows of $\mathbf{A}$

$$\mathbf{y}^T\mathbf{A} = \sum_{i=1}^{m} y_i a_{i*}.$$

A.I.2.2 The Linear Span of a Set of Vectors

Let $\mathbf{x}_1, \dots, \mathbf{x}_n$ be p-element column vectors, which form the $p \times n$ matrix

$$\mathbf{X} = [x_1, \dots, x_n].$$

Let $\mathbf{a} = [a_1, \ldots, a_n]^T$. Then, $\mathbf{y} = \displaystyle\sum_{i=1}^{n} a_i \mathbf{x}_i = \mathbf{Xa}$ is a p-element vector, which is a linear combination of the columns of $\mathbf{X}$. The linear span of vectors $\mathbf{x}_1, \ldots, \mathbf{x}_n$ is defined as the subspace in the $\mathbb{R}^p$ domain including all the possible linear combinations

$$\mathrm{span}\{\mathbf{x}_1, \ldots, \mathbf{x}_n\} = \{\mathbf{y} : \mathbf{y} = \mathbf{Xa}, \quad \mathbf{a} \in \mathbb{R}^n\}.$$

Therefore, when vector $\mathbf{a}$ extends to the entire $\mathbb{R}^p$ domain, vector $\mathbf{y}$ extends to the linear span of $\mathbf{x}_1, \ldots, \mathbf{x}_n$.

A.I.2.3 Rank of a Matrix

The rank (of the columns) of a matrix $\mathbf{A}$ equals the number of its linearly independent columns. The dimension of the column space of a matrix $\mathbf{A}$ of rank p is equal to p.
  If matrix $\mathbf{A}$ is of a full rank, then:

$$0 = \mathbf{Ax} = \sum_i x_i \mathbf{a}_{*i} \Leftrightarrow \mathbf{x} = 0.$$

Additionally, if $\mathbf{A}$ is a square matrix, then it is nonsingular.

A.I.2.4 Inverse Matrix

If $\mathbf{A}$ is a square nonsingular matrix and $\mathbf{A}^{-1}$ is its inverse, then $\mathbf{AA}^{-1} = \mathbf{I}$. In the specific case of a 2×2 matrix, the inverse matrix is given by the *Cramer* equation:

$$\begin{bmatrix} a & b \\ c & d \end{bmatrix}^{-1} = \frac{1}{ad - bc} \begin{bmatrix} d & -b \\ -c & a \end{bmatrix}, \quad ad \neq bc.$$

In some cases where a matrix has a specific type of structure, its inverse can be expressed in a general way.
  A typical example is the *Sherman-Morrison-Woodbury* property:

$$\left[\mathbf{A} + \mathbf{UV}^T\right]^{-1} = \mathbf{A}^{-1} - \mathbf{A}^{-1}\mathbf{U}\left[\mathbf{I} + \mathbf{V}^T\mathbf{A}^{-1}\mathbf{U}\right]^{-1}\mathbf{V}^T\mathbf{A}^{-1}, \tag{A.1}$$

which assumes that:

- $\mathbf{A}$, $\mathbf{U}$, $\mathbf{V}$ are compatible matrices (of appropriate dimensions to allow the multiplication between them)
- $[\mathbf{A} + \mathbf{UV}^T]^{-1}$ and $\mathbf{A}^{-1}$ exist.

Also, a useful expression is the partitioned matrix inverse identity:

$$\begin{bmatrix} \mathbf{A}_{11} & \mathbf{A}_{12} \\ \mathbf{A}_{21} & \mathbf{A}_{22} \end{bmatrix}^{-1} = \begin{bmatrix} \left[\mathbf{A}_{11} - \mathbf{A}_{12}\mathbf{A}_{22}^{-1}\mathbf{A}_{21}\right]^{-1} & -\mathbf{A}_{11}^{-1}\mathbf{A}_{12}\left[\mathbf{A}_{22} - \mathbf{A}_{21}\mathbf{A}_{11}^{-1}\mathbf{A}_{12}\right]^{-1} \\ -\mathbf{A}_{22}^{-1}\mathbf{A}_{21}\left[\mathbf{A}_{11} - \mathbf{A}_{12}\mathbf{A}_{22}^{-1}\mathbf{A}_{21}\right]^{-1} & \left[\mathbf{A}_{22} - \mathbf{A}_{21}\mathbf{A}_{11}^{-1}\mathbf{A}_{12}\right]^{-1} \end{bmatrix},$$

(A.2)

which assumes that all the inverse submatrices exist.

A.I.2.5 Orthogonal and Unitary Matrices

A square matrix $\mathbf{A}$ with real number elements is orthogonal if all of its columns are orthogonal:

$$\mathbf{A}^T\mathbf{A} = \mathbf{I}.$$

(A.3)

The condition for the generalization of the concept of orthogonality for complex matrices is that $\mathbf{A}$ must be unitary:

$$\mathbf{A}^H\mathbf{A} = \mathbf{I}.$$

Relation A.3 indicates that if $\mathbf{A}$ is orthogonal, then it is also invertible, so that:

$$\mathbf{A}^{-1} = \mathbf{A}^T.$$

A.I.2.6 Gramm-Schmidt Orthogonality Method

Let $\mathbf{x}_1, \ldots, \mathbf{x}_n$ be the column vectors of p linearly independent elements $(n \leq p)$, whose linear span belongs to the subspace H. The Gramm-Schmidt method is an algorithm applied to these vectors in order to create a new set of n vectors $\mathbf{y}_1, \ldots, \mathbf{y}_n$, which are produced in the same subspace. The algorithm is implemented as follows:

Step 1: Random choice of $\mathbf{y}_1$ as an arbitrary starting point in H. For example, let $\mathbf{a}_1 = [a_{11}, \cdots, a_{1n}]^T$, $\mathbf{y}_1 = \mathbf{X}\mathbf{a}_1$, where $\mathbf{X} = [\mathbf{x}_1, \ldots, \mathbf{x}_n]$.

Step 2: Derivation the rest of n-1 vectors $\mathbf{y}_2, \ldots \mathbf{y}_n$ through the following iterative procedure:

$$For \ \ j = 2, \cdots, n : \mathbf{y}_j = \mathbf{x}_j - \sum_{i=1}^{j} K_i \mathbf{y}_{i-1}, \ where \ \ Kj = \mathbf{x}_j^T\mathbf{y}_{j-1} / \mathbf{y}_{j-1}^T\mathbf{y}_{j-1}.$$

The above procedure can equally be expressed in the form of matrices as:

$$\mathbf{Y} = \mathbf{X}\mathbf{H},$$

where $\mathbf{Y} = [\mathbf{y}_1, \cdots, \mathbf{y}_n]$, and $\mathbf{H}$ is the Gramm-Schmidt matrix.

If, in every step $j = 1, \cdots, n$ of the process, the length $\mathbf{y}_j$ is normalized, that is, $\mathbf{y}_j \leftarrow \tilde{\mathbf{y}}_j = \mathbf{y}_j / \|\mathbf{y}_j\|$, the algorithm produces orthonormal vectors. This is called Gramm-Schmidt normalization and produces a matrix $\tilde{\mathbf{Y}}$ with orthonormal columns and identical column span as that of $\mathbf{X}$. This method is often used to create an orthonormal basis $[\mathbf{y}_1, \ldots, \mathbf{y}_p]$ in the $\mathbb{R}^p$ domain, arbitrarily starting from a vector $\mathbf{y}_1$. The matrix that is derived is of the following form:

$$\mathbf{Y} = \begin{bmatrix} \mathbf{y}_1 \\ \mathbf{v}_2 \\ \vdots \\ \mathbf{v}_n \end{bmatrix}$$

and

$$\mathbf{Y}^T \mathbf{Y} = \mathbf{I},$$

where $\mathbf{v}_2, \cdots, \mathbf{v}_n$ are orthonormal vectors that achieve completion of the basis with respect to the original vector $\mathbf{y}_1$.

A.I.2.7 Eigenvalues of a Symmetric Matrix

If the $n \times n$ matrix $\mathbf{R}$ is symmetric, that is, if $\mathbf{R}^T = \mathbf{R}$ is true, then there is a set of n orthonormal eigenvectors $\mathbf{v}_i$:

$$\mathbf{v}_i^T \mathbf{v}_i = \Delta_{ij} = \begin{cases} 1, & i = j \\ 0, & i \neq j \end{cases}$$

as well as a set of the corresponding eigenvalues, so that:

$$\mathbf{R}\mathbf{v}_i = \lambda_i \mathbf{v}_i, \quad i = 1, \cdots, n.$$

The above eigenvectors and eigenvalues satisfy the following conditions:

$$\mathbf{v}_i^T \mathbf{R} \mathbf{v}_i = \lambda_i$$
$$\mathbf{v}_i^T \mathbf{R} \mathbf{v}_j = 0, \quad i \neq j.$$

A.I.2.8 Diagonalization and Eigen-Decomposition of a Matrix

Let the $n \times n$ matrix $\mathbf{U} = [\mathbf{v}_1, \cdots, \mathbf{v}_n]$, which has been derived from the eigenvectors of a symmetric matrix $\mathbf{R}$. If matrix $\mathbf{R}$ is a real-symmetric matrix, then the matrix $\mathbf{U}$ is real-orthogonal. In the case that the matrix $\mathbf{R}$ is complex-symmetric, then the matrix $\mathbf{U}$ is complex-unitary:

$$\mathbf{U}^T\mathbf{U} = \mathbf{I}, \quad (\mathbf{U}: \text{orthogonal matrix})$$
$$\mathbf{U}^H\mathbf{U} = \mathbf{I}, \quad (\mathbf{U}: \text{unitary matrix}).$$

Since the Hermitian conjugate-transpose operator is general, next it will be used to denote an either real or a complex matrix ($\mathbf{A}^H = \mathbf{A}^T$, for real matrix $\mathbf{A}$).

The matrix $\mathbf{U}$ can be used for the diagonalization of $\mathbf{R}$:

$$\mathbf{U}^H\mathbf{R}\mathbf{U} = \mathbf{\Lambda}. \tag{A.4}$$

In cases of real and Hermitian symmetric $\mathbf{R}$, matrix $\mathbf{\Lambda}$ is diagonal with real values:

$$\mathbf{\Lambda} = \text{diag}(\lambda_i) = \begin{bmatrix} \lambda_1 & \cdots & 0 \\ \vdots & \ddots & \vdots \\ 0 & \cdots & \lambda_n \end{bmatrix},$$

where λ_i depict the eigenvalues of $\mathbf{R}$.

Expression A.4 suggests that:

$$\mathbf{R} = \mathbf{U}\mathbf{\Lambda}\mathbf{U}^H,$$

which is called the decomposition of matrix $\mathbf{R}$. Since the matrix $\mathbf{\Lambda}$ is diagonal, the above self-decomposition can be expressed as a sum of terms:

$$\mathbf{R} = \sum_{i=1}^{n} \lambda_i \mathbf{v}_i \mathbf{v}_i^H. \tag{A.5}$$

A.I.2.9 Square Form of Non-Negative Definite Matrix

For a square-symmetric matrix $\mathbf{R}$ and a compatible vector $\mathbf{x}$, the square form is a scalar coefficient defined as $\mathbf{x}^T\mathbf{R}\mathbf{x}$. The matrix $\mathbf{R}$ is non-negative definite if for each $\mathbf{x}$ it is true that:

$$\mathbf{x}^T\mathbf{R}\mathbf{x} \geq 0. \tag{A.6}$$

Matrix $\mathbf{R}$ is positive definite if the equality is excluded from Relation A.6, so that:

$$\mathbf{x}^T \mathbf{R} \mathbf{x} > 0, \quad \mathbf{x} \neq \mathbf{0}. \tag{A.7}$$

Examples of non-negative definite matrices:

- $\mathbf{R} = \mathbf{B}^T \mathbf{B}$, for any matrix $\mathbf{B}$.
- Symmetric matrix $\mathbf{R}$ only with non-negative (positive) eigenvalues.

Rayleigh Theorem:

If an $n \times n$ matrix $\mathbf{A}$ is non-negative definite with eigenvalues $\{\lambda_i\}_{i=1}^n$, it is true that:

$$\min(\lambda_i) \leq \frac{\mathbf{u}^T \mathbf{A} \mathbf{u}}{\mathbf{u}^T \mathbf{u}} \leq \max(\lambda_i),$$

where the lower limit is encountered when $\mathbf{u}$ is an eigenvector of $\mathbf{A}$ associated with the minimum eigenvalue of $\mathbf{A}$, while the upper limit is encountered by the eigenvector associated with the maximum eigenvalue of $\mathbf{A}$.

A.I.3 Positive Definiteness of Symmetric Partitioned Matrices

If a matrix $\mathbf{A}$ is symmetric and can be partitioned as described by Relation A.2, then it is true that:

$$\mathbf{A} = \begin{bmatrix} \mathbf{A}_{11} & \mathbf{A}_{12} \\ \mathbf{A}_{21} & \mathbf{A}_{22} \end{bmatrix} = \begin{bmatrix} \mathbf{I} & -\mathbf{A}_{12}\mathbf{A}_{22}^{-1} \\ \mathbf{0} & \mathbf{I} \end{bmatrix}^{-1} \begin{bmatrix} \mathbf{A}_{11} - \mathbf{A}_{12}\mathbf{A}_{22}^{-1}\mathbf{A}_{21} & \mathbf{0}^T \\ \mathbf{0} & \mathbf{A}_{22} \end{bmatrix} \begin{bmatrix} \mathbf{I} & \mathbf{0}^T \\ -\mathbf{A}_{22}^{-1}\mathbf{A}_{21} & \mathbf{I} \end{bmatrix}^{-1}$$

$$\tag{A.8}$$

as long as the submatrix $\mathbf{A}_{22}^{-1}$ exists. Therefore, if the matrix $\mathbf{A}$ is positive definite, then the submatrices $\mathbf{A}_{11} - \mathbf{A}_{12}\mathbf{A}_{22}^{-1}\mathbf{A}_{21}$ and $\mathbf{A}_{22}$ are positive definite. Similarly, the submatrices $\mathbf{A}_{22} - \mathbf{A}_{21}\mathbf{A}_{11}^{-1}\mathbf{A}_{12}$ and $\mathbf{A}_{11}$ are positive definite.

A.I.3.1 Determinant of a Matrix

Let $\mathbf{A}$ represent a square matrix. Its determinant is then defined as:

$$|\mathbf{A}| = \prod_i \lambda_i.$$

NOTE: A square matrix is non-singular if and only if its determinant is non-zero.

If the matrix **A** can be partitioned as in Relation A.2, and if matrices $\mathbf{A}_{11}^{-1}$ and $\mathbf{A}_{22}^{-1}$ exist, then:

$$|\mathbf{A}| = |\mathbf{A}_{11}||\mathbf{A}_{22} - \mathbf{A}_{21}\mathbf{A}_{11}^{-1}\mathbf{A}_{12}| = |\mathbf{A}_{22}||\mathbf{A}_{11} - \mathbf{A}_{12}\mathbf{A}_{22}^{-1}\mathbf{A}_{21}|,$$

which is derived from Relation A.8.

A.I.3.2 Trace of a Matrix

For any square matrix **A**, the trace is defined as:

$$\text{trace}\{\mathbf{A}\} = \sum_i a_{ii} = \sum_i \lambda_i.$$

The following properties are true:

- $\text{trace}\{\mathbf{AB}\} = \text{trace}\{\mathbf{BA}\}$.
- $\mathbf{x}^T\mathbf{Rx} = \text{trace}\{\mathbf{xx}^T\mathbf{R}\}$.

A.I.3.3 Differentiation of Vectors

Differentiating functions containing vectors is very common in signal detection and processing theory.

Let $\mathbf{h} = \left[h_1, \cdots, h_n\right]^T$ be an $n \times 1$ vector and $g(\mathbf{h})$ be a scalar function. Then the gradient of $g(\mathbf{h})$, which is noted as $\nabla g(\mathbf{h})$ or $\nabla_{\mathbf{h}} g(\mathbf{h})$, is defined as the column vector of the following partial derivatives:

$$\nabla g(\mathbf{h}) = \left[\frac{\partial g(\mathbf{h})}{\partial h_1}, \cdots, \frac{\partial g(\mathbf{h})}{\partial h_n}\right]^T.$$

Moreover, by introducing a constant, c, it is:

$$\nabla_{\mathbf{h}} c = \mathbf{0},$$

If $\mathbf{x} = [x_1, \cdots, x_n]^T$:

$$\nabla_{\mathbf{h}}(\mathbf{h}^T\mathbf{x}) = \nabla_{\mathbf{h}}(\mathbf{x}^T\mathbf{h}) = \mathbf{x},$$

and if **B** an $n \times n$ matrix:

$$\nabla_{\mathbf{h}}(\mathbf{h} - \mathbf{x})^T\mathbf{B}(\mathbf{h} - \mathbf{x}) = 2\mathbf{B}(\mathbf{h} - \mathbf{x}).$$

For a vector function $g(\mathbf{h}) = [g_1(\mathbf{h}), \cdots, g_m(\mathbf{h})]^T$, the gradient of $g(\mathbf{h})$ is an $m \times n$ matrix.

Appendix II: Solved Problems in Statistical Signal Processing

A.II.1 Compute the CDF $F_x(a)$, the expected value $\mathbb{E}[x]$, the second statistical moment $\mathbb{E}[x^2]$ and the variance σ_x^2 of a RV, which has the following PDF:

$$f_x(a) = \begin{cases} 1 - \dfrac{a}{2}, & \text{for } 0 \le a \le 2, \\ 0, & \text{otherwise.} \end{cases}$$

SOLUTION: The following graph illustrates the given PDF.

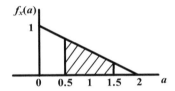

The CDF is equal to:

$$F_x(a) = \int_{-\infty}^{a} f_x(u)\,du = \int_{-\infty}^{+\infty}\left(1 - \frac{u}{2}\right)du = a - \frac{a^2}{4}, \quad 0 \le a \le 2.$$

The expected value of the RV x is expressed as:

$$\mathbb{E}[x] = \int_{-\infty}^{+\infty} a\left(1 - \frac{a}{2}\right)da = \frac{2}{3}.$$

The second statistical moment of the RV x is equal to:

$$\mathbb{E}[x^2] = \int_{-\infty}^{+\infty} a^2 f_x(a)\,da = \int_{-\infty}^{+\infty} a^2\left(1 - \frac{a}{2}\right)da = \frac{2}{3}.$$

The variance can be easily calculated from the above results:

$$\sigma_x^2 = \mathbb{E}[x^2] - \left(\mathbb{E}[x]\right)^2 = \frac{2}{3} - \left(\frac{2}{3}\right)^2 = \frac{2}{9}.$$

Moreover, from the CDF, we could easily calculate the probability of the RV x being within a range of values. For example, the probability $0.5 \le x \le 1.5$ is calculated as:

$$P[0.5 \le x \le 1.5] = F_x(1.5) - F_x(0.5)$$

$$= \frac{15}{16} - \frac{7}{16} = \frac{1}{2}.$$

This value corresponds to the shaded area of the figure.

A.II.2 Calculate the expected value and the variance of an RV x, the PDF of which is uniform and is illustrated in the following figure:

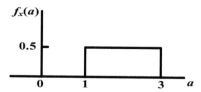

SOLUTION:

$$\mathbb{E}[x] = \int_{-\infty}^{+\infty} a\, f_x(a)\, da = \int_1^3 0.5\, a\, da = 0.5\, \frac{a^2}{2}\Big|_1^3 = \frac{1}{2}\left(\frac{9}{2} - \frac{1}{2}\right) = 2.$$

$$\mathbb{E}[x^2] = \int_{-\infty}^{+\infty} a^2 f_x(a)\, da = \int_1^3 0.5\, a^2\, da = 0.5\, \frac{a^3}{3}\Big|_1^3 = \frac{1}{2}\left(\frac{27}{3} - \frac{1}{3}\right) = \frac{26}{6} = \frac{13}{3}.$$

Thus:

$$\sigma_x^2 = \mathbb{E}[x^2] - \left(\mathbb{E}[x]\right)^2 = \frac{13}{3} - 2^2 = \frac{13}{3} - \frac{12}{3} = \frac{1}{3}.$$

A.II.3 Let the joint PDF of two RV x, y be:

$$f_{x,y}(a,b) = \begin{cases} A, & \text{for } -1 \le a \le 1, \quad 0 \le b \le 1, \\ 0, & \text{otherwise.} \end{cases}$$

Derive the value of constant A and calculate the probability of x and y being in the interval $0.5 \leq x \leq 1, 0 \leq y \leq 1$, respectively.

SOLUTION: From the basic property of the joint PDF, we have:

$$\int_{-\infty}^{+\infty}\int_{-\infty}^{+\infty} f_{x,y}(a,b)\,da\,db = 1$$

$$\Leftrightarrow A\int_{-1}^{1}\int_{0}^{1} da\,db = A\left[\int_{-1}^{1} da\right]\left[\int_{0}^{1} db\right] = A\,a\Big|_{-1}^{1}\,b\Big|_{0}^{1} = 1 \Leftrightarrow A\cdot 2\cdot 1 = 1$$

$$\Leftrightarrow 2A = 1.$$

Thus, $A = 1/2$.

The probability that x and y will be in the interval asked is illustrated in the shaded area of the following figure:

$$P\big[0.5 \leq x \leq 1, 0 \leq y \leq 1\big] = A\int_{0}^{1}\int_{0.5}^{1} da\,db = \frac{1}{2}\cdot 1\cdot\frac{1}{2} = \frac{1}{4}.$$

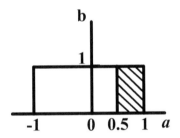

A.II.4 Let the random sinusoidal signal $x(n) = A\cos(\omega_0 n + \varphi)$, at the n point in time, with a PDF for A and Φ:

$$f_A(a) = \begin{cases} \dfrac{1}{4}, & 0 \leq a \leq 4, \\ 0, & \text{otherwise,} \end{cases} \qquad f_\varphi(\varphi) = \begin{cases} \dfrac{1}{2\pi}, & 0 \leq \varphi \leq 2\pi, \\ 0, & \text{otherwise.} \end{cases}$$

Calculate its statistical properties.

SOLUTION: The RVs A, φ are statistically independent (the amplitude and phase of a signal fluctuate independently), thus their joint PDF is equal to the product of the individual PDFs:

$$f_{A,\varphi}(a,\varphi) = \begin{cases} \dfrac{1}{8\pi}, & 0 \leq a \leq 4, \ 0 \leq \varphi \leq 2\pi, \\ 0, & \text{otherwise.} \end{cases}$$

The expected value of the signal is expressed as:

$$\mathbb{E}[x(n)] = \frac{1}{8\pi} \int_0^4 \int_0^{2\pi} a\cos(\omega_0 n + \varphi)\, da\, d\varphi$$

$$= \frac{1}{8\pi} \left(\int_0^4 a\, da \right) \left(\int_0^{2\pi} \cos(\omega_0 n + \varphi)\, d\varphi \right)$$

$$= \frac{1}{\pi} \left[\sin(\omega_0 n + 2\pi) - \sin(\omega_0 n) \right] = 0.$$

The second statistical moment of the signal is:

$$\mathbb{E}[x^2(n)] = \frac{1}{8\pi} \int_0^4 \int_0^{2\pi} a^2 \cos^2(\omega_0 n + \varphi)\, da\, d\varphi$$

$$= \frac{1}{8\pi} \left(\int_0^4 a^2\, da \right) \left(\int_0^{2\pi} \cos^2(\omega_0 n + \varphi)\, d\varphi \right)$$

$$= \frac{8}{3}.$$

The variance of the signal is equal to the second statistical moment since the expected value is 0:

$$\sigma_x^2 = \mathbb{E}[x^2] - \left(\mathbb{E}[x]\right)^2 = \mathbb{E}[x^2] - 0 = \frac{8}{3}.$$

The autocorrelation function is:

$$R_{xx}(m,n) = \mathbb{E}\left[x(m)x^*(n) \right]$$

$$= \frac{1}{8\pi} \left(\int_0^4 a^2\, da \right) \left(\int_0^{2\pi} \cos(\omega_0 m + \varphi)\cos(\omega_0 n + \varphi)\, d\varphi \right)$$

$$= \frac{8}{3}\cos(\omega_0(m-n)).$$

As shown in the following exercises, this sinusoidal signal is stationary in the wide sense, since it has a constant mean value $\left(\mathbb{E}[x(n)] = 0 \right)$, and the autocorrelation function depends only on the difference of the m and n points in time (since it is true that $R_{xx}(m,n) = (8/3) \cos(\omega_0(m-n))$).

A.II.5 Let:

$$x(n) = A\cos \omega n + B\sin \omega n,$$
$$y(n) = B\cos \omega n - A\sin \omega n,$$

where ω is a constant and A, B are the independent RVs of zero expected value and of variance σ^2. Calculate the cross-correlation of $x(n)$ and $y(n)$.

SOLUTION:

$$
\begin{aligned}
R_{xy}(m,n) &= \mathbb{E}[x(m)y^*(n)]\\
&= \mathbb{E}[(A\cos \omega m + B\sin \omega m)(B\cos \omega n - A\sin \omega n)]\\
&= \mathbb{E}\big[AB\cos \omega m \cdot \cos \omega n - A^2 \cos \omega m \cdot \sin \omega n\\
&\quad + B^2 \sin \omega m \cdot \cos \omega n - AB\sin \omega m \cdot \sin \omega n\big]\\
&= \mathbb{E}[(AB)]\cos \omega m \cdot \cos \omega n - \mathbb{E}[A^2]\cos \omega m \cdot \sin \omega n\\
&\quad + \mathbb{E}[B^2]\sin \omega m \cdot \cos \omega n - \mathbb{E}[AB]\sin \omega m \cdot \sin \omega n.
\end{aligned}
$$

However, because of the independence of the RVs, it is true that:

$$\mathbb{E}[AB] = \mathbb{E}[A]\cdot \mathbb{E}[B] = 0.$$

Also, it is given that:

$$\mathbb{E}[A^2] = \mathbb{E}[B^2] = \sigma^2.$$

According to the above relationships, the cross-correlation function becomes:

$$
\begin{aligned}
R_{xy}(m,n) &= \sigma^2(\sin \omega m \cdot \cos \omega n - \cos \omega m \cdot \sin \omega n)\\
&= \sigma^2 \sin(m-n).
\end{aligned}
$$

A.II.6 Calculate the expected value and the autocorrelation of the discrete-time random sequence $x(n) = A^n$, $n \geq 0$, where A is an RV with identical distribution in the interval $(0, 1)$.

SOLUTION:

$$f_A(a) = \begin{cases} 1, & 0 < a < 1,\\ 0, & \text{otherwise.} \end{cases}$$

$$\mathbb{E}[x(n)] = \mathbb{E}[A^n] = \int\limits_0^1 a^n \, da = \frac{a^{n+1}}{n+1}\bigg|_0^1 = \frac{1}{n+1}.$$

$$R_{xx}(n,m) = \mathbb{E}[x(n)x^*(m)] = \mathbb{E}[A^n A^m] = \mathbb{E}[A^{n+m}] = \int\limits_0^1 a^{n+m} \, da = \frac{1}{n+m+1}.$$

A.II.7   Prove that: $R_{xx}(n, n) \geq 0$.

SOLUTION:

$$R_{xx}(n,n) = \mathbb{E}\left[x(n)x^*(n)\right] = \mathbb{E}\left[|x(n)|^2\right] \geq 0.$$

A.II.8   Let a random sequence $x(n) = An + B$, where A, B are independent Gaussian RVs
with zero expected value and variances σ_A^2, σ_B^2, respectively. Calculate:
  a.   The expected value and the autocorrelation of the random sequence $x(n)$.
  b.   The second statistical moment of the random sequence $x(n)$.

SOLUTION:
  a.

$$\mathbb{E}[x(n)] = \mathbb{E}[An + B] = \mathbb{E}[A]n + \mathbb{E}[B] = 0, \text{ since } \mathbb{E}[A] = \mathbb{E}[B] = 0.$$

$$\begin{aligned}
R_{xx}(n,m) &= \mathbb{E}\left[x(n)x^*(m)\right] = \mathbb{E}\left[(An + B)(Am + B)\right] \\
&= \mathbb{E}\left[A^2 nm + ABn + BAm + B^2\right] \\
&= \mathbb{E}[A^2]nm + \mathbb{E}[AB]n + \mathbb{E}[BA]m + \mathbb{E}[B^2] \\
&= \sigma_A^2 nm + \mathbb{E}[A]\mathbb{E}[B]n + \mathbb{E}[B]\mathbb{E}[A]m + \sigma_B^2 \\
&= \sigma_A^2 nm + \sigma_B^2,
\end{aligned}$$

  since $\mathbb{E}[AB] = \mathbb{E}[A] \cdot \mathbb{E}[B] = 0$, because of the independence of the given RVs.
  b.   Following the same way of thinking as in the last relation and substituting
       $m \to n$, it is derived:

$$\mathbb{E}[x^2(n)] = \mathbb{E}\left[x(n)x^*(n)\right] = \cdots = \sigma_A^2 n^2 + \sigma_B^2.$$

A.II.9   Let the random signals $x(n) = A\cos(\omega n + \theta)$, $y(n) = A\sin(\omega n + \theta)$, where A, ω
are constants and θ is an RV with identical distribution in the interval $[0, 2\pi]$.
Calculate their cross-correlation:

SOLUTION:

$$R_{xy}(n+l,n) = \mathbb{E}\big[x(n+l)y^*(n)\big] = \mathbb{E}\big[A\cos(\omega(n+l)+\theta)\cdot A\sin(\omega n+\theta)\big]$$
$$= A^2\mathbb{E}\big[\cos(\omega(n+l)+\theta)\cdot\sin(\omega n+\theta)\big]$$
$$= A^2\frac{1}{2}\mathbb{E}\big[\sin(-\omega l)+\sin(2\omega n+\omega l+2\theta)\big]$$
$$= \frac{A^2}{2}\big[-\sin\omega l+\mathbb{E}\big[\sin(2\omega n+\omega l+2\theta)\big]\big].$$

Further analysis of the second part of the last equation yields:

$$\mathbb{E}\big[\sin(2\omega n+\omega l+2\theta)\big] = \int_0^{2\pi}\sin(2\omega n+\omega l+2\theta)f(\theta)\,d\theta$$
$$= \int_0^{2\pi}\sin(2\omega n+\omega l+2\theta)\frac{1}{2\pi}\,d\theta$$
$$= \frac{1}{2\pi}\frac{1}{2}\int_0^{2\pi}\sin\Big(\underbrace{2\omega n+\omega l}_{a}+2\theta\Big)d2\theta$$
$$= \frac{1}{4\pi}\int_0^{2\pi}\sin(a+2\theta)\,d(a+2\theta)$$
$$= -\frac{1}{4}\cos(a+2\theta)\Big|_0^{2\pi} = 0.$$

Consequently, the first relation is now written as:

$$R_{xy}(n+l,n) = -\frac{A^2}{2}\sin\omega l = R_{xy}(l).$$

Next, the cross-correlation $R_{yx}(n+l,n)$ is calculated:

$$R_{yx}(n+l,n) = \mathbb{E}\big[y(n+l)x^*(n)\big] = \mathbb{E}\big[A\sin(\omega(n+l)+\theta)\cdot A\cos(\omega n+\theta)\big]$$
$$= A^2\mathbb{E}\big[\sin(\omega(n+l)+\theta)\cdot\cos(\omega n+\theta)\big]$$
$$= A^2\frac{1}{2}\mathbb{E}\big[\sin\omega l+\sin(2\omega n+\omega l+2\theta)\big]$$
$$= \frac{A^2}{2}\big[\sin\omega l+\underbrace{\mathbb{E}\big[\sin(2\omega n+\omega l+2\theta)\big]}_{=0}\big]$$
$$= \frac{A^2}{2}\sin\omega l = R_{yx}(l) = -R_{xy}(l) = R_{xy}(-l).$$

Thus, we have proven that: $R_{yx}(l) = R_{xy}(-l)$.

A.II.10 Let the random signal $x(n) = A \cos(\omega n + \theta)$, where θ, ω are constants and A is an RV. Determine if the signal is wide-sense stationary (WSS).

SOLUTION: In order to determine whether the signal is WSS, we should examine whether its expected value is stationary and whether its autocorrelation is independent of time but depends only on the time difference $m - n = l$.

$$\mathbb{E}[x(n)] = \mathbb{E}[A\cos(\omega n + \theta)]$$
$$= \cos(\omega n + \theta)\mathbb{E}[A].$$

We can see that the expected value is not constant, unless $\mathbb{E}[A] = 0$.

$$R_{xx}(m, n) = \mathbb{E}[x(m)x^*(n)] = \mathbb{E}\left[A\cos(\omega m + \theta) \cdot A\cos(\omega n + \theta)\right]$$
$$= \mathbb{E}[A^2]\cos(\omega m + \theta)\cdot\cos(\omega n + \theta)$$
$$= \mathbb{E}[A^2]\frac{1}{2}\left[\cos(\omega(m-n)) + \cos(\omega(m+n) + 2\theta)\right].$$

We note that the autocorrelation is not dependent only on the difference $m - n$ and, therefore, the random signal is not WSS.

A.II.11 Let $x(n)$ be a WSS signal and $R_{xx}(l)$, its autocorrelation. Prove that:

$$\mathbb{E}\left[(x(n+l) - x(n))^2\right] = 2\left[R_{xx}(0) - R_{xx}(l)\right].$$

SOLUTION:

$$\mathbb{E}\left[(x(n+l) - x(n))^2\right] = \mathbb{E}\left[x^2(n+l) - 2x(n+l)x(n) + x^2(n)\right]$$
$$= \mathbb{E}\left[x^2(n+l)\right] - 2\mathbb{E}\left[x(n+l)x(n)\right] + \mathbb{E}\left[x^2(n)\right]$$
$$= R_{xx}(0) - 2R_{xx}(l) + R_{xx}(0)$$
$$= 2\left[R_{xx}(0) - R_{xx}(l)\right].$$

A.II.12 Let $x(n)$, $y(n)$ be two independent and random WSS signals. Determine if their sum is WSS.

SOLUTION: We define $s(n) \triangleq x(n) + y(n)$.

$$\mathbb{E}[s(n)] = \mathbb{E}[x(n) + y(n)] = \mathbb{E}[x(n)] + \mathbb{E}[y(n)].$$

Since the signals are WSS, their expected value will be constant. Therefore, their sum $\mathbb{E}[x(n)] + \mathbb{E}[y(n)]$ will also be constant.

The autocorrelation is calculated as follows:

$$
\begin{aligned}
R_{ss}(n+l,n) &= \mathbb{E}[s(n+l)s^*(n)] \\
&= \mathbb{E}\left[(x(n+l)+y(n+l))(x(n)+y(n))\right] \\
&= \mathbb{E}\left[x(n+l)x(n)+x(n+l)y(n)+y(n+l)x(n)+y(n+l)y(n)\right] \\
&= \mathbb{E}[x(n+l)x(n)]+\mathbb{E}[x(n+l)y(n)]+\mathbb{E}[y(n+l)x(n)] \\
&\quad +\mathbb{E}[y(n+l)y(n)] \\
&= R_{xx}(l)+\mathbb{E}[x(n+l)]\mathbb{E}[y(n)]+\mathbb{E}[y(n+l)]\mathbb{E}[x(n)]+R_{yy}(l) \\
&= R_{xx}(l)+R_{yy}(l)+2\mathbb{E}[x(n)]\mathbb{E}[y(n)].
\end{aligned}
$$

Thus, the signal $s(n)$ is WSS since its expected value is constant and its autocorrelation depends on the time difference, l.

A.II.13 Calculate the expected value and the variance of a WSS signal, the autocorrelation function of which is:

$$
R_{xx}(l) = \frac{10+21l^2}{1+3l^2}.
$$

SOLUTION: It is true that $\lim_{l\to+\infty} R_{xx}(l) = (\mathbb{E}[x(n)])^2$. Thus:

$$
(\mathbb{E}[x(n)])^2 = \lim_{l\to+\infty} R_{xx}(l) = \lim_{l\to+\infty} \frac{\dfrac{10}{l^2}+21}{\dfrac{1}{l^2}+3} = \frac{21}{3} = 7.
$$

Therefore, it is $\mathbb{E}[x(n)] = \pm\sqrt{7}$.

$$
\mathbb{E}[x^2(n)] = R_{xx}(0) \text{ yields } \mathbb{E}[x^2(n)] = 10.
$$

Finally, the variance is calculated from $\sigma_x^2 = R_{xx}(0)-\left(\mathbb{E}[x(n)]\right)^2 = 10-7 = 3$.

A.II.14 Let $x(n)$ be an identical distributed random signal, the PDF of which is illustrated in the figure below. Calculate the expected value and the variance of the output signal $y(n)$ when the input $x(n)$ is applied to a 3-point casual moving average filter with the impulse response $h(n) = \{1/3, 1/3, 1/3\}$.

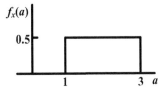

SOLUTION: For the given input signal, it is already known from a previous exercise that $\mathbb{E}[x] = 2$, $\sigma_x^2 = 1/3 \ \forall n$. Thus:

$$\mathbb{E}[y] = \mathbb{E}[x] \sum_{k=-\infty}^{+\infty} h(k) = \mathbb{E}[x](h(0) + h(1) + h(2)) = 2\left(\frac{1}{3} + \frac{1}{3} + \frac{1}{3}\right) = 2.$$

The variation of the output signal is:

$$\sigma_y^2 = \sigma_x^2 \sum_{k=-\infty}^{+\infty} |h(k)|^2 = \sigma_x^2 \left(h^2(0) + h^2(1) + h^2(2)\right) = \frac{1}{3}\left(\frac{1}{9} + \frac{1}{9} + \frac{1}{9}\right) = \frac{1}{9}.$$

A.II.15 A random WSS signal with expected value $1 - e^{-2}$ is applied to a linear time-invariant (LTI) system, the impulse response of which is:

$$h(n) = 3e^{-2n}u(n).$$

Calculate the output.

SOLUTION:

$$\mathbb{E}[y] = \mathbb{E}[x] \sum_{k=-\infty}^{+\infty} h(k) = \mathbb{E}[x]H(e^{j0}).$$

But:

$$H(e^{j\omega}) = \sum_{n=-\infty}^{+\infty} h(n)e^{-j\omega n} = 3\sum_{n=0}^{+\infty} e^{-2n}e^{-j\omega n}$$

$$= 3\sum_{n=0}^{+\infty} e^{-(2+j\omega)n} = 3\frac{1}{1 - e^{-(2+j\omega)}}.$$

For $\omega = 0$, it is:

$$H(e^{j0}) = 3\frac{1}{1 - e^{-2}}.$$

Finally, the expected value of the output is:

$$\mathbb{E}[y] = \mathbb{E}[x]H\left(e^{j0}\right) = \left(1 - e^{-2}\right)\frac{3}{\left(1 - e^{-2}\right)} = 3.$$

A.II.16 Let a stochastic signal have an expected value, $\mathbb{E}[x] = 2$, and variance, $\sigma_x^2 = 1/4$. This signal is applied to a casual filter with transfer function $H(z) = -\dfrac{1}{4} + \dfrac{1}{2} z^{-1} - \dfrac{1}{4} z^{-2}$. Calculate the expected value and the variance of the output signal.

SOLUTION:

$$H(z) = -\frac{1}{4} + \frac{1}{2} z^{-1} - \frac{1}{4} z^{-2} \overset{Z}{\leftrightarrow} h(n) = -\frac{1}{4}\delta(n) + \frac{1}{2}\delta(n-1) - \frac{1}{4}\delta(n-2)$$

$$\Rightarrow h(n) = \left\{ -\frac{1}{4}, \frac{1}{2}, -\frac{1}{4} \right\}.$$

$$\mathbb{E}[y] = \mathbb{E}[x] \sum_{k=-\infty}^{+\infty} h(k) = \mathbb{E}[x]\left[h(0) + h(1) + h(2) \right] = 2\left[-\frac{1}{4} + \frac{1}{2} - \frac{1}{4} \right] = 2 \cdot 0 = 0.$$

$$\sigma_y^2 = \sigma_x^2 \sum_{k=-\infty}^{+\infty} |h(k)|^2 = \sigma_x^2\left[|h(0)|^2 + |h(1)|^2 + |h(2)|^2 \right]$$

$$= \frac{1}{4}\left[\frac{1}{16} + \frac{1}{4} + \frac{1}{16} \right] = \frac{1}{4} \cdot \frac{3}{8} = \frac{3}{32}.$$

A.II.17 A white noise signal $x(n)$ of zero expected value and variance σ_x^2 is applied to an LTI system with impulse response $h(n) = (0.6)^n u(n)$ producing the output $v(n)$. The signal $v(n)$ is then applied to a second LTI system of impulse response $g(n) = (0.8)^n u(n)$, producing the output $y(n)$. Calculate the expected values and variances of the signals $v(n)$ and $y(n)$.

SOLUTION: For the first system with $h(n) = (0.6)^n u(n)$, it is:

$$\mathbb{E}[v] = \mathbb{E}[x] \sum_{n=-\infty}^{+\infty} h(n) = \mathbb{E}[x]H(e^{j0}) = 0, \text{ since } \mathbb{E}[x] = 0.$$

$$\sigma_v^2 = \sigma_x^2 \sum_{n=-\infty}^{+\infty} |h(n)|^2 = \underbrace{\sigma_x^2 \frac{1}{2\pi} \int_{-\pi}^{\pi} |H(e^{j\omega})|^2 \, d\omega}_{\text{by using Parseval's theorem}}$$

$$= \sigma_x^2 \frac{1}{2\pi} \int_{-\pi}^{\pi} H(e^{j\omega})H(e^{-j\omega}) \, d\omega = \sigma_x^2 \frac{1}{2\pi} \int_{-\pi}^{\pi} \frac{1}{1 - 0.6e^{-j\omega}} \frac{1}{1 - 0.6e^{j\omega}} \, d\omega$$

$$= \sigma_x^2 \frac{1}{2\pi} \int_{-\pi}^{\pi} \frac{1}{1.36 - 1.2\cos\omega} \, d\omega = \sigma_x^2 \frac{1}{2\pi} \frac{2\pi}{\sqrt{(1.36)^2 - (-1.2)^2}}$$

$$= \sigma_x^2 \frac{1}{0.64} = \frac{\sigma_x^2}{(0.8)^2} = 1.5625\sigma_x^2,$$

where the fourth equality is due to the fact that:

$$H(e^{j\omega}) = \text{DTFT}\{x(n)\} = \text{DTFT}\{0.6^n u(n)\} = 1/(1 - 0.6e^{-j\omega}),$$

while the last integral is solved by the use of the relation:

$$\int_0^{2\pi} \frac{d\omega}{a + b\cos\omega} = \frac{2\pi}{\sqrt{a^2 - b^2}}.$$

For the second system with $g(n) = (0.8)^n u(n)$, it is:

$$\mathbb{E}[y] = \mathbb{E}[v] \sum_{n=-\infty}^{+\infty} g(n) = \mathbb{E}[v]G(e^{j0}) = 0, \text{ since } \mathbb{E}[v] = 0.$$

$$\sigma_y^2 = \sigma_v^2 \sum_{n=-\infty}^{+\infty} |g(n)|^2 = \sigma_v^2 \frac{1}{2\pi} \int_{-\pi}^{\pi} |G(e^{j\omega})|^2 d\omega$$

$$= \sigma_v^2 \frac{1}{2\pi} \int_{-\pi}^{\pi} G(e^{j\omega})G(e^{-j\omega}) d\omega = \sigma_v^2 \frac{1}{2\pi} \int_{-\pi}^{\pi} \frac{1}{1 - 0.8e^{-j\omega}} \frac{1}{1 - 0.8e^{j\omega}} d\omega$$

$$= \sigma_v^2 \frac{1}{2\pi} \int_{-\pi}^{\pi} \frac{1}{1.64 - 1.6\cos\omega} d\omega = \sigma_v^2 \frac{1}{2\pi} \frac{2\pi}{\sqrt{(1.64)^2 - (-1.6)^2}}$$

$$= \frac{\sigma_v^2}{0.36} = \frac{\sigma_v^2}{(0.6)^2} = \frac{\frac{\sigma_x^2}{(0.8)^2}}{(0.6)^2} = \frac{\sigma_x^2}{(0.48)^2} = 4.34\sigma_x^2.$$

A.II.18 A WSS white noise signal $x(n)$ of zero expected value is applied to a casual LTI discrete-time system with an impulse response $h(n) = \delta(n) - a\delta(n-1)$ producing the WSS output $v(n)$. Calculate the power spectrum and the average power of the output $y(n)$. What is the effect of a on the average power of the output?

You are given the following formulas for the calculations.

• Power spectrum of the output: $P_{yy}(\omega) = |H(e^{j\omega})|^2 P_{xx}(\omega)$.

• Autocorrelation of white random process WSS: $P_{yy}(\omega) = |H(e^{j\omega})|^2 P_{xx}(\omega)$.

SOLUTION: It is true that:

$$H(e^{j\omega}) = \text{DTFT}\{h(n)\} = \sum_{n=-\infty}^{+\infty} h(n)e^{-j\omega n} = 1 - ae^{-j\omega}.$$

Thus,

$$P_{yy}(\omega) = \left|H(e^{j\omega})\right|^2 P_{xx}(\omega)$$
$$= H(e^{j\omega})H(e^{-j\omega})P_{xx}(\omega)$$
$$= (1 - a e^{-j\omega})(1 - a e^{j\omega})\sigma_x^2$$
$$= (1 - a e^{-j\omega} - a e^{j\omega} + a^2)\sigma_x^2.$$

The average power for a WSS signal is given by:

$$P_y = R_{yy}(0) = \sigma_y^2 + (\mathbb{E}[y])^2.$$

But:

$$\mathbb{E}[y] = \mathbb{E}[x] \sum_{n=-\infty}^{+\infty} h(n) = 0,$$

$$\sigma_y^2 = \sigma_x^2 \sum_{n=-\infty}^{+\infty} |h(n)|^2 = \sigma_x^2 \underbrace{\frac{1}{2\pi} \int_{-\pi}^{\pi} |H(e^{j\omega})|^2 \, d\omega}_{\text{by using Parseval's theorem}}$$

$$= \sigma_x^2 \frac{1}{2\pi} \int_{-\pi}^{\pi} |H(e^{j\omega})|^2 \, d\omega = \sigma_x^2 \frac{1}{2\pi} \int_{-\pi}^{\pi} (1 + a^2 - 2a\cos\omega) \, d\omega$$

$$= \sigma_x^2 \frac{1}{2\pi} \left[(1 + a^2) \int_{-\pi}^{\pi} d\omega - 2a \int_{-\pi}^{\pi} \cos\omega \, d\omega \right]$$

$$= \sigma_x^2 \frac{1}{2\pi} \left[(1 + a^2)2\pi - 2a \cdot 0 \right] = \sigma_x^2(1 + a^2).$$

Alternatively, the variance can be calculated as follows:

$$\sigma_y^2 = \sigma_x^2 \sum_{n=-\infty}^{+\infty} |h(n)|^2 = \sigma_x^2 \left(|h(0)|^2 + |h(1)|^2 \right) = \sigma_x^2(1 + a^2).$$

We can see that the average power is increased with the increase by a.

A.II.19 The input $x(n)$ to the discrete-time system of the figure is white noise with power spectrum σ^2, i.e., $P_{xx}(\omega) = \sigma^2$. Calculate the power spectrum and the average power of the output $y(n)$. It is given that $a < 1$.

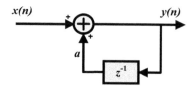

SOLUTION: The power spectrum of the output is $P_{yy}(\omega) = |H(e^{j\omega})|^2 P_{xx}(\omega)$. From the figure, it is $y(n) = x(n) + ay(n-1)$ and, taking the DTFT of the two parts of this relation, it is derived:

$$Y(e^{j\omega}) = X(e^{j\omega}) + ae^{-j\omega}Y(e^{j\omega})$$

$$\Rightarrow \frac{Y(e^{j\omega})}{X(e^{j\omega})} = H(e^{j\omega}) = \frac{1}{1-ae^{-j\omega}}, \quad a < 1, \ |\omega| < \pi.$$

Thus:

$$P_{yy}(\omega) = H(e^{j\omega})H(e^{-j\omega})P_{xx}(\omega)$$

$$= \frac{1}{1-ae^{-j\omega}} \cdot \frac{1}{1-ae^{j\omega}} \sigma^2$$

$$= \frac{\sigma^2}{1+\alpha^2 - 2\alpha\cos\omega}, \quad |\omega| < \pi.$$

The average power of the output $y(n)$ is given by:

$$\mathbb{E}\left[|y(n)|^2\right] = R_{yy}(0).$$

The autocorrelation $R_{yy}(l)$ is derived by the inverse Fourier transform of the power spectrum $P_{yy}(\omega)$:

$$R_{yy}(l) = \text{IFT}\{P_{yy}(\omega)\} = \frac{1}{2\pi}\int_{-\pi}^{\pi} P_{yy}(\omega)e^{j\omega l}\,d\omega$$

$$= \frac{1}{2\pi}\int_{-\pi}^{\pi} \frac{\sigma^2}{1+a^2-2a\cos\omega}e^{j\omega l}\,d\omega = \frac{\sigma^2}{1-a^2}\alpha^{|l|}.$$

Finally, for $l = 0$, it is:

$$R_{yy}(0) = \frac{\sigma^2}{1-a^2} = \mathbb{E}\left[|y(n)|^2\right].$$

A.II.20 A WSS white noise signal $x(n)$ with zero expected value and variance σ_x^2 is used to an LTI discrete-time system, of the impulse response $h(n) = (0.5)^n u(n)$ and produces the WSS output signal$y(n)$. Compute the power spectrum $P_{yy}(\omega)$ of the output.

SOLUTION:

$$P_{yy}(\omega) = H(e^{j\omega})H(e^{-j\omega})P_{xx}(\omega).$$

But $P_{xx}(\omega) = \sigma_x^2$, since $\mathbb{E}[x] = 0$.

Moreover:

$$
\begin{aligned}
\left|H(e^{j\omega})\right|^2 &= H(e^{j\omega})H(e^{-j\omega}) \\
&= \frac{1}{1-0.5e^{-j\omega}} \cdot \frac{1}{1-0.5e^{j\omega}} \\
&= \frac{1}{1-0.5e^{j\omega}-0.5e^{-j\omega}+(0.5)^2} \\
&= \frac{1}{1.25-\cos\omega}.
\end{aligned}
$$

Finally, by combining the above relations, it is:

$$P_{yy}(\omega) = \frac{\sigma_x^2}{1.25-\cos\omega}.$$

A.II.21 Stochastic WSS signal $x(n)$ with the expected value $\mathbb{E}[x(n)] = 2$ and second statistical moment $\mathbb{E}[x^2] = 7$ is applied to an LTI discrete-time system, producing the stochastic signal $y(n)$. The realization structure of the system is illustrated in the following figure. Calculate the expected value and the variance of the output signal.

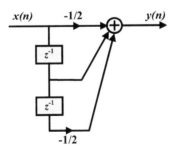

SOLUTION: For the input signal, it is:

$$\mathbb{E}[x(n)] = 2,$$
$$\sigma_x^2 = \mathbb{E}[x^2] - \mathbb{E}[x(n)] = 7 - 2^2 = 3.$$

Also, for the system, it is:

$$y(n) = -\frac{1}{2}x(n) + x(n-1) - \frac{1}{2}x(n-2)$$

$$\Leftrightarrow Y(z) = -\frac{1}{2}X(z) + z^{-1}X(z) - \frac{1}{2}z^{-2}X(z)$$

$$\Leftrightarrow \frac{Y(z)}{X(z)} = H(z) = -\frac{1}{2} + z^{-1} - \frac{1}{2}z^{-2}.$$

By taking the inverse z transform of $H(z)$, $h(n)$ is derived:

$$h(n) = -\frac{1}{2}\delta(n) + \delta(n-1) - \frac{1}{2}\delta(n-2) \Rightarrow h(n) = \left\{ \underset{n=0}{-\frac{1}{2}}, 1, -\frac{1}{2} \right\}.$$

For the output, it is:

$$\mathbb{E}[y(n)] = \mathbb{E}[x(n)] \sum_{n=-\infty}^{+\infty} h(n) = \mathbb{E}[x(n)][h(0) + h(1) + h(2)] = 2\left[-\frac{1}{2} + 1 - \frac{1}{2} \right] = 0,$$

$$\sigma_y^2 = \sigma_x^2 \sum_{n=-\infty}^{+\infty} |h(n)|^2 = 3\left[\frac{1}{4} + 1 + \frac{1}{4} \right] = \frac{9}{2}.$$

Bibliography

Abromowitz, M. and Stegun, I., *Handbook of Mathematical Functions*. Dover, 1965.

Adali, J. and Hayken, S., *Adaptive Signal Processing: Next Generation Solutions*. Wiley, 2010.

Antoniou, A., *Digital Filters: Analysis, Design, and Applications, 2nd Edition*. McGraw-Hill, 1993.

Bickel, P.-J. and Doksum, K.-A., *Mathematical Statistics: Basic Ideas and Selected Topics*. Holden-Day, 1977.

Bode, H. and Shannon, C., "A simplified derivation of linear least squares smoothing and prediction theory," *Proc. of the Institute of Radio Engineers (IRE)*, vol. 38, pp. 417–425, April 1950.

Bose, T., *Digital Signal and Image Processing*. Wiley, 2003.

Brillinger, D.-R., *Time Series: Data Analysis and Theory*. Springer-Verlag, 1981.

Cavicchi, T., *Digital Signal Processing*. Wiley, 2000.

Chaparro, L., *Signals and Systems Using MATLAB*. Elsevier, 2011.

Chassaing, R., *Digital Signal Processing with the C6713 and C6416 DSK*. Wiley, 2004.

Chassaing, R. and Reay, D., *Digital Signal Processing and Applications with the TMS320C6713 DSK*. Wiley-IEEE, 2008.

Chen, C.-T., *Digital Signal Processing*. Oxford University Press, 2001.

Clark, C., *LabVIEW Digital Signal Processing*. McGraw-Hill, 2005.

Corinthios, M., *Signals, Systems, Transforms, and Digital Signal Processing with MATLAB*. CRC Press, 2009.

Cristi, R., *Modern Digital Signal Processing*. Thompson, 2004.

Davenport, W. and Root, W., *An introduction to the theory of random signals and noise*. IEEE Press (reprint of 1958 McGraw-Hill edition), 1987.

Eaton, M.-L., *Multivariate Statistics—A Vector Space Approach*. Wiley, 1983.

Gelman, A., Carlin, J.-B., Stern, H.-S., and Rubin, D.-B., *Bayesian Data Analysis*. Chapman and Hall/CRC, 1995.

Gentleman, W.M. and G. Sande, "Fast fourier transforms—For fun and profit," in AFIPS Conference Proceedings, Nov. 1966, vol. 29, pp. 563–578.

Golub, G.-H. and Van Loan, C.-F., *Matrix Computations, 2nd Edition*. The Johns Hopkins University Press, 1989.

Graybill, F.-A., *Matrices with Applications in Statistics*. Wadsworth Publishing Co., 1983.

Grover, D. and Diller, J., *Digital Signal Processing*. Prentice Hall, 1999.

Hayes, M., *Schaum's Outline of Digital Signal Processing*. Schaum, 1999.

Hayes, M., *Statistical Digital Signal Processing and Modeling*. Wiley, 1996.

Haykin, S., *Array Signal Processing*. Prentice-Hall, 1985.

Heck, B. and Kamen, E., *Fundamentals of Signals and Systems Using the Web and MATLAB*. Prentice Hall, 2007.

Helstrom, C., *Elements of signal detection and estimation*. Prentice-Hall, 1995.

Hollander, M. and Wolfe, D.-A., *Nonparametric statistical methods, 2nd Edition*. Wiley, 1991.

Ifwachor, E. and Jervis, B., *Digital Signal Processing, 2nd Edition*. Addison Wesley, 2001.

Ingle, V. and Proakis, J., *Digital Signal Processing Using MATLAB*. Cengage Learning, 2007.

Jackson, L., *Digital Filters and Signal Processing*. Kluwer, 1989.

Johnson, N.-L., Kotz, S., and Balakrishnan, A., *Continuous Univariate Distributions: Vol. 2*. Wiley, 1995.

Kariya, T. and Sinha, B.-K., *Robustness of Statistical Tests*. Academic Press, 1989.

Kassam, S. and Thomas, J., *Nonparametric detection—Theory and Applications*. Dowden, Hutchinson and Ross, 1980.

Kay, S.-M., *Statistical Estimation*. Prentice-Hall, 1991.

Kehtarnavaz, N. and Kim, N., *Digital Signal Processing System Design Using LabVIEW*. Newnes, 2005.

Kuo, S. and Gan, W.-S., *Digital Signal Processing*. Prentice Hall, 2004.

Lehmann, E.-L., *Testing Statistical Hypotheses*. Wiley, 1959.

Lutovac, M., Tosic, D., and Evans, B., *Filter Design for Signal Processing, Using MATLAB and Mathematica*. Prentice Hall, 2001.

Lyons, R., *Understanding Digital Signal Processing*. Prentice Hall, 2004.

Madisetti, V., *The Digital Signal Processing Handbook, 2nd Edition*. CRC Press, 2009.

McClellan, J., Schafer, R., and Yoder, M., *Signal Processing First*. Prentice Hall, 1998.

Mendel, J.-M., *Lessons in Estimation for Signal Processing, Communications, and Control*. Prentice-Hall, 1995.

Mersereau, R. and Smith, M., *Digital Filtering*. Wiley, 1993.

Mitra, S., *Digital Signal Processing Laboratory Using MATLAB*. McGraw-Hill, 1999.

Mitra, S., *Digital Signal Processing: A Computer-Based Approach, 4th Edition*. McGraw-Hill, 2011.

Moon, T.-K. and Stirling, W.-C., *Mathematical Methods and Algorithms for Signal Processing*. Prentice Hall, 2000.

Moon, T. and Stirling, W., *Mathematical Methods and Algorithms for Signal Processing*. Prentice Hall, 2000.

Muirhead, R.-J., *Aspects of Multivariate Statistical Theory*. Wiley, 1982.

Oppenheim, A.-V. and Willsky, A.-S., *Signals and Systems*. Prentice-Hall, 1983.

Oppenheim, A. and Schafer, R., *Digital Signal Processing, 2nd Edition*. Prentice Hall, 1999.

Oppenheim, A. and Schafer, R., *Discrete-Time Signal Processing, 3rd Edition*. Prentice Hall, 2010.

Orfanidis, S., *Signal Processing*. Prentice Hall, 1996.

Papoulis, A., *Probability, Random Variables, and Stochastic Processes, 3rd Edition*. McGraw Hill, 1991.

Parhi, K., *VLSI Ditial Signal Processing and Implementation*. Wiley, 1995.

Parks, T. and Burrus, S., *Digital Filter Design*. Wiley, 1987.

Phillips, C., Parr, J., and Riskin, E., *Signals, System, and Transforms*. Prentice Hall, 2003.

Poral, B., *A Course in Digital Signal Processing*. Wiley, 1997.

Proakis J.-G. and Manolakis, D.-G., *Digital Signal Processing: Principles, Algorithms, and Applications*. Prentice-Hall, 1996.

Proakis, J. and Manolakis, D., *Digital Signal Processing, 4th Edition*. Prentice Hall, 2007.

Rao, P., *Signals and Systems*. McGraw-Hill, 2008.

Saben, W.E., *Discrete-Time Signal Analysis and Design*. Wiley, 2008.

Scharf, L.-L., *Statistical Signal Processing: Detection, Estimation, and Time Series Analysis*. Addison-Wesley, 1991.

Schilling, R. and Harris, S., *Fundamentals of Digital Signal Processing Using MATLAB*. Thomson, 2005.

Schlichtharle, D., *Digital Filters*. Springer, 2000.

Schuler, C., and Chugani, M., *Digital Signal Processing*. McGraw-Hill, 2004.

Searns, S. and Hush, D., *Digital Signal Processing with Examples in MATLAB*. CRC Press, 2011.

Shannon, C., "A Mathematical Theory of Communication," *Bell System Technical Journal*, vol. 27(3), pp. 379–423, 1948.

Singh, A. and Srinivansan, S., *Digital Signal Processing*. Thompson, 2004.

Smith, S., *The scientists and engineer's guide to digital signal processing*. www.DSPguide.com.

Tanner, M.-A., *Tools for Statistical Inference; Methods for the exploration of posterior distributions and likelihood functions*. Springer-Verlag, 1993.

Taylor, F., *Digital Filter Design Handbook*. Marcel Dekker, 1983.

Taylor, F. and Mellott, J., *Hands-On Digital Signal Processing*. McGraw-Hill, 1995.

Van-Trees, H.-L., *Detection, Estimation, and Modulation Theory: Part III*. Wiley, 2001.

Veloni, A., Miridakis, N., and Boucouvala, E., *Digital and Statistical Signal Processing*. CRC, 2018.

Whalen, A.-D., *Detection of Signals in Noise, 2nd Edition*. Academic Press, 1995.

Widrow, B. and Kollar, I., *Quantization Noise*. Cambridge, 2008.

Williams, A. and Taylor, F., *Electronic Filter Design Handbook, 4th Edition*. McGraw-Hill, 2006.

Yates, W., *Digital Signal Processing*. CSP, 1989.

Zelniker, G. and Taylor, F., *Advanced Digital Signal Processing*. Dekker, 1994.

Index

Page numbers followed by f and t indicate figures and tables, respectively.

A

Accumulator, 55
Accuracy, defined, 4
Acquisition time, 18
ADC, *see* Analog to digital converter (ADC)
ADC 3 digits, input-output relation of one,
    7–8, 8f
Adder component, block diagram, 179, 180f
Additivity, in discrete-time systems, 55, 56f
Affine estimator
    geometric interpretation (orthogonality
        condition and projection theorem),
        459
        minimum affine MSE estimation, 462
        minimum MSE linear estimation,
            460–461, 461f
        optimization for linear Gaussian model,
            462–463
        optimal affine vector estimator, 463–464
        examples, 464–467
        superposition property of, 459
Affine functions, 455; *see also* Linear estimation
Affine projection theorem, 462
Affine vector estimator, optimal, 463–464
Analog filters, 287; *see also* Digital filters
    vs. digital filters, 287
Analog signal processing, 3–4, 4f
Analog signal(s)
    digital processing of, 4, 4f
    digitization steps of, 5
        coding, 9–10, 9f–10f
        quantization, 7–9, 7f–8f
        sampling, 5–7, 6f
    non-accurate reconstruction of, 21–22, 22f
    real-time processing applications, 21
Analog to digital converter (ADC), 4, 5, 6, 287
Analog-to-digital signal conversion (A/D)
    conversion, 8, 9f
Analytical technique
    convolution calculation, 61
Associativity property, convolution, 60, 60t
Autocorrelation
    discrete-time signals, 63–64
    properties, 64

B

Backward difference method, digital IIR filter
    design, 295–296, 295f
Band-pass filter, 289, 289f; *see also* Digital
    filters
Band-stop filter, 289, 289f; *see also* Digital
    filters
Bartlett (triangular) window, 313, 314f
Bayes approach, signal detection, 488
    average risk minimization, 488–489
    disadvantages, 501
    minimax Bayes detector, 491–493, 492f
    minimum probability of error test, 490
    optimal Bayes test, 489–490
    performance evaluation of Bayes likelihood
        ratio test, 490–491, 491f
    priori probabilities, 488
Bayes estimator, 407
Bayesian estimation (examples), 413–420
    Gaussian signal
        estimation of, 415–417, 415f, 416f
        sign estimation, 419–420
        size estimation, 417–419, 418f
    uniform PDF amplitude estimation, 413–415,
        414f
Bayes likelihood ratio test, performance
    evaluation of, 490–491, 491f
Bayes risk criterion, 422
Bayes risk function, 407
Beta distribution, 391–392
BIBO stable (Bounded Input-Bounded Output
    stable), 58, 58f
Bilateral z-transform, 116
Bilinear (Tustin) method, digital IIR filters
    design, 297–298, 298f
Binary codes, 10
Binary likelihood ratios, 398
Blackman window, 314, 315f
Block diagrams
    branches, 179
    implementation components, 179, 180f
    nodes, 179
    realization structures, discrete-time systems,
        179–180, 180f

Branch(es)
    block diagrams, 179
    SFGs, 188, 188f
'Butterfly' graph, FFT, 224, 224f

C

Cartesian form, z-transform, 113
Casual discrete systems, 57–58
Cauchy distribution, 391
    position parameters of, minimal sufficiency
        of, 402
Cauchy's formula, 123
Cauchy's residue theorem, 122
Causality
    of discrete-time systems, 124–125
CDF, *see* Cumulative distribution function
        (CDF)
Central limit theorem (CLT), 387
Chebychev inequality, 423, 424
Chi-squared mixed distribution, 389–390
Chi-squared RV distribution, 388
    non-central, 389
Chunking method, 60–61
Circular convolution
    defined, 219
    DFT of, 219, 219f
Circular shift, DFT of, 218
    in frequency domain, 218
    in time domain, 218
CME, *see* Conditional mean estimator (CME)
Code, 9
Codeword, 9
Coding
    analog signals, 9–10, 9f–10f
    defined, 9
    voice coding system, 10, 10f
Commutativity property, convolution, 60, 60t
Complex integration method
    inverse z-transform calculation, 123–124
Complex system analysis, 58
Composite hypothesis, signal detection, 485
Computational complexity, 181
Conditional mean estimator (CME), 407–408,
        408f
    minimum mean absolute error estimation, 409
    MSE estimation, 407–408
    MUE estimation, 412
Conjugation of complex sequence, z-transform,
        120
Connected systems
    feedback connection of systems, 126f
    systems connected in parallel, 126f

    systems connected in series, 126f
    transfer function of, 126, 126f–127f
Constant (multiple non-random) parameters,
        estimation of, 441
    Cramer-Rao (CR) matrix bound in
        covariance matrix, 442–446
    criteria for estimator performance
        evaluation:, 441
    maximum likelihood vector estimation,
        447–451
    vector estimation methods, 446
Constant (non-random) parameters, estimation
        of, 422–423, 423f
    Cramer-Rao Bound (CRB) in estimation
        variance, 433–440
    maximum likelihood (ML), scalar estimators
        for, 429–433, 430f, 431f
    method of statistical moments for scalar
        estimators, 426–429, 428f
    scalar estimation criteria, 423–425,
        424f–425f
Continuous signal, 7
Continuous-time signal, 11
Convolution, 59–62, 60f, 60t
    associativity property, 60, 60t
    circular, DFT, 219, 219f
    commutativity property, 60, 60t
    distributivity property, 60, 60t
    fast, FFT of, 228
        overlap and add method, 228–229, 228t
        overlap and save method, 229
    identity property, 60, 60t
    matrix, use of, 60, 60f
    periodic, DFS, 215
    properties, 60t
    of two finite-length signals, 60–61, 60f
    z-transform, 119
Correlation
    discrete-time signals, 63–64
    of two sequences, z-transform, 121
Covariance matrix
    Cramer-Rao (CR) matrix bound in, 442–446
Cramer-Rao bound (CRB)
    defined, 433–434
    in estimation variance, 433–440
    examples, 439–440
    predictions for non-random nuisance
        parameters, 453–454
    scalar CRB, properties, 436–439
Cramer-Rao (CR) matrix bound, in covariance
        matrix, 442–446
    proof of, 445–446
    properties, 443–445

CRB, *see* Cramer-Rao bound (CRB)
Cross-correlation
  discrete-time signals, 63–64
  properties, 64
Cumulative distribution function (CDF)
  of Gaussian RV, 384–385
Cut-off frequency, 11

D

DAC, *see* Digital to analog converter (DAC)
DCT, *see* Discrete cosine transform (DCT)
Decimation in frequency (DIF) technique, 227
Decimation in time (DIT) process, 224, 226, 227
Decision function, signal detection, 486–487,
    486f, 487f
Deconvolution, 62
Delay component, block diagram, 179, 180f
Dependent variable
  conversion of, 54
Difference equations (D.E.), 113
  discrete-time systems, 64–65
  solution via *z*-transform, 113–114, 114f
Digital filters; *see also* Analog filters
  advantages, 287–288
  band-pass filter, 289, 289f
  band-stop filter, 289, 289f
  design specifications, 288–290, 289f–290f
  digital FIR filters, 288
  digital IIR filters, 288
    frequency transformations, 301–303
  digital IIR filters, design, 290–291
    direct methods, 300–301
    indirect methods, 292–300
  FIR filters, 303–304
    stability of, 307
  FIR filters, design, 307
    comparison of methods, 319, 319t
    optimal equiripple FIR filter design,
      317–319, 318f
    using frequency sampling method,
      309–311
    using window method, 311–317, 311f, 312t,
      313f–316f, 317t
  FIR linear phase filters, 304–307
  high-pass filter, 289, 289f
  low-pass filter, 289, 289f
  moving average filters, 307–308, 309f
  overview, 287–288
  pass band, 289, 290f
  signal filtering process using, 287, 288f
  stop band, 290, 290f
  transition band, 290, 290f

types of, 288, 289f
  vs. analog filters, 287
Digital FIR filters, 288; *see also* Finite impulse
    response (FIR) filters
Digital IIR filters, 288; *see also* Infinite impulse
    response (IIR) filters
  design, 290–291
    direct methods, 291, 300–301
    indirect methods, 291, 292–300
  direct design methods
    design of $H(e^{jw})^2$ method, 300–301
    $h[n]$, method of calculating, 301
  frequency transformations, 301–303
  indirect design methods, 291, 292–300
    backward difference method, 295–296, 295f
    bilinear (Tustin) method, 297–298, 298f
    forward difference method, 296
    impulse invariant method, 292
    matched pole-zero method, 299–300
    step invariant method (or *z*-transform
      method with sample and hold),
      293–294, 293f–294f
  optimization methods, 291
Digital signal processing, 4f
  advantages, 4–5
  analog signal, 3–4, 4f
  applications, 3, 4–5, 6
  objectives, 3
  overview, 3
Digital signal processor (DSP), 4, 5, 287
Digital signal(s), 45
  analog signal conversion into, 5
    coding, 9–10, 9f–10f
    quantization, 7–9, 7f–8f
    sampling, 5–7, 6f
  defined, 3
Digital to analog converter (DAC), 4, 5, 6, 287
  features, 4
Digital voice coding system, 10, 10f
Digitization
  analog signals, 5
    coding, 9–10, 9f–10f
    quantization, 7–9, 7f–8f
    sampling, 5–7, 6f
  defined, 5
Direct methods, digital IIR filters design
  design of $H(e^{jw})^2$ method, 300–301
  $h[n]$, method of calculating, 301
Discrete cosine transform (DCT), 212, 229–231,
    231f
Discrete Fourier series (DFS), 211, 214–216
  coefficients of, 215
  periodic convolution, 215

relation of components and DTFT over a
period, 216
Discrete Fourier transform (DFT), 211, 216–220;
see also Fast Fourier transform (FFT)
circular convolution, 219, 219f
circular shift, 218
complex multiplications for calculations,
227t
linearity, 218
multiplication of sequences, 220
N-point DFT, 217
Parseval's theorem, 220, 220t
properties, 218–220
sampling at DTFT frequency, 217, 217f
time dependent, 216
Discrete likelihood ratios, 398
Discrete-time Fourier transform (DTFT), 211,
212, 213t–214t
defined, 212
of fundamental signals, 213t
properties, 214t
relation of DFS components and, 216
sampling at DTFT frequency, 217, 217f
Discrete-time moving average system, 55
Discrete-time signals, 54f; *see also* Discrete-time
systems
autocorrelation, 63–64
correlation, 63–64
cross-correlation, 63–64
defined, 45, 46f
energy of, 53
even and odd, 51–53, 52f
exponential function, 48–50, 49f–50f
impulse function, 45, 46f, 46t
power of, 53
ramp function, 47, 48f
sinusoidal sequence, 50–51, 51f
transfer function of, 124
unit rectangular function (pulse function),
48, 48f
unit step function, 47, 47f
Discrete-time systems, 54–55, 54f; *see also*
Discrete-time signals
additivity in, 55, 56f
block diagrams, 179–180, 180f
casual, 57–58
categories, 55–58
causality of, 124–125
convolution, 59–62, 60f, 60t
associativity property, 60, 60t
commutativity property, 60, 60t
distributivity property, 60, 60t
identity property, 60, 60t

matrix, use of, 60, 60f
properties, 60t
of two finite-length signals, 60–61, 60f
deconvolution, 62
difference equations, 64–65
discrete-time moving average system, 55
examples, 55
of finite impulse response, 65–66
homogeneity in, 55, 56f
ideal discrete time delay system, 55
implementation, 179
using hardware, 179
using software, 179
impulse response, 59
invertible, 57
linear, 55–56, 56f
with memory, 57
overview, 179
realization structure, 181–183, 181f–182f
FIR discrete systems, 186–187, 186f, 187f
IIR discrete systems, 183–186, 184f–185f
signal flow graphs (SFGs), 188, 188f–189f
Mason's gain formula, 189, 189f–190f
stability of, 125–126
stable, 58, 58f
system connections, 58
mixed connection, 58, 59f
parallel connection, 58, 59f
in series connection, 58, 59f
time-invariant, 56–57
transfer function of, 127–128
unstable, 58, 58f
Discrete wavelets transform (DWT), 212,
231–236, 232f
Distribution(s); *see also* Gaussian distribution
reproducing, 392
Distributivity property, convolution, 60, 60t
Division method
inverse z-transform calculation, 122
DSP, *see* Digital signal processor (DSP)
DTFT, *see* Discrete-time Fourier transform
(DTFT)
Duty cycle, 19
DWT, *see* Discrete wavelets transform (DWT)
DWT coefficients, 232
Dynamic systems, 57

E

8-point FFT algorithm, 224–227, 225f–227f
Energy
of discrete-time signals, 53
Even discrete-time signal, 51–53, 52f

Expected value, 393–395
of Gaussian distribution, minimal
sufficiency of, 400–401
and variance of Gaussian distribution,
minimal sufficiency of, 401–402
Exponential distributions category, statistical
sufficiency, 402–404
PDF and, 404
Exponential function, 48–50, 49f–50f
Exponential sequence
multiplication by, 120

F

Fast convolution, FFT of, 228
overlap and add method, 228–229, 228t
overlap and save method, 229
Fast Fourier transform (FFT), 212, 221; *see also*
Discrete Fourier transform (DFT)
'butterfly' graph, 224, 224f
complex multiplications for calculations,
227t
decimation in frequency (DIF) technique,
227
decimation in time (DIT) process, 224, 226,
227
8-point FFT algorithm, 224–227, 225f–227f
equations, 221–227, 224f–226f, 227t
fast convolution, 228–229
Fourier transform estimation through, 229
fundamental equations of radix-2 FFT
algorithm, 223
IDFT computation using FFT, 228
iterative FFT algorithm, 223–224, 224f
Feedback connection of systems, 126f
Final value theorem, 121
Finite energy signals, 53
Finite impulse response (FIR)
discrete-time systems of, 65–66
Finite impulse response (FIR) filters, 288,
303–304; *see also* Digital FIR filters
design, 307
comparison of methods, 319, 319t
optimal equiripple FIR filter design,
317–319, 318f
using frequency sampling method,
309–311
using window method, 311–317, 311f, 312t,
313f–316f, 317t
ideal impulse responses, 312t
stability of, 307
Finite-length signals
convolution of, 60–61, 60f

Finite power signals, 53
Finite word length effects, 181
FIR, *see* Finite impulse response (FIR)
FIR discrete systems, implementation
structures, 186–187
in cascade form, 187, 187f
in direct form I, 186, 186f
FIR filters, *see* Finite impulse response (FIR)
filters
FIR linear phase filters, 304–307
Fisher-Cochran theorem, 392–393, 394
Fisher factorization, 397
Fisher-Snedecor F-distribution, 390–391
Formula tables
z-transform, 129, 129t–130t
Forward difference method, digital IIR filters
design, 296
Forward path, SFGs, 188, 189f
Fourier analysis, 211
Fourier transform; *see also* Discrete Fourier
transform (DFT); Fast Fourier
transform (FFT)
estimation through FFT, 229
Frequency domain, 211
DFT of circular shift in, 218
Frequency domain analysis
discrete cosine transform, 229–231, 231f
discrete Fourier series (DFS), 214–216
periodic convolution, 215
relation of components and DTFT over a
period, 216
discrete Fourier transform (DFT), 216–220
circular convolution, 219, 219f
circular shift, 218
linearity, 218
multiplication of sequences, 220
Parseval's theorem, 220, 220t
properties, 218–220
sampling at DTFT frequency, 217, 217f
discrete-time Fourier transform (DTFT), 212,
213t–214t
fast Fourier transform (FFT), 221
equations, 221–227, 224f–226f, 227t
fast convolution, 228–229
IDFT computation using FFT, 228
Fourier transform, estimation through FFT,
229
overview, 211–212
wavelet transform, 231–236, 232f
theory, 233–236, 235f
Frequency sampling method
FIR filters design using, 309–311
Frequency transformations, IIR filters, 301–303

Frequentist approach, signal detection, 502
    defined, 502
    simple hypothese, case of, 502–505
FSR, *see* Full scale range (FSR)
Full scale range (FSR), 8

G

Gamma distribution, 388–389
Gaussian distribution, 383–392
    beta distribution, 391–392
    Cauchy distribution, 391
    central limit theorem, 387
    chi-squared mixed distribution, 389–390
    chi-squared RV distribution, 388
    Fisher-Snedecor F-distribution, 390–391
    gamma distribution, 388–389
    minimal sufficiency of expected value and
        variance of, 401–402
    minimal sufficiency of expected value of,
        400–401
    multivariate, 385–387
    non-central chi-squared RV distribution, 389
    overview, 383–385
    reproducing distributions, 392
    student's t-distribution, 390
Gaussian models
    linear, affine estimator optimization for,
        462–463
    LMMSE optimization in, 477–478
Gaussian random variable X (RV); *see also*
        Gaussian distribution
    cumulative distribution function (CDF),
        384–385
    independent and identically distributed
        (IID), 385
    probability density function (PDF) of, 383
    reproducing distribution, 392
Gaussian signal, Bayesian estimation, 415–417,
        415f, 416f
    sign estimation, 419–420
    size estimation, 417–419, 418f
Gramm-Schmidt orthonormalization method, 394
Graphical technique
    convolution calculation, 61

H

Hamming window, 314, 315f
Hanning window, 313, 314f
Hardware
    discrete-time system implementation using,
        179

Heaviside formula, 122
$H(e^{jw})^2$ method, digital IIR filters design,
        300–301
High-pass filter, 289, 289f; *see also* Digital filters
$H[n]$, method of calculating, digital IIR filters
        design, 301
Holding time, 20
Homogeneity, in discrete-time systems, 55,
        56f

I

IDCT, *see* Inverse discrete Cosine transform
        (IDCT)
Ideal discrete time delay system, 55
Ideal impulse responses, 312t
Identical system, 54
Identity property, convolution, 60, 60t
IDFT, *see* Inverse discrete Fourier transform
        (IDFT)
IFFT, *see* Inverse Fast Fourier Transform (IFFT)
IIR, *see* Infinite impulse response (IIR)
IIR discrete systems, implementation
        structures, 183–186
    in canonical form, 184–185, 185f
    in direct form I, 184, 184f
    intermediate stage, 184, 184f
IIR filters, *see* Infinite impulse response (IIR)
        filters
Impulse function, 45
    defined, 45, 46f
    properties, 46t
Impulse invariant method, digital IIR filters
        design, 292
Impulse responses, 59; *see also* Convolution;
        Deconvolution
    ideal, 312t
Independent and identically distributed (IID)
        Gaussian RVs, 385
Independent variable
    conversion of, 54
    lag, 54
Indirect methods, digital IIR filters design, 291,
        292–300
    backward difference method, 295–296, 295f
    bilinear (Tustin) method, 297–298, 298f
    forward difference method, 296
    impulse invariant method, 292
    matched pole-zero method, 299–300
    step invariant method (or z-transform
        method with sample and hold),
        293–294, 293f–294f
Infinite impulse response (IIR), 66

Infinite impulse response (IIR) filters, 288;
   see also Digital IIR filters
  frequency transformations, 301–303
Initial value theorem, 121
Input node, SFGs, 188, 188f
Input-output relation, of one ADC 3 digits,
   7–8, 8f
Inverse discrete Cosine transform (IDCT), 230
Inverse discrete Fourier transform (IDFT), 221
  computation, using FFT, 228
Inverse Fast Fourier Transform (IFFT), 221
Inverse z-transform, 121–122
  complex integration method, 123–124
  defined, 122
  partial fraction expansion method, 122–123
  power series expansion method (division
    method), 122
Invertible discrete systems, 57
Iterative FFT algorithm, 223–224, 224f

K

Kaiser window, 315–316, 316f
Karush-Kuhn-Tucker method, 503

L

Lag, 63
Laplace transforms, 113
  defined, 113–114
  to z-transform, 113–116, 114f
Linear/affine estimator, *see* Affine estimator
Linear discrete systems, 55–56, 56f
Linear estimation; *see also* Parametric
   estimation
  affine functions, 455
  constant MSE minimization (linear and
    affine estimation), 455
   optimal constant estimator of scalar RV,
    456
  geometric interpretation (orthogonality
    condition and projection theorem),
    459
   affine estimator optimization for linear
    Gaussian model, 462–463
   minimum affine MSE estimation, 462
   minimum MSE linear estimation,
    460–461, 461f
  LMWLS optimization in Gaussian models,
    477–478
  non-statistical least squares technique
    (linear regression), 467–472, 468f, 470f,
    472f, 473f

optimal affine vector estimator, 463–464
  examples, 464–467
  overview, 455
  scalar random variable
   optimal affine estimator of, 458–459
   optimal constant estimator of, 456
   optimal linear estimator of, 456–458
   superposition property of linear/affine
    estimators, 459
  theory of, 455
  of weighted LLS, 473–477
Linear Gaussian model, affine estimator
   optimization for, 462–463
Linearity
  DFT, 218
  z-transform, 118
Linear least squares theory (LLS), 455
Linear minimum mean square error (LMMSE),
   455
  estimator, 455, 457–458; *see also* Linear
    estimation
  optimization, in Gaussian models,
    477–478
Linear projection theorem, 460–461, 461f
Linear regression, 467–472, 468f, 470f, 472f,
   473f
Linear time invariant (LTI) systems, 55, 57, 116,
   124; *see also* Discrete-time systems
  impulse response of, 59–60, 60t; *see also*
    Convolution
Linear WSSE estimator, 477
LMMSE, *see* Linear minimum mean square
   error (LMMSE)
Loop
  SFGs, 188, 188f–189f
Low-pass filter, 289, 289f; *see also* Digital filters

M

Mason, Samuel J., 188
Mason's gain formula, 189, 189f–190f
Matched pole-zero method, digital IIR filters
   design, 299–300
Matrix
  convolution computation, 60, 60f
  covariance, Cramer-Rao (CR) matrix bound
    in, 442–446
Maximum-likelihood (ML) method, 425
  scalar estimators for, 429–433, 430f, 431f
  vector estimation, 447–451
Mean square error (MSE)
  conditional mean estimator (CME),
    407–408

constant MSE minimization (linear and
    affine estimation), 455
  optimal constant estimator of scalar RV,
    456
  estimation of, 407–409, 408f
  minimum affine MSE estimation, 462
  minimum MSE linear estimation, 460–461,
    461f
Mean uniform error (MUE)
  CME, 412
  estimation of, 411–412, 411f–413f
Memory, 181
  discrete systems with, 57
Memory-based system, 57
Memory-less system, 57
MIMO systems (multiple inputs, multiple
    outputs), 54
Minimal sufficiency, 399–402
  defined, 399–400
  examples, 400–402
  of expected value and variance of Gaussian
    distribution, 401–402
  of expected value of Gaussian distribution,
    400–401
  of position parameters of Cauchy
    distribution, 402
  proof, 400
  theorem, 400
Minimax Bayes detector, 491–493, 492f
Minimum mean absolute error
  CME, 409
  estimation of, 409–411, 410f
MISO system, *see* Multiple inputs, single output
    (MISO) system
Mixed connection, discrete-time systems, 58, 59f
Mixed node, SFGs, 188, 188f
ML method, *see* Maximum-likelihood (ML)
    method
Model validation, 55
Moving average filters, 307–308, 309f
MSE, *see* Mean square error (MSE)
Multiple hypotheses tests, signal detection,
    496–498, 497f
  average risk minimization, 498–501, 501f
  Bayes approach, disadvantages, 501
  priori probabilities, 498
Multiple inputs, multiple outputs (MIMO
    systems), 54
Multiple inputs, single output (MISO) system, 54
Multiple non-random (constant) parameters,
    estimation of, 441
  Cramer-Rao (CR) matrix bound in
    covariance matrix, 442–446

  criteria for estimator performance
    evaluation:, 441
  maximum likelihood vector estimation,
    447–451
  vector estimation methods, 446
Multiplication of sequences, DFT, 220
Multiplier component, block diagram, 179, 180f
Multivariate Gaussian distribution, 385–387

N

Neyman-Pearson strategy, 502, 503
Nodes
  block diagrams, 179
Non-accurate reconstruction, of analog signals,
    21–22, 22f
Non-central chi-squared RV distribution, 389
Non-random nuisance parameters
  CRB predictions for, 453–454
  handling, 452–453
Non-random (constant) parameters, estimation
    of, 422–423, 423f
  Cramer-Rao Bound (CRB) in estimation
    variance, 433–440
  maximum likelihood (ML), scalar estimators
    for, 429–433, 430f, 431f
  method of statistical moments for scalar
    estimators, 426–429, 428f
  scalar estimation criteria, 423–425, 424f–425f
Non-statistical least squares technique (linear
    regression), 467–472, 468f, 470f, 472f,
    473f
N-order difference equation, 64
Normal distribution, 383; *see also* Gaussian
    distribution
N-point DFT, 217
Nuisance parameters, handling, 452
  CRB predictions for non-random, 453–454
  non-random, 452–453
  random, 452
Nyquist frequency, 11
Nyquist theorem, 211

O

Odd discrete-time signal, 51–53, 52f
Optimal affine estimator, of scalar RV, 458–459
Optimal affine vector estimator, 463–464
Optimal Bayes test, 489–490
Optimal constant estimator, of scalar RV, 456
Optimal equiripple FIR filter design, 317–319,
    318f
Optimal linear estimator, of scalar RV, 456–458

Overlap and add method, 228–229, 228t
Overlap and save method, 229

P

PAM, *see* Pulse amplitude modulation (PAM)
PAM signal, 21
Parallel connection, discrete-time systems, 58, 59f
Parametric estimation; *see also* Linear estimation
  components, 405–406, 406f
  of multiple non-random (constant) parameters, 441
    Cramer-Rao (CR) matrix bound in covariance matrix, 442–446
    maximum likelihood vector estimation, 447–451
    vector estimation methods, 446
  of non-random (constant) parameters, 422–423, 423f
    Cramer-Rao Bound (CRB) in estimation variance, 433–440
    maximum likelihood (ML), scalar estimators for, 429–433, 430f, 431f
    method of statistical moments for scalar estimators, 426–429, 428f
    scalar estimation criteria, 423–425, 424f–425f
  nuisance parameters, handling, 452
    CRB predictions for non-random, 453–454
    non-random, 452–453
    random, 452
  overview, 405
  of random vector parameters, 421
    squared vector error, 421–422, 422f
    uniform vector error, 422
  of scalar random parameters, 406–407, 407f
    Bayesian estimation (examples), 413–420
    mean square error (MSE), 407–409, 408f
    mean uniform error (MUE), 411–412, 411f–413f
    minimum mean absolute error, 409–411, 410f
Parseval's theorem, 220, 220t
Partial fraction expansion method
  inverse z-transform calculation, 122–123
Pass band, digital filters, 289, 290f
Path, SFGs, 188, 189f
PDF, *see* Probability density function (PDF)
PDF amplitude (uniform), Bayesian estimation, 413–415, 414f

Periodic convolution
  DFS, 215
Physical sampling, 17–20, 18f–19f, 20f
  pulse format of sampled signal, 18, 19f
  signal spectra in, 19, 20f
Polar form, z-transform, 113
Pole-zero matching method, digital IIR filters design, 299–300
Position parameters of Cauchy distribution, minimal sufficiency of, 402
Power
  of discrete-time signals, 53
Power series expansion method
  inverse z-transform calculation, 122
Probability density function (PDF)
  Cauchy distribution, 391
  chi-squared RV distribution, 388
  defined, 388–389
  exponential distribution category and, 404
  Fisher/Snedecor–F, 390–391
  gamma distribution, 388–389
  of Gaussian RV, 383
  non-central chi-squared RV distribution, 389
  student's t-distribution, 390
Pulse amplitude modulation (PAM), 21
Pulse function (unit rectangular function), 48, 48f

Q

Quantization, 10
  analog signals, 7–9, 7f–8f
  defined, 7
  error, 7, 7f
  noise, 8

R

Radar application, signal detection and, 479–481, 480f
  error analysis, 481–483, 481f–482f, 484f
  examples, 493–499
Radix-2 FFT algorithm, fundamental equations of, 223
Ramp function, 47, 48f
Random nuisance parameters, handling, 452
Random vector parameters, estimation of, 421
  squared vector error, 421–422, 422f
  uniform vector error, 422
Realization structures, discrete-time systems, 181–183, 181f–182f
  block diagrams, 179–180, 180f
  Cascade form, 181, 182f

direct form I, 181, 181f
direct form II or canonical form, 181, 182f
FIR discrete systems, 186–187, 186f, 187f
IIR discrete systems, 183–186, 184f–185f
implementation, 179
overview, 179
parallel form, 181, 182f
realization structure, 181–183, 181f–182f
signal flow graphs (SFGs), 188, 188f–189f
    Mason's gain formula, 189, 189f–190f
Real-time processing and reconstruction
    analog signals, 21
Receiver operating characteristic (ROC) curve
    properties, 506
    for threshold testing, 506–516, 506f–509f
        examples, 510–516
Reconstruction
    non-accurate reconstruction of analog
        signals, 21–22, 22f
    sinusoidal signals, 10–11
Rectangular window, 313, 313f
Reduction ratio (RR), statistical sufficiency and,
    396, 396t
Region of convergence (ROC), 116
    s- and z-planes into, comparison of, 116–117,
        117f
Residue theorem, 123
ROC, *see* Region of convergence (ROC)
ROC curve, *see* Receiver operating
        characteristic (ROC) curve
RR, *see* Reduction ratio (RR)

S

Samples, variance of, 393–395
Sampling, 5
    analog signals, 5–7, 6f
    example, 7
    frequency, 5, 6, 11
    and holding device, 20–21, 20f
    and holding method, 20–21, 20f
    period, 5, 10
    rate, 6
    sinusoidal signals, 10–11
    uniform, 5, 6f
Sampling theorem, 11
    ideal sampling using impulses, 12f
    low-pass filter, 15, 16f
    proof and detailed discussion, 12–17, 12f–16f
    sampling system, 13, 14f
    signals, 13, 13f
    sinusoidal signals, 12–17, 12f–16f
    spectrum, 13–14, 14f–15f, 16f

Scalar estimation criteria, for non-random
        parameters, 423–425, 424f–425f
Scalar estimators
    for maximum likelihood (ML), 429–433, 430f,
        431f
    statistical moments method for, 426–429
Scalar random parameters, estimation of,
        406–407, 407f
    Bayesian estimation (examples), 413–420
    mean square error (MSE), 407–409, 408f
    mean uniform error (MUE), 411–412,
        411f–413f
    minimum mean absolute error, 409–411, 410f
Scalar random variable
    optimal affine estimator of, 458–459
    optimal constant estimator of, 456
    optimal linear estimator of, 456–458
    superposition property of linear/affine
        estimators, 459
Series connection, discrete-time systems in, 58, 59f
SFGs, *see* Signal flow graphs (SFGs)
Shifted impulse function, 45
Signal(s); *see also* Analog signal(s); Digital
        signal(s)
    analog form, 3–4
    analog signal processing, 3–4, 4f
    defined, 3
Signal detection, 479–516
    Bayes approach, 488
        average risk minimization, 488–489
        minimax Bayes detector, 491–493, 492f
        minimum probability of error test, 490
        optimal Bayes test, 489–490
        performance evaluation of Bayes
            likelihood ratio test, 490–491, 491f
        priori probabilities, 488
    decision function, 486–487, 486f, 487f
    Frequentist approach, 502
        defined, 502
        simple hypotheses, case of, 502–505
    general problem, 484–485, 485f
    multiple hypotheses tests, 496–498, 497f
        average risk minimization, 498–501, 501f
        Bayes approach, disadvantages, 501
        priori probabilities, 498
    radar application, 479–481, 480f
        error analysis, 481–483, 481f–482f, 484f
        examples, 493–499
    ROC curves for threshold testing, 506–516,
        506f–509f
        examples, 510–516
        properties, 506
    simple and composite hypotheses, 485

Signal filtering process, using digital filters, 287, 288f
Signal flow graphs (SFGs), 188, 188f–189f
    branch, 188, 188f
    canonical form implementation structure, 189, 190f
    of difference equation, 189, 189f
    forward path, 188, 189f
    loop, 188, 189f
    Mason's gain formula, 189, 189f–190f
    mixed node, 188, 188f
    path, 188, 189f
    sink node, 188, 188f
    source or input node, 188, 188f
Signal spectra
    in physical sampling, 19, 20f
Signal-to-noise ratio (SNR), 8–9, 19
Simple hypothesis, signal detection, 485
    case of, 502–505
Single-input, single-output system (SISO), 54
Sink node, SFGs, 188, 188f
Sinusoidal sequence, discrete-time, 50–51, 51f
Sinusoidal signals
    reconstruction, 10–17
    sampling, 10–17
        sampling frequency, 11
        sampling period, 10
        sampling theorem, 11
            ideal sampling using impulses, 12f
            proof and detailed discussion, 12–17, 12f–16f
            sampling system, 13, 14f
            signals, 13, 13f
            spectrum, 13, 14f
SISO, *see* Single-input, single-output system (SISO)
SNR, *see* Signal-to-noise ratio (SNR)
Software
    discrete-time system implementation using, 179
Source node, SFGs, 188, 188f
s-plane *vs.* *z*-plane, into region of convergence, 116–117, 117f
Squared vector error, estimation of, 421–422, 422f
Stability
    of discrete-time systems, 125–126
Stable discrete systems, 58
Stable system, 59f
Static system, 57
Statistical models
    expected value and variance of samples, 393–395

Fisher-Cochran theorem, 392–393
Gaussian distribution, 383–392
    beta distribution, 391–392
    Cauchy distribution, 391
    central limit theorem, 387
    chi-squared mixed distribution, 389–390
    chi-squared RV distribution, 388
    Fisher-Snedecor F-distribution, 390–391
    gamma distribution, 388–389
    multivariate, 385–387
    non-central chi-squared RV distribution, 389
    overview, 383–385
    student's t-distribution, 390
    overview, 383
    reproducing distributions, 392
    statistical sufficiency, 395–396
        exponential distributions category, 402–404
        minimal sufficiency, 399–402
        PDF and exponential distribution category, 404
        reduction ratio and, 396, 396t
        sufficient condition, definition, 397–399
Statistical moments method
    for scalar estimators, 426–429
    vector estimation through, 446
Statistical sufficiency, 395–396
    exponential distributions category, 402–404
    minimal sufficiency, 399–402
    PDF and exponential distribution category, 404
    reduction ratio and, 396, 396t
    sufficient condition, definition, 397–399
Step invariant method (z-transform method with sample and hold), digital IIR filters design, 293–294, 293f–294f
Step-size, 8
Stop band, digital filters, 290, 290f
Student's t-distribution, 390, 393
Sufficient condition
    binary likelihood ratios, 398
    defined, 397–399
    discrete likelihood ratios, 398
    examples, 397–399
    Fisher factorization, 397
    likelihood ratio trajectory, 399
System analysis, 55
    z-transform in, 124
        causality of discrete-time systems, 124–125
        stability of discrete-time systems, 125–126

transfer function of connected systems, 126, 126f–127f
transfer function of discrete-time signal, 124
transfer function of discrete-time systems, 127–128
System connections, discrete-time systems, 58
mixed connection, 58, 59f
parallel connection, 58, 59f
in series connection, 58, 59f
System identification, 55
System modeling, 55

T

Time Dependent Discrete Fourier Transform, 216
Time domain
DFT of circular shift in, 218
Time-invariant discrete systems, 56–57
Time reversal, 54
mirroring, 54
z-transform, 118–119
Time scaling, 54
Time shift, 54
z-transform, 117–118
Tracking time, 18
Transfer function
of connected systems, 126, 126f–127f
defined, 124
of discrete-time signals, 124
of discrete-time systems, 127–128
Transition band, digital filters, 290, 290f
Triangular (Bartlett) window, 313, 314f
Tustin (bilinear) method, digital IIR filters design, 297–298, 298f

U

UMVU estimator, *see* Uniform minimum variance unbiased (UMVU) estimator
Uniform minimum variance unbiased (UMVU) estimator, 424–425, 425f
Uniform PDF amplitude, Bayesian estimation, 413–415, 414f
Uniform sampling, 5, 6f
Uniform vector error, estimation of, 422
Unilateral z-transform, 116
Unit rectangular function (pulse function), 48, 48f
Unit step function, 47, 47f
Unstable system, 59f

V

Variance
expected value and, of Gaussian distribution, minimal sufficiency of, 401–402
of samples, 393–395
Vector estimation
maximum likelihood, 447–451
through statistical moments, 446
Vector estimator, optimal affine, 463–464

W

Wavelets, defined, 232
Wavelet transform, 212, 231–236, 232f
theory, 233–236, 235f
Weighted LLS, linear estimation of, 473–477
Window functions
Bartlett (triangular), 313, 314f
Blackman, 314, 315f
election, 316, 317t
Hamming, 314, 315f
Hanning, 313, 314f
Kaiser, 315–316, 316f
rectangular, 313, 313f
Window method, FIR filters design using, 311–317, 311f, 312t, 313f–316f, 317t

Z

Zero-order hold filter (ZOH), 293
operation of, 293–294, 293f
signal reconstruction with, 294, 294f
ZOH, *see* Zero-order hold filter (ZOH)
z-plane
differentiation in, 119–120
vs. s-planes into region of convergence, 116–117, 117f
z-Transform
bilateral, 116
Cartesian form, 113
defined, 113
formula tables, 129, 129t–130t
of fundamental functions, 129t
inverse, 121–122
complex integration method, 123–124
partial fraction expansion method, 122–123
power series expansion method (division method), 122
from Laplace transform to, 113–116, 114f
method with sample and hold, digital IIR filters design, 293–294, 293f–294f

Polar form, 113
properties, 130t
    conjugation of complex sequence, 120
    convolution, 119
    correlation of two sequences, 121
    differentiation in z-plane, 119–120
    initial and final value theorem, 121
    linearity, 118
    multiplication by exponential sequence,
      120
    time reversal, 118–119
    time shift, 117–118
solution of D.E. via, 113–114, 114f

s- vs. z-planes into region of convergence,
    116–117, 117f
in system analysis, 124
    causality of discrete-time systems,
      124–125
    stability of discrete-time systems, 125–126
    transfer function of connected systems,
      126, 126f–127f
    transfer function of discrete-time signal,
      124
    transfer function of discrete-time
      systems, 127–128
unilateral, 116

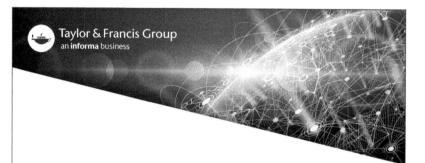

Milton Keynes UK
Ingram Content Group UK Ltd.
UKHW052027071024
449327UK00027B/2463